Advanced Technologies for Rechargeable Batteries

This volume focuses on alkaline metal ion, redox flow, and metal sulfur batteries and provides details about the various kinds of advanced rechargeable batteries. It explains magnesium-ion batteries, sodium-ion batteries, metal sulfur batteries, and redox flow batteries with an introduction to rechargeable batteries and major upcoming batteries (magnesium-/sodium-ion batteries). Various kinds of redox flow batteries from introduction extending to the recent progress in redox flow batteries have been extensively discussed.

Features:

- Covers recent battery technologies in detail, from chemistry to advances in post-lithium-ion batteries.
- Reviews magnesium-ion batteries, sodium-ion batteries, metal sulfur batteries, and redox flow batteries.
- Explains various metal sulfur batteries.
- Explores different types of redox flow batteries for large-scale energy storage application.
- Provides authoritative coverage of scientific contents via global contributing experts.

This book is aimed at graduate students, researchers, and professionals in materials science, chemical and electrical engineering, and electrochemistry.

Advanced Technologies for Rechargeable Batteries

Alkaline Metal Ion, Redox Flow, and Metal Sulfur Batteries

Volume 1

Edited by
Prasanth Raghavan, Akhila Das,
and Jabeen Fatima M. J.

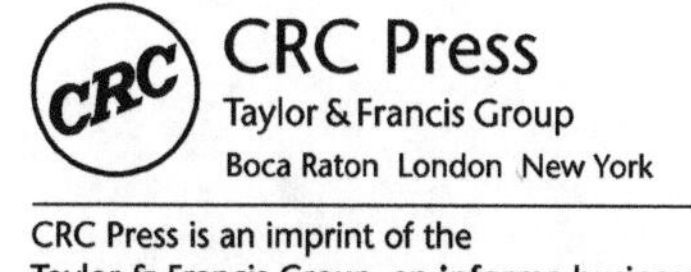

CRC Press is an imprint of the
Taylor & Francis Group, an **informa** business

Designed cover image: © Shutterstock Images

First edition published 2025
by CRC Press
2385 NW Executive Center Drive, Suite 320, Boca Raton FL 33431

and by CRC Press
4 Park Square, Milton Park, Abingdon, Oxon, OX14 4RN

CRC Press is an imprint of Taylor & Francis Group, LLC

Library of Congress Cataloging-in-Publication Data
Names: Raghavan, Prasanth, editor.
Title: Advanced technologies for rechargeable batteries /
edited by Prasanth Raghavan, Akhila Das, and Jabeen Fatima M. J.
Description: First edition. | Boca Raton : CRC Press, 2025.
Identifiers: LCCN 2024009475 (print) | LCCN 2024009476 (ebook) |
ISBN 9781032315348 (v. 1 ; hardback) | ISBN 9781032315355 (v. 1 ; paperback) |
ISBN 9781032315362 (v. 2 ; hardback) | ISBN 9781032315379 (v. 2 ; paperback) |
ISBN 9781003310167 (v. 1 ; ebook) | ISBN 9781003310174 (v. 2 ; ebook)
Subjects: LCSH: Storage batteries.
Classification: LCC TK2941 .A33 2025 (print) | LCC TK2941 (ebook) |
DDC 621.31/2424–dc23/eng/20240404
LC record available at https://lccn.loc.gov/2024009475
LC ebook record available at https://lccn.loc.gov/2024009476

ISBN: 9781032315348 (hbk)
ISBN: 9781032315355 (pbk)
ISBN: 9781003310167 (ebk)

DOI: 10.1201/9781003310167

Typeset in Times
by codeMantra

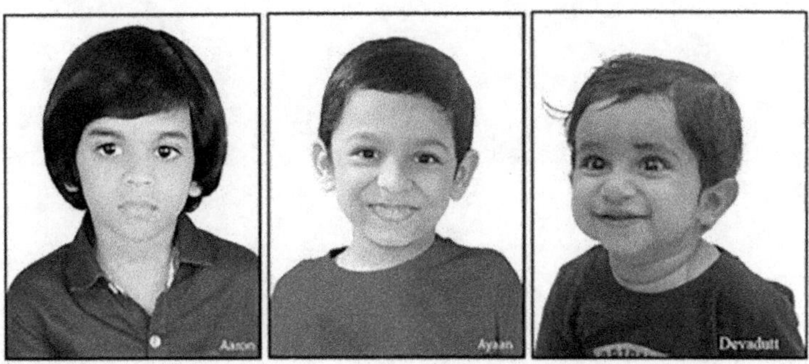

Dedicated to the Little Lanterns, who gave the best support for the successful completion of this book, but if they are understanding, then the book must be completed much earlier than expected.

Contents

Preface

Energy storage systems have emerged as an inevitable part of modern life ever since the commercialization of batteries began. An era of portable devices was initiated when lithium-ion batteries (LIBs) were commercialized by Sony in 1991, which led to a revolution of electronic gadgets like mobile phones, laptops, palmtops, and tablets. In the current scenario, such storage devices have also been used commercially for electric vehicles to reduce pollution and the exploitation of severely depleted stocks of fossil fuels. Energy storage devices capable of satisfying the present energy demands of portable devices consist mainly of LIBs, supercapacitors, and fuel cells. The performance of these storage devices is assessed using two main parameters: the energy density and the power density. The first parameter defines the amount of energy that can be stored in a given volume or weight, whereas the second parameter describes the speed at which energy is stored in or discharged from the device. An ideal storage device should simultaneously deliver high energy density and high power density. The commercially available LIBs are capable of releasing high energy density, whereas supercapacitors can liberate high density power. Present-day research and development of new and innovative component materials are progressing well to address the demands of super gadgets. The ideal energy storage devices for long-range applications are still in their infancy, and hence, there is still plenty of room available in the exploratory domain of materials.

Even though LIBs have already been widely used in different areas, they are still facing a lot of challenges, including poor safety, short performance life, and relatively low specific energy. To address these issues, the new format of batteries, namely solid-state Li batteries, has been developed. Improving the efficiency of any storage device depends directly on the performance of its components, in particular, the behavior of its electrodes and electrolytes during charging and discharging. Material selection is the primary concern in developing advanced energy storage applications. The electrolyte is considered to be the heart of the energy storage devices, and its properties greatly affect the energy capacity, rate performance, cyclability, and safety of these devices. Safety is a prime concern, since portability is the major requirement for these devices, demanding special electrolytic systems capable of replacing conventional liquid electrolyte systems. The organic chemicals used in the liquid electrolytes initiate the solid electrolyte interface formation at the anode of the LIBs, which, on continuous cycling, formed dendritic projections extending toward cathode, leading to explosions in the devices.

The present book offers a detailed explanation of recent evidence of progress and challenges in electrolyte research for energy storage devices. The influences of electrolyte properties on the performances of different energy storage devices are discussed in detail. The detailed explanation has been classified under two major categories, which include a general introduction to energy storage devices and a history of LIBs followed by a detailed investigation on polymer gel electrolytes for energy storage devices. In total, this book consists of 12 chapters. The introduction chapter (Chapter 1) deals with types of energy storage systems and the function of these devices. A detailed view on the history of LIBs has been carefully presented in Chapter 2. These chapters are followed by a detailed review of polymer-based gel electrolytes for LIBs, consisting of seven chapters based on different types of polymeric systems like PVdF, PVdF-co-HFP, PAN, blend polymeric systems, composite polymeric systems (silica, clay, etc.), and polymeric ionic liquid gel electrolytes for LIBs. A detailed review on biopolymer electrolytes for energy storage applications was also included in Chapter 12.

The present book covers a wide range of polymer-based electrolytes and includes useful information for the development of the various electrolytes. We truly believe that this book will be very useful not only for researchers but also for engineers in the development of next-generation energy storage devices. As the battery industry has grown so much over the past 10 years, there have been many new entrants into the battery world from other industries. So, whether you are looking to learn something about one aspect of energy storage devices to reinforce your current knowledge, or are entirely neoteric and are looking to learn all the basics, this book will be a good manual to add to your dictionaries of science, engineering, and technology.

Past few decades have seen the exponential growth in the optimization and enhancement of LIB technology. LIB has become a benign technology for powering portable devices, electronic gadgets, and smart watches to electrifying electric vehicles. However, consistency regarding the raw material prices and availability of materials has always been interrogated. As greater advancements in electronics sector and electric vehicle technology arose, demand of high-energy-density batteries increased. Many battery technologies have been introduced in the past few decades, of which other batteries were one of popular among the post-LIB technology. The book *Advanced Technologies for Rechargeable Batteries* highlights developments in the battery technology. This book introduces general considerations and then dives deep into the mechanism and electrochemistry of each battery technology; hence, it unveils the complexities of battery chemistry. In a nutshell, this book sheds light into the topics of different types of batteries, which ubiquitously enhance the knowledge on the future of post-LIB era batteries. This book will serve as a reference material for those who work in the areas of energy storage technology and materials for battery technology.

About the Editors

Prasanth Raghavan is a Professor at Materials Science and NanoEngineering Lab (MSNE-Lab), Department of Polymer Science and Rubber Technology, Cochin University of Science and Technology (CUSAT) and a Visiting Professor at the Department of Materials Engineering and Convergence Technology, Gyeongsang National University, Republic of Korea, International and Inter University Centre for Nanoscience and Nanotechnology (IIUCNN), Mahatma Gandhi University, Kottayam, Kerala, India, and at Biorefining and Advanced Materials Research Center, SRUC (Scotland's Rural College), Edinburgh, UK. He received his PhD in Engineering under the guidance of Prof. Jou-Hyeon Ahn from the Department of Chemical and Biological Engineering, Gyeongsang National University, Republic of Korea, in 2009, under prestigious Brain Korea (BK21) Fellowship. He completed his BTech and MTech from CUSAT, India. After a couple of years of attachment stint as a Project Scientist at Indian Institute of Technology (IIT-D), New Delhi, he moved to abroad for his PhD studies in 2007. His PhD research was focused on fabrication and investigation of nanoscale fibrous electrolytes for high-performance energy storage devices. He completed his Engineering doctoral degree in less than 3 years, and still it's an unbroken record in Republic of Korea. After the PhD, Dr. Prasanth joined as a Research Scientist at Nanyang Technological University (NTU), Singapore in collaboration with Energy Research Institute at NTU (ERI@N) and TUM CREATE, a joint electromobility research center between Germany's Technische Universität München (TUM) and NTU, where he was working with Prof. Rachid Yazami, the one who has successfully introduced graphitic carbon as anode for commercial lithium-ion batteries, and received Draper Prize along with the Nobel Laurates Prof. J. B. Goodenough and Prof. Akira Yoshino. After 4 years in Singapore, Dr. Prasanth moved to Rice University, USA as a Research Scientist where he was working with Prof. Pulickal M. Ajayan, the co-inventor of carbon nanotubes, and lucky to work with 2019 Chemistry Nobel Laurate Prof. J. B. Goodenough. Dr. Prasanth was selected for Brain Korea Fellowship (2007), SAGE Research Foundation Fellowship, Brazil (2009), Estonian Science Foundation Fellowship, European Science Foundation Fellowship (2010), Faculty Recharge, UGC (2015), etc. He received many international awards including Young Scientist award, Korean Electrochemical Society (2009) and was selected for Bharat Vikas Yuva Ratna Award (2016). He developed many products such as high-performance breaking parachute, flex wheels for space shuttles, and high-performance lithium-ion batteries for leading portable electronic device and automobile industries. He has a general research interest in polymer synthesis and processing, nanomaterials, green/nanocomposites, and electrospinning. His current research focuses on nanoscale materials and polymer composites for printed and lightweight charge storage solutions including high-temperature supercapacitors and batteries. He has published a good number of research papers and book/book chapters in high-impact-factor journals and have more than 5000 citations and an h-index of 45 plus. Apart from science and technology, Dr. Prasanth is a poet, activist, and a columnist in online portals and printed media.

Akhila Das is a graduate student pursuing PhD at Materials Science and NanoEngineering Lab (MSNE-Lab), Department of Polymer Science and Rubber Technology (PSRT), Cochin University of Science and Technology (CUSAT), India. She completed her Master's degree in Chemistry from St. Thomas College, Thrissur after receiving her Bachelor's degree in Chemistry from the University of Calicut (Christ College, Irinjalakuda). She qualified IIT JAM exam. She has expertise in extraction of natural oils and medicinal components from medicinal plants. She was an intern/project student in Dr. Ajay Ghosh's group (Winner of Shanti Swarup Bhatnagar prize). Her MS project was entitled "An efficient supramolecular approach for CO_2 detection". Her research area focused on the "Development of alternative battery technology other than Li ion technology". More than 25 peer-reviewed publications of international publications and couple of international conference publications are into her credits. Her current area of interest includes batteries, supercapacitors, and e-waste management for a sustainable future and a green environment.

Jabeen Fatima M. J. recently joined Carborundum Universal Limited as Manager of Technology. She was previously engaged as a research scientist at Materials Science and NanoEngineering Lab (MSNE-Lab), Department of Polymer Science and Rubber Technology (PSRT), Cochin University of Science and Technology (CUSAT), India. Before joining MSNE-Lab, she was working as a tentative Assistant Professor at the Department of Nanoscience and Technology, University of Calicut, India. She received her PhD in Nanoscience and Technology from the University of Calicut in 2016, India, with prestigious National Fellowship JRF/SRF from the Council of Scientific and Industrial Research (CSIR), under the Ministry of Science and Technology, Government of India. Her research area was focused on the synthesis of nanostructures for photoelectrodes for photovoltaic applications, energy storage devices, photoelectrochemical water splitting, catalysis, etc. She received her MS degree in Applied Chemistry (University First Rank) after receiving her BSc degree in Chemistry from Mahatma Gandhi University (MGU), Kottayam, India. She received many prestigious fellowships like Junior/Senior Research Fellowship (JRF/SRF) from the Centre for Science and Research, Department of Science and Technology, Ministry of India, Post-doctoral/research scientist fellowship from Kerala State Council for Science, Technology and Environment (KSCSTE), and InSc research excellence award. She has published a good number of full-length research articles in peer-reviewed international journals and book chapters with international publishers. She is serving as reviewer for many STM journals published by Wiley International, Elsevier, and Springer Nature. Her current area of interest includes the development of flexible and free-standing electrodes for printable and stretchable energy storage solutions and development of novel nanostructured materials and ternary composite electrodes and electrolytes for sustainable energy applications like supercapacitors, fuel cells, and lithium-ion batteries.

Contributors

Jou-Hyeon Ahn
Department of Materials Engineering and Convergence Technology
Gyeongsang National University
Jinju, Republic of Korea

Sreenivasan Anandhan
Metallurgical & Materials Engineering
NITK Surathkal
Mangaluru, India

Neethu T. M. Balakrishnan
Materials Science and NanoEngineering Lab (MSNE-Lab), Department of Polymer Science and Rubber Technology
Cochin University of Science and Technology
Cochin, India

Madhushri Bhar
Department of Chemistry
Indian Institute of Technology Hyderabad
Sangareddy, India

Ankur Bhattacharjee
Department of Electrical and Electronics Engineering
BITS Pilani
Secunderabad, India

Udita Bhattacharjee
Department of Chemistry
Indian Institute of Technology Hyderabad
Sangareddy, India

K. Biswas
Metallurgical & Materials Engineering
IIT Kharagpur
Kharagpur, India

Gurdial Blugan
Swiss Federal Laboratories for Material Science and Technology
Empa, Switzerland

Anita Cymann-Sachajdak
Department of Energy Conversion and Storage, Faculty of Chemistry
Gdańsk University of Technology
Narutowicza, Poland

Akhila Das
Materials Science and NanoEngineering Lab (MSNE-Lab), Department of Polymer Science and Rubber Technology
Cochin University of Science and Technology
Cochin, India

Zhao Deng
Soochow Institute for Energy and Materials Innovations, College of Energy
Soochow University, Suzhou, China.
Key Laboratory of Advanced Carbon Materials and Wearable Energy Technologies of Jiangsu Province
Soochow University
Suzhou, China

Thomas George
Materials Science and NanoEngineering Lab (MSNE-Lab), Department of Polymer Science and Rubber Technology
Cochin University of Science and Technology
Cochin, India

Shuvajit Ghosh
Department of Chemistry
Indian Institute of Technology Hyderabad
Sangareddy, India

Magdalena Graczyk-Zajac
Fachbereich Material und Geowissenschaften
Technische Universität Darmstadt
Darmstadt, Germany
EnBW Energie Baden-Württemberg AG
Karlsruhe, Germany

Jabeen Fatima M. J.
Materials Science and NanoEngineering Lab (MSNE-Lab), Department of Polymer Science and Rubber Technology
Cochin University of Science and Technology
Cochin, India

S. Janakiraman
Metallurgical & Materials Engineering
IIT Kharagpur
Kharagpur, India

Jishnu N. S.
Institute of Materials Science, Faculty of Mechanical Science and Engineering
Technical University of Dresden
Dresden, Germany

Roshny Joy
Materials Science and NanoEngineering Lab (MSNE-Lab), Department of Polymer Science and Rubber Technology
Cochin University of Science and Technology
Cochin, India

Alexander Kempf
Division of Molecular Chemistry, Materials and Catalysis
Fachbereich Material und Geowissenschaften
Technische Universität Darmstadt
Darmstadt, Germany

Mohammed Khalifa
Metallurgical & Materials Engineering
NITK Surathkal
Mangaluru, India

Irshad U. Khan
Department of Chemical Engineering
National Institute of Technology
Warangal, India

Devika S. Lal
School of Energy Materials
Mahatma Gandhi University
Kottayam, India

Surendra K. Martha
Department of Chemistry
Indian Institute of Technology Hyderabad
Sangareddy, India

Nikhil Medhavi
Materials Science and NanoEngineering Lab (MSNE-Lab), Department of Polymer Science and Rubber Technology
Cochin University of Science and Technology
Cochin, India

Chinmaya Mirle
Department of Chemistry
Indian Institute of Technology Madras
Chennai, India

Anandu M. Nair
School of Energy Materials
Mahatma Gandhi University
Kottayam, India

Vijayamohanan Pillai
Department of Chemistry
Indian Institute of Science Education and Research (IISER)
Tirupati, India

Abhilash Pullanchiyodan
Materials Science and NanoEngineering Lab (MSNE-Lab), Department of Polymer Science and Rubber Technology
Cochin University of Science and Technology
Cochin, India

Nawin Ra
Department of Electrical and Electronics Engineering
BITS Pilani
Hyderabad, India

Rajmohan, K. S.
Department of Chemical Engineering
National Institute of Technology
Warangal, India

Prasanth Raghavan
Materials Science and NanoEngineering Lab (MSNE-Lab), Department of Polymer Science and Rubber Technology
Cochin University of Science and Technology
Cochin, India
Department of Materials Engineering and Convergence Technology
Gyeongsang National University
Jinju, Republic of Korea
International and Inter University Centre for Nanoscience and Nanotechnology (IIUCNN)
Mahatma Gandhi University
Kerala, India
Biorefining and Advanced Materials Research Center
SRUC (Scotland's Rural College)
Edinburgh, United Kingdom

Kothandaraman Ramanujam
Department of Chemistry
Indian Institute of Technology Madras
Chennai, India
DST-IITM Solar Energy Harnessing Centre
Indian Institute of Technology Madras
Chennai, India

Leya Rose Raphael
Materials Science and NanoEngineering Lab (MSNE-Lab), Department of Polymer Science and Rubber Technology
Cochin University of Science and Technology
Cochin, India

Amir Abdul Razzaq
Soochow Institute for Energy and Materials Innovations, College of Energy
Soochow University
Suzhou, China
Key Laboratory of Advanced Carbon Materials and Wearable Energy Technologies of Jiangsu Province
Soochow University
Suzhou, China

M. V. Reddy
Nouveau Monde Graphite (NMG)
Saint-Michel-des-Saints, Canada

Hiranmay Saha
Centre of Excellence for Green Energy and Sensor Systems
IIEST
Howrah, India

Pradeep Vallachira Warriam Sasikumar
Swiss Federal Laboratories for Material Science and Technology
Empa, Switzerland

Murali Mohan Seepana
Department of Chemical Engineering
National Institute of Technology
Warangal, India

Rupali Singh
Metallurgical & Materials Engineering
IIT Kharagpur
Kharagpur, India

Vijay Kumar Thakur
Biorefining and Advanced Materials Research Center
SRUC (Scotland's Rural College)
Edinburgh, UK

Anusree Thilak
Materials Science and NanoEngineering Lab (MSNE-Lab), Department of Polymer Science and Rubber Technology
Cochin University of Science and Technology
Cochin, India

Anjumole P. Thomas
Materials Science and NanoEngineering Lab (MSNE-Lab), Department of Polymer Science and Rubber Technology
Cochin University of Science and Technology
Cochin, India

A. Venimadhav
Cryogenic Engineering Center
IIT Kharagpur
Kharagpur, India

S. K. Vineeth
Department of Polymer and Surface Engineering
Institute of Chemical Technology
Mumbai, India

Alexandru Vlad
Division of Molecular Chemistry, Materials and Catalysis
Institute of Condensed Matter and Nanosciences
Université catholique de Louvain
Ottignies-Louvain-la-Neuve, Belgium

Monika Wilamowska-Zawłocka
Department of Energy Conversion and Storage, Faculty of Chemistry
Gdańsk University of Technology
Gdańsk, Poland

Zuzanna Zarach
Department of Chemistry and Technology of Functional Materials, Faculty of Chemistry
Gdańsk University of Technology
Gdańsk, Poland

Xiaohui Zhao
Soochow Institute for Energy and Materials Innovations, College of Energy
Soochow University
Suzhou, China
Key Laboratory of Advanced Carbon Materials and Wearable Energy Technologies of Jiangsu Province
Soochow University
Suzhou, China

Abbreviations

AEM	anion exchange membrane
AIB	aluminum-ion battery
AMSA	aminomethylsulfonic acid
AMV	apical membrane vesicles
APS	Asahi polysulfone
AQ	anthraquinone
AQDS	9,10-anthraquinone-2,7-disulfonic acid
AQDS	anthraquinone-2,6-disulfonic acid
As	arsenic
BCF3EPT	3,7 bis(trifluoromethyl)-n-ethylphenothiazine
Bi	bismuth
Bi NT	bismuth nanotube
Bi_2O_3	bismuth oxide
BiF	bismuth fluoride
BiOF	bismuth oxynanofluoride
Black P	black phosphorous
B-Mg	bulk magnesium
BP	black phosphorous
CE	coulombic efficiency
CEM	cation exchange membrane
CNT	carbon nanotube
CNTs	carbon nanotubes
CoO_3	cobalt oxide
CP	Chevrel phase
CTAB	hexadecyl trimethyl ammonium bromide
CuCP	copper-based Chevrel phase
CuHCF	copper hexacyanoferrate
CV	cyclic voltammetry
CVD	chemical vapor deposition
DBBB	2,5-di-tert-butyl-1,4-bis(2-methoxyethoxy)benzene
DEC	diethyl carbonate
DFT	density functional theory
DHAQ	2,6-dihydroxyanthraquinone
DI	deionized water
DMC	dimethyl carbonate
DME	dimethoxyethane
DME	dimethyl ethylene
DMF	dimethyl formamide
DMSO	dimethyl sulfoxide
DPSO	dipropyl sulfone
EC	ethylene carbonate
EC/PC	ethylene carbonate/propylene carbonate
EDTA	ethylenediamine tetraacetic acid
EE	energy efficiency
EES	electrical energy storage
EtMgBr	ethyl magnesium bromide
EV	electrical vehicle

EW	equivalent weight
FEC	fluoroethylene carbonate
FeFB	iron flow battery
FIB	fluoride-ion battery
G2	diglyme
GF	graphite felts
GO	graphene oxide
GPE	gel polymer electrolyte
HC	hard carbon
HER	hydrogen evolution reaction
HEV	hybrid electrical vehicle
HMDS	hexamethyldisilazide magnesium chloride
HRSEM	high-resolution scanning electron micrographs
HRTEM	high-resolution tunneling electron microscope
HTFIB	high-temperature FIB
4-HO-TEMPO	4-hydroxy-2,2,6,6-tetramethyl-piperidin-1-oxyl
ICB	iron-chromium battery
IEC	ion exchange capacity
In	indium
IVB	iron-vanadium battery
KCl	potassium chloride
KIT	Karlsruhe Institute of Technology
LBL	layer by layer
LCO	lithium cobalt oxide
LFP	lithium iron phosphate
LIB	lithium-ion battery
Li-Mg battery	lithium–magnesium hybrid ion batteries
LTO NPs	spinel lithium titanium oxide
MA	mechanical activation
MACC	magnesium–aluminum chloride complex
MB	magnesium-bismuth
Mg	magnesium
Mg(TF)	magnesium trifluoromethanesulfonate
$Mg(TFSI)_2$	magnesium bis(trifluoromethanesulfonate)
$MgCl_2$	magnesium chloride
MgO	magnesium oxide
MIB	magnesium-ion batteries
MOFS	metal-organic structures
MOLICEL	Moli energy to recall
MSA	methane sulfonic acid
MSS	molten solid-state synthesis technique
MV	methyl viologen
$NaNO_2$	sodium nitrite
NASA	National Aeronautics and Space Administration
NASICON	Na super ionic conductor
NaTFSI	sodium salt of bis(trifluoromethanesulfonyl) imide
NCs	nanocrystals
NIBs	sodium-ion batteries
NiHCF	nickel hexacyanoferrate
N-Mg	nano magnesium
NMO	nickel manganese oxide

NMR	nuclear magnetic resonance
OCV	open-circuit voltage
OER	oxygen evolution reaction
P(VDF-co-HFP)	poly(vinylidene fluoride-co-hexafluoropropylene)
PAN	polyacrylonitrile
PB	Prussian blue
Pb	lead
PBA	Prussian blue analog
PBAs	Prussian blue analogs
PC	propylene carbonate
PCET	proton-coupled electron transfer
PE	polymer electrolyte
PEC	poly(ethylene carbonate)
PEGDE	poly(ethylene glycol) dimethyl ether
PEO	polyethylene oxide
PMMA	poly(methyl methacrylate)
PP	polypropylene
PSSS	poly(sodium-4-styrenesulfonate)
PTFE	polytetrafluoroethylene
PTIO	2-phenyl-4,4,5,5-tetramethylimidazoline 1oxyl 3-oxide
PVA	poly vinyl alcohol
PVDF	poly(vinylidene fluoride)
PVP	polyvinyl pyrrolidone
PWA	phosphotungstic acid
R&D	research and development
RAHC	radially aligned hierarchical columnar (structures)
RFB	redox flow battery
RGO	reduced graphene oxide
RMB	rechargeable magnesium battery
RP	red phosphorous
RTFIB	room-temperature fluoride-ion battery
RTIL	room-temperature ionic liquid
SAPE	solid acid polymer electrolyte
SAXS	small angle X-ray scattering
Sb	antimony
SEI	solid electrolyte interphase
SEM	scanning electron microscopy
SFPAE	stable fluorinated sulfonated poly(arylene ether)
SHE	standard hydrogen electrode
SHMP	sodium hexametaphosphate
SIB	sodium-ion battery
Sn	tin
SN	succinonitrile
SOC	state of charge
SOD	state of discharge
SPE	solid polymer electrolyte
SPEEK	sulfonated poly(ether ether ketone)
SPEU	segmented poly(ether urethane)
TEM	transmission electron microscopy
TG	tetraglyme
THF	tetrahydrofuran

$Ti(TFSI)_2Cl_2$	titanium(bis(trifluoromethanesulfonyl)imide)chloride
TM	transition metals
TMQ	2,3,6-trimethylquinoline
TMV	tobacco mosaic virus
2D	two dimensional
VE	voltage efficiency
VRB	vanadium redox battery
VRFB	vanadium redox flow batteries
XPS	X-ray photoelectron spectroscopy
XRD	X-ray diffraction
ZBB	zinc-bromine flow battery
15D3GAQ	1,5-bis(2-[2-(2-methoxyethoxy) ethoxy] ethoxy) anthracene-9,10-dione

Symbols

$AlCl_3$	aluminum chloride
$AlCl_3$	aluminum trichloride
BF_4	tetrafluoroborate
BF-rGO	benzoic acid-functionalized rGO
Bi_3Sn_2	bismuth-tin alloy
BiF_3	bismuth fluoride
BMIMBF4	1-n-butyl 3-methyl imidazolium tetrafluoroborate
BMImTFSI	1-butyl-3-methylimidazolium TFSI
BMP	1-butyl 1-methyl pyrrolidinium
$CB_{11}H_{12}$-	carborane
CH_3SO_3H	methanesulfonic acid
CoO_3	cobalt oxide
Cr_2S_4	chromium sulfate
EMImTFSI	1-ethyl-3-methylimidazolium bis(trifluoromethylsulfonyl)imide
EtMgBr	ethylene magnesium bromide
$Fe_2(SO_4)_3$	ferrous sulfate
FeS_2	iron disulfide
H_2SO_4	sulfuric acid
H_3PO_4	phosphoric acid
HCl	hydrochloric acid
K_2SO_4	potassium sulfate
K_3PO_4	potassium phosphate
$Mg\ (ClO_4)_2$	magnesium perchlorate
$Mg\ (PF_6)$	magnesium hexafluoro phosphate
$Mg(B(Bu_2RPh_2))_2$THF-DMF	magnesium organohaloaluminate in tetrahydro furan and dimethyl sulfoxide
Ah_{charge}	charge ampere hours
$Ah_{discharge}$	discharge ampere hours
c	electrolyte concentration
E	electrode potential
E_eq^	equilibrium potential
E_c	number of watt hours the system charges on DC side
E_d	number of watt hours the system discharges on the DC side
F	Faraday's constant
I	terminal current
I_0	exchange current
I_{charge}	charge current
$I_{discharge}$	discharge current
$LiClO_4$	lithium per chlorate
$LiFePO_4$/LFP	lithium iron phosphate
$LiMnO_2$	lithium manganese oxide
$Mg\ (BBu_2Ph_2)_2$	magnesium dibutyldiphenylborate
$Mg(BF_4)_2$	magnesium tetrafluoroborate
$Mg(TFSI)_2$	magnesium bis(trifluoro sulfonyl)imide
$MgBH_4$	magnesium borohydride

$\mathbf{MgCl_2}$	magnesium dichloride
$\mathbf{MgCl_2/AlCl_3}$	magnesium chloride/aluminum chloride
$\mathbf{MgClO_4}$	magnesium perchlorate
$\mathbf{MgHf(WO_4)_3}$	magnesium hafnium tungstate
$\mathbf{MgNO_3}$	magnesium nitrate
$\mathbf{Mg(B(Bu_2RPh_2))_2}$	B-boron, R-alkyl, Bu-butyl, Ph-phenyl
$\mathbf{MgSc_2Se_4}$	magnesium scandium selenide
$\mathbf{MgSO_4}$	magnesium sulfate
$\mathbf{MnO_2}$	manganese dioxide
$\mathbf{Mo_6S_8}$	Chevrel phase molybdenum sulfide
$\mathbf{MoS_2}$	molybdenum disulfide
$\mathbf{Na_2SO_4}$	sodium sulfate
$\mathbf{NaCF_3SO_3}$	trifluoromethanesulfonic acid sodium salt
NaCl	sodium chloride
$\mathbf{NaClO_4}$	sodium perchlorate
$\mathbf{NaIO_4}$	sodium periodate
$\mathbf{NaPF_6}$	sodium hexafluorophosphate
$\mathbf{NaZr_2(PO_4)_2}$	sodium zirconium phosphate
$\mathbf{(NH_4)_2SO_4}$	ammonium sulfate
$\mathbf{(NH_4)_3PO_4}$	ammonium phosphate
ne	number of electrons transferred per mole of electrolyte solution
$\mathbf{NiMnO_4}$	nickel manganese oxide
NiTFSI	sodium bis(trifluoromethylsulfonyl)imide
Pyr13FSI	N-methyl-N-propylpyrrolidinium bis-(fluorosulfonyl)imide
Pyr14FSI	N-butyl-N-methylpyrrolidinium bis(fluorosulfonyl)imide)
R	universal gas constant
T	ambient temperature
$\mathbf{t_{charge}}$	charge time
$\mathbf{t_{discharge}}$	discharge time
$\mathbf{Tf_2N^-}$	bis(trifluoromethanesulfonyl)imide
TFSI	trifluorosulfonylimide
$\mathbf{TiO_2}$	titanium dioxide
$\mathbf{TiS_2}$	titanium disulfide
$\mathbf{UF_4}$	uranium tetrafluoride
$\mathbf{\underline{V}}$	electrolyte volume
$\mathbf{\bar{v}}$	average voltage during discharge
$\mathbf{V_{\ charge}}$	average discharged voltage
$\mathbf{V_2O_5}$	vanadium pentoxide
$\mathbf{V_{charge}}$	charge voltage
$\mathbf{V_{discharge}}$	average discharged voltage
$\mathbf{V_{discharge}}$	discharge voltage
$\mathbf{VOSO_4}$	vanadyl sulfate
$\mathbf{W_c}$	auxiliary consumption during charge
$\mathbf{W_d}$	auxiliary consumption during discharge
$\mathbf{Wh_{actual}}$	actual capacity discharged from the battery
$\mathbf{Wh_{charge}}$	charge watt hours
$\mathbf{Wh_{discharge}}$	discharge watt hours
$\mathbf{Wh_{theoretical}}$	theoretical capacity discharged from the battery
α _ **a**	charge transfer coefficient in the anolyte side

α _ c	charge transfer coefficient in the catholyte side
η	energy efficiency of the system
η utilization	electrolyte utilization
ηC	coulombic efficiency
ηE	energy efficiency
ηV	voltage efficiency

Units

Ah g^{-1}	ampere hour per gram
A g^{-1}	ampere per gram
Å	angstrom
A	ampere
C	mole m^{-3} or mol L^{-1}
°C	degree Celsius
cm	centimeter
cP	centipoise
E	volt
g cm^{-3}	gram per cubic centimeter
g L^{-1}	gram per liter
g mol^{-1}	grams per mole
g	gram
GJ m^{-3}	giga joule per meter cube
GPa	gigapascal
h	hour
I	A
K	kelvin
kg	kilogram
kJ	kilojoule
kJ mol^{-1}	kilojoule per mol
kW	kilowatt
M	molar
mA	milliampere
mAh cc^{-1}	milliampere hour per cubic centimeter
mAh cm^{-3}	milliampere hour per centimeter cube
mAh g^{-1}	milliampere hour per gram
mAh mL^{-1}	milliampere hour per milliliter
mA-h	milliampere-hour
mF cm^{-2}	millifarad per centimeter square
mol	mole
MPa	megapascal
mS cm^{-1}	millisiemens per centimeter
mV s^{-1}	millivolt per second
nm	nanometer
P	pressure
Pa	pascal
pH	negative decadic logarithm of the relative activity of H_3O^+ ions
ppm	parts per million
S	siemens
S cm^{-1}	siemens per centimeter
K	kelvin
μm	micrometer
ton^{-1}	ton inverse
V	voltage
W	watt

Wh kg^{-1}	watt-hour per kilogram
Wh L^{-1}	watt-hour per liter
wt	weight
Ω	ohm
%	percent

1 Current Rechargeable Batteries
Issues and Challenges

Akhila Das, Neethu T. M. Balakrishnan, S. K. Vineeth, Jou-Hyeon Ahn, Alexandru Vlad, Vijayamohanan Pillai, and Prasanth Raghavan

1.1 INTRODUCTION

The global energy demand is steadily increasing day by day, and it is predicted to increase in the future as well. The U.S. energy administration reports that there will be about 50% of increase in the global energy consumption by the year of 2050. The overdependence of the non-renewable resources such as fossil fuel leads to a serious concern such as pollution which can impact the environment dangerously. The hazardous impact and the concern of extinction of fossil fuel resources on environment were facilitated by the use of renewable resources, such as wind and solar, which can also rectify the extinction of non-renewables. The major challenge faced during the exploration of the renewable was the proper management of energy generation and its consumption at the required time. In order to meet the balancing of the supply and consumption of energy harvested from the renewable resources, the role of energy storage system is considered to be noteworthy. The development of energy storage systems was started in 1990s in order to overcome the enhanced oil price. Later, an energy storage programme called Annex IX was initiated that highly influenced the growth and development of energy storage systems.

Different energy storage systems are employed for the proper storage of the energy harvested from the renewable resources. The storage systems such as batteries, supercapacitors and fuel cells are highly recommended for the proper supply of stored energy for various applications. Among them, batteries that can effectually convert the chemical energy to electrical energy are most promising and widely accepted system. The wide consumer acceptance of batteries owing to its flexibility in size, shape and most importantly its portability has attracted its convenient application in vast variety of areas. The vision of zero-emission automobiles is made true by the proper utilization of batteries that ensure an environmentally friendly technology. For the application of the batteries, it is important to understand its history, issues and challenges. This chapter discusses the detailed examination of current rechargeable technologies, as well as their issues and future challenges in different sections.

1.2 THE EMERGENCE OF BATTERIES

Batteries are the energy storage system that effectively converts the chemical energy to electrical energy, which are broadly classified as primary and secondary batteries. The term primary batteries are meant for the non-reusable energy storage systems that cannot be recharged, hence they are also termed as non-rechargeable batteries. The mere dependence of primary batteries can lead to the formation of electrical waste that outcomes as a result of the systems throw out after single use. As an alternative, secondary batteries were introduced that can be reused by charging again and hence termed as "rechargeable batteries" [1]. The birth of batteries happened long time ago, predating even before the recognition of their relevance in energy storage. History also sheds light on the ancient Agastya Samhitha battery which resembles the modern rechargeable battery (Figure 1.1a).

DOI: 10.1201/9781003310167-1

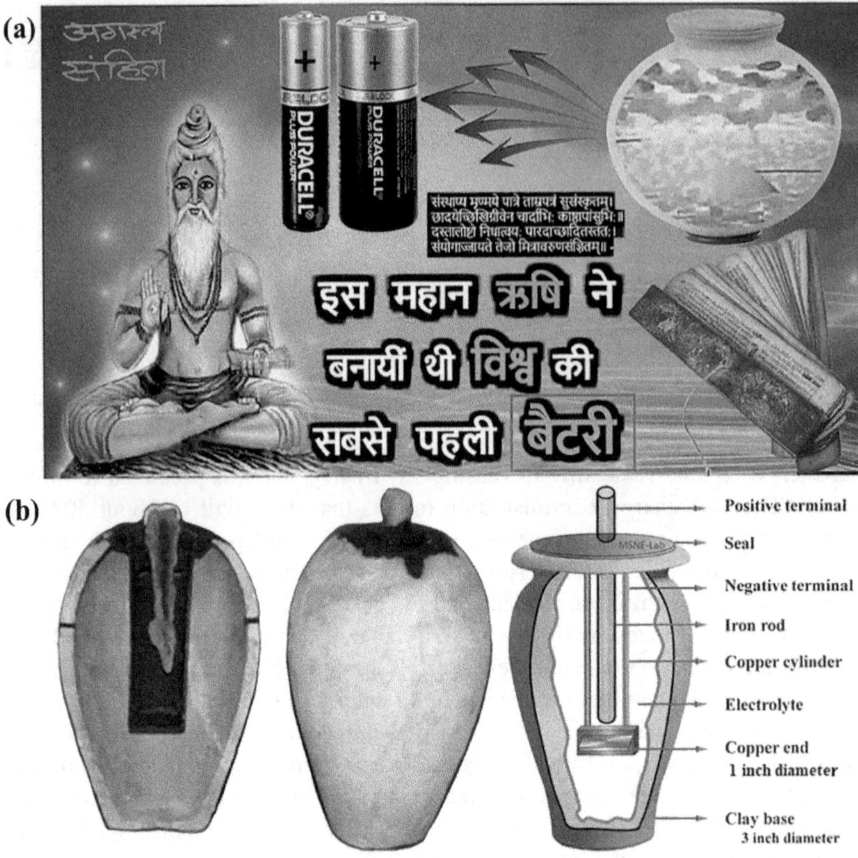

FIGURE 1.1 (a) Images of Agastya battery. (b) Schematic illustration of Baghdad battery.

Baghdad batteries or Parthian batteries that are found by the workers near the Baghdad railway were considered to be the first battery type. Baghdad battery constitutes an iron rod surrounded by copper cylinder which is submerged in a vinegar solution (Figure 1.1b). The acceptance of this system as a battery still remains a controversial topic [2]. As a technological advancement, the first practical application of static current was invented by Alessandro Volta in 1745 [3], and later in 1791, Luigi Galvani [4] discovered the animal electricity by observing the muscle contraction in frog after touched by a metallic object. Inspired from this successful experiment, he continued to examine the electrical energy formation using zinc, lead, tin and iron as the cathode and copper, silver, gold, and graphite as the anode, and it gathered much interest as Galvanic electricity. In 1800, the first voltaic cell which could be termed as the "battery" [5] has been discovered. It was a crucial initiative in the development of the current battery technology that got much scientific acceptance. From the voltaic cell, an improved cell structure was developed by John F. Daniell, an English chemist in 1836 [6].

In 1859, the first rechargeable battery was developed by French physician Gaston Planté using a lead system [7,8]. It is noteworthy that this system is still in use. Followed by it, the rechargeable system was in the path of steady development and advancement in technology. In 1899, the nickel-cadmium battery was invented by Waldemar Jungner from Sweden by making use of nickel oxyhydroxide and cadmium as the cathode and anode, respectively [9]. The material cost of cadmium (Cd) was a serious issue for the execution of this battery technology, and later, Edison substituted cadmium with iron (Fe) and developed the Nickel-Iron (Ni-Fe) battery which is also known as nickel-alkaline battery or Edison battery [10]. This rechargeable system was attractive because of its durability, discharge rate and the ability to withstand the overcharge discharge and short circuit. However, the application was restricted in terms of its cost, efficiency, size and difficulty in maintenance. So, Ni-Cd was widely explored in portable electronics [11] which was first commercialized in 1960 by Panasonic Corporation in the USA and Japan, and it later started to be produced in enormous quantities. Although cadmium remains toxic even after the disposal of batteries as an E-waste, environmentalists have raised serious concern about this [12]. As an alternative for the technology, nickel-metal hydride battery was proposed which is far more environmentally friendly as compared to that of Ni-Cd battery. In 1990s, the drastic development of nickel-metal hydride batteries by Panasonic and Sanyo Electric Co., Ltd promoted the small portable electronic devices like mobile phones, notebook computers, camcorders and digital cameras [13,14].

1.3 NEXT-GENERATION RECHARGEABLE SYSTEMS

The rechargeable lithium-ion battery (LIB) technology was a greatest discovery that made a prominent influence in the transformation of lives of common people. The major components that give LIBs their best performance are the electrodes comprising anode, cathode and electrolyte (Figure 1.2). Each of these components is crucial since the damage of the single component can affect the proper functioning of the battery. The significance of lithium ions (Li^+ ions) as a battery component was recognized in 1958 when Harris et al. [15] examined the solubility of lithium in various electrolytes. This investigation revealed the capability of lithium to form a passivation film that prevents the direct reaction of lithium metal and electrolyte by allowing the ionic transport. Later, Selis et al. [16] recognized the importance of lithium ions while examining the battery made up of calcium and silver electrodes, and in 1965, Casey et al. [17] found that the lithium can explore in electrolytes that could be operable above the temperature of 55°C in nickel hydroxide-cadmium battery. Lithium-based primary batteries were available in commercial market even in the 1960s, which are considered to be the 3 V primary batteries including lithium sulphur dioxide $Li//SO_2$ [18], lithium polycarbon monofluoride ($Li//(CF_x)n$) [19], lithium manganese oxide [20] and lithium copper oxide [21].

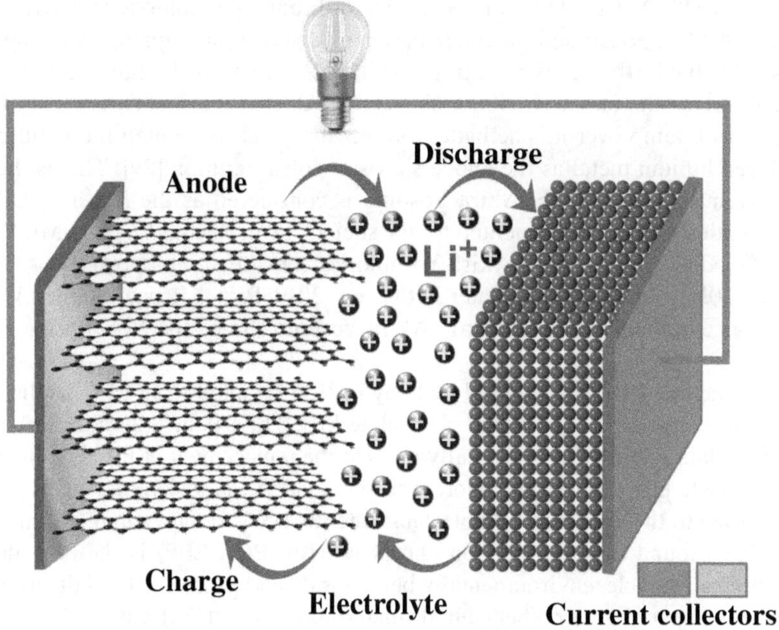

FIGURE 1.2 Schematic illustration of components of battery.

The emergence of first rechargeable lithium-based system was made possible with the introduction of the intercalating mechanism of Whittingham in 1978 [22]. He explored the intercalation mechanism in titanium disulphide (TiS_2) and fabricated the cell using TiS_2 as the cathode and lithium as the anode. TiS_2 was attracted for its high intercalation capacity (240 mAh g^{-1}) and high cyclic stability. However, the use of TiS_2 at ambient condition was hazardous owing to the release of the toxic gases such as H_2S on reaction with the moisture. Further, the use of lithium metal as anode also led to serious safety hazards that were significant from the safety issues caused by the MOLICEL batteries. MOLICEL explored lithium metal anode along with the molybdenum disulphide (MoS_2) as the cathode. In 1980s, the fire issues reported with the cell phones forced the company Moli Energy to recall all their products from the market [23].

The intercalation mechanism proposed by Whittingham initiated the investigation of different layered materials by different research groups that can produce the best and safer intercalation mechanism for LIBs. The new material, lithium cobalt oxide ($LiCoO_2$, simply LCO), studied by Godshall et al. and Mizushima revealed another leap in the history of LIBs, which exhibits a layered structure similar to TiS_2. In 1981, Goodenough et al. [24] recognized LCO as the best material for the cathode for LIBs that can exhibit a better stability and high lithium-ion insertion voltage of 3.5–4 V vs Li/Li^+ as compared to that of TiS_2. However, toxic nature of cobalt content in LCO and the requirement of high voltage for the complete delithiation are some of the effects that deteriorate its performance in LIBs. Later in 1989, Thackeray proposed the $LiMnO_2$ cathodes for LIBs that can deliver a safer electrochemical mechanism. Even though lithium manganese oxide was a successful positive electrode proposed, its structural variations by the Jahn–Teller distortion during the electrochemical mechanism led the search for another intercalation compounds for LIBs. After that, again in 1989, John Goodenough and team reported the effectiveness of cathodes containing polyanions as the best substitute for oxides that are capable of producing higher voltage than oxide owing to their inductive effect. It was a remarkable finding in the development of safer LIBs.

In the search of anodes for LIBs, Basu et al. [25,26] in 1978 proposed the lithium intercalation mechanism in graphite, and in 1980, Besenhard et al. [27] scrutinized the intercalation and de-intercalation mechanism of Li^+ ions in graphite. By exploring the solid polymer electrolyte, polyethylene oxide (PEO) activated with electrolyte of lithium per chlorate ($LiClO_4$); Yazami and Touzain [28] in 1983 demonstrated the intercalation mechanism in graphite. These inventions and observations led to the birth of LIB by Akira Yoshino using LCO as the cathode and vapour phase grown carbon fibre as the anode. In 1986, Yoshino conducted the first safety test of the battery by dropping an iron lump over it. The battery successfully self-prevented the ignition, while the battery employed lithium metal as the anode shows a violent ignition [29]. This is the moment at which LIBs originated, and hence, Akira Yoshino is considered as the father of LIBs. For this remarkable invention that greatly benefited for the society, Prof. Rachid Yazami, Mr. Yoshio Nishi, Prof. John B. Goodenough and Prof. Akira Yoshino were awarded with prestigious Charles Stark Draper prize in 2014 (Figure 1.3), and later in the year 2019, Prof. Michael Stanley Whittingham, Prof. John Bannister Goodenough and Prof. Akira Yoshino shared the Nobel prize in Chemistry (Figure 1.4).

The first commercial LIB was released by Sony in 1991 by employing LCO as the cathode and graphite as the anode along with a liquid electrolyte [31]. The use of liquid electrolyte was risky owing to its high flammability that eventually leads to the replacement of these liquid counterparts with thermally stable polymer system. Moreover, the volume expansion and safety issues caused by LCO further led to the search of a stable substitute. In 1996, Goodenough, Akshaya Padhi and co-workers [32,33] found that lithium iron phosphate ($LiFePO_4$, LFP) is the best alternative for LCO as it is thermally stable, environmentally benign, safe and possesses high theoretical capacity (170 mAh g^{-1}) as compared to any other cathode materials proposed [34]. Currently, most of the LIB technologies explore LFP as the cathode along with the carbonaceous anode and a stable polymer electrolyte separating them.

FIGURE 1.3 Photograph of Draper Prize winners in lithium-ion battery of the year 2014. (From right to left: Prof. Rachid Yazami, Mr. Yoshio Nishi, Prof. John B. Goodenough and Prof. Akira Yoshino.) Adapted and Reproduced from National Academy of Sciences for the National Academy of Engineering.

FIGURE 1.4 Images of Nobel laureates 2019. (From right to left: John. B. Goodenough, Stanley M. Wittingham and Akira Yoshino.) Adapted and reproduced from Ref. [30]. Copyright © Nobel Media 2019. Illustration: Niklas Elmehed.

1.4 THE ERA OF POST-LITHIUM-ION BATTERIES: INTERNAL ISSUES AND CHALLENGES

The advent of LIBs significantly influenced the consumer markets and its various attractions such as portability owing to its light weight and small size enhanced the production of miniaturized electronic gadgets. However, the major challenge with the LIB technology is the low abundance of lithium in the earth crust that could substantially raise serious concern about the supply of lithium ions. The sources of lithium are limited in regions of Australia, Portugal and Zimbabwe, North and South America, and further, the expensiveness of mining and import can significantly affect the cost of the material [35]. This scenario leads to the essentiality of proper alternative for the

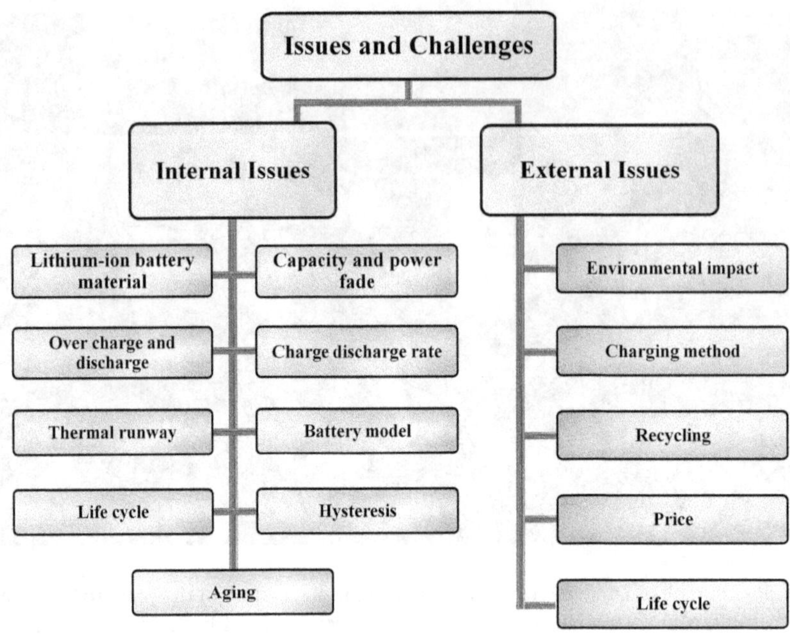

FIGURE 1.5 Classification of issues and challenges.

LIBs technology. Apart from that, there are many internal and external issues that challenge the application of LIBs as shown in Figure 1.5 [36]. The internal issues such as battery materials and thermal runway ageing mechanism and the external issues such as charging methods, environmental impact, and recycling were the essentiality of alternate technologies that could overcome these issues without compromising the final performance. In search of it, different battery technologies such as magnesium-ion batteries (MIBs) [37], lithium magnesium hybrid ion batteries (Li–Mg battery) [38,39], sodium-ion batteries (SIBs) [40,41] and potassium-ion batteries (KIBs) [42,43] were explored in current rechargeable systems. This section summarizes the issues and challenges of current LIBs including their external and internal issues as well as different post-LIBs.

1.5 ISSUES OF CURRENT LITHIUM-ION BATTERIES

1.5.1 Battery Components

The internal issues of LIBs are mostly raised from its components including anode, cathodes, electrolytes and the separators. The anodes mostly explored in LIBs include graphitic anodes, lithium metal and lithium-based alloys, while the cathodes examined are LCO [44,45], lithium iron phosphate (LFP) [46,47], lithium manganese oxide (LMO) [48], etc. Cathodes like LCO were toxic material owing to the presence of cobalt and there was an essentiality of careful precaution owing to its high-level heating and stress. Apart from it, they have a short life span that can limit their performance in LIBs [49]. Compared to LCO, LFP is better in properties and battery performance. They are non-toxic, inexpensive and also reveal better thermal stability as compared to that of LCO [49]. Even though the service LFP is affected by the thermal variation, it can cause serious issue with the battery [50]. Cathodes like nickel manganese oxide (NMO) are good in overall performance, but the poor stability of nickel and the low specific energy of the manganese can further make concerns [51].

Considering the performance anodes, lithium metal initially explored in LIBs was limited by its poor stability and the safety issues aroused from it. While shifting towards the graphitic anodes, the safety concern in the aspects of anode stability was achieved to some extent, but their electrochemical performance was limited. Considering other anodes such as silicon and lithium titanium oxides can reveal an enhanced lithium capturing capacity, but limited by high volume expansion [52]. The major challenge with the use of LIB is the safety concern caused by the electrolyte used in it. During initial stages, the liquid electrolytes such as organic solvents with the lithium salt were extensively examined. The continuous charging and discharging process cause the heating of the battery that can eventually heat the liquid electrolyte inside [35]. This liquid electrolytes with low vapour pressure can boil to cause the swelling of the battery that could finally burst out and cause explosion at extreme condition [53]. Replacing the liquid counterparts with the polymer electrolytes assisted to resolve this issue to great extent, but still the research is progressing to meet the better performance and to resolve further limitation in performance [54,55].

1.5.2 Thermal Runway

The temperature has a crucial role in the performance of LIBs. The heat development inside the battery can accelerate the exothermic reaction that can outcome dangerous issues with the battery. Normally, the thermal reaction inside the battery is considered to follow a three-step process including the anodic reaction that starts at 90°C followed by the decomposition of the solid electrolyte interface (SEI) layer above 120°C. Enhancing the temperature above 140°C initiates the cathodic exothermic reaction that can trigger the evolution of oxygen. The cathodic decomposition and the electrolyte oxidation take place with high-rate exothermic reaction that occurs when the temperature reaches above 180°C. The thermal decomposition in different battery materials is illustrated in Figure 1.6 [36]. Different electrode systems exhibit varying temperature stability. The thermal decomposition of LCO is observed at 150°C, while that of $LiNi_{0.8}Co_{0.15}Al_{0.05}O_2$ is at 160°C. $LiNi_xCo_yMn_zO_2$ and $LiMn_2O_4$ decompose at 210°C and 265°C, respectively, while $LiFePO_4$ decomposes at a higher temperature which is about 310°C. Considering the low temperature, the deposition of metallic lithium over the anodes can cause poor performance of the battery, and at this temperature, cathodes can break down and result in short circuit. So, it is extremely important to consider the thermal issue of LIBs.

1.5.3 Ageing

The ageing of LIBs occurs due to the innate chemical reaction between the electrodes and electrolyte. The ageing effect of LIBs can happen in two different conditions either during the storage or by the working of the battery. The ageing during the storage can significantly affect the calendar life of the battery. The common change that occurs in the anode during the ageing effect is the thickening of the SEI layer that will increase the internal impedance of the cell. The composition and morphological changes in the SEI that formed during the first few charge discharge cycles generate the electrode impedance that will inhibit the battery performance. The major source for ageing at the anode is the reaction that occurs at the electrode–electrolyte interface. During the charged state, the electrolyte decomposition and the irreversible consumption of the lithium ions form a protective layer over the anode that comprises both SEI and non-SEI layers. They are observed to be essential for the protection of electrodes from the electrolyte degradation, but owing to the ageing effect, they cause the restriction of battery performance. The ageing mechanism is a temperature-dependent phenomenon. It has been reported that the resistance and capacitance vary with ageing cycling over a temperature range of 25°C–55°C.

1.5.4 Other Internal Issues with Lithium-Ion Batteries

Other internal issues normally observed with the LIBs include capacity fade and power fade, life cycle, charge/discharge rate and battery model. The capacity fade and power fade of LIBs are associated with changes that occur in the electrodes. The loss of lithium ions from the SEI layer of the

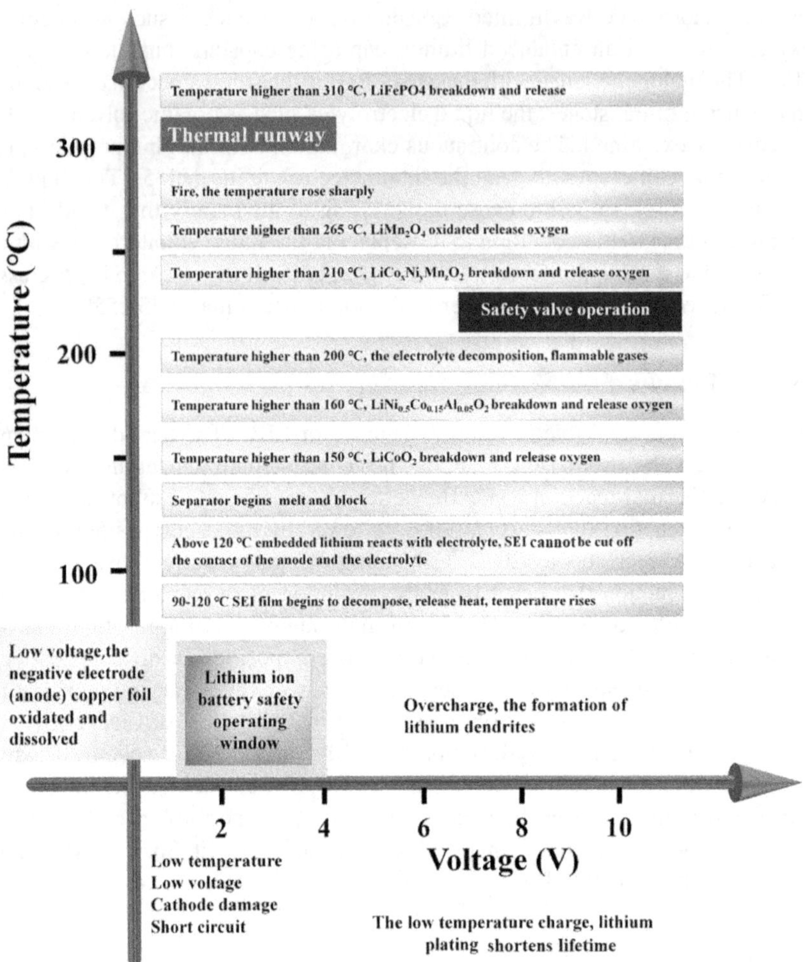

FIGURE 1.6 Lithium-ion cell operating window. Adapted and reproduced with permission from Ref. [36]. Copyright © 2018 Elsevier.

anode and the dissolution of the active material and the formation of a surface layer over the positive electrodes expressively affect the performance of commercial LIBs. Similarly, overcharge and overdischarge are another phenomena that have a serious negative impact on the performance of the battery. Overcharge can lead to the upright of the internal pressure of the battery that may cause leakage, which finally leads to the explosion of the battery. Overdischarge on the other hand can affect the life span of battery. The high current can significantly reduce the battery life.

The life of LIBs can also influenced by the temperature at which they are getting operated. It has been reported that the LIBs are best performing at room temperature without any deterioration in performance. At low temperature, the performance and life of the battery get reduced owing to the formation of lithium plating, and at high temperature of above 60°C, the chemical breakdown can cause the damage of the battery. Another state that should be considered seriously is the charge/discharge rate of the battery. It has been observed that the LIBs should be operated in a range of voltage and current to avoid any type of issues during its operation. Similarly, they can determine the degradation of the battery and hence the battery life as well. The fast charging can cause explosion of the battery. The thermal runway triggered by the fast charging can occur through the three-stage process in which the reaction between the lithium ions and electrolyte is vital to cause this serious situation.

1.5.4.1 Thermal Exploitation

When portable electronics are connected by low resistance paths, external short circuit appears. In addition to this, if liquid electrolytes are used, the chance of leakage is high leading to the internal side reactions. Water immersion can also enhance side reactions, and when an external electrical connection was made in contact with the terminals of the electrodes, the chance of electrical short circuiting and thereby ignition arises. Overheating is another major external issue which arises either due to external short circuiting or internal short circuiting. When overheating increases, the formation of SEI layer on the anode materials and the degradation of cathode materials occur, which leads to the capacity loss. This can also increase the pressure inside and swell the battery to explode.

1.5.4.2 Mechanical Exploitation

A cell generally consists of anode, cathode and electrolyte combined in jelly roll covered by thin aluminium foil. Hence, the chances of ignition mainly on the weakest part due to the mechanical abuse such as collision are extreme. The outer shell is designed in such a way to withstand any type of mechanical force and hence protects its cell components [56]. However, small deformation in the shell casing produces a trigger in the internal components mainly current collectors. Insufficient flexibility of these current collectors and separators leads to the direct contact among electrodes which heads to short circuit and hence explosion. Several reports [57,58] on nail penetration tests reveal that the passage of air near the nail penetrates by piercing leading to the exothermic oxidation reactions with the liberation of excess heat. LIB penetrates by small notch or nail head for the generation of heat sink [59]. If the same experiment is performed with higher capacity and voltage, significant damage occurs in the battery. During penetration, large amounts of discharge current overflow from the battery to produce large amount of joule heat [60]. Figure 1.7 shows reasons for the thermal runway of the cell.

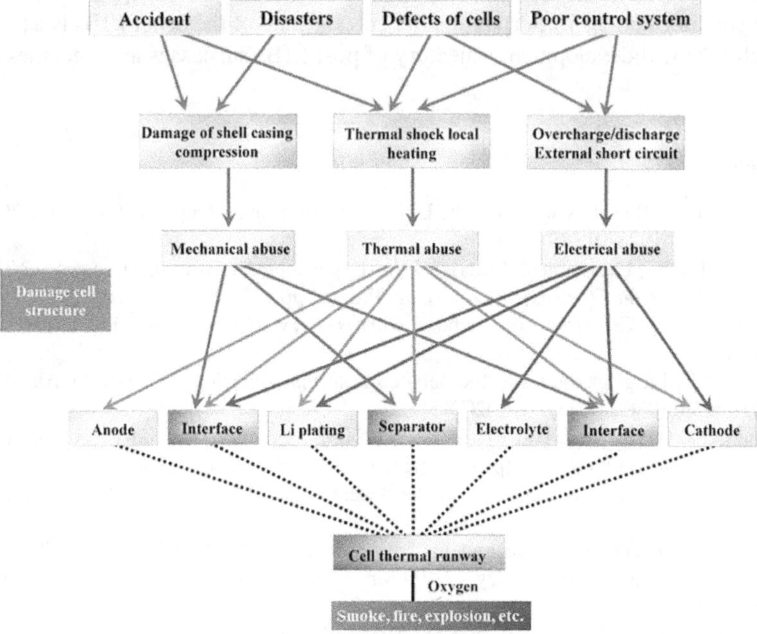

FIGURE 1.7 An overview of battery safety issues. Battery accidents, disasters, defects and poor control systems (a) lead to mechanical, thermal and/or electrical abuse (b and c), which can trigger side reactions in battery materials (d). Broken separators and oxygen released from cathodes are the main reasons for cell thermal runaway, which can generate smoke, fires and explosions with the help of oxygen from air (e). Adapted and reproduced with permission from [63]. Copyright © 2021 Elsevier.

1.5.4.3 Electrical Exploitation

Electrical exploitation occurs mainly by overdischarging or overcharging. The reasons for the overcharging include undesirable electrochemical reaction, inconsistency of battery cell and manufacturing defect. An electrolyte decomposition in the cathode phase will occur during overcharging which governs the release of high temperature. The unstable cathode materials due to the excessive lithium-ion intercalation release huge amount of oxygen resulting in the dendrite formation. Subsequently, more heat will be generated leading to the explosion of batteries. Similar to overcharging, overdischarge can also affect the degradation and explosion of batteries. Some of the battery cells touch the state of discharge (SOD) in advance [61]. During this state, if a cell is forced to discharge, then overdischarge occurs and releases excess Li ions from the anodes. This can lead to the deformation in the graphite structure and hence the destruction of SEI. Short circuiting occurs with the successive deposition of Cu ions from the current collector at the cathode material [62]. Among these two issues (overcharging and overdischarging), overcharging is more serious than overdischarging.

1.6 CONCLUSION

On growing demand of electrical storage devices, lithium shortage and safety concerns paved to a cost-effective, safer post-LIBs. The chapter also discusses the emergence of batteries and the brief history of current LIBs. This chapter mainly focused on the limitation of LIBs such as ageing, hysteresis, thermal runway, battery model and electrical abuse. Some of the inadequacy of LIBs are surmount by multivalent and monovalent ion batteries. Researchers are focusing on these post-LIBs which play a pivotal role in meeting these demands. The major post-LIBs comprise sodium-, potassium-, calcium-, chloride- and magnesium-ion batteries. Ironically, development of these batteries is still at infancy and researchers concentrated on the development and optimization of suitable electrode and electrolyte materials for possible commercialization. Among these different batteries, emergence of superior battery by surpassing the current state of the art of LIBs is a tedious process. Certainly, technological development trajectory of post-LIBs surpasses and becomes successors of current LIBs.

REFERENCES

1. Goodenough JB (2018) How we made the Li-ion rechargeable battery. *Nat Electron* 1:204–204. https://doi.org/10.1038/s41928-018-0048-6.
2. Křepelková M (2017) Evolution of Batteries: From Experiments to Everyday Usage. In: *21th International Student Conference on Electrical Engineering,* Prague, pp 1–7.
3. Chilton JP (1971) Corrosion of metals. *Corros Prev Control* 18:8–15. https://doi.org/10.1049/jipe.1951.0004.
4. Kipnis N (1987) Luigi galvani and the debate on animal electricity, 1791-1800. *Ann Sci* 44:107–142. https://doi.org/10.1080/00033798700200151.
5. Pop V, Bergveld HJ, Notten PHL, Regtien PPL (2005) State-of-the-art of battery state-of-charge determination. *Meas Sci Technol* 16. https://doi.org/10.1088/0957-0233/16/12/R01.
6. Bandyopadhyay R (2018) A transition in the development of batteries. *IJRAR – Int J Res Anal Rev* 5(2):19–21.
7. Kurzweil P (2010) Gaston planté and his invention of the lead-acid battery – The genesis of the first practical rechargeable battery. *J Power Sources* 195:4424–4434. https://doi.org/10.1016/j.jpowsour.2009.12.126.
8. Placke T, Kloepsch R, Dühnen S, Winter M (2017) Lithium ion, lithium metal, and alternative rechargeable battery technologies: The odyssey for high energy density. *J Solid State Electrochem* 21:1939–1964. https://doi.org/10.1007/s10008-017-3610-7.
9. Johansson LE (1999) End-of-life and Swedish environment policy of the 20th century. In: *Ecological Technology and Management Kalmar,* Sweden, pp 79–82.

10. Halpert G (1984) Past developments and the future of nickel electrode cell technology. *J Power Sources* 12:177–192.
11. Justine Marie E. Abarro (2023) A tale of nickel-iron batteries: Its resurgence in the age of modern batteries. *Batteries* 9(7):383. https://doi.org/10.3390/batteries9070383.
12. Chen X, Shen W, Vo TT, et al (2012) An overview of lithium-ion batteries for electric vehicles. In *10th Int Power Energy Conf IPEC 2012*. pp. 230–235. https://doi.org/10.1109/ASSCC.2012.6523269.
13. Winter M, Barnett B, Xu K (2018) Before Li ion batteries. *Chem Rev* 118:11433–11456. https://doi.org/10.1021/acs.chemrev.8b00422.
14. Winter M, Bessenhard JO (1999) Rechargeable batteries. *Chem Our Time* 33:252–266.
15. Harris WS (1958) *Electrochemical Studies in Cyclic Esters.* Univ California, Lawrence Radiat Lab, California, 8381.
16. Sells SM, Wondowski JP, Justus RF (1964) A high-rate, high-energy thermal battery system. *J Electrochem Soc* 111:6–13. https://doi.org/10.1149/1.2426065.
17. Lake PE, Dubois AR, Moroz WJ (1965) Effects of foreign ions on nickel hydroxide and cadmium electrodes. *J Electrochem Soc* 112:371–383. https://doi.org/10.1149/1.2423552.
18. Linden D, McDonald B (1980) The lithium-sulfur dioxide primary battery – Its characteristics, performance and applications. *J Power Sources* 5:35–55. https://doi.org/10.1016/0378-7753(80)80094-2.
19. Ritchie AG, Giwa CO, Bowles PG, et al (2001) Further development of lithium/polycarbon monofluoride envelope cells. *J Power Sources* 96:180–183. https://doi.org/10.1016/S0378-7753(01)00563-8.
20. Park MS, Yoon WY (2003) Characteristics of a Li/MnO_2 battery using a lithium powder anode at high-rate discharge. *J Power Sources* 114:237–243. https://doi.org/10.1016/S0378-7753(02)00581-5.
21. Reddy MV, Mauger A, Julien CM, et al (2020) Brief history of early lithium-battery development. *Materials (Basel)* 13:1–9. https://doi.org/10.3390/MA13081884.
22. Whittingham MS (1978) Chemistry of intercalation compounds: Metal guests in chalcogenide hosts. *Progress in Solid State Chemistry* 12, 1:41–99. https://doi.org/10.1016/0079-6786(78)90003-1.
23. Laman FC, Brandt K (1988) Effect of discharge current on cycle life of a rechargeable lithium battery. *J Power Sources* 24:195–206. https://doi.org/10.1016/0378-7753(88)80115-0.
24. Mizushima KJ, Jones PC, Wiseman PJ, Goodenough JB (1981) $LixCoO_2$ ($0<x<-1$): A new cathode material for batteries of high energy density. *Solid State Ionics* 4:171–174.
25. Basu S, Zeller C, Flanders PJ, et al (1979) Synthesis and properties of lithium-graphite intercalation. *Mater Sci Eng* 38:275–283.
26. Zanini M, Basu S, Fischer JE (1978) Alternate synthesis and reflectivity spectrum of stage 1 lithium-graphite intercalation compound. *Carbon N Y* 16:211–212. https://doi.org/10.1016/0008-6223(78)90026-X.
27. Besenhard JO, Theodoridou E (1982) Electrochemical applications of graphite intercalation compounds. *Synth Met* 4:211–223.
28. Yazami R, Touzain P (1983) A reversible graphite-lithium negative electrode for electrochemical generators. *J Power Sources* 9:365–371. https://doi.org/10.1016/0378-7753(83)87040-2.
29. Yoshino A (2012) The birth of the lithium-ion battery. *Angew Chemie – Int Ed* 51:5798–5800. https://doi.org/10.1002/anie.201105006.
30. Prize N History (2019) In: *Nobel Media* https://www.nobelprize.org/prizes/chemistry/2019/press-release/.
31. Dawn T, Batteries L (2016) The dawn of lithium-ion batteries. *Electrochem Soc Interface Fall* 25:71–74.
32. Mandal BK, Padhi AK, Shi Z, et al (2006) New low temperature electrolytes with thermal runaway inhibition for lithium-ion rechargeable batteries. *J Power Sources* 162:690–695. https://doi.org/10.1016/j.jpowsour.2006.06.053.
33. Bachtin K, Kaus M, Pfaffmann L, et al (2016) Comparison of electrospun and conventional $LiFePO_4$ /C composite cathodes for Li-ion batteries. *Mater Sci Eng B Solid-State Mater Adv Technol* 213:98–104. https://doi.org/10.1016/j.mseb.2016.04.006.
34. Julien CM (2017) Lithium iron phosphate: Olivine material for high power li-ion batteries. *Res Dev Mater Sci* 2:3–6. https://doi.org/10.31031/rdms.2017.02.000545.
35. Liu B, Jia Y, Li J, et al (2018) Safety issues caused by internal short circuits in lithium-ion batteries. *J Mater Chem A* 6:21475–21484. https://doi.org/10.1039/C8TA08997C.
36. Lipu MSH, Hannan MA, Hussain A, et al (2018) A review of state of health and remaining useful life estimation methods for lithium-ion battery in electric vehicles: Challenges and recommendations. *J Clean Prod* 205:115–133.

37. Su S, Huang Z, Nuli Y, et al (2015) A novel rechargeable battery with a magnesium anode, a titanium dioxide cathode, and a magnesium borohydride/tetraglyme electrolyte. *Chem Commun* 51:2641–2644. https://doi.org/10.1039/c4cc08774g.
38. Cheng Y, Chang HJ, Dong H, et al (2016) Rechargeable Mg-Li hybrid batteries: Status and challenges. *J Mater Res* 31:3125–3141. https://doi.org/10.1557/jmr.2016.331.
39. Fan X, Gaddam RR, Kumar NA, Zhao XS (2017) A hybrid Mg_{2+}/Li_+ battery based on interlayer-expanded MoS_2/graphene cathode. *Adv Energy Mater* 7:2–11. https://doi.org/10.1002/aenm.201700317.
40. Slater MD, Kim D, Lee E, Johnson CS (2013) Sodium-ion batteries. *Adv Funct Mater* 23:947–958. https://doi.org/10.1002/adfm.201200691.
41. Hwang JY, Myung ST, Sun YK (2017) Sodium-ion batteries: Present and future. *Chem Soc Rev* 46:3529–3614. https://doi.org/10.1039/c6cs00776g.
42. Ji B, Zhang F, Song X, Tang Y (2017) A novel potassium-ion-based dual-ion battery. *Adv Mater* 29: 1700519. https://doi.org/10.1002/adma.201700519.
43. Luo W, Wan J, Ozdemir B, et al (2015) Potassium ion batteries with graphitic materials. *Nano Lett* 15:7671–7677. https://doi.org/10.1021/acs.nanolett.5b03667.
44. Blomgren GE (2017) The development and future of lithium ion batteries. *J Electrochem Soc* 164:A5019–A5025. https://doi.org/10.1149/2.0251701jes.
45. Jow TR, Delp SA, Allen JL, et al (2018) Factors limiting Li+charge transfer kinetics in Li-ion batteries. *J Electrochem Soc* 165:A361–A367. https://doi.org/10.1149/2.1221802jes.
46. Naoki N, Feixiang W, Tae LJ, Gleb Y (2015) A new cathode material for batteries of high-energy density. *Mater Today* 18:252–264. https://doi.org/10.1016/j.mattod.2014.10.040.
47. Kim HJ, Bae GH, Lee SM, et al (2019) Properties of lithium iron phosphate prepared by biomass-derived carbon coating for flexible lithium ion batteries. *Electrochim Acta* 300:18–25. https://doi.org/10.1016/j.electacta.2019.01.057.
48. Zhou H, Ding X, Liu G, et al (2015) Preparation and characterization of ultralong spinel lithium manganese oxide nanofiber cathode via electrospinning method. *Electrochim Acta* 152:274–279. https://doi.org/10.1016/j.electacta.2014.11.147.
49. Choi SH, Son JW, Yoon YS, Kim J (2006) Particle size effects on temperature-dependent performance of $LiCoO_2$ in lithium batteries. *J Power Sources* 158:1419–1424. https://doi.org/10.1016/j.jpowsour.2005.10.076.
50. Eliakim R, Karmeli F (2003) Divergent effects of nicotine administration on cytokine levels in rat small bowel mucosa, colonic mucosa, and blood. *Isr Med Assoc J* 5:178–180.
51. Omar N, Verbrugge B, Mulder G, et al (2010) Evaluation of performance characteristics of various lithium-ion batteries for use in BEV application. In: *2010 IEEE Vehicle Power and Propulsion Conference*. IEEE, Lille, France, pp 1–6.
52. Liu K, Liu Y, Lin D, et al (2018) Materials for lithium-ion battery safety. *Sci Adv* 4. https://doi.org/10.1126/sciadv.aas9820.
53. Raghavan P, Choi JW, Ahn JH, et al (2008) Novel electrospun poly(vinylidene fluoride-co-hexafluoropropylene)-in situ SiO_2 composite membrane-based polymer electrolyte for lithium batteries. *J Power Sources* 184:437–443. https://doi.org/10.1016/j.jpowsour.2008.03.027.
54. Prasanth R, Aravindan V, Srinivasan M (2012) Novel polymer electrolyte based on cob-web electrospun multi component polymer blend of polyacrylonitrile/poly(methyl methacrylate)/polystyrene for lithium ion batteries – Preparation and electrochemical characterization. *J Power Sources* 202:299–307. https://doi.org/10.1016/j.jpowsour.2011.11.057.
55. Prasanth R, Shubha N, Hng HH, Srinivasan M (2013) Effect of nano-clay on ionic conductivity and electrochemical properties of poly(vinylidene fluoride) based nanocomposite porous polymer membranes and their application as polymer electrolyte in lithium ion batteries. *Eur Polym J* 49:307–318. https://doi.org/10.1016/j.eurpolymj.2012.10.033.
56. Greve L, Fehrenbach C (2012) Mechanical testing and macro-mechanical finite element simulation of the deformation, fracture, and short circuit initiation of cylindrical Lithium ion battery cells. *J Power Sources* 214:377–385. https://doi.org/10.1016/j.jpowsour.2012.04.055.
57. Zhao R, Liu J, Gu J (2016) Simulation and experimental study on lithium ion battery short circuit. *Appl Energy* 173:29–39. https://doi.org/10.1016/j.apenergy.2016.04.016.
58. Li X, Qian K, He YB, et al (2017) A dual-functional gel-polymer electrolyte for lithium ion batteries with superior rate and safety performances. *J Mater Chem A* 5:18888–18895. https://doi.org/10.1039/c7ta04415a.
59. Zhao W, Luo G, Wang C-Y (2015) Modeling nail penetration process in large-format Li-ion cells. *J Electrochem Soc* 162:A207–A217. https://doi.org/10.1149/2.1071501jes

60. Ali MY, Lai WJ, Pan J (2013) Computational models for simulations of lithium-ion battery cells under constrained compression tests. *J Power Sources* 242:325–340. https://doi.org/10.1016/j.jpowsour.2013.05.022.
61. Ouyang D, Chen M, Liu J, et al (2018) Investigation of a commercial lithium-ion battery under overcharge/over-discharge failure conditions. *RSC Adv* 8:33414–33424. https://doi.org/10.1039/C8RA05564E.
62. Li HF, Gao JK, Zhang SL (2008) Effect of overdischarge on swelling and recharge performance of lithium ion cells. *Chinese J Chem* 26:1585–1588. https://doi.org/10.1002/cjoc.200890286.
63. Chen Y, Kang Y, Zhao Y, et al (2021) A review of lithium-ion battery safety concerns: The issues, strategies, and testing standards. *J Energy Chem* 59:83–99. https://doi.org/10.1016/j.jechem.2020.10.017.

2 Ion Batteries
An Overview

Akhila Das, Neethu T. M. Balakrishnan, Leya Rose Raphel, Anandu M. Nair, Alexandru Vlad, Jabeen Fatima M. J., and Prasanth Raghavan

2.1 INTRODUCTION

The limitations of lithium-ion batteries (LIBs) paved a way for the tremendous development of post-LIBs such as multivalent and monovalent batteries. LIBs are limited by their abundance, cost, safety issues and performance. Researchers intrude the competences of different monovalent batteries such as sodium ion, potassium ion and multivalent ions such as magnesium ion and calcium ion which could reconnoitre new battery technology [1]. These batteries could perform better and support industrial application in smart grid applications and hybrid electric vehicles. Low cost and safety are the two main mandatory requirements for the large grid industrial development. Hence, researchers focused on post-LIBs with enhanced performance [2,3]. This chapter also discusses major post-LIBs and their development directions. The volumetric capacity, abundance and gravimetric capacity of different batteries were depicted in Figure 2.1.

2.2 SODIUM-ION BATTERIES

Sodium-ion batteries are at a quick pace owing to the chemical analogues between lithium- and sodium-ion batteries. The major similarities of lithium and sodium metals rely on: (i) similar chemical kinetics and (ii) similar chemistry behind its mechanism. Sodium-ion batteries are cost effective since Na metal is abundant in the earth's crust compared to LIBs [4]. However, there is a difference in the size of ionic radius and it is estimated that the ionic radius of lithium is 50% lesser than Na and hence its electrochemical properties will be different. There are several reports on hard carbon

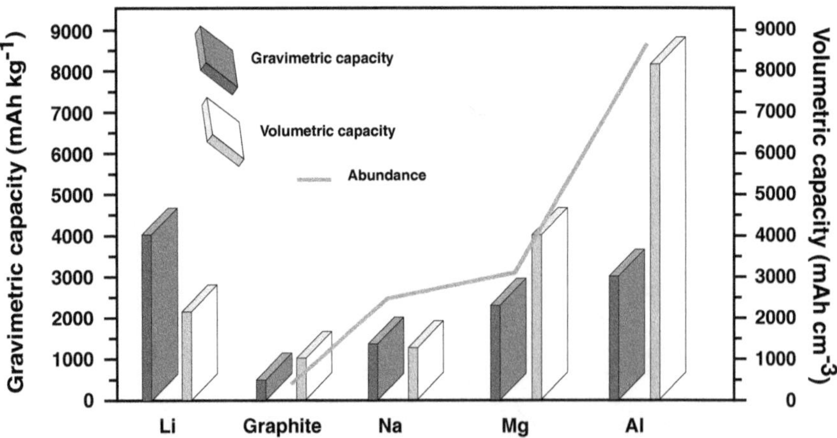

FIGURE 2.1 Comparison of gravimetric capacity, volumetric capacity and abundance of different types of batteries.

DOI: 10.1201/9781003310167-2

which has been used as anodes showing better battery performance [5–7]. However, sodium-ion batteries with hard carbon anode have poor rate capability owing to the larger size of Na metal which cannot easily intercalate in hard carbon host. At the same time, the introduction of graphene as anode material enhances cycling stability and rate capability owing to the larger void space in graphene material and hence faster intercalation of Na metal ion. In addition, rate capability and ionic conductivity have been improved with the incorporation of composite materials and carbonaceous substance to the Na metal. Types of anodes incorporated in Na metal were classified as carbonaceous materials such as graphite, alloying anodes and conversion anodes [8]. The major anode materials with their capacity range are described in Figure 2.2.

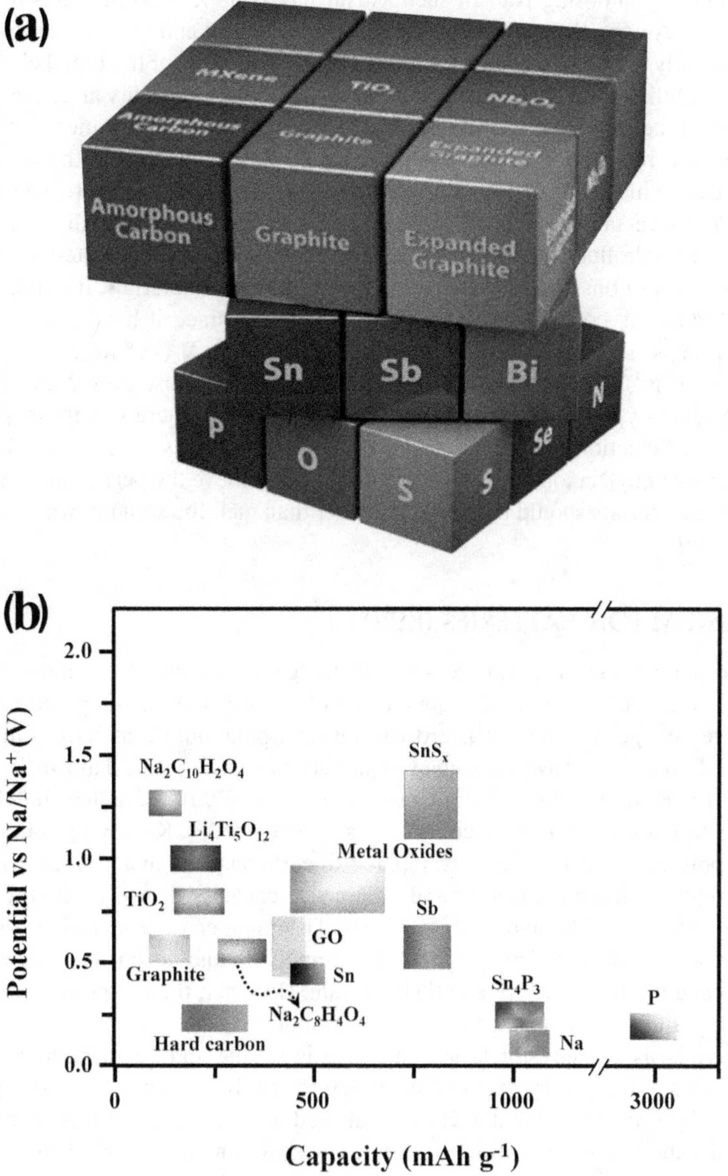

FIGURE 2.2 (a) A summary of elements and materials that are potential anodes of SIBs. (b) Some typical anodes along with their working potential range and capacity. Adapted and reproduced with permission from [8]. Copyright © 2019 Elsevier.

Electrolytes are the crucial part of the battery since it should be both electronically insulating and ionically conducting. The major criteria for Na ion conduction are as follows: (i) Thermal stability and wide liquid range, (ii) chemical inertness, (iii) environmentally benign and non-toxic and (iv) wide electrochemical stability window. Different types of electrolytes were fabricated for the application of SIB including carbonate ester types, polymer electrolytes, ionic liquid based and NASICON-type electrolytes. Lewis acidity of sodium is lower than lithium hence desolvation energy (40–70 kJ mol^{-1}) of sodium ion is lower than lithium. Carbonate-based solvents such as EC, PC, DMC and DME have compared the conductivity and viscosity, and they were optimized. EC:DME ratio has highest ionic conductivity among DMC, Triglyme, PC, DMC, EC:DMC and so on. However, viscosity was lower. Gebrekidian et al. [9] reported the effect of different salts on sodium-ion battery by choosing Na salt such as $NaClO_4$, $NaPF_6$, NaTFSI and NaFSI dissolved in carbonate solvents. Among these inorganic solvents, TFSI based and PF_6 based show better performance [10]. Similarly, polymer electrolytes are the best alternative for the liquid electrolytes in SIB owing to their flexibility, transparency, high chemical and thermal stability and so on [11,12]. Nimah et al. [12] reported the improved electrochemical performance of gel polymer electrolyte with the addition of inorganic fillers such as TiO_2 and SiO_2 along with plasticizers. Cathodes are key components in batteries in which sodium ions are interacted with the anode host lattice during the process of charging and vice versa [13]. Paula Serus et al. [13] reported the major high voltage type cathode materials which include fluorophosphate-type framework structure, sulphate-based structure and phosphate-based polyanions. Among different types of cathode materials, fluorophosphates-based cathode (Na_2FePO_4F) is considered as the suitable cathode since it has high operating voltage. Among these groups, sodium vanadium fluorophosphates ($Na_3V_2(PO_4)_2F_3$) show better specific capacity (128 mAh g^{-1}). Other cathode materials that have been used are $NaMnPO_4$, $NaFePO_4$, $Na_2Ti_3O_7$, NASICON-type framework and so on having similar structure to those of lithium-based cathode materials. Selection and optimization of suitable cathodes, anode and electrolyte is the major challenge faced by Na ion battery since it could not achieve the performance of rechargeable LIBs. New anode materials should be fabricated other than metallic sodium, which has tendency to form dendrites [14].

2.3 POTASSIUM-ION BATTERIES (KIBs)

In recent decades, potassium-ion batteries (KIBs) emerge as a potential alternative to sodium- and lithium-ion batteries. Potassium is also abundant in the earth's crust compared to LIBs and has a high operative voltage. Due to the fast diffusion rate, potassium ions have a high power density. Similar to LIBs, KIBs have rock-chair type batteries and the mechanism is similar to that of LIBs and SIBs. However, the mass to charge ratio of K^+ (39.10) is higher than Li^+ (6.94). The major anode materials used in KIBs include oxides, chalcogenides, K alloying compounds, organic materials and polyanionic compounds. Compared to carbonaceous material, alloying compounds often exhibit high gravimetric capacity and volumetric capacity. However, high operating range was expected for chalcogenides and oxide materials. The major cathode materials used are Prussian blue analogues $KFe[Fe(CN)_6]$ [15], polyanionic compounds and so on. Polyanionic compounds have structural and thermal stability and they are safe. However, these types of cathodes have low material capacity and specific capacity. Layered cathode materials are another class of cathodes in which the synthesis strategy for large-scale grid is simple and has a high material capacity. These compounds are dragged by potential polarization, low K ion content and high phase changes. Hexacyanometallates are another cathode material used in KIB, which have high thermal stability, rapid interstitial stability and so on. Still these materials have low volumetric density and high water content [16]. The best alternative solution for this was enhancing the performance by doping with efficient dopants. Different electrolytes are employed in KIB including aqueous electrolytes, ionic electrolytes, organic electrolytes and polymer electrolytes. Research focuses on the development of suitable electrolytes that are compatible with anodes and cathodes [17–19]. Though KIB is almost similar to rechargeable LIBs, inherent difference between K and Li metals including ionic radius

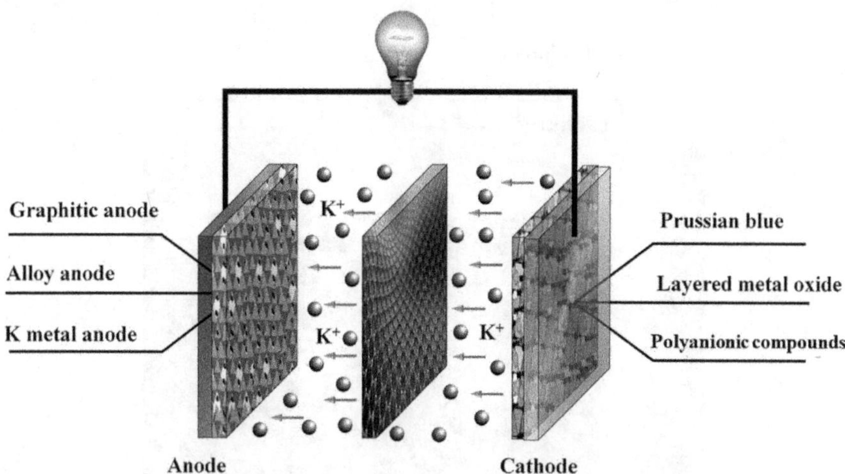

FIGURE 2.3 Research status of potassium-ion battery. The anode materials mainly include carbon-based anode, alloys anode and K metal anode (left). The types of cathode electrodes are Prussian blue (PB) and its analogues, layered metal oxides and polyanionic compounds (right). Adapted and reproduced with permission from [16]. Copyright © 2020 John Wiley and sons.

should be taken into consideration. Low volumetric capacity and gravimetric capacity is the major hurdle faced by KIB. Other hindrance faced during commercialization of KIB is the high reactivity of K ion and the formation of dendrites, which decline the electrochemical performance of the battery. Research and development focuses on cracking these issues related to rechargeable KIB, and it should be established as future battery in the scientific community. The major components of KIB and the types of anode and cathod materials were illustrated in Figure 2.3.

2.4 CALCIUM-ION BATTERIES

Several multivalent ions have been introduced as an alternative to the rechargeable LIBs such as Mg^{2+}, Ca^{2+} and Zn^{2+}. It was noted that divalent metals such as Ca and Mg have uniform distribution on the electrode/electrolyte system and the chance of forming dendrites is feasible. Hence, calcium-ion battery has been investigated by a group of researchers. Comparing the abundance of calcium with Mg and Na, calcium is the fifth most abundant mineral in the earth's crust and hence they are cost effective. Ca is environmental benign material and it has higher deposition potential (0.17 V) which is higher than lithium metal. Since reversible stripping and deposition was introduced only in the year 2016, calcium-ion batteries are considered to be in infant stage. The volumetric capacity of Ca is 2,072 mAh mL^{-1} which is higher than Li metal ion, and metallic Ca itself can act as anode material [20–22]. Since Ca metal belongs to alkali earth metal, a spontaneous solid electrode interface was formed in nonaqueous electrolyte which allows ions through the electrolytes and not the electrons. Earlier studies of calcium thionyl chloride reveal the formation of thicker SEI layer which even blocks the ion migration pathway and thus increases the polarization [23]. The major drawback of using Ca itself as metallic anode includes voltage instability, higher impedance and poor reversibility. Hence, an alternative to anode materials is alloys such as Ca-Si intermetallic alloys. Specific cathode materials employed calcium-ion batteries that include layered structures like V_2O_5 layer, 3D tunnel structures, Chevrel phases ($CaMo_6S_8$), etc. Among these cathode materials, V_2O_5-based layered materials have tendency to form co-intercalation of water from these electrolytes in which the host lattice and intercalating ions effectively shield the interaction. Prussian blue analogues are dragged by the poor specific capacity and low capacity retention which are not suitable for high-energy-density applications. A suitable electrolyte is necessary for

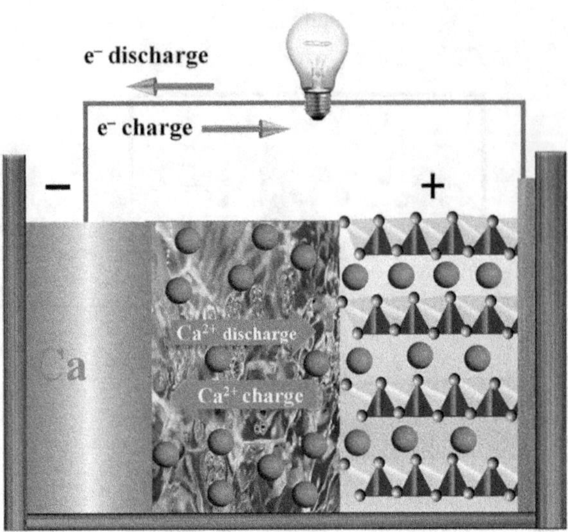

FIGURE 2.4 Schematic of a Ca battery using a Ca metal anode and an intercalation cathode. Adapted and reproduced with permission from [26]. Copyright © 2020 American Chemical Society.

the better electrochemical performance which should be both ionically conducting and electronically insulating, forming a stable SEI layer. Electrolytes in EC:PC dissolving THF solvent could not achieve this criterion, and hence, Muldoon et al. [24], who investigated Grignard-based electrolytes in magnesium-ion batteries (MIBs), developed similar type of electrolytes in calcium-ion batteries. Moreover, these electrolytes have low anodic stability and are not applicable for high voltage applications. Later, several research on heavy Grignard has been investigated and its studies are at early stages. The general working principle of calcium-ion battery is depicted in Figure 2.4 [25].

2.5 MAGNESIUM-ION BATTERIES (MIBs)

The major barrier that declines the development of LIB is its cost due to the shortage of raw material. Many of the researchers in this decade are trying to find a non-lithium metal in which they can improve most of the defects caused by lithium-based batteries. Most lithium-based batteries' fires and explosion come down due to the needle-like structure that developed between the electrodes in it resulting in short circuiting. Availability of the raw material is an important factor. In contrast to lithium, magnesium has attained an attractive attention in the research and development field since it is abundant in the earth's crust and safer. Magnesium batteries also possess high theoretical volumetric capacity (32.731 GJ m^{-3}), which is almost double the volumetric capacity of LIBs (22.569 GJ m^{-3}) [5]. The basic working principle behind MIBs is similar to that of LIBs. Magnesium is found to be a better alternative since the magnesium metal that carries +2 charge shuttles between magnesium metal as anode and cathode that carries charge. The first rechargeable magnesium-ion battery (rMIB) was discovered by Aurbach's group in 2000 in which magnesium metal is set as anode, Chevrel phase Mo_6S_8 as cathode and Mg $(AlCl_2BuEt)_2$ as electrolyte. Later on, several improvements occurred in electrolyte in which organoborates were first used (1.9 V), which are substituted by Grignard reagents. The cathode used by Aurbach was found to be most successful one. Even though several works were done in this field, it took 15 years for the discovery of practical rechargeable MIB that was done by Muldoon who synthesized a non-nucleophilic electrolyte and electrochemically active species from Hexamethyldisilazide Magnesium Chloride (HMDS) and $AlCl_3$. This battery shows wide electrochemical window and low corrosive nature. The Chevrel phase has low operational voltage and many works were progressing for the modification of cathode.

There are several challenges faced to find suitable cathodes that should satisfy the requirements such as cyclability, reversibility and energy density. Similar to MIB, investigation on magnesium metal batteries [27] mainly magnesium sulphur (Mg-S) batteries was carried out, which have a high theoretical capacity than Li-S battery (2062 vs 3832 mAh cm^{-3}) [27]. In Mg-S batteries, redox reaction occurs between Mg metal which serves as anode and sulphur which serves as cathode. During oxidation, Mg releases 2 electrons and Mg^{2+} moves towards the cathodes through the separator and electrolytes. This Mg ion will be combined with S ion and transformed to Polysulfone chains to form MgS. Mg^{2+} is deposited at the anode during the charging process and hence continues. Low sulphur utilization from the cathodes, severe overcharge behaviour and poor electrochemical behaviour are some of the drawbacks faced by them [28].

2.6 FLUORIDE-ION BATTERIES (FIBs)

Besides cation batteries, anion batteries are emerged as a new alternative to the rechargeable LIBs. Among anion batteries, fluoride-ion batteries (FIBs) are prominent owing to the higher electronegative element in the periodic table, which have high electrochemical stability window and are stable. When the electrochemical couple is either discharged or charged, transfer of fluoride ions takes place by charge compensation mechanism. The theoretical working principle of FIB is simple considering Mg as the anode and BiF_3 as the cathode as an example (Figure 2.5) [29–31]. During discharge process, oxidation of anode (Mg) occurs with the release of two electrons to form MgF_2. At cathode, reduction takes place in which electron reduces to form Bi from BiF_3. Migration of F^- through the electrolyte confirms the charge neutrality. The basic reaction in the FIB is given in the following equations:

$$\text{Cathode reaction: } MF_x + xe^- \rightarrow M + xF^- \tag{2.1}$$

$$\text{Anode reaction: } M' + yF^- \rightarrow MF_y + ye^- \tag{2.2}$$

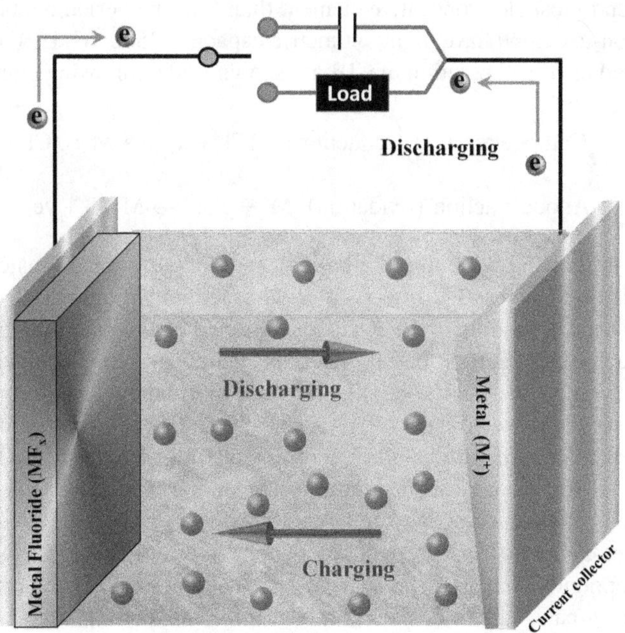

FIGURE 2.5 Schematic illustration of fluoride-ion battery.

TABLE 2.1
Major Breakthrough and Roadblock in FIB Development

Type of Battery	Breakthrough	Principal Obstacle
HTFIB	• Proof of principle for a second FIB	• Battery beaks apart due to volume changes
	• In-depth analysis of a superionic conductor as electrolyte	• Fading of capacity
	• Use of Mg anode in HTFIB	• Low fluoride conduction
	• Thin film for HTFIB	
RTFIB	• Proof of principle for a functional FIB at RT with a liquid electrolyte	• No cycling due to passivation of the anode
	• Liquid electrolytes based on fluoride-doped PEG polymer	• Low ionic conductivity in the electrolyte
	• Mg as anode	• Solubility of the fluoride compound must be improved
	• Optimization of cathode preparation	• Reproducibility

The major advantage of using metal fluoride as cathode materials includes higher gravimetric capacity and volumetric capacity. Like standard anodes and cathodes in rechargeable LIBs, no such materials were established in FIBs. Generally, FIBs are divided into two: Room temperature FIB (RTFIB), which are liquid electrolytes, and high-temperature FIB (HTFIB), which are all solid-state FIBs. The major pros and cons of these two types of batteries are shown in Table 2.1; the major road block of marketing rechargeable FIBs is its safety since fluorine atom is highly reactive.

2.7 CHLORIDE-ION BATTERIES (CIBs)

Chloride-ion batteries (CIBs) are another type of anion batteries similar to that of FIB in which chlorine is the second most electronegative element than F in the periodic table. In addition, chlorine is abundant, non-toxic and have high volumetric capacity (2500 Wh L^{-1}) than LIBs. The schematic illustration and chemical reaction in CIB are shown in the following equations:

$$\text{Cathode reaction (reduction): } MCl_x + xe^- \rightarrow M + xCl \tag{2.3}$$

$$\text{Anode reaction (oxidation): } M' + yCl^- \rightarrow M'Cl_y + ye^- \tag{2.4}$$

The first CIB was introduced with alkali, alkaline earth metal as anode material and Lewis acid metal chloride as cathode material [20] with chloride-based ionic liquid as electrolyte. However, it experiences severe capacity fading and FeOCl-based doped conducting polymer electrolyte offers better specific capacity [32]. Carbon-based materials such as graphite as anode have been widely explored in CIBs owing to their controllability, surface morphology and so on. Different cathode materials were investigated including BiOCl, $CoCl_2$ and $CuCl_2$. However, these types of cathode materials have inadequate structure stability and less cycling performance. Optimization of suitable cathode material for CIBs is major hindrance for its development. Chloride-ion conduction (CC) is mainly due to the anion vacancy and Schottky effect. $SrCl_2$, $BrCl_2$, etc. were exploited as chloride-ion conductors but have good conductivity behaviour only at high temperature [33]. Later, solid inorganic compounds such as cryptands and crown ethers were also reported with good transference number and greater lattice defects [34]. Recently, polymer electrolytes were investigated [35,36] which have apparent chemical and thermal stability, flexibility and good ionic conductivity. Some of the cation conducting polymers conduct anions governing the total ionic conductivity. Ionic

TABLE 2.2
Overview of the Existing Literature on CIBs

Cathode	Anode	Electrolyte	Best Observed Cycling	Best Observed Capacity (First Discharge, mAh g⁻¹)	Estimated Nominal Voltage	OCV	Discussed Problems
$CoCl_2$							Dissolution of cathode
VCl_3	Li	OMMCl	3	140	2.4		material
$BiCl_3$		$BMMBF_4$					
FeOCl		$Ni_{116}Cl$					Postulated transformation of
BiOCl	Li	Ni_{114} TFSI	6	60	2.2		BiOCl to bioxide was not clearly observed
FeOCl	Mg	PP14Cl	30	125	1		Formation of a $MgCl_2$
BiOCl		PP15 TFSI					passivation layer
VOCl	Mg	PP14Cl	50	80	1-1.2		No clear evidence of VO
	$MgCl_2$	PP15 TFSI					
VOCl	Li	PP14Cl In PC	50	150	1.6		Possible intercalation of cations in VOCl
PPy on CNT	Li	PP14Cl PP15TFSI	40	118	2.2		Using polymer as cathode
$BiCl_3$ PANI	Zn	TBACl in Gelatine, PVdF-HF	-	150	0.8		Very rudimental battery setup, built in air. Only discharge tests

Source: Adapted and reproduced with permission from [37]. Copyright © 2017 Wiley.

liquids are the major class of CIB electrolytes [20], whereas studies on polymer electrolytes are few. Several other combinations were introduced by researchers; Table 2.2 depicts major anodes, cathodes and electrolytes employed in CIB and the major drawbacks faced by these CIBs along with its electrochemical properties.

2.8 ALUMINIUM-ION BATTERIES (AIBs)

Aluminium-based batteries are another class of post-lithium batteries explored attributable to the advantage of abundance, safer, inertness, ease of handling and high volumetric capacity as they are sixfold denser than lithium (8.04 vs 2.06 mAh cm^{-3} for lithium) [15]. Aluminium can exchange three electrons at a time compared to Li^+ which can only exchange one electron per cation and has similar ionic radii (0.54A° for Al and 0.76 A° for Li) [38]. Aluminium batteries are generally classified as aluminium-ion batteries (AIBs) and aluminium air batteries [39]. The pioneer work of AIB was started in the year 1855; however, it remained unnoticed due to its least reversible stripping/plating. During the initial stage, pure $AlCl_3$ was chosen as electrolyte, and later, its performance was improved by the addition of small amount of LiCl. In the meantime, research were also focused on cathode materials to substitute toxic Cl_2 by FeS_2. Later, Al-ion battery was fabricated with FeS_2 as cathode material, and a mixture of $AlCl_3$ and NaCl as electrolyte exhibits high-temperature electrochemical performance and cycling stability up to 200 cycles with ~100% columbic efficiency [40]. Different other combinations were also experimented by researchers to attain high energy density and lifespan. However, Al-ion battery works only at high operational temperature, and later, a breakthrough on rechargeable Al-ion batteries working at ambient temperature was accomplished in 2010. The complete knowledge as well as behaviour of cathodes and electrolytes in AIBs acquired only by the stimulation analysis and experimental strategies. In addition, the complete mechanism of electrode–electrolyte interface was not revealed yet [41]. In future, researchers would discover a fruitful result which could narrow the gap of limitations and create a new opportunity for rechargeable AIB.

2.9 CONCLUSION

The development and optimization of different cathode, anode and electrolyte materials in LIBs leads to the commercialization of electric vehicle in technological world. Although the development of LIBs is cutting edge, the technological boom is handicapped by its safety measurement and its cost. Many post-LIBs such as potassium-, sodium-, chloride-, calcium-, fluoride- and magnesium-ion batteries compete with conventional LIBs in battery research community. This chapter debates the working principle and state of art of current post-LIBs. From the schematic representation, it is obvious that the working principle and chemistry in the next-generation batteries are almost similar. Hence, these batteries are the powerful tool for the future scientific and technological innovations. The energy density of post-LIBs could not stand in line with conventional LIBs, and today, R&D department mainly focuses on the fabrication of suitable cathode materials which can contend with rechargeable LIBs.

REFERENCES

1. Walter M, Kovalenko MV, Kravchyk K V (2020) Challenges and benefits of post-lithium-ion batteries. *New J Chem* 44:1677–1683. https://doi.org/10.1039/c9nj05682c.
2. Kulova TL, Fateev VN, Seregina EA, Grigoriev AS (2020) A brief review of post-lithium-ion batteries. *Int J Electrochem Sci* 15:7242–7259. https://doi.org/10.20964/2020.08.22.
3. Sawicki M, Shaw LL (2015) Advances and challenges of sodium ion batteries as post lithium-ion batteries. *RSC Adv* 5:53129–53154. https://doi.org/10.1039/c5ra08321d.
4. Chayambuka K, Mulder G, Danilov DL, Notten PHL (2018) Sodium-ion battery materials and electrochemical properties reviewed. *Adv Energy Mater* 8:1–49. https://doi.org/10.1002/aenm.201800079.
5. Hwang JY, Myung ST, Sun YK (2017) Sodium-ion batteries: Present and future. *Chem Soc Rev* 46:3529–3614. https://doi.org/10.1039/c6cs00776g.
6. Shadike Z, Zhao E, Zhou YN, et al (2018) Advanced characterization techniques for sodium-ion battery studies. *Adv Energy Mater* 8:1–29. https://doi.org/10.1002/aenm.201702588.
7. Wang Y, Wang C, Wang Y, et al (2016) Boric acid assisted reduction of graphene oxide: A promising material for sodium-ion batteries. *ACS Appl Mater Interfaces* 8:18860–18866. https://doi.org/10.1021/acsami.6b04774.
8. Zhang W, Zhang F, Ming F, Alshareef HN (2019) Sodium-ion battery anodes: Status and future trends. *EnergyChem* 100012. https://doi.org/10.1016/j.enchem.2019.100012.
9. Eshetu GG, Elia GA, Armand M, et al (2020) Electrolytes and interphases in sodium-based rechargeable batteries: Recent advances and perspectives. *Adv Energy Mater* 10. https://doi.org/10.1002/aenm.202000093.
10. Eshetu GG, Grugeon S, Kim H, et al (2016) Comprehensive insights into the reactivity of electrolytes based on sodium ions. *ChemSusChem* 9:462–471. https://doi.org/10.1002/cssc.201501605.
11. Yang YQ, Chang Z, Li MX, et al (2015) A sodium ion conducting gel polymer electrolyte. *Solid State Ion* 269:1–7. https://doi.org/10.1016/j.ssi.2014.11.015.
12. Ni'Mah YL, Cheng MY, Cheng JH, et al (2015) Solid-state polymer nanocomposite electrolyte of $TiO_2/PEO/NaClO_4$ for sodium ion batteries. *J Power Sources* 278:375–381. https://doi.org/10.1016/j.jpowsour.2014.11.047.
13. Chen M, Zhang Y, Xing G, Tang Y (2020) Building High power density of sodium-ion batteries: Importance of multidimensional diffusion pathways in cathode materials. *Front Chem* 8:1–7. https://doi.org/10.3389/fchem.2020.00152.
14. Palomares V, Serras P, Villaluenga I, et al (2012) Na-ion batteries, recent advances and present challenges to become low cost energy storage systems. *Energy Environ Sci* 5:5884–5901. https://doi.org/10.1039/c2ee02781j.
15. Levy NR, Ein-Eli Y (2020) Aluminum-ion battery technology: A rising star or a devastating fall? *J Solid State Electrochem* 24:2067–2071. https://doi.org/10.1007/s10008-020-04598-y.

16. Liu Y, Gao C, Dai L, et al (2020) The features and progress of electrolyte for potassium ion batteries. *Small* 2004096:1–13. https://doi.org/10.1002/smll.202004096.
17. Rajagopalan R, Tang Y, Ji X, et al (2020) Advancements and challenges in potassium ion batteries: A comprehensive review. *Adv Funct Mater* 1909486:1–35. https://doi.org/10.1002/adfm.201909486.
18. Pramudita JC, Sehrawat D, Goonetilleke D, Sharma N (2017) An initial review of the status of electrode materials for potassium-ion batteries. *Adv Energy Mater* 1602911:1–21. https://doi.org/10.1002/aenm.201602911
19. Zhang W, Liu Y, Guo Z (2019) Approaching high-performance potassium-ion batteries via advanced design strategies and enneering *Sci Adv* 5:eaav7412.
20. Zhao X, Ren S, Bruns M, Fichtner M (2014) Chloride ion battery : A new member in the rechargeable battery family. *J Power Sources* 245:706–711. https://doi.org/10.1016/j.jpowsour.2013.07.001.
21. Ponrouch A, Palacin MR (2018) On the road toward calcium-based batteries. *Curr Opin Electrochem* 9:1–7. https://doi.org/10.1016/j.coelec.2018.02.001.
22. Stievano L, De Meatza I, Bitenc J, et al (2021) Emerging calcium batteries. *J Power Sources* 482:228875. https://doi.org/10.1016/j.jpowsour.2020.228875.
23. Meitav A, Peled E (1982) Calcium-Ca(AICI4)2-thionyl chloride cell: Performance and safety. *J Electrochem Soc* 129:451.
24. Muldoon J, Bucur CB, Gregory T (2017) Fervent hype behind magnesium batteries: An open call to synthetic chemists - electrolytes and cathodes needed. *Angew Chemie* 129:12232–12253. https://doi.org/10.1002/ange.201700673.
25. Dompablo MEA, Ponrouch A, Johansson P, Palac MR (2019) Achievements, challenges, and prospects of calcium batteries. https://doi.org/10.1021/acs.chemrev.9b00339.
26. Arroyo-De Dompablo ME, Ponrouch A, Johansson P, Palacín MR (2020) Achievements, challenges, and prospects of calcium batteries. *Chem Rev* 120:6331–6357. https://doi.org/10.1021/acs.chemrev.9b00339.
27. Vinayan BP, Zhao-Karger Z, Diemant T, et al (2016) Performance study of magnesium-sulfur battery using a graphene based sulfur composite cathode electrode and a non-nucleophilic Mg electrolyte. *Nanoscale* 8:3296–3306. https://doi.org/10.1039/c5nr04383b.
28. Wang P, Buchmeiser MR (2019) Rechargeable magnesium-sulfur battery technology: State of the art and key challenges. *Adv Funct Mater* 29. https://doi.org/10.1002/adfm.201905248.
29. Rongeat C, Reddy MA, Witter R, Fichtner M (2013) Nanostructured fluorite-type fluorides as electrolytes for fluoride ion batteries. *J Phys Chem C* 117:4943–4950.
30. Fabienne S (2014) A fluoride-doped PEG matrix as an electrolyte for anion transportation in a room-temperature fluoride ion battery. *J Mater Chem A* 2:1214–1218. https://doi.org/10.1039/c3ta13881j.
31. Gschwind F, Rodriguez-garcia G, Sandbeck DJS, et al (2016) Fluoride ion batteries: Theoretical performance, safety, toxicity, and a combinatorial screening of new electrodes. *J Flu Chem* 182:76–90. https://doi.org/10.1016/j.jfluchem.2015.12.002.
32. Zhao X, Zhao-karger Z, Wang D, Fichtner M (2013) Metal oxychlorides as cathode materials for chloride ion batteries. *Angew Chem* 125:13866–13869. https://doi.org/10.1002/ange.201307314.
33. Okamoto K, Imanaka N, Adachi G (2002) Chloride ion conduction in rare earth oxychlorides. *Solid State Ionics* 155:577–580.
34. Euchner H (2017) Chloride ion battery review: Theoretical calculations, state of the art, safety, toxicity and an outlook towards future developments. *Eur J Inorg Chem* 2017:2784–2799. https://doi.org/10.1002/ejic.201700288
35. Ogawa K, Hirano S (1984) Chloride ion conductivity in a plasticized quaternary ammonium polymer. *Macromolecules* 17:975–977.
36. Qiao J, Zhang J, Zhang J (2013) Anion conducting poly(vinyl alcohol)/poly(diallyldimethylammonium chloride) membranes with high durable alkaline stability for polymer electrolyte membrane fuel cells. *J Power Sources* 237:1–4. https://doi.org/10.1016/j.jpowsour.2013.02.059.
37. Gschwind, F, Euchner R-G (2017) Chloride ion battery review: Theoretical calculations, state of the art, safety, toxicity and an outlook towards future developments. *Eur J Inorg Chem* 2017:2784–2799.
38. Das SK, Mahapatra S, Lahan H (2017) Aluminium-ion batteries: Developments and challenges. *J Mater Chem A* 5:6347–6367. https://doi.org/10.1039/c7ta00228a.
39. Leisegang T, Meutzner F, Zschornak M, et al (2019) The aluminum-ion battery: A sustainable and seminal concept? *Front Chem* 7. https://doi.org/10.3389/fchem.2019.00268.
40. Holleck GL (1972) The reduction of chlorine on carbon in AlCl[sub 3]-KCl-NaCl melts. *J Electrochem Soc* 119:1158. https://doi.org/10.1149/1.2404432.
41. Hu Y, Sun D, Luo B, Wang L (2019) Recent progress and future trends of aluminum batteries. *Energy Technol* 7:86–106. https://doi.org/10.1002/ente.201800550.

3 Magnesium-Ion Batteries
An Overview

Anjumole P. Thomas, Akhila Das, Anusree Thilak, Abhilash Pullanchiyodan, Jou-Hyeon Ahn, M. V. Reddy, and Prasanth Raghavan

3.1 INTRODUCTION

Electrochemical energy storage is considered a prominent innovation in human history. From the invention of Alessandro Volta's voltaic pile, an early electric battery, which produced a steady electric current [1], to lithium batteries that drive electric cars, has great importance in our society. Many electrochemical energy storage systems such as lithium-ion batteries (LIBs) have been invented and commercialized to make life more convenient [2]. Li-ion secondary batteries are one of the most powerful achievements among other energy storage devices of their high energy and power density, which allow users to operate their devices for longer times and find applications in military and space applications [3]. Due to great concern over worldwide environmental issues, such as global warming, more focus is deviating in the direction of a new candidate with environmental friendliness and compatibility. Even though LIBs are the present leading technology for electric energy storage around the world, LIBs have disadvantages such as relatively high costs, safety limitations, and scarcity of lithium. The magnesium battery is an alternative prospective candidate, conceivably offering a cost-effective and safe option for future energy storage devices [4].

The role of energy production and consumption plays an important role in our society. Trends towards reduced fossil fuel dependence are encouraged worldwide on global concern regarding pollution, climate change, and resource scarcity. It will not be possible for many of the current policies to continue moving forward to depend on fossil fuels to power the nation [5]. Renewable energies are not that much exploited up to our needs but will need to grow rapidly to keep pace with world energy consumption, which is projected to grow by 56% between 2010 and 2040 [6]. There are substantial efforts that have been made to develop and install renewable energy harvesting technology [7]. However, the establishment of reliable and robust storage devices since harvesting renewable energy and electricity is inherently intermittent and does not stand out in the final outcome [8].

Therefore, the need for more economically viable and sustainable energy storage devices to electrify our society is a great concern nowadays. LIBs are the most reliable technology in the current market. But at the same time, improved electrochemical energy storage applications should come from other sources with a reduction in cost and improved safety which is more appreciable. Sodium-ion (Na) and magnesium-ion (Mg) batteries are two new budding technologies that might resolve the problems causing LIBs in terms of safety and efficiency [8]. Both metals are cheaper and more abundant than Li, while divalent magnesium has the added bonus of passing twice as much charge per atom. On the other side, both are still in research labs with challenges to overcome. Magnesium-ion battery technology is promising for several reasons. First of all, due to the higher abundance of magnesium in the earth's crust, approximately 104 times that of lithium, its incorporation into electrode materials is inexpensive [10]. Figure 3.1 shows the relative abundance and electrode potential of different elements in metal ion battery technology. Second, magnesium is more atmosphere stable and has a higher melting point than lithium, making it safer relative to

DOI: 10.1201/9781003310167-3

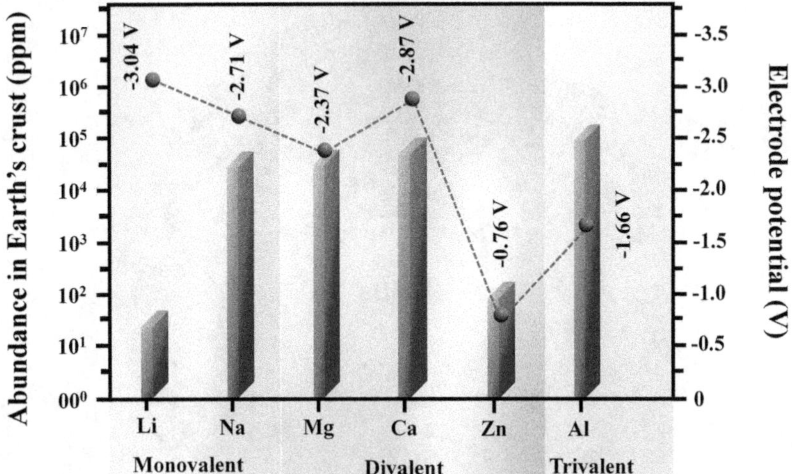

FIGURE 3.1 Schematic illustrating the abundance in the earth's crust (bar chart, left y-axis) and electrode potentials (line chart, right y-axis) of elements that are of interest to the rechargeable metal ion battery industry. Adapted and Reproduced from Ref. [9] Copyright © 2019 Elsevier.

lithium. Due to the divalent nature of magnesium ions, it possesses a high volumetric capacity (3833 mAh cm^{-3} for Mg vs 2046 mAh cm^{-3} for Li) [10].

The rechargeable magnesium batteries (RMBs) consist of cathode, anode, and electrolyte, and the working principle is demonstrated in Figure 3.2. The de-intercalation reaction of Mg^{2+} ions can be realized from Figure 3.2 where the Mg^{2+} ions are released from the anode and inserted into the cathode in the discharge process; the process is reversed during charging, and the Mg^{2+} ions are released from the cathode and deposited on the anode.

This chapter will overview the key materials of RMBs: electrodes (cathodes and anode) and electrolytes. Figure 3.3 shows different types of cathodes, anode, and electrolyte used in rechargeable magnesium-ion battery. According to different storage mechanisms, the cathode materials of RMBs, such as intercalation-type cathodes, are explained in detail. The modification and interface issues of Mg anode materials are also explained. The crucial developments of RMB electrolytes and their future perspective with recent research reports are analysed and summarized.

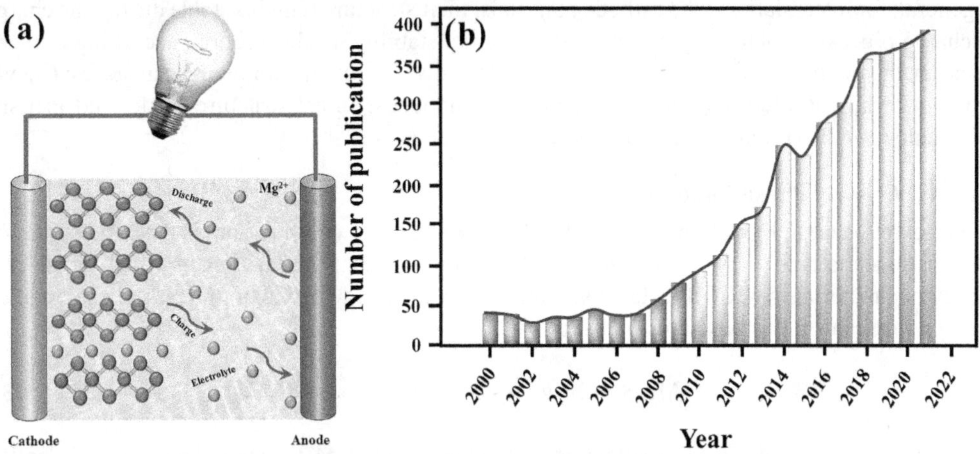

FIGURE 3.2 (a) The principle of RMBs. (b) The number of published papers related to magnesium batteries from 2000 to 2019. Adapted and reproduced from Ref. [11]. Copyright © 2019 Elsevier.

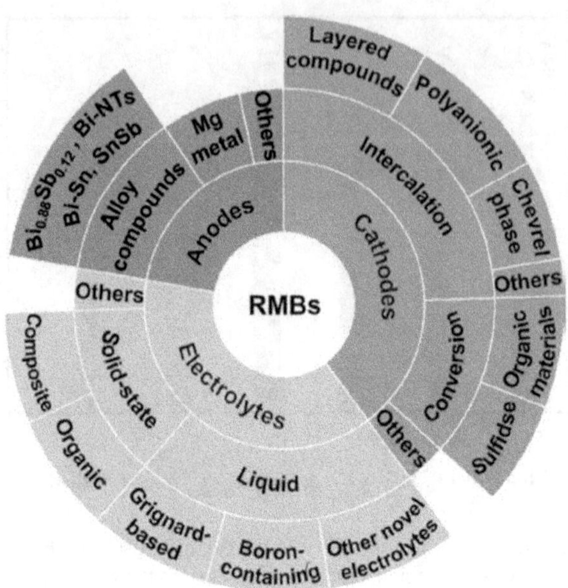

FIGURE 3.3 An overview of key materials for rechargeable magnesium batteries (RMBs). Adapted and reproduced from Ref. [11] Copyright 2021 Wiley.

3.2 CATHODE MATERIALS

The first challenge in producing RMBs is finding suitable high-energy cathode materials which have a high reversible capacity and adequate operating voltage under appropriate power output conditions. Due to the high valency of Mg^{2+} ions, the ion diffusion is very slow in inorganic cathode material resulting in low reversible capacity and reduced power input. In order to solve these issues, different strategies were adopted like the use of large surface area nano- and mesoporous materials as cathode materials [12]. With the recent development in cathode materials for magnesium batteries, different types of cathode compositions and various phases were explained in detail.

3.2.1 Intercalation-Type Cathode

In general, for intercalation-type cathode materials, their structure remains stable during the charge/discharge process, which can guarantee good cycling stability. In this section, according to recent research, the intercalation-type cathode materials are divided into the following categories: Chevrel phase materials, layered materials, polyanionic compounds, spinel structure oxide, and Prussian blue analogue (PBA) were discussed in detail.

3.2.1.1 Chevrel Phase Materials

The Chevrel phase (CP) with the formula Mo_6T_8 (T=S, Se, or a combination) is the earliest intercalated cathode material for Mg^{2+} ions storage which can be seen as the stacking of Mo_6S_8 blocks [13]. As cathode materials for RMBs, Mg^{2+} ions can insert and occupy two main interstitial sites, and the reaction mechanism can be expressed as follows:

$$Mo_6S_8 + 2Mg^{2+} + 4e^- \rightarrow Mg_2Mo_6S_8. \tag{3.1}$$

Mo_3S_4 with CP structure is an excellent cathode material for RMBs; when the discharge depth is 100%, it could be cycled more than 2000 times with the capacity retention exceeding 85% Mo_6S_8 [14]. Synthesized by acid leaching of copper from $Cu_2Mo_6S_8$, it exhibits good electrochemical performance such as long cycle (150 cycles), rate capability, and high discharge capacity (≈76 mAh g^{-1} at 20 mA g^{-1}) [15]. However, during the first charge, only 60%–80% Mg^{2+} ions are extracted from the Mo_6S_8 cathode, and this is attributed to the partial charge entrapment. The other approach is to reduce the particle size of the Chevrel Mo_6S_8 powder through mechanical milling [16]. The smaller particles can provide shorter Mg^{2+} diffusion lengths, leading to enhanced rate capabilities, and the sub-micro-sized CP (S-CP) exhibits higher discharge capacity than that of the micro-sized CP (M-CP).

3.2.1.2 Layered Material

The layered structure material offers 2D transmission channels and active insertion sites for Mg^{2+} ions; moreover, its high electronic conductivity is beneficial for enhancing the kinetics of the Mg^{2+} ion, thus improving the Mg storage capacity.

3.2.1.2.1 Transition-Metal Oxides

There is great importance given to discovering new cathode materials for rechargeable Mg batteries. High-capacity materials are essential for increasing the energy density of rechargeable Mg batteries. Various metal oxide materials are suggested as cathode material for magnesium batteries as shown in Figure 3.4. By theory, transition metals in the oxygen framework would allow a higher operating voltage and excellent structural stability [4].

3.2.1.2.1.1 Vanadium Oxides (V_2O_5) V_2O_5 is considered as a high-energy cathode material for use in LIBs due to low cost, facile synthesis, and high energy densities. Theoretically, V_2O_5 can accommodate more than 3 mol of Li^+ into its layered structure and has a charge capacity of about 400 mAh g^{-1}. Because of these promising features, V_2O_5 has been studied as a cathode material for rechargeable Mg batteries. Novak et al. conducted the study on single-crystal V_2O_5 and found that V_2O_5 can host very small amounts of Mg^{2+}, preferably on the surface of crystals surface [18]. Zhang and Yu found that V_2O_5 a high discharge capacity (159 mAh g^{-1}) of V_2O_5 could be attained in 0.1 M Mg $(ClO_4)_2$ + 1.79 M H_2O/propylene carbonate electrolytes [19].Yuan et al. [20,21] found out that vanadium oxide nanotubes (VO-NTs) and Cu-doped VO-NTs to be Mg^{2+} hosting materials that allowed reversible electrochemical Mg^{2+} insertion and extraction. The Cu-doped VO-NTs exhibited a significantly improved discharge capacity of about 120.2 mAh g^{-1} at a current density of 10 mAg^{-1}

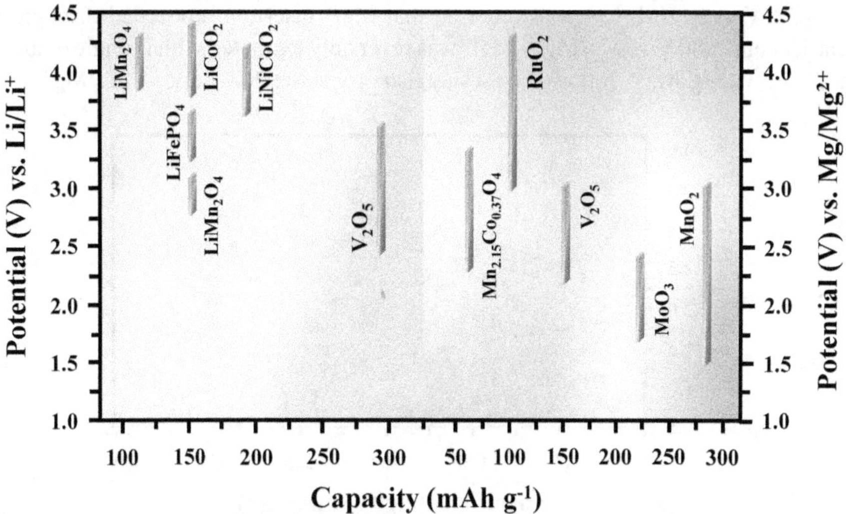

FIGURE 3.4 Comparison of reaction voltages for Li^+ or Mg^{2+} insertion and extraction of selected transition-metal oxide cathodes. Adapted and reproduced from Ref. [17].

because of their enhanced electronic conductivity. Stojkovic et al. [22] successfully synthesized amorphous V_2O_5 xerogel from a solution of V_2O_5 in H_2O_2. Xerogel V_2O_5 showed a reversible capacity of 107 mAh g^{-1}, whereas crystalline V_2O_5 yielded 30 mAh g^{-1} in a solution of $Mg(NO_3)_2$.

3.2.1.2.1.2 Manganese Oxide (MnO_2) MnO_2 is considered as a promising material for rechargeable Mg batteries because of its low cost, environmental friendliness, and nontoxicity. MnO_2 has different polymorphs, which can be roughly classified into a tunnel (1D), layer phase (2D), and spinel (3D) structures. Zhang et al. [23] studied the reversible Mg^{2+} insertion and extraction behaviour of hollandite-MnO_2. Hollandite-MnO_2, with a size of 20 nm, showed a reversible capacity of about 240 mAh g^{-1} in the first cycle with a high Coulombic efficiency of 86%. The galvanostatic cycling of hollandite MnO_2 at 0.015 C is shown in Figure 3.5.

Rasul et al. [25] also reported on the electrochemical performance of nanosized hollandite-MnO_2 with a 2×2 tunnel structure synthesized by sol–gel and hydrothermal methods. The discharge and charge capacities of hollandite-MnO_2 were estimated to be about 85 and 49 mAh g^{-1}, which corresponded to the Mg^{2+} insertion of 0.16 Mg/Mn and Mg^{2+} extraction of 0.09 Mg/Mn, respectively. Todorokite-MnO_2, which has the largest tunnel structure (3×3) with triple chains of edge-sharing MnO_6 octahedra, was synthesized through a hydrothermal treatment of layered MnO_2. The crystallinity of todorokite-MnO_2 depends on the hydrothermal temperature. Todorokite-MnO_2 displayed a relatively higher capacity than that of hollandite-MnO_2 with its smaller tunnel. The discharge capacity of todorokite-MnO_2 was estimated to be 85 mAh g^{-1} during Mg^{2+} insertion, corresponding to the transfer of 0.3 mol of Mg^{2+}, whereas Mg^{2+} could barely be inserted into hollandite-MnO_2 due to the smaller tunnel size.

3.2.1.2.1.3 Molybdenum Oxide MoO_3 Orthorhombic MoO_3 is considered an intercalation material for both mono- and multivalent cations. It consists of edge- and corner-sharing MoO_6 octahedra as a double layer. The layered structure of MoO_3 can easily accommodate guest ions, such as magnesium ion. Gregory et al. [26] found that 0.5 mol of Mg^{2+} could be chemically inserted using a solution of dibutyl magnesium. The charge capacity of $Mg_{0.5}MoO_3$ was estimated to be about 143 mAh g^{-1}. Gregory et al. [27] studied the reaction mechanism of MoO_3 during Mg^{2+} insertion and extraction, and crystalline MoO_3 thin films with a thickness of 100 nm were prepared by electrodeposition. The MoO_3 thin films had an orthorhombic structure with a preferentially oriented (010) direction. The charge capacity of crystalline MoO_3 thin films was about 220 mAh g^{-1} in a voltage range of 1.7–2.8 V vs Mg/Mg^{2+}. It was reported that Mg^{2+} insertion proceeded through two-stage reactions at 1.74 and 1.80 V vs Mg/Mg^{2+}. Mg^{2+} was reversibly extracted with a single oxidation peak, centred at 2.15 V vs Mg/Mg^{2+}, followed by a sluggish process at 2.2–2.8 V vs Mg/Mg^{2+}.

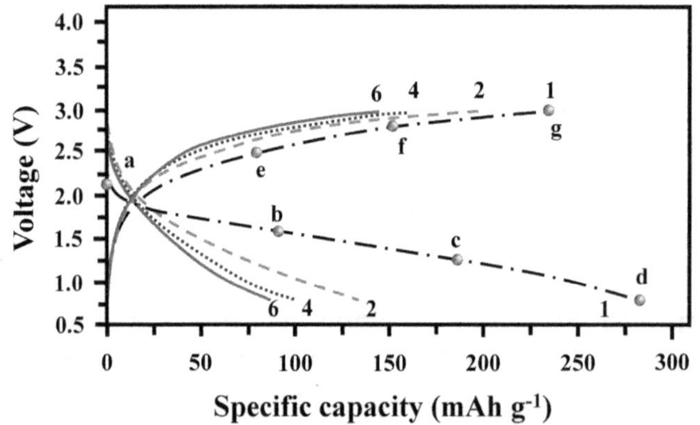

FIGURE 3.5 Galvanostatic cycling of hollandite MnO_2 at 0.015 C. Adapted and reproduced from Ref. [24] copyright 2012 Elsevier.

3.2.1.2.2 Sulphides: MoS_2, TiS_2, and NbS_3

Two-dimensional layered transition-metal disulphides, MX_2 (M=Ti, Zr, Hf, Nb, Ta, Mo, W, V, and X=S), are well-known materials that allow reversible intercalation of cations [28–30]. Liu et al. [31] proposed a cathode material of a graphene-like MoS_2/C composite, which had a reversible capacity of about 200 mAh g^{-1}. The report confirmed the potential of a nanostructural MoS_2 cathode with an expanded capacity for reversibility compared with the conventional CP Mo_6S_8 cathode. TiS_2 has also been proposed as a potential cathode material for rechargeable Mg batteries, with the advantages of a single homogeneous intercalation phase and a high theoretical capacity. In practice, TiS_2 nanotubes exhibited a high capacity for reversibility of about 236 mAh g^{-1} with a moderate reaction voltage of 1.4 V vs Mg/Mg^{2+}. Based on these characteristics, it is expected that more than twice the energy density of current Chevrel Mo_6S_8 cathodes can be attained using TiS_2 nanotubes. Apart from the advantages of TiS_2 nanotubes, the great loss of capacity during cycling remains an unresolved issue in this cathode material.

3.2.1.3 Polyanion-Based Material

3.2.1.3.1 Silicates: $MgMnSiO_4$ and $MgFeSiO_4$

Polyanionic compounds are 3D network structures of transition metals and polyanions with strong interconnected covalent bonds. These materials have several advantages over other cathodes such as variety, high voltage, and stable structure. The research of polyanionic compounds mainly include olivine-structured silicates and NASICON-structured phosphates. Olivine-type $MgMnSiO_4$ (M=Fe, Mn, Co, or Ni) possesses a high theoretical capacity and high theoretical redox potential, and the presence of multivalent transition metals can reduce the damage to the structure during de-intercalation of Mg^{2+} ions. NuLi et al. [32,33] reported on various sizes and morphologies of $MgMnSiO_4$. Orikasa et al. [34] reported that $MgFeSiO_4$ was synthesized by an ion-exchange method for Mg^{2+} host material. Ion-exchanged $MgFeSiO_4$ has a reversible capacity of 300 mAh g^{-1} after 5 cycles in the electrolyte containing a 0.5 M solution of Mg $(TFSI)_2$.

3.3 ANODES

Unlike lithium metal, pure Mg metal could be used directly as an anode rechargeable battery because it's less sensitive to the atmospheric environment and dendrite-free characteristic during recharging [35]. In RMBs, during the Mg deposition and dissolution process, a passivation film formed has an insulating effect on Mg^{2+} ions in most polar organic electrolytes. Innovative ideas in developing efficient anodes that can reduce the formation of passivation layers, which is crucial for RMBs, are encouraged. Mainly, the materials used for the anode of RMBs are mainly modified: Mg metal anodes, alloy-based materials, carbon-based materials, and so on.

3.3.1 Mg Metal Anodes

Magnesium metal is promising anodes (Li and Na) from a thermodynamic point of view, with properties such as low standard potential (–2.38 V vs NHE), natural abundance, and a high theoretical capacity (2205 Ahk g^{-1}) [36]. The first RMBs with magnesium as anode are reported by Aurbach et al. [14] where the system consists of Mg organohaloaluminate salts in solvents of THF and polyethers of the glyme family. The major drawback of RMBs is the existence of formation of the passivation film, which hinders the passage of Mg^{2+} ions and thus leads to poor cycling stability. In order to solve this problem, micro–nanostructured Mg metal was used as an anode material, which can effectively decrease the thickness of the passivating films, promote ion diffusion, and improve the performance of Mg storage. Chen's group [37] reported the various Mg nano/mesoscale structures that were showing excellent electrochemical properties for application in Mg/air batteries [38]. In RMBs, as an anode, the ultra-small N-Mg nanoparticles with an average diameter of 2.5 nm have better cycle performance than bulk Mg.

3.3.2 Mg^{2+} Insertion Materials

Insertion-type materials were primarily introduced to overcome the issues generated from the reaction between magnesium metal and electrolytes. In insertion-type anodes, electrolytes derived from Mg ionic salts in polar aprotic solvents are used. Bismuth, antimony, and Bi-Sb alloys were considered as insertion anode materials for Mg rechargeable batteries [39]. Antimony (Sb) showed poor cycling stability, while Bismuth and Bi-Sb showed reasonable cycling performances over 100 cycles at a rate of 1 C. The galvanostatic discharge of Bi, Sb, $Bi_{0.88}Sb_{0.12}$, and $Bi_{0.55}Sb_{0.45}$ anodes at a 0.01 C rate is shown in Figure 3.6. The largest capacity of 298 mAh g^{-1} was obtained from $Bi_{0.88}Sb_{0.12}$ at a rate of 1 C, which dropped to 215 mAh g^{-1} after 100 cycles. Bismuth showed the best capacity retention of 89% from the 20th (247 mAh g^{-1}) to the 100th (222 mAh g^{-1}) cycles. Because the cell voltage is one of the factors that determine the energy density, it is necessary to develop more kinds of RMBs anode materials with lower Mg^{2+} insertion/extraction voltages and higher capacities. Sn was considered as another candidate for anode, with the Mg^{2+} insertion/extraction voltage (+0.15/0.20 V) being taken into consideration. Combining with the advantages of Bi and Sn, Niu et al. [40] developed a novel high-performance dual-phase NP-Bi-Sn alloys, and the high phase boundary density could provide more Mg^{2+} transmission channels; the unique nanopore structure eases large volume changes; as an anode for RMBs, the NP-Bi_6Sn_4 exhibited an excellent electrochemical performance.

3.4 ELECTROLYTES IN RECHARGEABLE MIB

The electrolyte plays a role in transferring ions between the cathode and anode electrodes of the battery. For RMBs, as exhibited in Figure 3.7, [42] the electrolytes should have the following characteristics: (i) reversible deposition and dissolution of Mg; (ii) high ionic conductivity; and (iii) wide electrochemical window. In addition, safety properties such as good thermal stability, low volatility, low flammability, and low toxicity are also critical for Mg electrolytes. In this section, according to their phase states, Mg electrolytes can be classified into liquid electrolytes and solid-state electrolytes. Figure 3.7 shows the schematic illustration of main prerequisites of RMB electrolytes.

3.4.1 Liquid Electrolytes

Mg metal anode reacts with oxidizing compounds to form Mg^{2+} nonconductive passivation layers such as in conventional polar solvents common commercial magnesium salts ($Mg(ClO_4)_2$), which makes the poor performance of RMBs [43]. According to the ingredients in electrolytes, they can be classified into four categories: Grignard-based electrolytes, boron-containing electrolytes, $(HMDS)_2$ Mg-based electrolytes, and other novel magnesium electrolytes.

3.4.1.1 Grignard-Based Electrolytes

Grignard reagent, expressed as RMgX (R may be alkyl or aryl; X is Cl, Br, or other halides), is used in RMBs, because it enables reversible Mg deposition-stripping and prevent the formation of passivation film. Aurbach et al. [14] developed first Grignard reagent-based Mg $(AlCl_2R)_2$/tetrahydrofuran (THF) electrolyte. However, the narrow electrochemical stability window (≈2.2 V vs Mg/Mg^{2+} on Pt) of this electrolyte limits its use in RMBs with high energy density. The properties of the electrolyte are related to the R group in RMgX, so some researchers used amines or phenolates to replace it. Wang et al. synthesized an air-stable electrolyte system through the reaction between organic Mg salt (ROMgCl) and $AlCl_3$ that has shown excellent ionic conductivity (2.56×10^{-3} S cm^{-1}) and anodic stability (2.6 V vs Mg). Yoshimto et al. reported electrolyte system of Grignard reagent and ionic liquid (IL) which displayed high ionic conductivity of 7.44×10^{-3} S cm^{-1} (at 25°C). Adding IL additives is an effective strategy to enhance anodic stability and ionic conductivity of Grignard-based electrolytes.

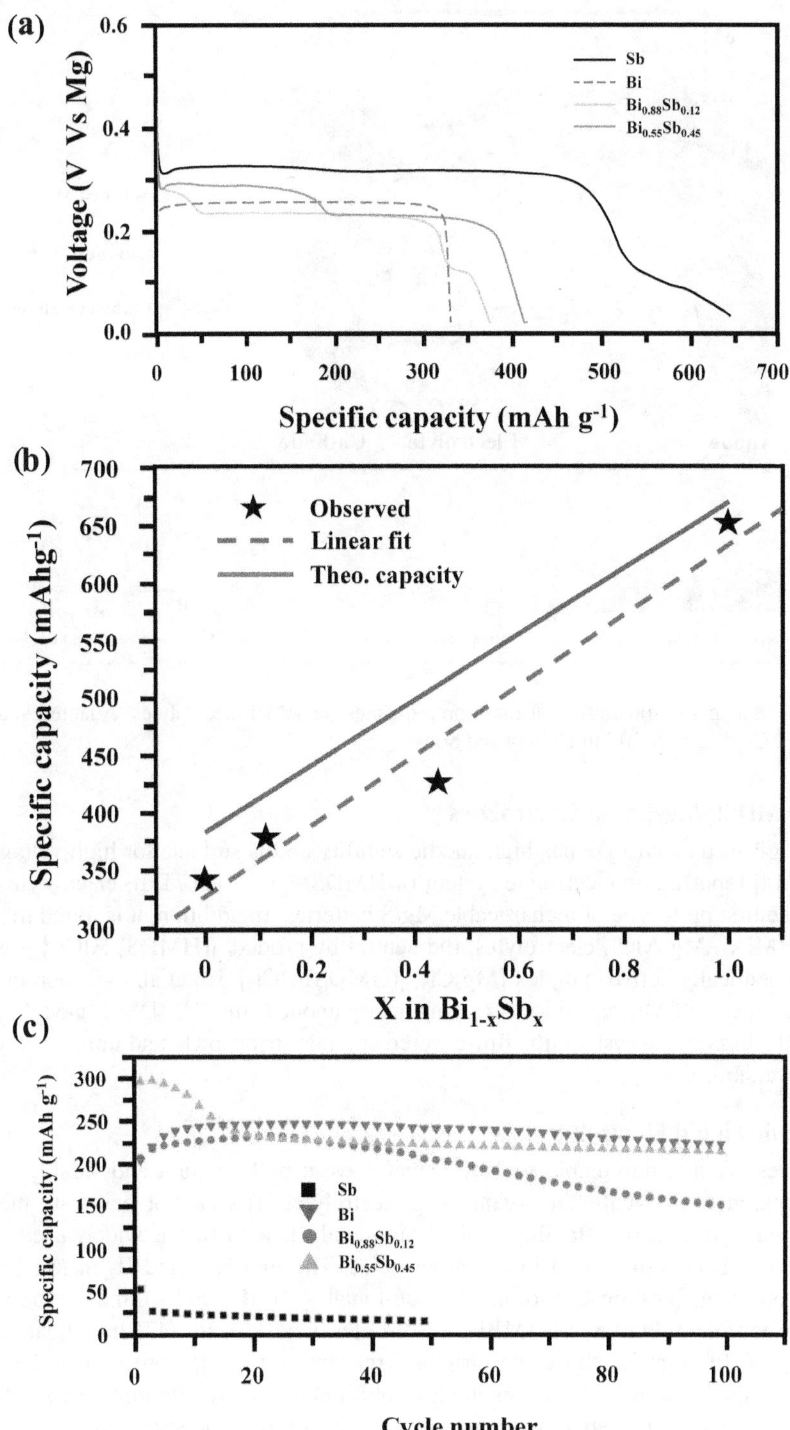

FIGURE 3.6 (a) Galvanostatic discharge of Bi, Sb, $Bi_{0.88}Sb_{0.12}$, and $Bi_{0.55}Sb_{0.45}$ anodes at a 0.01 C rate. (b) Observed capacity of the alloy anodes with a linear fit of these capacities (dashed line) compared with the theoretical capacity (solid line). (c) Cycling capabilities of Bi, Sb, $Bi_{0.88}Sb_{0.12}$, and $Bi_{0.55}Sb_{0.45}$ anodes at a 1 C rate. Adapted and reproduced from Ref. [41] copyright 2012 Elsevier.

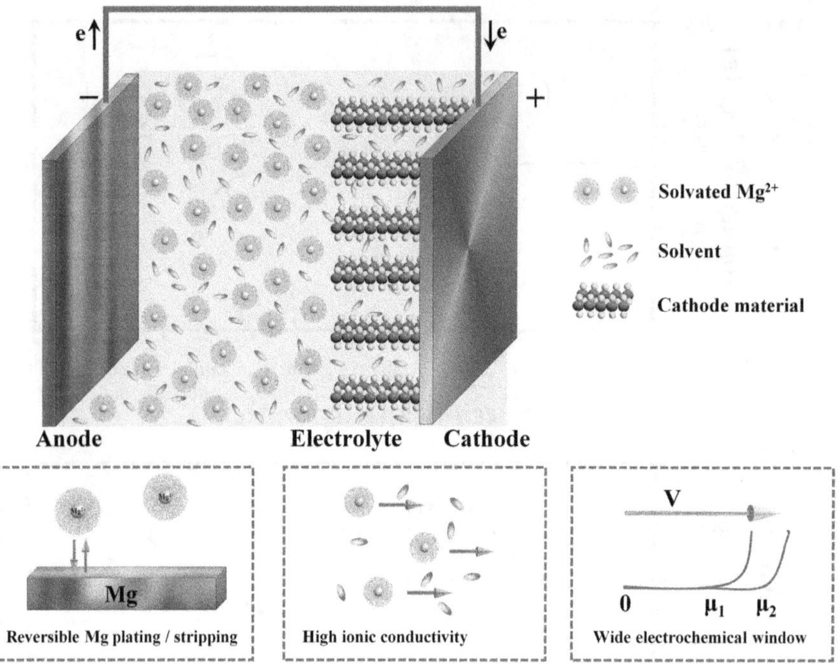

FIGURE 3.7 Schematic illustration of main prerequisites for RMB electrolytes. Adapted and reproduced from Ref. [11]. Copyright 2020 John Wiley and Sons.

3.4.1.2 $(HMDS)_2$ Mg-Based Electrolytes

$(HMDS)_2$ Mg-based electrolyte has high anodic stability and is suitable for high-voltage cathodes. Kim et al. [44] reported an electrolyte system of HMDSMgCl-$AlCl_3$/THF electrolyte and used it to prepare the first prototype of rechargeable Mg/S batteries. In addition, it is found that by adding $MgCl_2$ to $(HMDS)_2$Mg-$AlCl_3$ electrolytes, the neutral by-product [(HMDS) $AlCl_2$] was converted into electrochemically active complex [Mg_2Cl_3] [(HMDS)$AlCl_3$]. Hu et al. [45] thoroughly studied the interface process of Mg deposition/stripping on Mg anode using $(HMDS)_2$Mg-$MgCl_2$/ether electrolytes. In the tetraglyme system, the full-covered crystals strip slowly and uniformly, which leads to a good reversibility.

3.4.1.3 Ionic Liquid Electrolytes

IL electrolytes are non-flammable solvents which have a bulky cation and weakly coordinating anion. Compared to conventional organic Mg electrolytes, ILs cannot passivate metal surface and possess enough reductive stability against Mg metal. IL with BF^{4-} is widely used one because of its high ionic conductivity and low melting point. The first reported IL in Mg batteries was a combination of magnesium trifluoromethanesulfonate (Mg $(CF_3SO_3)_2$) and 1-n-butyl-3-methyl-imidazolium tetrafluoroborate ($BMIMBF_4$) [46]. Later, Cheek et al. [47] investigated the reductive stability of different ILs by using Grignard reagent as the Mg source. It is found that Mg can be deposited from 1-butyl-1-methylpyrrolidinium bis (trifluoromethylsulphonyl)imide (BMPTFSI) but not from 1-ethyl-3-methylimidazolium tetrafluoroborate ($EMImBF_4$). Following these reports, Morita et al. reported a number of papers on IL in Mg battery electrolytes using Grignard reagent as the Mg source [48–50]. In their first report, they developed an electrolyte mixture with ethyl magnesium bromide in THF (EtMgBr/THF) and an IL of quaternary ammonium salt, N,N-diethyl-N-methyl-N-(2-methoxyethyl) ammonium-bis(trifluoromethanesulphonyl)imide (DEMETFSI). The ionic conductivity of mixture to be 7.4 mS cm^{-1} at room temperature, and it is noted that the solution was capable of Mg deposition/dissolution on a silver substrate.

Attempts to add IL as additive in Mg electrolyte are also investigated by Cho et al. [51] yielding interesting results. They found that the addition of 1-allyl-1-methylpyrrolidinium to a Grignard reagent improves its oxidation stability by 1.0 V. Hui et al. [52] carried out a detailed investigation of ILs with different cations (pyridinium, imidazolium, piperidinium, and pyrrolidinium) with $Mg(TFSI)_2$ salt and acetonitrile solvent. The authors observed a correlation with high conductivity IL and the presence of unsaturated rings combined with short carbon chain lengths. The most desirable properties were obtained for the composition with $Mg(TFSI)_2$ dissolved in 40% 1-ethyl-1-methylpyrrolidinium bis(trifluoromethanesulphonyl) imide (EMPTFSI) and 60% acetonitrile, which had a conductivity of 40 mS cm^{-1} and a voltage of 5.23 V vs Mg^0/Mg^{2+}.

3.4.2 Solid-State Electrolytes

Solid-state electrolytes have attracted many researchers' attention because of its excellent properties such as good safety performance, excellent mechanical properties, wide voltage window, and high energy density. There are three categories of solid-state electrolytes: inorganic solid-state electrolytes, organic solid-state electrolytes, and organic–inorganic composite solid-state electrolytes. Most magnesium inorganic solid-state electrolytes such as $MgZr_4(PO_4)_6$ [53] and $Mg(BH_4)(NH_2)$ have low ionic conductivity. Organic solid-state electrolytes can be called solid-state polymer electrolytes, which are formed by complexing organic polymer and Mg salt. Common organic polymer substrates include polyoxyethylene (PEO) [54,55] polyvinylidene fluoridehexafluoropropylene (P(VDF-HFP)) [56,57], polyvinylidene fluoride (PVDF) [58], and polyvinyl alcohol (PVA) [59]. Chusid et al. [58] prepared a magnesium polymer electrolyte by using PEO or PVDF as polymer matrices and mixing $Mg(AlCl_2EtBu)_2$/ether solution. The PVdF-$Mg(AlCl_2EtBu)_2$-tetraglyme system possessed a high conductivity (3.7×10^{-3} S cm^{-1} at 25°C) and achieved reversible Mg intercalation in the Mo_6S_8 cathode.

To recapitulate, the research on Mg solid-state electrolyte remains at the first stage, and the main factors restricting its development are: (i) the non-Mg^{2+} conductive passivation film on Mg anodes and (ii) poor magnesium mobility in solid. Therefore, the development of solid electrolytes with reversible magnesium deposition on magnesium anodes and high ion conductivity is the focus of the solid-state Mg battery research in the future.

3.5 CONCLUSION

RMBs are considered as the future candidate in the field of energy storage beyond LIBs in recent years. However, owing to the relatively poor compatibility of electrolyte and electrode materials, the electrochemical performances of the RMBs are at low level so far. In summary, the research on the key materials of RMBs should mainly focus on the following aspects in the future: designing and synthesizing new cathode material to reduce the polarization and enhance the diffusion kinetics of Mg^{2+} ions, optimizing and modifying the Mg anode to reduce or eliminate the passivation film as soon as possible, developing efficient electrolyte materials that could achieve reversible deposition/dissolution of Mg without corroding the current collector, and enhancing the interface stability the electrode/electrolyte. Studies on Mg storage mechanism in electrode materials, as well as understanding and prevention of failure mechanisms will definitely improve the future RMBs performance. By rationally designing the structure of the material and optimizing the electrode/electrolyte system, the construction of a high-capacity, long-life, and high-safety RMBs is finally achieved, which will further promote its widespread application in the field of large-scale energy storage.

REFERENCES

1. Scrosati B (2011) History of lithium batteries. *Journal of Solid State Electrochemistry* 15:1623–1630. https://doi.org/10.1007/s10008-011-1386-8.
2. Geng P, Zheng S, Tang H, et al (2018) Transition metal sulfides based on graphene for electrochemical energy storage. *Advanced Energy Materials* 8:1–26. https://doi.org/10.1002/aenm.201703259.
3. Diouf B, Pode R (2015) Potential of lithium-ion batteries in renewable energy. *Renewable Energy* 76:375–380. https://doi.org/10.1016/j.renene.2014.11.058.
4. Park MS, Kim JG, Kim YJ, et al (2015) Recent advances in rechargeable magnesium battery technology: A review of the field's current status and prospects. *Israel Journal of Chemistry* 55:570–585. https://doi.org/10.1002/ijch.201400174.
5. IEA (2020) *World Energy Investment 2020.* Paris. https://doi.org/10.1787/6f552938-en.
6. US EIA (2013) International Energy Outlook 2013- DOE/EIA-0484(2013). *Outlook* 2013:312.
7. Ellabban O, Abu-Rub H, Blaabjerg F (2014) Renewable energy resources: Current status, future prospects and their enabling technology. *Renewable and Sustainable Energy Reviews* 39:748–764. https://doi.org/10.1016/j.rser.2014.07.113.
8. Massé RC, Uchaker E, Cao G (2015) Beyond Li-ion: Electrode materials for sodium- and magnesium-ion batteries. *Science China Materials* 58:715–766. https://doi.org/10.1007/s40843-015-0084-8.
9. Deivanayagam R, Ingram BJ, Shahbazian-Yassar R (2019) Progress in development of electrolytes for magnesium batteries. *Energy Storage Materials* 21:136–153.
10. Huie MM, Bock DC, Takeuchi ES, et al (2015) Cathode materials for magnesium and magnesium-ion based batteries. *Coordination Chemistry Reviews* 287:15–27. https://doi.org/10.1016/j.ccr.2014.11.005.
11. Liu F, Wang T, Liu X, Fan LZ (2021) Challenges and recent progress on key materials for rechargeable magnesium batteries. *Advanced Energy Materials* 11:2000787–2000815.
12. Peng B, Liang J, Tao Z, Chen J (2009) Magnesium nanostructures for energy storage and conversion. *Journal of Materials Chemistry* 19:2877–2883. https://doi.org/10.1039/b816478a.
13. Levi E, Gershinsky D, Aurbach, O. Isnard GC (2015) New insight on the unusually high ionic mobility in chevrel phases. *Chemistry of Materials* 21:1390–1399. https://doi.org/10.1007/s10832.
14. Aurbach D, Lu Z, Schechter A, et al (2010) ChemInform abstract: Prototype systems for rechargeable magnesium batteries. *ChemInform* 32. https://doi.org/10.1002/chin.200103014.
15. Saha P, Jampani PH, Datta MK, et al (2014) A convenient approach to mo 6 s 8 chevrel phase cathode for rechargeable magnesium battery. *Journal of the Electrochemical Society* 161:A593–A598. https://doi.org/10.1149/2.061404jes.
16. Cho W, Moon B, Woo SG, et al (2015) Size effect of chevrel $Mg_xMo_6S_8$ as cathode material for magnesium rechargeable batteries. *Bulletin of the Korean Chemical Society* 36:1209–1214. https://doi.org/10.1002/bkcs.10231.
17. Park MS, Kim JG, Kim YJ, et al (2015) Recent advances in rechargeable magnesium battery technology: A review of the field's current status and prospects. *Israel Journal of Chemistry* 55:570–585.
18. Yu L, Zhang X (2004) Electrochemical insertion of magnesium ions into V_2O_5 from aprotic electrolytes with varied water content. *Journal of Colloid and Interface Science* 278:160–165. https://doi.org/10.1016/j.jcis.2004.05.028.
19. Muldoon J, Bucur CB, Gregory T (2014) Quest for nonaqueous multivalent secondary batteries: Magnesium and beyond. *Chemical Reviews* 114:11683–11720. https://doi.org/10.1021/cr500049y.
20. Jiao LF, Yuan HT, Si YC, et al (2006) Synthesis of Cu0.1-doped vanadium oxide nanotubes and their application as cathode materials for rechargeable magnesium batteries. *Electrochemistry Communications* 8:1041–1044. https://doi.org/10.1016/j.elecom.2006.03.043.
21. Jiao L, Yuan H, Wang Y, et al (2005) Mg intercalation properties into open-ended vanadium oxide nanotubes. *Electrochemistry Communications* 7:431–436. https://doi.org/10.1016/j.elecom.2005.02.017.
22. Kanatzidis MG, Wu C-G, Liu Y-J, et al (1991) Intercalation of layered V_2O_5 Xerogel with polymers. *MRS Proceedings* 233. https://doi.org/10.1557/proc-233-183.
23. Manuscript A Liu Q, Hsu C-W, et al. (2020) A fluorine-substituted pyrrolidinium-based ionic liquid for high-voltage Li-ion batteries. *Chemical Communications.* https://doi.org/10.1039/D0CC02184A.
24. Zhang R, Yu X, Nam KW, et al (2012) α-MnO_2 as a cathode material for rechargeable Mg batteries. *Electrochemistry Communications* 23:110–113. https://doi.org/10.1016/j.elecom.2012.07.021.
25. Rasul S, Suzuki S, Yamaguchi S, Miyayama M (2012) Synthesis and electrochemical behavior of hollandite MnO_2/acetylene black composite cathode for secondary Mg-ion batteries. *Solid State Ionics* 225:542–546. https://doi.org/10.1016/j.ssi.2012.01.019.

26. Gregory TD, Hoffman RJ, Winterton RC (1990) Nonaqueous electrochemistry of magnesium: Applications to energy storage. *Journal of the Electrochemical Society* 137:775–780. https://doi.org/10.1149/1.2086553.
27. Gershinsky G, Yoo HD, Gofer Y, Aurbach D (2013) Electrochemical and spectroscopic analysis of Mg2+ intercalation into thin film electrodes of layered oxides: V_2O_5 and MoO_3. *Langmuir* 29:10964–10972. https://doi.org/10.1021/la402391f.
28. Bissessur R, Heising J, Hirpo W, Kanatzidis M (1996) Toward pillared layered metal sulfides. Intercalation of the chalcogenide clusters Co_6 $Q_8(PR_3)_6$ (Q=S, Se, and Te and R=Alkyl) into MoS_2 materials with open frameworks and void spaces are important in the field of separation science and ca. *Chemistry of Materials* 8:318–320.
29. Benavente E, Santa Ana MA, Mendizábal F, González G (2002) Intercalation chemistry of molybdenum disulfide. *Coordination Chemistry Reviews* 224:87–109. https://doi.org/10.1016/S0010-8545(01)00392-7.
30. Enyashin AN, Gemming S, Bar-Sadan M, et al (2007) Structure and stability of molybdenum sulfide fullerenes. *Angewandte Chemie – International Edition* 46:623–627. https://doi.org/10.1002/anie.200602136.
31. Kuang C, Zeng W, Li Y (2019) A review of electrode for rechargeable magnesium ion batteries. *Journal of Nanoscience and Nanotechnology* 19, 12–25. https://doi.org/10.1166/jnn.2019.16435.
32. Nuli Y, Yang J, Li Y, Wang J (2010) Mesoporous magnesium manganese silicate as cathode materials for rechargeable magnesium batteries. *Chemical Communications* 46:3794–3796. https://doi.org/10.1039/c002456b.
33. Feng Z, Yang J, Nuli Y, Wang J (2008) Sol-gel synthesis of Mg1.03Mn0.97SiO4 and its electrochemical intercalation behavior. *Journal of Power Sources* 184:604–609. https://doi.org/10.1016/j.jpowsour.2008.05.021.
34. Orikasa Y, Masese T, Koyama Y, et al (2014) High energy density rechargeable magnesium battery using earth-abundant and non-toxic elements. *Scientific Reports* 4:1–6. https://doi.org/10.1038/srep05622.
35. Aurbach D, Lu Z, Schechter A, et al (2000) Prototype Mg batteries. *Nature* 407:724–727.
36. Besenhard JO, Winter M (2002) Advances in battery technology: Rechargeable magnesium batteries and novel negative-electrode materials for lithium ion batteries. *ChemPhysChem* 3:155–159. https://doi.org/10.1002/1439-7641(20020215)3:2<155::AID-CPHC155>3.0.CO;2-S.
37. Liang Y, Feng R, Yang S, et al (2011) Rechargeable Mg batteries with graphene-like MoS2 cathode and ultrasmall Mg nanoparticle anode. *Advanced Materials* 23:640–643. https://doi.org/10.1002/adma.201003560.
38. Li W, Li C, Zhou C, et al (2006) Metallic magnesium nano/mesoscale structures: Their shape-controlled preparation and Mg/air battery applications. *Angewandte Chemie – International Edition* 45:6009–6012. https://doi.org/10.1002/anie.200600099.
39. Ling C, Banerjee D, Matsui M (2012) Study of the electrochemical deposition of Mg in the atomic level: Why it prefers the non-dendritic morphology. *Electrochimica Acta* 76:270–274. https://doi.org/10.1016/j.electacta.2012.05.001.
40. Niu J, Gao H, Ma W, et al. (2018) Dual phase enhanced superior electrochemical performance of nanoporous bismuth-tin alloy anodes for magnesium-ion batteries. *Energy Storage Materials* 14:351–360. https://doi.org/10.1016/j.ensm.2018.05.023.
41. Arthur TS, Singh N, Matsui M (2012) Electrodeposited Bi, Sb and Bi1-xSbx alloys as anodes for Mg-ion batteries. *Electrochemistry Communications* 16:103–106. https://doi.org/10.1016/j.elecom.2011.12.010.
42. Liu F, Wang T, Liu X, Fan L (2020) Challenges and recent progress on key materials for rechargeable magnesium batteries. *Advanced Energy Materials* 2000787:1–28. https://doi.org/10.1002/aenm.202000787.
43. Lu Z, Schechter A, Moshkovich M, Aurbach D (1999) On the electrochemical behavior of magnesium electrodes in polar aprotic electrolyte solutions. *Journal of Electroanalytical Chemistry* 466:203–217. https://doi.org/10.1016/S0022-0728(99)00146-1.
44. Kim HS, Arthur TS, Allred GD, et al. (2011) Structure and compatibility of a magnesium electrolyte with a sulphur cathode. *Nature Communications* 2. https://doi.org/10.1038/ncomms1435.
45. Hu XC, Shi Y, Lang SY, et al (2018) Direct insights into the electrochemical processes at anode/electrolyte interfaces in magnesium-sulfur batteries. *Nano Energy* 49:453–459. https://doi.org/10.1016/j.nanoen.2018.04.066.
46. NuLi Y, Yang J, Wu R (2005) Reversible deposition and dissolution of magnesium from BMIMBF4 ionic liquid. *Electrochemistry Communications* 7:1105–1110. https://doi.org/10.1016/j.elecom.2005.07.013.
47. Cheek GT, O'Grady WE, el Abedin SZ, et al (2008) Studies on the electrodeposition of magnesium in ionic liquids. *Journal of the Electrochemical Society* 155:D91. https://doi.org/10.1149/1.2804763.

48. Yoshimoto N, Matsumoto M, Egashia M, Morita M (2010) Mixed electrolyte consisting of ethylmagnesiumbromide with ionic liquid for rechargeable magnesium electrode. *Journal of Power Sources* 195:2096–2098. https://doi.org/10.1016/j.jpowsour.2009.10.073.
49. Kakibe T, Yoshimoto N, Egashira M, Morita M (2010) Optimization of cation structure of imidazolium-based ionic liquids as ionic solvents for rechargeable magnesium batteries. *Electrochemistry Communications* 12:1630–1633. https://doi.org/10.1016/j.elecom.2010.09.012.
50. Kakibe T, Hishii JY, Yoshimoto N, et al (2012) Binary ionic liquid electrolytes containing organo-magnesium complex for rechargeable magnesium batteries. *Journal of Power Sources* 203:195–200. https://doi.org/10.1016/j.jpowsour.2011.10.127.
51. Cho JH, Ha JH, Lee SH, et al (2017) Effect of 1-allyl-1-methylpyrrolidinium chloride addition to ethylmagnesium bromide electrolyte on a rechargeable magnesium battery. *Electrochimica Acta* 231:379–385. https://doi.org/10.1016/j.electacta.2017.02.062.
52. Huie MM, Cama CA, Smith PF, et al (2016) Ionic liquid hybrids: Progress toward non-corrosive electrolytes with high-voltage oxidation stability for magnesium-ion based batteries. *Electrochimica Acta* 219:267–276. https://doi.org/10.1016/j.electacta.2016.09.107.
53. Adamu M, Kale GM (2016) Novel sol-gel synthesis of $MgZr_4P_6O_{24}$ composite solid electrolyte and newer insight into the $Mg2^+$-ion conducting properties using impedance spectroscopy. *Journal of Physical Chemistry C* 120:17909–17915. https://doi.org/10.1021/acs.jpcc.6b05036.
54. Dissanayake MAKL, Bandara LRAK, Karaliyadda LH, et al (2006) Thermal and electrical properties of solid polymer electrolyte PEO_9 $Mg(ClO_4)_2$ incorporating nano-porous Al_2O_3 filler. *Solid State Ionics* 177:343–346. https://doi.org/10.1016/j.ssi.2005.10.031.
55. Shao Y, Gu M, Li X, et al (2014) Highly reversible Mg insertion in nanostructured Bi for Mg ion batteries. *Nano Letters* 14:255–260. https://doi.org/10.1021/nl403874y.
56. Oh JS, Ko JM, Kim DW (2004) Preparation and characterization of gel polymer electrolytes for solid state magnesium batteries. *Electrochimica Acta* 50:903–906. https://doi.org/10.1016/j.electacta.2004.01.099.
57. Pandey GP, Agrawal RC, Hashmi SA (2011) Magnesium ion-conducting gel polymer electrolytes dispersed with fumed silica for rechargeable magnesium battery application. *Journal of Solid State Electrochemistry* 15:2253–2264. https://doi.org/10.1007/s10008-010-1240-4.
58. Chusid O, Gofer Y, Gizbar H, et al (2003) Solid-state rechargeable magnesium batteries. *Advanced Materials* 15:627–630. https://doi.org/10.1002/adma.200304415.
59. Manjuladevi R, Thamilselvan M, Selvasekarapandian S, et al. (2017) Mg-ion conducting blend polymer electrolyte based on poly(vinyl alcohol)-poly (acrylonitrile) with magnesium perchlorate. *Solid State Ionics* 308:90–100. https://doi.org/10.1016/j.ssi.2017.06.002.

4 Potential Anode Materials for Magnesium-Ion Batteries

Anjumole P. Thomas, Akhila Das, S. K. Vineeth, Abhilash Pullanchiyodan, M. V. Reddy, Jou-Hyeon Ahn, and Prasanth Raghavan

4.1 INTRODUCTION

Energy storage devices occupy a central role in driving electrification of our societies due to an ever-increasing demand of electrical energy to power numerous aspects of human life [1]. It plays a major role in harvesting energy from different sources and stored it for different applications such as industry, transportation, utility and building. Energy storage systems have been used for centuries and tremendous improvements take place in this field [2]. There is a growing demand for zero-emission energy sources for future energy needs, and in this regard, batteries are considered to be a prospective candidate. Nowadays, with the alarming environmental issues related to the utilization of fossil fuels, there is a great demand for efficient energy storage devices that store renewable energy. Researching and using new energy sources has become a common thing around the world and also considered as the future trend in irreversible energy development [3]. Renewable energy is regarded as an important solution for human energy demand in future due to environmental concerns [4].

Battery energy storage devices are the promising candidate for harvesting renewable energy in future, and among these, lithium batteries are finding application in electric vehicles, cell phones, laptops, computers and energy power stations nowadays [5,6]. Lithium-ion batteries (LIBs) are considered to be most successful batteries among rechargeable battery technologies and find application in various sectors due to its high energy density and portability [7,8]. However, they possess some demerits such as tendency of formation of dendrite during continued charge–discharge cycle and scarcity of lithium metal on earth, which in turn encourage the development of other battery technologies for future [9]. In a post-lithium era, there is tremendous effort put into developing a new battery technology with suitable characteristics. Among the possible secondary battery technologies, magnesium-ion batteries (MIBs) are gaining more attention due to its low cost and high energy density [10].

Magnesium metal possesses some superior properties compared to other anode material used in energy storage devices. It possesses (i) high theoretical capacity of 3833 mAh cm^{-1} (vs lithium 2046 mAh cm^{-1}), (ii) low reduction potential (–2.37 V vs SHE) and (iii) relatively low abundance which thereby reduce the production cost of RMBs [11]. In addition to this, magnesium is low density metal (1.74 g cm^{-1}) that is expected to find applications in portable electronics [12]. Magnesium and its compounds are mainly non-toxic and environmentally friendly. For these reasons, RMBs are considered to be potential candidates for large-scale energy storage applications. RMBs consist of cathode, anode and electrolytes where Mg^{2+} ions are released from the anode and inserted into the cathode during discharge, and Mg^{2+} ions are released from the cathode and deposited on the anode during charging.

RMB is first reported in the 1990s in which magnesium metal as the anode, cobalt oxide CoO_3 as the cathode and 0.25 mol L^{-1} Mg $(B(Bu_2RPh_2))_2$THF-DMF as the electrolyte were employed [13]. Later, Aurbach et al. [14] made the first rechargeable magnesium battery technology, in which magnesium metal as the anode, chevrel phase (Mo_6S_8) as the cathode and organohaloaluminate salt as the electrolyte were used. Afterward, several reports came on rechargeable magnesium batteries,

DOI: 10.1201/9781003310167-4

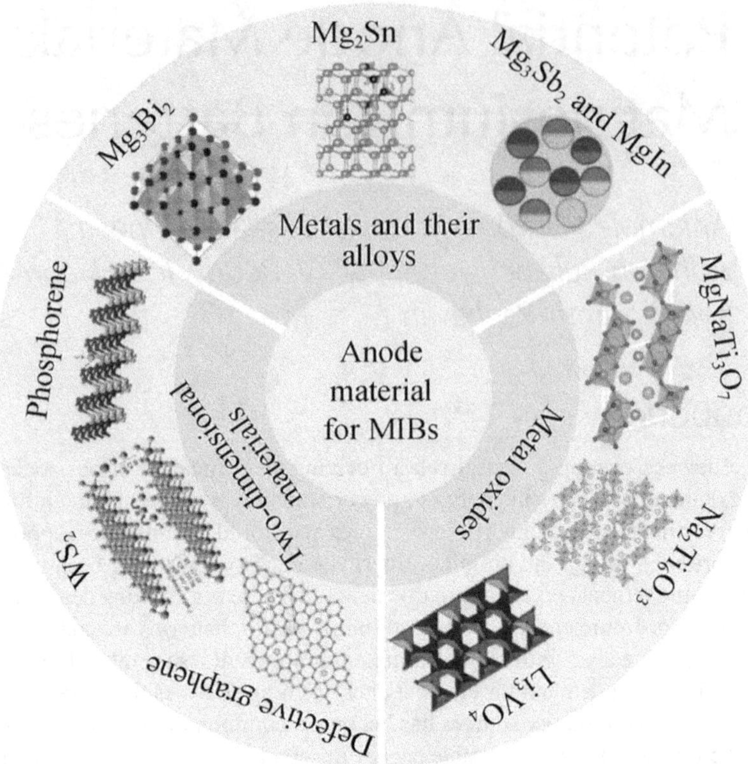

FIGURE 4.1 Anode materials for MIBs. Adapted and reproduced from Ref. [21]. Copyright 2020 Springer Nature.

but due to several reasons, the development of efficient RMBs with required parameters is not accomplished yet. The main reason for this is attributed to the passivation film formed on the surface of Mg metal which prevents the reversible deposition of magnesium ion and reduces the columbic efficiency of the Mg anode [15]. And also, the divalent nature of Mg ion makes it less mobile in the electrolyte system, which results in poor ionic conductivity [16]. Therefore, exploring and obtaining suitable electrode materials and electrolytes is the key to achieving high-performance and practical RMBs in the future.

As compared to LIBs, the development of efficient electrode materials is crucial to commercialize magnesium batteries [17]. In the case of MIBs, most of the conventional electrolytes such as $Mg(ClO_4)_2$ and $Mg(PF_6)$ trigger the formation of passivation layer on metal surface during cycling and this eventually diminishes the Coulombic efficiency of MIBs [18]. Finding novel electrolyte that is compatible with the magnesium metal is the only solution to this problem. The electrolyte that is not forming passivation layer on metal anodes has some other limitations such as high volatility, low operating voltage and corrosive nature [19,20]. Therefore, developing alternative anode material may help in overcoming this issue in MIBs. This chapter reviews the up-to-date progress in alternative anode materials in MIBs and classified as metal oxides, alloys and two-dimensional (2D) materials (Figure 4.1). The chapter also explains mechanism, problems and its limitations in detail.

4.2 RECHARGEABLE MAGNESIUM-ION BATTERY (MIB)

Over these years, LIBs are considered to be the most promising choice for portable energy storage device due to its high specific energy density, longer cycle life, slow self-discharge rate and wide

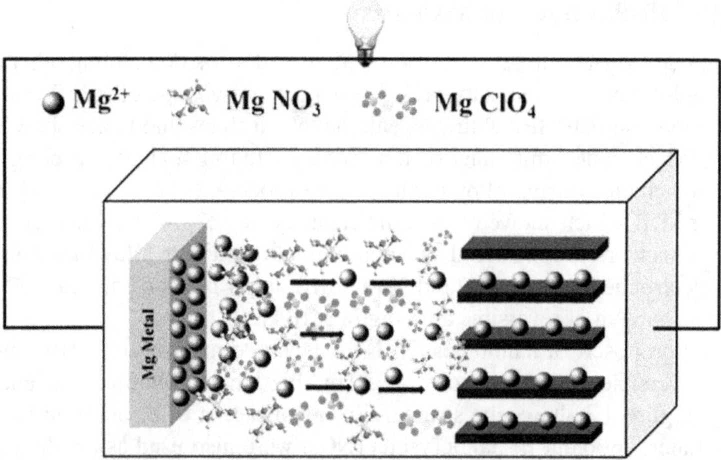

FIGURE 4.2 Working principle of magnesium-ion batteries. Adapted and reproduced from Ref. [26]. Copyright 2019 Elsevier.

range of operating temperatures [22]. However, LIBs have disadvantages such as high cost, safety issues and scarcity of lithium metal [23]. In this aspect, magnesium battery is an alternative candidate, offering cost effective and safety option for future energy storage applications. Magnesium metal (Mg) is abundant in the earth's crust and it is electrochemically active in nature (equivalent weight of 12.15/faraday). Mg has a negative electrode potential of –2.37 V and offers two electrons per redox centre possessing a theoretical capacity of 2205Ah Kg^{-1} [19]. Moreover, Mg is non-toxic and stable towards moisture, container materials and the environment compared to lithium and sodium [24]. Mg does not form dendrite during charging/discharging unlike lithium, which is important because it assures safety in all aspects [25].

A conventional MIB system consists of a cathode, anode and electrolyte. The physical and chemical properties of each component influence the overall functioning of the battery. Figure 4.2 shows the working principle of MIBs. The cathode material possesses high reversible intercalation/deintercalation capacity of Mg^{2+} ion and has good structural stability. An electrolyte acts as medium for transporting ion; in MIB, conventional electrolytes form passivation layer on metal surface, which leads to reduced stability of the battery system. Due to this reason, searching of compatible anode material with electrolyte is a wise method to avoid the passivation layer. An ideal anode material should possess a good specific capacity (mAh g^{-1}), high volumetric capacity (mAh cm^{-1}), low working voltage and chemical inertness to electrolytes. The working principle of MIB is almost similar to that of LIB and based on the reversible intercalation and deintercalation of Mg^{2+} ions on the anode and cathode. In the discharge process, oxidation takes place in anode (Mg metal), where Mg^{2+} ion moves towards cathode through electrolyte. In charging, external voltage is applied and redox reactions were reversed.

4.3 ALLOY-BASED ANODE MATERIAL FOR MAGNESIUM BATTERIES

Alloy-based anode materials are widely explored in LIBs, which are inspired from this Mg metal replaced with metal alloy that was also reported in MIBs. However, in contradiction to the expectation, the alloys give poor performance in MIB in comparison to LIB. To date, there are several P-block elements (Bi, Sn, Pb, Sb and In) that were reported in MIBs as alternatives to Mg metal as anode [27].

4.3.1 Bismuth (Bi)-Based Anode Materials

Bi-based alloy anode (Mg_3Bi_2) material garnered increased attraction among other alloy anodes in MIB due to its high theoretical capacity of 385 mAh g^{-1} and voltage of 0.25 V vs Mg/Mg^{2+}. This alloy possesses good compatibility with conventional electrolytes and hence shows reversibility in MIBs. However, it has some limitations such as capacity fading and poor cycling stability attributed by the volume change during alloying/dealloying process. Arthur et al. [28] first reported Bi alloy electrode in MIB which showed a specific capacity of 222 mAh g^{-1} and good compatibility with electrolyte. This work is considered as a major breakthrough in alloy-based anode material in MIBs. Various electrochemical studies and X-ray diffraction methods are carried out in order to understand the mechanism of reversible alloying in MIBs [29].

Shao et al. [30] proposed Bi nanotubes (Bi-NTs) as anode material in MIBs, and from studies, this Bi NT has a reversible specific capacity of 350 mAh g^{-1}, high columbic efficiency and excellent cycling stability. Figure 4.3 shows the structural transformation of bismuth during the discharge/charge process. Later, colloidal Bi nanocrystals (NCs) were also used as anode material in MIBs [31]. These Bi NCs showed specific capacity of 325 mAh g^{-1} with excellent capacity retention. The magnesiation of Bi was studied using crystal-structure prediction methodologies and total-energy density functional theory (DFT) calculations. It is found that magnesiation occurs through an alloying mechanism and it is found that there is a formation of two stable phases. The first one is the low-temperature trigonal structure (Mg_3Bi_2), and the second is high-temperature cubic structure (Mg_3Bi_2). From the small angle X-ray scattering (SAXS) measurements, it is found that there is a formation of monodispersed Bi NCs anode during the electrochemical cycling. Some active Bi particles were converted into smaller nanostructures after the first discharge.

There are several reports on different synthetic routes for developing Bi alloy anode materials. Tan et al. [32] reported nano-clustered Bi-Mg alloy via one-step solid-state alloying setup. This is prepared by mixing Mg and Bi powder under different alloying temperatures. Figures 4.4 and 4.5 show the nanoscale morphologies characterization of products obtained at different alloying temperatures [32]. The anode alloyed at temperatures of 650°C (MB-650) has the highest reversible

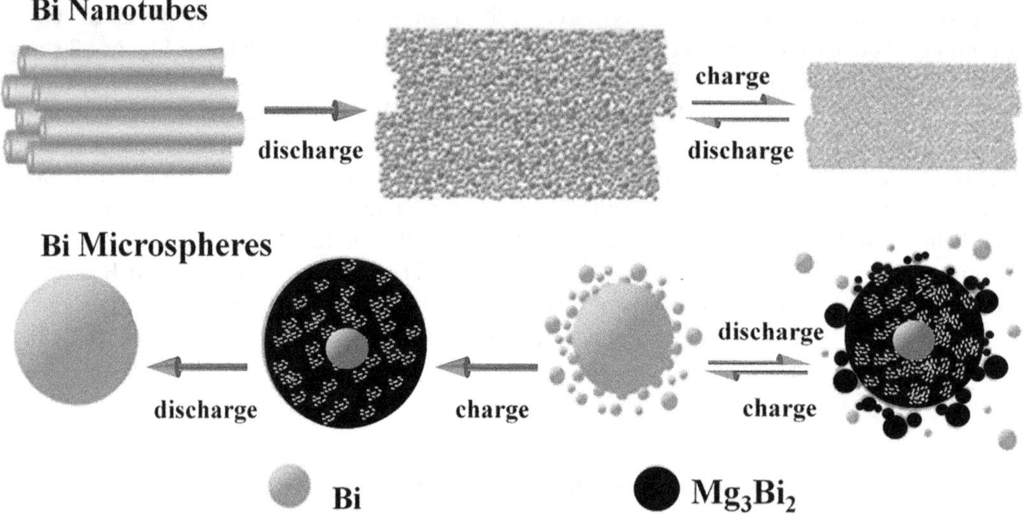

FIGURE 4.3 Structural transformation of bismuth during the discharge/charge process. Bi-NTs may be fully magnesiated, and the structure evolves into porous interconnected nanoparticles which keeps electric contact; however, Bi large particles can only be partially magnesiated, and the structure pulverizes, losing electric contact. This leads to high performance of Bi-NTs and low performance of Bi-Micro in terms of rate capability and cycling stability. Adapted and reproduced from Ref. [30]. Copyright 2014 American Chemical Society.

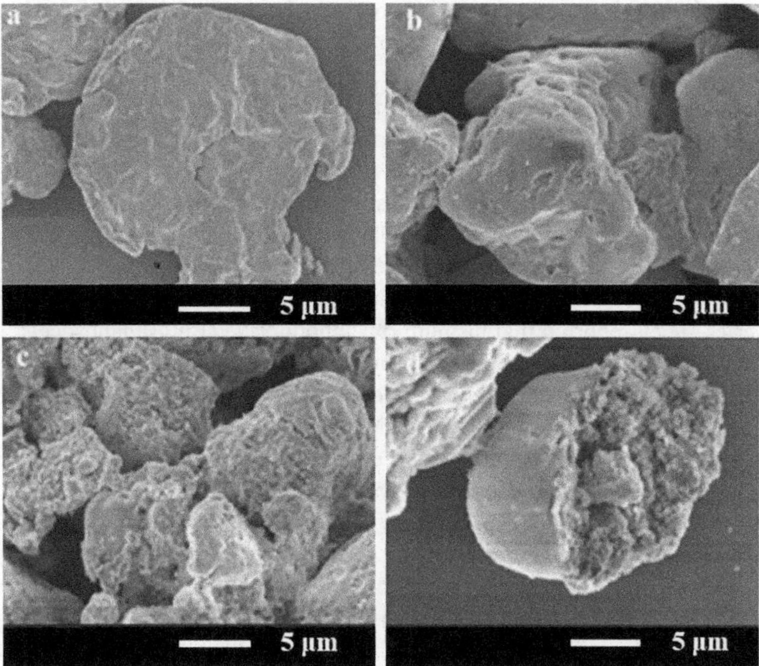

FIGURE 4.4 Nanoscale morphologies characterization of products obtained at different alloying temperatures. (a–d) SEM image of MB-300, MB-500, MB-650 and MB-750, respectively. Adapted and reproduced from Ref. [32]. Copyright 2018 American Chemical Society.

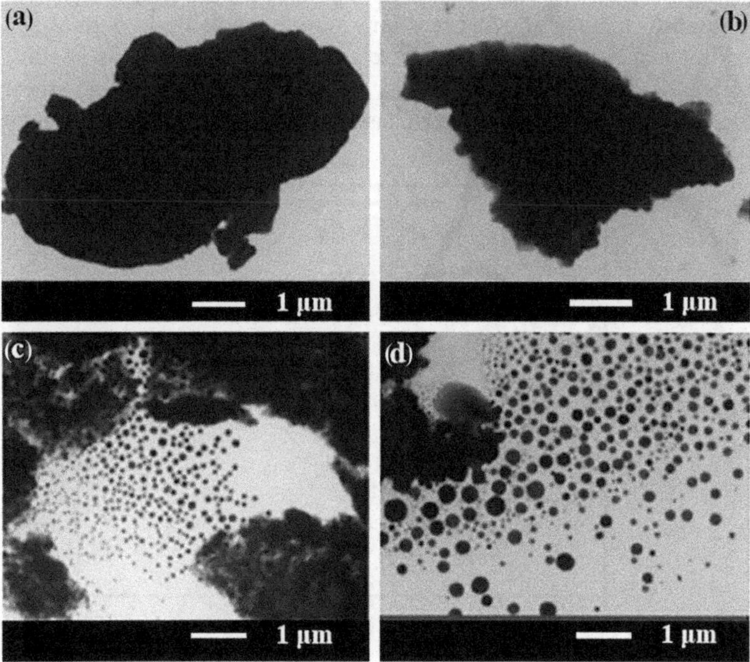

FIGURE 4.5 Nanoscale morphologies characterization of products obtained at different alloying temperatures. (a–d) TEM image of MB-300, MB-500, MB-650 and MB-750, respectively. Adapted and reproduced from Ref. [32]. Copyright 2018 American Chemical Society.

specific capacity of 360 mAh g^{-1}, along with good Coulombic efficiency (average 98.4% in 200 cycles) and excellent capacity retention (90.7% of total capacity was retained over 200 cycles). In addition to this, Bi-based composite materials are also employed as anode materials in MIBs. For instance, the nanocomposite prepared from Bi and reduced graphene oxide (RGO) with different component ratios was used as the anode material for MIBs. The nanocomposite of 60% Bi:40% RGO (Bi_{60}) exhibits good cycle stability with a discharge capacity of 372 mAh g^{-1} and relatively good high-rate capability. The presence of RGO strengthens the electrical conductivity of the composite and also acts as a buffer to accommodate large volume changes during the charge/discharge cycling.

Wang et al. [17] developed a novel bismuth-based anode material for rechargeable magnesium-ion batteries. They reported that bismuth oxynanofluoride (BiOF) nanosheets were prepared by solvothermal method and analysed its performance as anode for the first time (Figure 4.6) [17]. Mg-ion storage mechanism was studied by coupled ex situ X-ray diffraction (XRD) measurements and X-ray photoelectron spectroscopy (XPS) analysis with accuracy. In this experiment, conversion reaction from BiOF to metallic bismuth occurs during the Mg^{2+} ion insertion/extraction process, and two-dimensional bismuth metallic anode was developed. This anode material showed high reversible capacity (335 mAh g^{-1}), superior rate capability and excellent cycle life with capacity retention >96% after 100 cycles.

Research on the electrochemical reaction process of the BiOF electrode in rechargeable Mg battery was investigated in which the phase transformation of BiOF starts at discharging (0.25 V). In between that, a minor reaction from the irreversible reaction of equation 4.1 took place:

$$Bi_2O_3 + 3\,Mg^{2+} + 6\,e^- \rightarrow 3\,MgO + 2Bi \tag{4.1}$$

The irreversible reaction was confirmed by the disappearance of the Bi_2O_3 in the as-prepared sample. Bi, MgO and MgF_2 are generated. O-Mg and F-Mg bonds in the ex situ XPS spectrum did not change; this implies that MgO and MgF_2 did not participate in the subsequent reactions. After the phase transformation, alloying reactions between metallic Bi with Mg^{2+} ions dominate the electrochemical magnesium storage processes. As the metallic Bi was in situ generated during the conversion reaction, it may show superior stability during the alloying processes due to the space confinement of in situ reaction. After phase transformation, alloying reactions between Bi

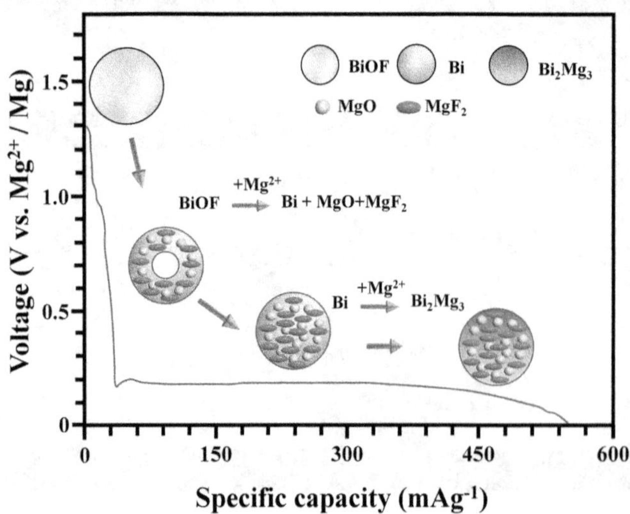

FIGURE 4.6 Schematic illustration of the phase evolution reaction during discharge process. Adapted and reproduced from Ref. [17]. Copyright © 2015 John Wiley and Sons.

and magnesium ion dominate the electrochemical magnesium storage processes. Later, Niu et al. [33] prepared porous Bi-Sn (P-Bi-Sn) composites with varying compositions/sizes. Figure 4.7 shows the schematic illustration of design strategy of P-Bi-Sn alloys. It is inevitable that optimum alloy composition results in the porous structure which is beneficial for improving the electrochemical performance of the P-Bi-Sn anode. As an anode for MIBs, the P–Bi_3Sn_2 anode has better electrochemical properties such as high discharge capacity (367 mAh g^{-1} even at a high current density of 1000 mA g^{-1}).

To recapitulate all of these, Bi-Mg alloy is the most explored material as anode in MIBs. Incorporating nanostructured materials is one of the effective strategies for improving the performance of Bi anodes. Bi alloys possess several electrochemical characteristics such as discharge capacity, Coulombic efficiency, cycling stability and rate performance. These Bi-Mg alloys have limitations too; anode has a limited theoretical capacity in MIBs. Hence, synthesizing composites with other materials is another promising way to further improve the performance of MIBs.

4.3.2 Tin (Sn)-Based Anode Materials

Tin (Sn) is another material which finds application in MIBs, where Singh et al. [34] first reported the high energy Sn as anode in the same. Sn possesses lower reaction voltage (0.15 V) and high theoretical capacity (903 mAh g^{-1}). Malyi et al. [35] reported the DFT calculation of structure and electrochemical characteristics of Mg-Sn alloy anodes. It is found that within a certain range of doping concentrations of Sn in Mg, the Mg atoms can effectively diffuse through Sn enhancing the electrochemical property. Later, Nguyen and Song [35] prepared Mg_2Sn micro-sized anode materials for MIBs via simple ball-milling method. Its electrochemical performance was analysed in a THF electrolyte. This anode shows a discharge capacity of 182 mAh g^{-1} during the first cycle, and later, capacity increased with cycles. The Mg-ion full cell, assembled with a V_2O_5 cathode and $Mg(TFSI)_2$/diglyme electrolyte, delivered good cycling ability in the early cycles at room temperature.

Yaghoobnejad et al. [36] reported a low-cost and facile method for preparing nanostructured Sn material as anode for MIBs. It showed the specific capacity of 300 mAh g^{-1} over 150 cycles. From electrochemical studies, it is observed that the formation of passivation layer has huge impact on the performance of MIBs. It is found that the removal of passivation layer by the conventional constant-current cycle and an oxidation pulse is beneficial for achieving good cycling stability of the Sn electrode. Later, Parent et al. [37] reported morphology, crystal structure and local composition of Sn-Sb NPs anode during the (de)magnesiation processes. Electrochemical studies reveal that Sb played a sterling role in the formation of stable Sn NPs in the alloy system. Cheng et al. [38] reported the potential use of an Sn-Sb alloy in MIBs, and studies showed that the anode has a high reversible specific capacity of 420 mAh g^{-1}. This is mainly attributed to the stable interface formed between

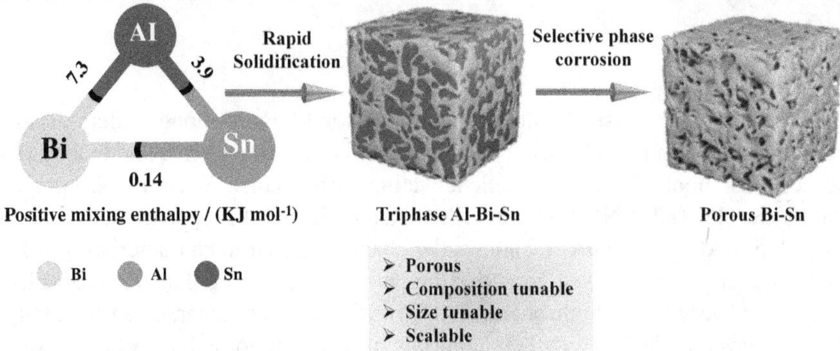

FIGURE 4.7 Schematic illustration of design strategy of P-Bi-Sn alloys. Adapted and reproduced from Ref. [33]. Copyright 2019 Royal Society of Chemistry.

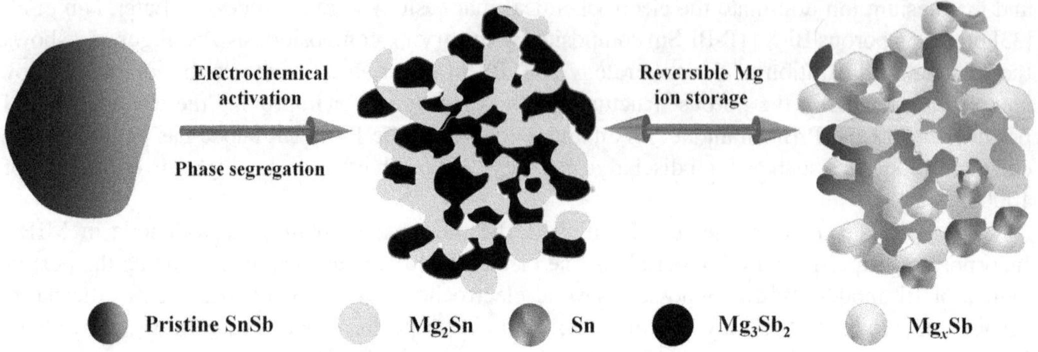

FIGURE 4.8 Schematic illustration of the electrochemical reaction mechanism of Sn-Sb particles with Mg ions. Pristine Sn-Sb particle experiences breakdown and phase segregation during electrochemical activation, and reversible Mg-ion storage for activated Sn-Sb alloys. Adapted and Reproduced from Ref. [38]. Copyright 2015 Wiley.

Sb and Sn. Figure 4.8 elucidates the electrochemical reaction mechanism of Sn-Sb particles with Mg ions. In addition to this, the Sn-Sb anode has been found to exhibit good compatibility with conventional Mg salt electrolytes.

4.3.3 Other Metal-based Anode Materials

The P-block elements such as lead (Pb), antimony (Sb) and indium (In) were also employed as anode material in MIBs. Murgia et al. [39] prepared Sb-Bi via high energy mechanochemical alloying. Sb and Bi facilitated the electrochemical magnesiation of the electrode due to the synergy created through their chemical combination. Later, Blondeau et al. [39] reported In-Sb alloy during the Mg alloying reaction. It is found that In-Sb reacts reversely with Mg leading to better electrochemical performance over a few tens of cycles. Periyapperuma et al. [40] prepared sputtered Pb films as anode in MIBs. This showed quasi-reversible Mg electrochemistry in Grignard-based electrolytes, resulting in a high discharge capacity of 450 mAh g^{-1}. However, due to the toxicity of Pb, it is not considered a viable candidate for high-performance rechargeable MIBs when effective recovery methods exist.

In conclusion, Bi and Sn are the most studied anode materials for MIBs; Sb is limited by its low electrochemical activity, while Pb and In suffer from toxicity and scarcity, respectively. Different nanostructured materials have been developed as anode for MIBs; however, their high cost and complexity of synthesis halt their commercialization in the development of production routes. Table 4.1 summarizes the electrochemical performance of the alloy-based anode materials for MIB.

4.4 METAL OXIDES

Metal oxides have been extensively studied as anode for MIBs. Among widely accepted metal oxides, titanium-based compounds have been considered to be prospective anode for MIBs [41,42]. It possesses low cost, nontoxicity and excellent cycling performance. Wu et al. [45] prepared spinel $Li_4Ti_5O_{12}$ nanoparticles (LTO NPs) with excellent electrochemical properties such as good reversible capacity (175 mAh g^{-1}). Figure 4.9 shows the charge–discharge characteristics of LTO in Mg batteries. LTO NP electrode showed good cycling performance (more than 95% capacity retention over 500 cycles) and very low volume change (0.8%) during cycling. Later, Chen et al. [44] proposed the possibility of layered $Na_2Ti_3O_7$ (NTO) nanoribbons as anode for MIBs. NTO showed the theoretical capacity of 88 mAh g^{-1} and exhibited a good reversible capacity (75 mAhg^{-1}).

TABLE 4.1
Summary of Electrochemical Performance of Alloy-Based Anode for Magnesium-Ion Battery [21]

Anode Materials	Discharge Capacity/($mAhg^{-1}$)	Operating Voltage (V)	Cycling Performance/ ($mAhg^{-1}$)/Cycles	Rate Performance ($mAh.g^{-1}$)
Bi nanotubes	350 at 0.05 C	0–0.6	303/200	216 at 5 C
Colloidal Bi nanocrystals	350 at 2 C	0–0.6	325/150; 189/4500	335 at 0.5 C; 350 at 2 C
Nanocluster Mg_3Bi_2	360 at 100 mAg^{-1}	0.05–1.60	263/200	103 at 100 mAg^{-1}; 58 at 2000 mAg^{-1}
60%Bi:40%RGO nanocomposites	413 at 39 mAg^{-1}	0.005–0.600	372/50	381, 372, 354, 295 and 238 at 53, 100, 200, 500 and 700 mAg^{-1}
P–Bi_3Sn_2 electrode	416 at 100 mAg^{-1}	0–0.8	301/200 (1000 mAg^{-1})	367 at 1000 mAg^{-1}
Micronized bulk Mg_2Sn anode	182 at 0.025 C	0.01–0.60	270/15; 197/50	-
Nanostructured Sn-Sb Alloys	420 at 50 mAg^{-1}	0–0.8	270/200	350 at 1000 mAg^{-1}
Bi_6Sn_4 anode	434 at 50 mAg^{-1}	0–0.8	280/200	362 at 1000 mAg^{-1}
Activated Mg_2Sn anode	382 at 0.05 C	0.05–0.60	305/50; 264/60	-

To summarize all these, the investigation on the possible use of metal oxides as anodes for MIBs has been mainly focused on Ti-based composites, which showed excellent cycling performance because of the small volumetric change during the intercalation/deintercalation process. However, they can provide relatively low reversible capacities as compared to other anode materials. A suitable morphology would be beneficial for enhancing their performance in MIBs.

4.5 TWO-DIMENSIONAL MATERIALS

Over the years, there is an increasing interest in the exploration of two-dimensional (2D) materials after the discovery of graphene and its derivatives. 2D materials have superior physicochemical properties compared to its bulk; based on this, mainly these 2D materials are studied at the atomic level by carrying out computational simulations.

4.5.1 Carbon-Based 2D Materials as Anode in MIBs

The superior properties of graphene compared to its other allotropes make it an attractive candidate for anode materials in energy storage devices. Er et al. [45] proposed using graphene as anode material in MIBs using computational simulations. It is found that the open-circuit voltage (OCV) and charge transfer have direct relation to the concentration of the introduced defect on the material. The low defect fraction on the material was unfavourable for adsorption of Mg on all possible sites. Later, Zhang et al. [46] reported the three types of carbon-nitrogen structures used as anodes for MIBs and the calculated results of the band structure three possess semiconductor characteristics. Figure 4.10 shows the structures of C_2N and g-C_3N_4 having theoretical capacities of 671.7 and 319.2 mAh g^{-1}, respectively, in MIB.

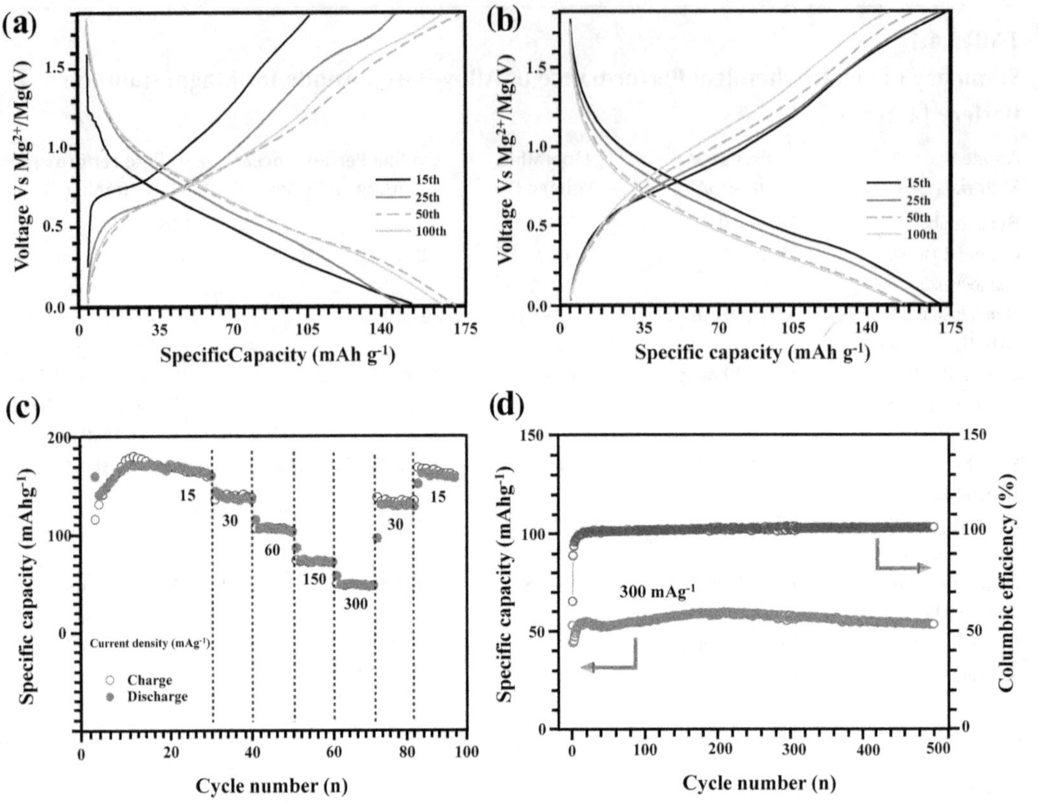

FIGURE 4.9 Charge–discharge characteristics of LTO electrodes in rechargeable Mg batteries (a and b). Typical galvanostatic discharge–charge voltage profiles of the cell cycled before (a) and after (b) 15 cycles at a current density of $15\,mA\,g^{-1}$. (c) Comparison of the rate capabilities of the cell cycled at different current densities. (d) Cycling performance of the cells using LTO electrodes cycled at a current density of $300\,mA\,g^{-1}$. Adapted and Reproduced from Ref. [43]. Copyright 2014 Springer Nature.

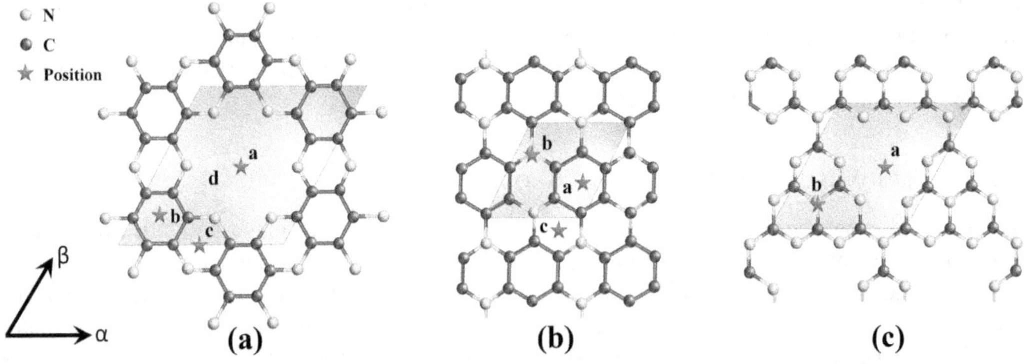

FIGURE 4.10 Schematic of fully relaxed (a) C_2N, (b) C_3N and (c) g-C_3N_4 structures. The purple balls represent carbon atoms and the yellow balls represent nitrogen atoms. The dashed box in each figure indicates the primitive cell, and the angle between vectors α and β is 60°. The lattice constants are 8.247, 4.934 and 7.133 Å for C_2N, C_3N and g-C_3N_4, respectively. Adapted and reproduced from Ref. [46]. Copyright 2019 Elsevier.

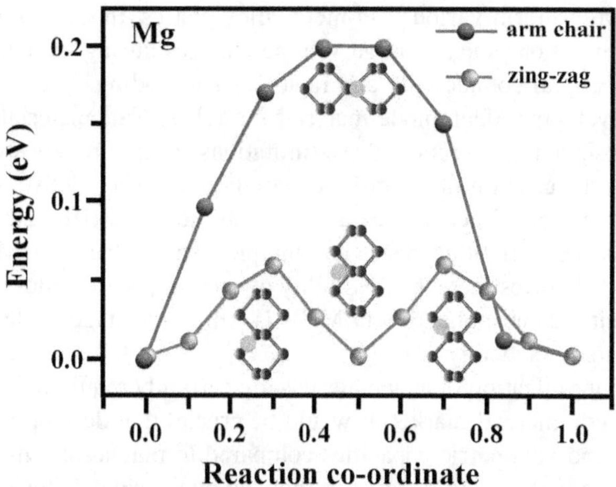

FIGURE 4.11 Energy profiles of Mg diffusion along armchair and zigzag directions. Adapted and reproduced from Ref. [49]. Copyright 2018 Elsevier.

4.5.2 Phosphorous- and Arsenic-Based Anode Material in MIB

Phosphorene has been widely explored material in various fields of chemistry in MIBs. Jin et al. [47] proposed the single-layer phosphorene that is derived from black phosphorus by DFT-based assumptions. The OCV of insertion of Mg^{2+} into the layers of phosphorene is a lower value (0.89 V). In addition, single-layer phosphorene was observed to exhibit a small volume expansion during ion insertion and good structural stability in MIB. Sibari et al. [48] reported that when the concentration of Mg atom increased, both sides of the phosphorene were covered by Mg atoms (X = 1). The specific capacity for $Mg_{0.5}P$ was calculated to be 310.71 mAh g^{-1}. Figure 4.11 explains the anisotropic diffusion of Mg atom in phosphorene along the armchair direction; this process requires jumping over the P atom in the upper puckering of phosphorene, which is required to overcome a high barrier that is not conducive to diffusion. Han et al. [49] observed that the Mg atoms tend to be absorbed on one side of phosphorene, and hence, the excessive accumulation of Mg on one side can cause significant electrostatic repulsion and finally instability of the monolayer structure. Figure 4.11 shows the energy profiles of Mg diffusion at different reaction coordinates.

Arsenic is an element found in the same group of phosphorous, where a single layer of arsenic exhibits good stability and promising electronic properties. Benzidi et al. [50] predicted the possibility of using arsenene as an anode in MIBs using theoretical calculations. In the case of arsenic, an increase in Mg adsorption introduced a new state within the Fermi level, which made the material from semiconductor to metallic nature. Theoretical specific capacities of 1430.90 mAh g^{-1} could be obtained with Mg_2As compositions in MIBs. Ye et al. [51] calculated theoretical capacity of the arsenene monolayer and found that it was similar to that of the previous study by Benzidi and Sibari. Improvement of the conductivity of arsenene and graphene was integrated with it to form a hybrid heterostructure; this combination makes the lone-pair electrons on the arsenene transfer to graphene, thus enhancing the bonding strength of Mg to arsenene. The specific capacities of the arsenene/graphene composites were 409.9 mAh g^{-1}, which were much smaller than those of the arsenene monolayer.

4.6 CONCLUSION

In this chapter, the progress of research on alternative anode materials (metals and their alloys, metal oxides and 2D materials) for MIBs has been explained in detail. Alloy-based materials are the most promising alternative anode materials for MIBs, especially Bi- and Sn-based anode materials.

They can react with Mg to form various Mg-metal alloy phases, thus resulting in high capacities. However, they have limitations such as large volume changes during the cycling process, accompanied by a loss in electrical connectivity and rapid capacity fading. Therefore, various strategies were employed for developing ideal anode material for MIBs. Nanomaterials and heterostructure materials are being designed to overcome these limitations. There are few studies on metal oxides and most of them have focused on titanium-based compounds. They deliver excellent cycling performance; however, their specific capacities are generally low in MIBs. 2D materials are widely explored due to their large interstitial spaces and unique layered structure. Many researchers have also begun to theoretically investigate the possibility of using them as anodes in MIBs. Despite their high theoretical capacity, their application in MIBs is still at the stage of theoretical research and needs to be demonstrated practically.

The discovery and optimization of new materials are certainly required for the successful establishment of RMBs in commercial market. It would be crucial that developing anodes that possess very low gravimetric and volumetric capacities compared to magnesium metal is the first preference. The second relates to the sluggish kinetics induced by the slow diffusion of magnesium ions. The third point relates to examining the presence and nature of possible insertion anode–electrolyte interfaces that may form electrochemically/chemically. Not only does this impact the rate of magnesium ions insertion/extraction, but also provides valuable insight into potential interfaces that may enable facile magnesium ion diffusion for the future RMB.

REFERENCES

1. Gür TM (2018) Review of electrical energy storage technologies, materials and systems: Challenges and prospects for large-scale grid storage. *Energy and Environmental Science* 11:2696–2767.
2. Koohi-Fayegh S, Rosen MA (2020) A review of energy storage types, applications and recent developments. *Journal of Energy Storage* 27:101047–101070.
3. Liu F, Wang T, Liu X, Fan LZ (2021) Challenges and recent progress on key materials for rechargeable magnesium batteries. *Advanced Energy Materials* 11:2000787.
4. Goodenough JB (2014) Electrochemical energy storage in a sustainable modern society. *Energy and Environmental Science* 7:14–18.
5. Goodenough JB, Park KS (2013) The Li-ion rechargeable battery: A perspective. *Journal of the American Chemical Society* 135:1167–1176.
6. Goodenough JB, Kim Y (2010) Challenges for rechargeable Li batteries. *Chemistry of Materials* 22:587–603.
7. Vikström H, Davidsson S, Höök M (2013) Lithium availability and future production outlooks. *Applied Energy* 110:252–266. https://doi.org/10.1016/j.apenergy.2013.04.005.
8. Ortiz-Vitoriano N, Drewett NE, Gonzalo E, Rojo T (2017) High performance manganese-based layered oxide cathodes: Overcoming the challenges of sodium ion batteries. *Energy and Environmental Science* 10:1051–1074.
9. Kim H, Jeong G, Kim YU, et al (2013) Metallic anodes for next generation secondary batteries. *Chemical Society Reviews* 42:9011–9034.
10. Mohtadi R, Mizuno F (2014) Magnesium batteries: Current state of the art, issues and future perspectives. *Beilstein Journal of Nanotechnology* 5:1291–1311.
11. Yoo HD, Shterenberg I, Gofer Y, et al. (2013) Mg rechargeable batteries: An on-going challenge. *Energy and Environmental Science* 6:2265–2279.
12. Rashad M, Asif M, Wang Y, et al (2020) Recent advances in electrolytes and cathode materials for magnesium and hybrid-ion batteries. *Energy Storage Materials* 25:342–375. https://doi.org/10.1016/j.ensm.2019.10.004.
13. Aurbach D, Pour (2011) Nonaqueous electrochemistry of magnesium. Woodhead Publishing Series in Metals and Surface Engineering, pp. 484–515.
14. Aurbach D, Lu Z, Schechter A, et al. (2000) Prototype systems for rechargeable magnesium batteries. *Nature* 407:724–727.
15. Huie MM, Bock DC, Takeuchi ES, et al. (2015) Cathode materials for magnesium and magnesium-ion based batteries. *Coordination Chemistry Reviews* 287:15–27.

16. Yoo HD, Liang Y, Dong H, et al (2017) Fast kinetics of magnesium monochloride cations in interlayer-expanded titanium disulfide for magnesium rechargeable batteries. *Nature Communications* 8. https://doi.org/10.1038/s41467-017-00431-9.
17. Wang W, Liu L, Wang PF, et al (2018) A novel bismuth-based anode material with a stable alloying process by the space confinement of an: In situ conversion reaction for a rechargeable magnesium ion battery. *Chemical Communications* 54:1714–1717. https://doi.org/10.1039/c7cc08206a.
18. Lu Z, Schechter A, Moshkovich M, Aurbach D (1999) On the electrochemical behavior of magnesium electrodes in polar aprotic electrolyte solutions. *Journal of Electroanalytical Chemistry* 466:203–217.
19. Muldoon J, Bucur CB, Oliver AG, et al (2012) Electrolyte roadblocks to a magnesium rechargeable battery. *Energy and Environmental Science* 5:5941–5950.
20. Zhang Z, Cui Z, Qiao L, et al (2017) Novel design concepts of efficient Mg-ion electrolytes toward high-performance magnesium-selenium and magnesium-sulfur batteries. *Advanced Energy Materials* 7. https://doi.org/10.1002/aenm.201602055.
21. Guo Q, Zeng W, Liu SL, et al (2021) Recent developments on anode materials for magnesium-ion batteries: A review. *Rare Metals* 40:290–308.
22. Manthiram A (2017) An outlook on lithium ion battery technology. *ACS Central Science* 3:1063–1069. https://doi.org/10.1021/acscentsci.7b00288.
23. Park MS, Kim JG, Kim YJ, et al (2015) Recent advances in rechargeable magnesium battery technology: A review of the field's current status and prospects. *Israel Journal of Chemistry* 55:570–585. https://doi.org/10.1002/ijch.201400174.
24. Imhof R, Haas O, Novák P (1999) Magnesium insertion electrodes for rechargeable nonaqueous batteries Ð a competitive alternative to lithium? *Electrochimica Acta* 45:351–367.
25. Matsui M (2011) Study on electrochemically deposited Mg metal. *Journal of Power Sources* 196:7048–7055. https://doi.org/10.1016/j.jpowsour.2010.11.141.
26. Deivanayagam R, Ingram BJ, Shahbazian-Yassar R (2019) Progress in development of electrolytes for magnesium batteries. *Energy Storage Materials* 21:136–153. https://doi.org/10.1016/j.ensm.2019.05.028.
27. Niu J, Yin K, Gao H, et al (2019) Composition- and size-modulated porous bismuth-tin biphase alloys as anodes for advanced magnesium ion batteries. *Nanoscale* 11:15279–15288. https://doi.org/10.1039/c9nr05399a.
28. Arthur TS, Singh N, Matsui M (2012) Electrodeposited Bi, Sb and Bi1-xSbx alloys as anodes for Mg-ion batteries. *Electrochemistry Communications* 16:103–106. https://doi.org/10.1016/j.elecom.2011.12.010.
29. Murgia F, Stievano L, Monconduit L, Berthelot R (2015) Insight into the electrochemical behavior of micrometric Bi and Mg_3Bi_2 as high performance negative electrodes for Mg batteries. *Journal of Materials Chemistry A* 3:16478–16485. https://doi.org/10.1039/C5TA04077A.
30. Shao Y, Gu M, Li X, et al (2014) Highly reversible Mg insertion in nanostructured Bi for Mg ion batteries. *Nano Letters* 14:255–260. https://doi.org/10.1021/nl403874y.
31. Kravchyk KV, Piveteau L, Caputo R, et al (2018) Colloidal bismuth nanocrystals as a model anode material for rechargeable Mg-ion batteries: Atomistic and mesoscale insights. *ACS Nano* 12:8297–8307. https://doi.org/10.1021/acsnano.8b03572.
32. Tan YH, Yao WT, Zhang T, et al (2018) High voltage magnesium-ion battery enabled by nanocluster Mg_3Bi_2 alloy anode in noncorrosive electrolyte. *ACS Nano* 12:5856–5865. https://doi.org/10.1021/acsnano.8b01847.
33. Niu J, Yin K, Gao H, et al (2019) Composition- and size-modulated porous bismuth-tin biphase alloys as anodes for advanced magnesium ion batteries. *Nanoscale* 11:15279–15288. https://doi.org/10.1039/c9nr05399a.
34. Singh N, Arthur TS, Ling C, et al (2013) A high energy-density tin anode for rechargeable magnesium-ion batteries. *Chemical Communications* 49:149–151. https://doi.org/10.1039/c2cc34673g.
35. Malyi OI, Tan TL, Manzhos S (2013) In search of high performance anode materials for Mg batteries: Computational studies of Mg in Ge, Si, and Sn. *Journal of Power Sources* 233:341–345. https://doi.org/10.1016/j.jpowsour.2013.01.114.
36. Yaghoobnejad Asl H, Fu J, Kumar H, et al (2018) In situ dealloying of bulk Mg_2Sn in Mg-ion half cell as an effective route to nanostructured sn for high performance Mg-ion battery anodes. *Chemistry of Materials* 30:1815–1824. https://doi.org/10.1021/acs.chemmater.7b04124.
37. Parent LR, Cheng Y, Sushko PV, et al (2015) Realizing the full potential of insertion anodes for Mg-ion batteries through the nanostructuring of Sn. *Nano Letters* 15:1177–1182. https://doi.org/10.1021/nl5042534.
38. Cheng Y, Shao Y, Parent LR, et al (2015) Interface promoted reversible Mg insertion in nanostructured tin-antimony alloys. *Advanced Materials* 27:6598–6605. https://doi.org/10.1002/adma.201502378.

39. Murgia F, Laurencin D, Weldekidan ET, et al (2018) Electrochemical Mg alloying properties along the Sb1-xBix solid solution. *Electrochimica Acta* 259:276–283. https://doi.org/10.1016/j.electacta.2017.10.170.
40. Periyapperuma K, Tran TT, Purcell MI, Obrovac MN (2015) The reversible magnesiation of Pb. *Electrochimica Acta* 165:162–165. https://doi.org/10.1016/j.electacta.2015.03.006.
41. Fu S, Ni J, Xu Y, et al (2016) Hydrogenation driven conductive $Na_2Ti_3O_7$ Nanoarrays as robust binder-free anodes for sodium-ion batteries. *Nano Letters* 16:4544–4551. https://doi.org/10.1021/acs.nanolett.6b01805.
42. Subramanian V, Karki A, Gnanasekar KI, et al (2006) Nanocrystalline TiO_2 (anatase) for Li-ion batteries. *Journal of Power Sources* 159:186–192. https://doi.org/10.1016/j.jpowsour.2006.04.027.
43. Wu N, Lyu YC, Xiao RJ, et al (2014) A highly reversible, low-strain Mg-ion insertion anode material for rechargeable Mg-ion batteries. *NPG Asia Materials* 6. https://doi.org/10.1038/am.2014.61.
44. Chen C, Wang J, Zhao Q, et al (2016) Layered $Na_2Ti_3O_7$/$MgNaTi_3O_7$/$Mg0.5NaTi_3O_7$ nanoribbons as high-performance anode of rechargeable Mg-ion batteries. *ACS Energy Letters* 1:1165–1172. https://doi.org/10.1021/acsenergylett.6b00515.
45. Er D, Detsi E, Kumar H, Shenoy VB (2016) Defective graphene and graphene allotropes as high-capacity anode materials for Mg ion batteries. *ACS Energy Letters* 1:638–645. https://doi.org/10.1021/acsenergylett.6b00308.
46. Zhang J, Liu G, Hu H, et al (2019) Graphene-like carbon-nitrogen materials as anode materials for Li-ion and mg-ion batteries. *Applied Surface Science* 487:1026–1032. https://doi.org/10.1016/j.apsusc.2019.05.155.
47. Jin W, Wang Z, Fu YQ (2016) Monolayer black phosphorus as potential anode materials for Mg-ion batteries. *Journal of Materials Science* 51:7355–7360. https://doi.org/10.1007/s10853-016-0023-4.
48. Sibari A, el Marjaoui A, Lakhal M, et al (2018) Phosphorene as a promising anode material for (Li/Na/Mg)-ion batteries: A first-principle study. *Solar Energy Materials and Solar Cells* 180:253–257. https://doi.org/10.1016/j.solmat.2017.06.034.
49. Han X, Liu C, Sun J, et al (2018) Density functional theory calculations for evaluation of phosphorene as a potential anode material for magnesium batteries. *RSC Advances* 8:7196–7204. https://doi.org/10.1039/c7ra12400g.
50. Benzidi H, Lakhal M, Garara M, et al (2019) Arsenene monolayer as an outstanding anode material for (Li/Na/Mg)-ion batteries: Density functional theory. *Physical Chemistry Chemical Physics* 21:19951–19962. https://doi.org/10.1039/c9cp03230d.
51. Ye XJ, Zhu GL, Liu J, et al (2019) Monolayer, bilayer, and heterostructure arsenene as potential anode materials for magnesium-ion batteries: A first-principles study. *Journal of Physical Chemistry C*. https://doi.org/10.1021/acs.jpcc.9b02399.

5 Cathode Materials for Rechargeable Magnesium-Ion Batteries

Akhila Das, Devika S. Lal, Anandu M. Nair, Vijay Kumar Thakur, Abhilash Pullanchiyodan, and Prasanth Raghavan

5.1 INTRODUCTION

Over the last few decades, the battery technology focuses on the electric vehicles and advanced portable energy storage devices in which lithium-ion batteries (LIBs) worked at the forefront, which helps to eliminate or diminish the fossil fuel consumption [1]. Since the demand on large-scale batteries increases, battery technology pushes LIBs to leap beyond the limitation of safety hazards and lack of energy density and volumetric density [2–4]. However, developing cost effective and safer LIBs is still challenging. A switching from LIBs to other batteries such as magnesium-ion batteries (MIBs) [5] and sodium-ion batteries (NIBs) [6] is inevitable. Among these different types, MIBs are found to be a promising option for next-generation battery technology owing to its advantages such as earth abundance, large volumetric density (3833 mAh cc^{-1}), and environmentally and economically benign. Annual production of Mg increased by 20-fold than that of lithium, with a current market price of \$2700 ton^{-1} for Mg and \$64,800 ton^{-1} for lithium [7]. Detailed examination of MIBs revealed that Mg metal itself acts as suitable anode materials due to its high electrochemical performance and no dendrite formation compared to LIB is another highlight of magnesium-based battery system. The basic electrochemistry behind rechargeable MIBs is similar to that of LIBs, and the schematic illustration of mechanism behind MIB is given in Figure 5.1.

Aurbach's group [8] was the first to synthesize a prototype system for advanced MIBs. Mo_6S_8, also known as chevrel phase (CP), Mg metal, and organo-halo aluminum were used as cathode, anode, and electrolyte. The compatibility between Mg electrodes and electrolytes is very crucial for the better electrochemical performance of MIBs. Hence, the potential cathode is a matter of discussion and the performance of battery is closely related to certain factors related to anode/electrolyte interface, intercalation behavior of cathode material, compatibility of electrode materials, etc. Kinetic feasibility of cathode divalent Mg^{2+} ion is comparatively less than that of monovalent Li^+ ion owing to the slower solid-state diffusion mechanism of divalent cation. Thus research and development focuses on the better compatible cathode materials that can offer higher electrochemical performance [9]. Different types of cathodes successfully investigated include chalcogenides, CP, and vanadium-based compounds, which are discussed in the upcoming sections.

5.2 CHEVREL PHASE AS CATHODES FOR RMIBS

The general formula for CP is $M_xMo_6T_8$, where M is the metal and T is S, Se, etc. [10]. CP was earlier investigated as cathode materials in monovalent metal such as Li in LIBs [11] owing to the rapid reversible insertion of metal ion. The general reaction mechanism behind CP is shown in equation 5.1 [12]. The basic crystal structure of Mo_6S_8 consists of two vacant sites in which the metals are

DOI: 10.1201/9781003310167-5

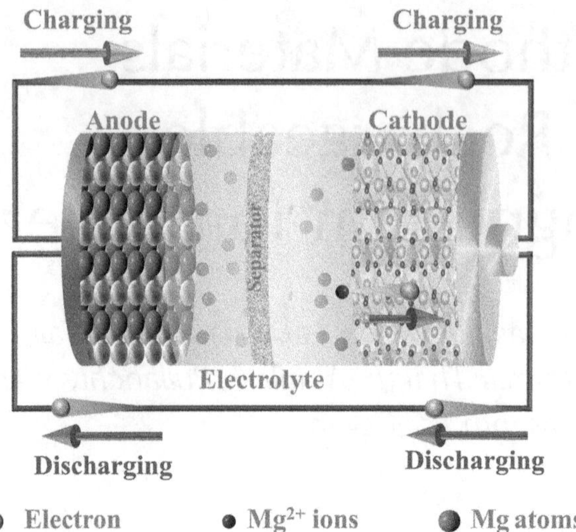

FIGURE 5.1 Schematic illustration of the structure and operating principles of magnesium-ion batteries including the movement of ions between electrodes during charge (forward arrow) and discharge (backward arrow) states.

localized with an open three-dimensional channel. After the successful outcome of the CP in LIBs, researchers explored this material in MIBs. The larger, immobile cations that typically lie in cavity 1 and cavity 2 are generally occupied by smaller cation. Burdett et al. [13] reported the structural examination of CP by molecular orbital theory. Hence, the phase diagram of CP including its crystal structures was reported by Levi et al. [14] who was the first to introduce CP in rechargeable MIBs. Magnesium dissolution mechanism of inorganic halide-based electrolytes on the CP was reported by Wan et al. [15] (Figure 5.2). CP crystal structure was chosen as cathode materials owing to the fast kinetic reaction, along with organo magnesium aluminum halide as electrolyte and Mg metal as anode material, which exhibit a maximal charge capacity of 122 mAh g^{-1}. Prolonged cycling stability (2000 cycles) was observed owing to the excellent compatibility of CP in MIBs. The general equation by the insertion of Mg ions in Mo_6S_8 in the two vacant sites is given in the following equations:.

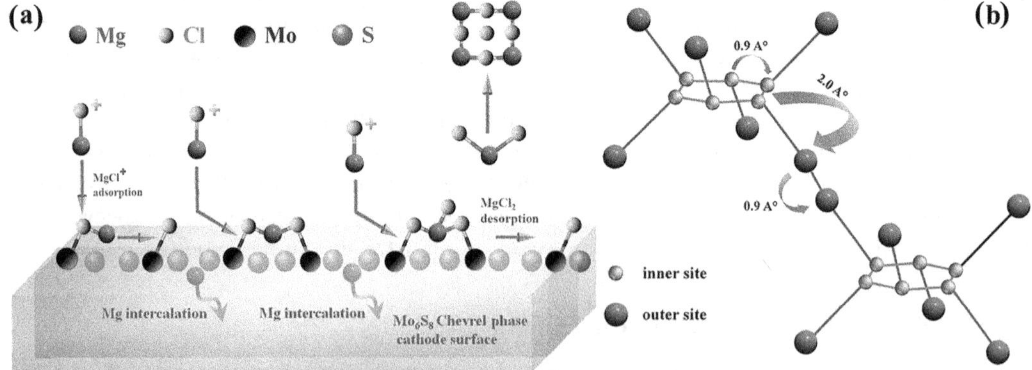

FIGURE 5.2 (a) Schematic illustration of the interaction of chevrel phase and $MgCl_2$ electrolyte. (b) Arrangement of cation sites in the $Mg_xMo_6S_8$ crystal structure. Adapted and reproduced from Ref. [14] Copyright © 2006 American Chemical Society. Adapted and reproduced from Ref. [15]. Copyright © 2015 American Chemical Society.

$$xMn^{+} + M¢yMo_6T_8 + 4e^{-} \rightarrow MxMo_6T_8 + yM¢ \text{ (metal)} \tag{5.1}$$

$$Mg^{2+} + 2e^{-} + Mo_6X_8 \rightarrow MgMo_6X_8 \tag{5.2}$$

$$Mg^{2+} + 2e^{-} + MgMo_6X_8 \rightarrow Mg_2Mo_6X_8 \tag{5.3}$$

Thermodynamic factor regarding the CP insertion reveals that reversible Mg insertion and fast ion diffusion occur at a faster rate [16]. CP can be synthesized by the acid treatment of $CuMo_6S_8$ (CuCP) [17]. At C/6 rate, Mo_6S_8 cathode has a specific capacity of 76 mAh g^{-1} with high capacity retention (95%). The higher capacitive retention of CP is due to the higher wetting capacity of the electrode materials with electrolytes owing to the formation of cuboidal structure of Mo_6S_8. Since milling alters the particle size and crystal deformation of CuCP, Aurbach et al. [18] examined the milling process of CP, which improved the intercalation kinetics with the decrease in the particle size. An enhancement in surface electronic conduction increased compared to that of pristine CP electrodes. Levi et al. [19] also reported the kinetic and thermodynamic factors of CP in which both S and Se were investigated. Substituting sulfur to selenium, an increase in the Mg-ion interaction was observed owing to the high polarization effect of Se than S atom. Hence, Aurbach et al. [20] investigated the partial substitution of S to Se to form $Mo_6Se_{8-n}S_n$ CPs (n = 0, 1, 2) for the reversible interaction of Mg ions.

5.2.1 Copper-based Chevrel Phase as Cathode Materials

Incorporation of Cu ions in the CP to form $CuMo_6S_8$ (CuCP) was explored in MIBs providing improved electrochemical properties. CuCP also acts as a precursor for the synthesis of CP cathode material [17]. CuCP is less reactive compounds, similar to that of chalcogenide materials, which can be synthesized by molten solid-state synthesis (MSS) technique. Hence, Aurbach et al. [21] closely examined the electrochemical and physical properties of CuCP in rechargeable MIBs. The major synthetic strategy for the preparation suitable CuCP was milling the pristine CuCP followed by an annealing technique. The surface morphological images of pristine material reveal the presence of large spherical agglomerates on the electrode surface, which has been deformed after the milling process of CuCP. The surface analysis of pristine and milled CuCP at 5 μm is shown in Figure 5.3.

5.3 PRUSSIAN BLUE ANALOGS IN rMIBs

Prussian blue analogs (PBA) have an open framework structure, with the general formula $A_xM_y[M'(CN)_6]$, where CN is the ligand and M and M′ refer to transition metals [22]. Smaller cations can be easily inserted through the open structure due to the presence of stronger triple bonded CN, which paves the easy separation of transition metals. Pata et al. [23] reported the morphological and electrochemical properties of this open frame structure in advance energy storage systems (Figure 5.4). Manganese hexacyanomanganate (MnII–NiC–MnIII/II) has been used as an anode material providing a zero redox potential, whereas this material combined with its cathode material (CuII–NiC–FeIII/II) provides a potential range of 1 V. Wang et al. [24] investigate the behavior of alkaline earth metals such as Ca, Mg, and Ba in PBAs. Among different PBAs, copper and nickel hexacyanoferrate (CuHCF, NiHCF) exhibit high cycling stability compared to other monovalent metals. NiHCF has a high cycling stability (~2000 cycles) and its Mg^{2+} exhibits better specific capacity (65 mAh g^{-1}) compared to other alkaline earth metals. Later, the same group [25] investigated the reversible insertion of trivalent, divalent, and monovalent ions in Prussian blue structures. Electrochemical performance is conducted by choosing CuHCFe as cathode material. Since Rb^{+}, Y^{3+}, and Pb^{2+} were inserted in PBA sites through different pathways, it is concluded that the insertion

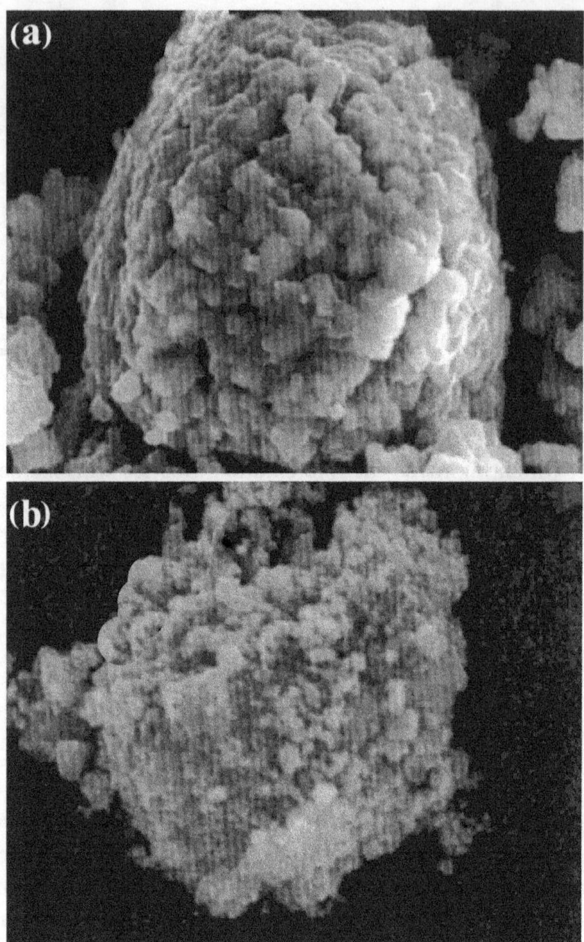

FIGURE 5.3 Scanning electron microscope (SEM) image on the morphology of the (a) Pristine and (b) milled CuCP at different stages with magnification 5 μm. Adapted and reproduced from [21]. Copyright © 2002 American Chemical Society.

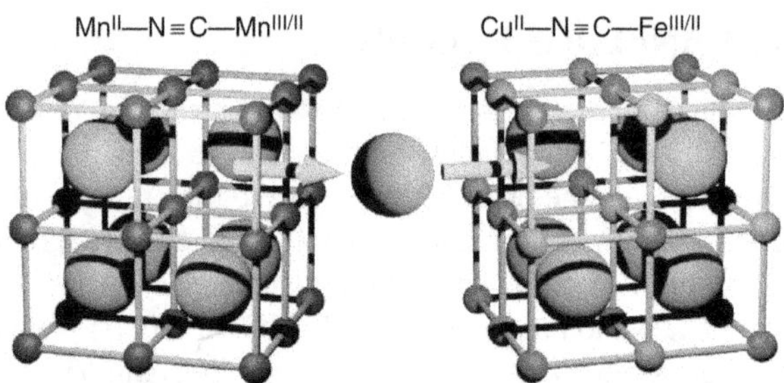

FIGURE 5.4 Symmetric open framework cell schematic consisting of insertion of sodium ions into the copper hexacyanoferrate (CuII–NC–FeIII/II) cathode and a newly developed manganese hexacyanomanganate (MnII–NC–MnIII/II) anode, each of which has the same open framework crystal structure.

of cations can be of different pathways, and hence, reversible insertion is applicable in Mg-ion, Al ion, and other rechargeable batteries. Lipson et al. [26] chose NiHCF as PBA cathode material in rMIBs and its electrochemical performance was conducted with $Mg(TFSI)_2$ in propylene carbonate as electrolytes and Carbon as anode material. Galvanostatic cycling studies shows that NiHCF cathode has a specific capacity of ~40 mAh g^{-1} with cycling stability up to 50 cycles. However, specific capacity and cycling stability are very low. Chen et al. [27] reported a polyamide cathode material ($Na_{1.4}N_{i1.3}Fe(CN)_6{\cdot}5H_2O$) in 1 M $MgSO_4$ aqueous electrolyte with a specific capacity of 150 mAh g^{-1} (1A g^{-1}) and a cycling stability of ~2000 cycles. $K_{0.1}Cu[Fe(CN)_6]_{0.7-3}{\cdot}6H_2O$ cathode (CuFe–PBA) was synthesized by Mizuno et al. [28], and its cyclic performance was conducted by 1M $MgNO_3$ and Pt as counter electrode material. The chemical reaction behind CuFe–PBA is given in equation 5.4. However, the performance of CuFe-PBA system was insignificant since it has cycling stability up to 20 cycles and a specific capacity of ~50 mAh g^{-1} at 0.1 A g^{-1}.

$$xMg^{2+} + 2xe^- + CuFe^-PBA \rightarrow Mg_xCuFe^-PBA\ (0 < x < 0.3) \quad (5.4)$$

5.4 VANADIUM COMPOUNDS AS CATHODE MATERIALS

Vanadium oxides are considered a potential candidate for cathode materials in secondary batteries owing to the advantages of the layered structure, earth abundance, and cost effectiveness. V_2O_5-based cathode materials are widely used in LIBs [29], and researchers explored this cathode material in MIBs as well. Novak et al. [30] investigated the effect of different metal oxides and sulfides on cathode materials such as V_2O_5, ZrS_2, and TiS_2 in $MgClO_4$/THF electrolyte solution and concluded that among other cathodes, V_2O_5 cathode material offers high reversibility and specific capacity (170 mAh g^{-1}). Comparing oxides and sulfides, oxides are stable than sulfides which prevent oxidative decomposition during magnetization. V_2O_5 exists as layered structure similar to TiO_2, and it exists in square pyramidal configuration. The major intercalation reaction in V_2O_5 spinel is given in the following equation [31]:

$$V_2O_5 + xMg^{2+} + 2xe^- \rightarrow Mg_xV_2O_5 \quad (5.5)$$

The intercalation of V_2O_5 crystals in Mg^{2+} is a sluggish process, and hence, researchers altered its physical structure to rod-type structure as evidenced by SEM and TEM images (Figure 5.5) [32,33], enabling a fast intercalation mechanism. Morphology clearly depicts the nanotube-like structure of V_2O_5, and the main advantage of nanotube-like V_2O_5 is the forthright insertion of Mg ion which is a redox-active material synthesized by sol–gel method [33]. In the same year, the same group reported its electrochemical performance with a full battery setup, using Mg ribbon as anode and $Mg(AlBu_2Cl_2)_2$/THF as electrolytes.

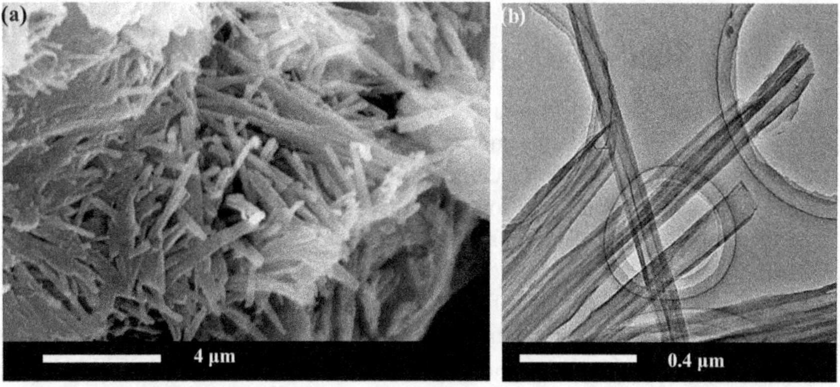

FIGURE 5.5 Morphological analysis of SEM and TEM images of nanorod V_2O_5. Adapted and reproduced from [32]. Copyright 2012 Elsevier.

V_2O_5 cathode material was synthesized by mainly sol–gel process [34], microwave plasma method (sulfur) [35], and metal-doped V_2O_5 [36], which alters the electrochemical performance of rMIBs. $Mg_{0.1}V_2O_5$ prepared by sol–gel method offers a high specific capacity of >270 mAh g^{-1} with an operating voltage of ~3.2 V [34]. Doping sulfur, Cu, etc. enhances the structural stability of the cathode and improves the ionic conductivity and electrochemical performance. Cu-doped V_2O_5 offers rate capability and kinetic behavior [36], whereas S-doped V_2O_5 along with the addition of small amount of MnO_2 exhibits very high specific capacity of ~420 mAh g^{-1} [35]. Slow state diffusion is the major reason for the decrease in the performance of the battery, and hence, this can be lowered by introducing V_2O_5-based aerogels which have short diffusion path followed by high rate capability [37]. However, the development of vanadium-based aerogels is restricted by its size constraints. Daichi et al. [38,39] reported V_2O_5/carbon-based composite electrode material that can intercalate more Mg ions exhibiting a specific capacity of 540 mAh g^{-1}.

5.5 MANGANESE DIOXIDE-BASED CATHODE MATERIALS FOR rMIBs

MnO_2-based cathode materials are potential candidates in MIBs since they are widely used in different types of secondary batteries, especially LIBs [40,41]. MnO_2 can be classified in different ways such as based on their structural framework and polymorphs. Based on their oxide framework system, MnO_2 has different phases such as tunnel, spinel, and layered phases. The tunnel structure is again classified as todorokite (3 × 3), hollandite (2 × 2), and pyrolusite (2 × 2). Todorokite has the largest tunnel structure, and hence, its electrochemical properties in LIBs and MIBs were examined by Kumagai et al. [42]. Mg insertion is possible in todorokite structure with a discharge capacity of 85 mAh g^{-1}, whereas Mg insertion was tedious in hollandite structure. However, hollandite $K_{0.14}MnO_2 \cdot x\ H_2O$ along with acetylene black was reported by Rasul et al. [43] owing to the structural stability of the cathode. Compared to pristine hollandite material, the hollandite/acetylene black composition with the addition of 0.39 Mg/Mn has a high specific capacity of 210 mAh g^{-1} at a current rate of 100 mA g^{-1}, which is twofold higher than pristine hollandite system. Later, the same group [44] (Figure 5.6) compared the electrochemical performance of tunnel structured hollandite–MnO_2 and layered structured birnessite–MnO_2. Specific capacities of 109 and 310 mAh g^{-1} were observed for Bis-MnO_2/AB electrode (0.20 Mg/Mn) and Hol-MnO_2/AB electrode (0.57 Mg/Mn), inferring that hollandite is better than layered structure. Based on different phases of MnO_2, these can classified as α, β, γ, and λ. Yuan et al. [45] investigated the intercalation mechanism of λ-MnO_2 in different divalent cations. Among different divalent cations, λ-MnO_2 in $MgCl_2$ offers a high specific capacity of 545 mAh g^{-1} at 13.6 mA g^{-1}. Zhang et al. [46] and Ling et al. [47] investigated α MnO_2 as cathode material and its electrochemical behavior. The tunnel structure of α-MnO_2 exhibits a specific capacity of 240 mAh g^{-1} [46] at 0.015 C rate.

5.6 SPINELS AS CATHODE MATERIALS IN rMIBs

Magnesium-based spinels have the general formula of AM_2X_4 (A = Mg, M = metal cations, X = O or S) and a tetrahedrally coordinated structure. Since several spinels were introduced as cathode materials in LIBs attributing to the enhancement in cycle life and thermal stability [48–50], similar types of spinels were attracted its attention in Mg-ion batteries. Leng et al. [51] reported that the effect of Li-, Mg-, and Na-based spinels on AMn_2O_4 structure is known as post-spinel compounds. The major attraction of post-spinel compounds is its fast kinetics throughout the crystal structure with one-dimensional channels. Among these three metals, Mg^{2+} is found to have high mobility and is a potential candidate for rechargeable MIBs. Thus, Goutham et al. [52] investigate the inversion and mobility of Mg in Mn_2O_4 spinel structure. The general structure of Mg-based spinels is in face-centered packing with octahedral configuration and concluded that the inversion of spinel structure alters the electrochemical properties of the cathode. Kim et al. [53] reported the intercalation mechanism of Mg spinel having Mg Mn_2O_4 structure. Mg^{2+} was directly intercalated into

FIGURE 5.6 Microstructure of MnO_2: (a) K^+-hollandite tunnel structure and (b) Na^+-birnessite layered structure. Adapted and reproduced with permission from Ref. [44]. Copyright © 2012 Elsevier.

the tetrahedral sites of the spinel having a specific capacity of 190 mAh g^{-1} and a voltage range of 2.9 V. $MgMn_2O_4$ spinel usually exists in tetrahedral structure owing to the presence of Jahn–Teller distortion. Knight et al. [54] investigated possible extraction methods such as acid treatment for demagnesiation of Mg from the spinel host structure of $MgMn_2O_4$ and concluded that easy Mg demagnesiation by altering the structure of this spinel structure. Similar to Mn-based spinels, Liu et al. [55] investigated the effect of Sulfur-based spinels in multivalent cathode materials. By choosing divalent cations such as Mg^{2+} and Ca^{2+} ions, the best combinations such as $MgTi_2S_4$, $CaCr_2S_4$, $MgCr_2S_4$, and $MgMn_2S_4$ possess high electrochemical performance. Among these spinel compounds, $MgTi_2S_4$ has a higher specific capacity of 165 mAh g^{-1} with an average voltage range of 1.2 V. Recently, ternary spinels are explored by Wang et al. [56] with Sc and Se to form $MgSc_2Se_4$. According to his findings, Mg tends to exist in tetrahedral site rather than octahedral site owing to the reduction in the activation energy, which facilitates the easy migration of Mg cation. Doping cerium or titanium in $MgSc_2Se_4$ spinels reduces the electronic conduction behavior and improves the ionic conductivity. Earlier, nickel-doped Mn spinel was fabricated as cathode material in LIBs [57] showing excellent cycling stability owing to the presence of Ni atom which

FIGURE 5.7 High-resolution scanning electron microscopic image (secondary electron contrast) of the Mo_6S_8 electrode: (a) a pristine electrode and (b) the electrode after a charge experiment of Mg-ion spinel-chevrel cell (as shown in a). Adapted and reproduced from [58]. Copyright © 2019 Springer.

reduces the cation disorder. $MgNiMnO_4$ exists in cubic octahedral geometry with the space group *Fd3m*. However, Mg-ion dissolution is energetically not favorable, which was obtained from DFT calculation. Figure 5.7 depicts the surface morphology images of pristine Mo_6S_8 electrode and the morphological change of electrode after charging [58].

5.7 CONCLUSION

Cathode materials are the major part of batteries since compatibility of cathodes is a matter of discussion. Major cathode materials for Mg-based rechargeable batteries include spinels, CP, PBAs, and vanadium compounds. Aurbach's group was the first to use CP Mo_6S_8 owing to its lower cutoff voltage and structural stability. Many modifications to CP, such as partial and fully substitution of sulfur to selenium and doping of transition metals such as Cu and Ni, enhance the electrochemical properties of rMIBs. Open framework model was also introduced by researchers, which mainly includes Prussian blue type analog. However, these compounds have very less cycling stability and specific capacity except for very few compounds. Vanadium compounds possess high specific capacity, easy Mg^{2+} transport pathway, and operational voltage, making them a promising cathode material for rMIBs. Manganese oxides exist in different polymorphs and structural frameworks, which distinctly differ in electrochemical properties of the material. Comparing different structural framework systems such as layered, hollandite, and tunnel structures, layered structures have high specific capacity and cycling stability compared to tunnel and hollandite structures. Spinels are another class of cathode material that has high structural stability and fast kinetics. Thus, different cathode materials have their own properties and better compatible cathode materials with their electrolyte and anode materials preferring high electrochemical properties.

REFERENCES

1. Brodd RJ, Bullock KR, Leising RA, et al (2004) Batteries, 1977 to 2002. *J Electrochem Soc* 151:K1. https://doi.org/10.1149/1.1641042.
2. Bansal D, Meyer B, Salomon M (2008) Gelled membranes for Li and Li-ion batteries prepared by electrospinning. *J Power Sources* 178:848–851. https://doi.org/10.1016/j.jpowsour.2007.07.070.
3. Manthiram A, Chemelewski K, Lee E-S (2014) A perspective on the high-voltage $LiMn_{1.5}Ni_{0.5}O_4$ spinel cathode for lithium-ion batteries. *Energy Environ Sci* 7:1339.
4. Scrosati B, Hassoun J, Sun YK (2011) Lithium-ion batteries. A look into the future. *Energy Environ Sci* 4:3287–3295. https://doi.org/10.1039/c1ee01388b.
5. Zhao-Karger Z, Gil Bardaji ME, Fuhr O, Fichtner M (2017) A new class of non-corrosive, highly efficient electrolytes for rechargeable magnesium batteries. *J Mater Chem A* 5:10815–10820. https://doi.org/10.1039/c7ta02237a.
6. Eftekhari A, Kim DW (2018) Sodium-ion batteries: New opportunities beyond energy storage by lithium. *J Power Sources* 395:336–348. https://doi.org/10.1016/j.jpowsour.2018.05.089.
7. Saha P, Datta MK, Velikokhatnyi OI, et al (2014) Rechargeable magnesium battery: Current status and key challenges for the future. *Prog Mater Sci* 66:1–86. https://doi.org/10.1016/j.pmatsci.2014.04.001.
8. Aurbach D, Schechter A, Gofer Y (2000) Prototype systems for rechargeable. 407:10–14. https://doi.org/10.1038/35037553.
9. Zhang R, Ling C (2016) Status and challenge of Mg battery cathode. *MRS Energy Sustain* 3:E1. https://doi.org/10.1557/mre.2016.2.
10. Peña O (2015) Chevrel phases: Past, present and future. *Phys C Supercond Appl* 514:95–112.
11. Pan J, Bai S, Ma Y, et al (2017) Carbon coated chevrel phase of Mo_6S_8 as anode material for improving electrochemical properties of aqueous lithium-ion batteries. *Electrochim Acta* https://doi.org/10.1016/j.electacta.2017.10.015.
12. Levi E, Gershinsky G, Aurbach D, Isnard O, Ceder G (2015) New insight on the unusually high ionic mobility in chevrel phases. *Am Chem Soc* 182:4–7. https://doi.org/10.1007/s10832.
13. Burdett JK, Lin J (1982) Structures of chevrel phases. *Inorg Chem* 21:5–10.
14. Levi E, Lancry E, Mitelman A, et al (2006) Phase diagram of Mg insertion into chevrel phases, $Mg_xMo_6T_8$ (T: S, Se). Part 2. The crystal structure of triclinic $MgMo_6Se_8$. *ChemInform* 37:1975–1983. https://doi.org/10.1002/chin.200643008.
15. Wan LF, Perdue BR, Apblett CA, Prendergast D (2015) Mg desolvation and intercalation mechanism at the Mo_6S_8 chevrel phase surface. *Chem Mater* 27:5932–5940. https://doi.org/10.1021/acs.chemmater.5b01907.
16. Ling C, Suto K (2017) Thermodynamic origin of irreversible magnesium trapping in chevrel phase Mo_6S_8: Importance of magnesium and vacancy ordering. *Chem Mater* 29:3731–3739. https://doi.org/10.1021/acs.chemmater.7b00772.
17. Saha P, Jampani PH, Datta MK, et al (2014) A convenient approach to Mo_6S_8 chevrel phase cathode for rechargeable magnesium battery. *J Electrochem Soc* 161:A593–A598. https://doi.org/10.1149/2.061404jes.
18. Lancry E, Levi E, Gofer Y, et al (2005) The effect of milling on the performance of a Mo_6S_8 chevrel phase as a cathode material for rechargeable Mg batteries. *J Solid State Electrochem* 9:259–266. https://doi.org/10.1007/s10008-004-0633-7.
19. Levi MD, Lancri E, Levi E, et al (2005) The effect of the anionic framework of Mo_6X_8 chevrel phase (X=S, Se) on the thermodynamics and the kinetics of the electrochemical insertion of Mg^{2+} ions. *Solid State Ionics* 176:1695–1699. https://doi.org/10.1016/j.ssi.2005.04.019.
20. Suresh GS, Levi DA (2008) Effect of chalcogen substitution in mixed $Mo_6S_{8-n}Se_n$ (n=0,1,2) chevrel phases on the thermodynamics and kinetics of reversible Mg ions insertion. *Electrochim Acta* 53:3889–3896.
21. Levi E, Gofer Y, Vestfreed Y, et al (2002) $Cu_2Mo_6S_8$ chevrel phase, a promising cathode material for new rechargeable Mg batteries: A mechanically induced chemical reaction. *Chem Mater* 14:2767–2773. https://doi.org/10.1021/cm021122o.

22. You Y, Wu XL, Yin YX, Guo YG (2014) High-quality Prussian blue crystals as superior cathode materials for room-temperature sodium-ion batteries. *Energy Environ Sci* 7:1643–1647. https://doi.org/10.1039/c3ee44004d.
23. Pasta M, Wessells CD, Liu N, et al (2014) Full open-framework batteries for stationary energy storage. *Nat Commun* 5:1–9. https://doi.org/10.1038/ncomms4007.
24. Wang RY, Wessells CD, Huggins RA, Cui Y (2013) Highly reversible open framework nanoscale electrodes for divalent ion batteries. *Nano Lett* 13:5748–5752. https://doi.org/10.1021/nl403669a.
25. Wang RY, Shyam B, Stone KH, et al (2015) Reversible multivalent (monovalent, divalent, trivalent) ion insertion in open framework materials. *Adv Energy Mater* 5:1–10. https://doi.org/10.1002/aenm.201401869.
26. Lipson AL, Han SD, Kim S, et al (2016) Nickel hexacyanoferrate, a versatile intercalation host for divalent ions from nonaqueous electrolytes. *J Power Sources* 325:646–652. https://doi.org/10.1016/j.jpowsour.2016.06.019.
27. Chen L, Bao JL, Dong X, et al (2017) Aqueous Mg-ion battery based on polyimide anode and prussian blue cathode. *ACS Energy Lett* 2:1115–1121. https://doi.org/10.1021/acsenergylett.7b00040.
28. Mizuno Y, Okubo M, Hosono E, et al (2013) Electrochemical Mg^{2+} intercalation into a bimetallic CuFe Prussian blue analog in aqueous electrolytes. *J Mater Chem A* 1:13055–13059. https://doi.org/10.1039/c3ta13205f.
29. Cheah YL, Aravindan V, Madhavi S (2012) Electrochemical lithium insertion behavior of combustion synthesized V_2O_5 cathodes for lithium-ion batteries. *J Electrochem Soc* 159:A273–A280. https://doi.org/10.1149/2.071203jes.
30. Novák P (1993) Electrochemical insertion of magnesium in metal oxides and sulfides from aprotic electrolytes. *J Electrochem Soc* 140:140. https://doi.org/10.1149/1.2056075.
31. Shklover V, Haibach T (1996) Crystal structure of the product of Mg^{2+} insertion into V_2O_5 single crystals. *J Solid State Chem* 323:317–323.
32. Jiao L, Yuan H, Wang Y, et al (2005) Mg intercalation properties into open-ended vanadium oxide nanotubes. *Electrochem Commun* 7:431–436. https://doi.org/10.1016/j.elecom.2005.02.017.
33. Jiao L, Yuan H, Si Y, et al (2006) Electrochemical insertion of magnesium in open-ended vanadium oxide nanotubes. *J Power Sources* 156:673–676. https://doi.org/10.1016/j.jpowsour.2005.06.004.
34. Lee SH, DiLeo RA, Marschilok AC, et al (2014) Sol gel based synthesis and electrochemistry of magnesium vanadium oxide: A promising cathode material for secondary magnesium ion batteries. *ECS Electrochem Lett* 3:A87–A90. https://doi.org/10.1149/2.0021408eel.
35. Inamoto M, Kurihara H, Yajima T (2013) Vanadium pentoxide-based composite synthesized using microwave water plasma for cathode material in rechargeable magnesium batteries. *Materials (Basel)* 6:4514–4522. https://doi.org/10.3390/ma6104514.
36. Jiao LF, Yuan HT, Si YC, et al (2006) Synthesis of Cu0.1-doped vanadium oxide nanotubes and their application as cathode materials for rechargeable magnesium batteries. *Electrochem Commun* 8:1041–1044. https://doi.org/10.1016/j.elecom.2006.03.043.
37. Le DB, Passerini S, Coustier F, et al (2004) Intercalation of polyvalent cations into V_2O_5 aerogels. *J Colloid Interface Sci* 278:682–684. https://doi.org/10.1021/cm9705101.
38. Imamura D, Miyayama M (2003) Characterization of magnesium-intercalated V_2O_5/carbon composites. *Solid State Ionics* 161:173–180. https://doi.org/10.1016/S0167-2738(03)00267-4.
39. Imamura D, Miyayama M, Hibino M, Kudo T (2003) Mg intercalation properties into V_2O_5 gel/carbon composites under high-rate condition. *J Electrochem Soc* 150:A753. https://doi.org/10.1149/1.1571531.
40. Zhao X, Hou Y, Wang Y, et al (2017) Prepared MnO_2 with different crystal forms as electrode materials for supercapacitors: Experimental research from hydrothermal crystallization process to electrochemical performances. *RSC Adv* 7:40286–40294. https://doi.org/10.1039/c7ra06369e.
41. Sheha E (2009) Ionic conductivity and dielectric properties of plasticized $PVA_{0.7}(LiBr)_{0.3}(H_2SO_4)_{2.7M}$ solid acid membrane and its performance in a magnesium battery. *Solid State Ionics* 180:1575–1579. https://doi.org/10.1016/j.ssi.2009.10.008.
42. Kumagai N, Komaba S, Sakai H, Kumagai N (2001) Preparation of todorokite-type manganese-based oxide and its application as lithium and magnesium rechargeable battery cathode. *J Power Sources* 97-98:515–517. https://doi.org/10.1016/S0378-7753(01)00726-1.
43. Rasul S, Suzuki S, Yamaguchi S, Miyayama M (2012) Synthesis and electrochemical behavior of hollandite MnO_2/acetylene black composite cathode for secondary Mg-ion batteries. *Solid State Ionics* 225:542–546. https://doi.org/10.1016/j.ssi.2012.01.019.
44. Rasul S, Suzuki S, Yamaguchi S, Miyayama M (2012) High capacity positive electrodes for secondary Mg-ion batteries. *Electrochim Acta* 82:243–249. https://doi.org/10.1016/j.electacta.2012.03.095.

45. Yuan C, Zhang Y, Pan Y, et al (2014) Investigation of the intercalation of polyvalent cations (Mg^{2+}, Zn^{2+}) into λ-MnO_2 for rechargeable aqueous battery. *Electrochim Acta* 116:404–412. https://doi.org/10.1016/j.electacta.2013.11.090.
46. Zhang R, Yu X, Nam KW, et al (2012) α-MnO_2 as a cathode material for rechargeable Mg batteries. *Electrochem Commun* 23:110–113.
47. Ling C, Zhang R, Arthur TS, Mizuno F (2015) How general is the conversion reaction in mg battery cathode: A case study of the magnesiation of α-MnO_2. *Chem Mater* 27:5799–5807. https://doi.org/10.1021/acs.chemmater.5b02488.
48. Zhu YR, Yin LC, Yi TF, et al (2013) Electrochemical performance and lithium-ion intercalation kinetics of submicron-sized $Li_4Ti_5O1_2$ anode material. *J Alloys Compd* 547:107–112. https://doi.org/10.1016/j.jallcom.2012.08.113.
49. Wu N, Yin YX, Guo YG (2014) Size-dependent electrochemical magnesium storage performance of spinel lithium titanate. *Chem Asian J* 9:2099–2102. https://doi.org/10.1002/asia.201402286.
50. Hashem AM, Abbas SM, Hou X, et al (2019) Facile one step synthesis method of spinel $LiMn_2O_4$ cathode material for lithium batteries. *Heliyon* 5:e02027. https://doi.org/10.1016/j.heliyon.2019.e02027.
51. Ling C, Mizuno F (2013) Phase stability of post-spinel compound AMn_2O_4 (A=Li, Na, or Mg) and its application as a rechargeable battery cathode. *Chem Mater* 25:3062–3071. https://doi.org/10.1021/cm401250c.
52. Sai Gautam G, Canepa P, Urban A, et al (2017) Influence of inversion on Mg mobility and electrochemistry in spinels. *Chem Mater* 29:7918–7930. https://doi.org/10.1021/acs.chemmater.7b02820.
53. Kim C, Phillips PJ, Key B, et al (2015) Direct observation of reversible magnesium ion intercalation into a spinel oxide host. *Adv Mater* 27:3377–3384. https://doi.org/10.1002/adma.201500083.
54. Knight JC, Therese S, Manthiram A (2015) On the utility of spinel oxide hosts for magnesium-ion batteries. *ACS Appl Mater Interfaces* 7:22953–22961. https://doi.org/10.1021/acsami.5b06179.
55. Liu M, Jain A, Rong Z, et al (2016) Evaluation of sulfur spinel compounds for multivalent battery cathode applications. *Energy Environ Sci* 9:3201–3209. https://doi.org/10.1039/c6ee01731b.
56. Wang LP, Zhao-Karger Z, Klein F, et al (2019) $MgSc_2Se_4$-A magnesium solid ionic conductor for all-solid-state Mg batteries? *ChemSusChem* 12:2286–2293. https://doi.org/10.1002/cssc.201900225.
57. Zhong Q, Bonakclarpour A, Zhang M, Dahn JR (1997) Synthesis and Electrochemistry of $LiNi_xMn_{2-x}O_4$. *J Electrochem Soc* 144:205.
58. Shasha H, Yatom N, Prill M, et al (2019) Unveiling ionic diffusion in $MgNiMnO_4$ cathode material for Mg-ion batteries via combined computational and experimental studies. *Short Commun* 23:3209–3216.

6 Electrolytes for Storage Magnesium-Ion Batteries
An Overview

Akhila Das, Abhilash Pullanchiyodan, Vijay Kumar Thakur, Nikhil Medhavi, Alexandru Vlad, and Prasanth Raghavan

6.1 INTRODUCTION

Every day, we rely on different types of energy, and non-renewable energy sources such as fossil fuels are at serious risk due to their adverse effect on global climate change by emitting CO_2 to the atmosphere. A gradual switching from non-renewable to renewable energy sources is necessary [1,2]. The necessity of storing these renewable energies is a matter of discussion. In this context, rechargeable energy storage devices such as batteries, supercapacitors, and fuel cells have emerged. After the invention of the first battery called Baghdad battery, many discoveries took place. The first rechargeable battery was Ni-Cd alkaline battery which was discovered in 1899. With the development of portable electronic devices such as laptops and mobile phone, conventional batteries such as Ni-Cd and lead acid were curtailed, since the reduction of size was possible only to an extent [3,4]. Lithium-ion batteries exceed the performance of all other batteries that existed due to their low molecular weight, low reduction potential, and high energy density [5]. Sony was the first commercialized lithium-ion battery in 1990. The present lithium-ion batteries were retarded by some practical issues. The major challenge faced by those batteries is their lack of efficient energy storage. Lithium-ion battery does not arrive at an affordable price due to its lack of raw materials. Lithium-ion batteries catch fire if handled at abnormal conditions. 'Dendrite' formation is another major challenge faced by lithium-ion batteries, which affects rate capability and cyclability [6–8]. Limitations of such batteries can be reduced by replacing lithium-ions with some metals, such as lithium-sulfur batteries, which are promising candidate because of its long cycle life, high energy density, and high rate capability as well as being economically cheaper. Sulfur is used as the best metal due to its readiness to undergo catenation. To compete between lithium-ion batteries, lithium-sulfur batteries have to improve their cycle life and their higher energy discharge [9,10]. Some of the battery parameters can be enhanced to bring them to an industrial scale. Recently, several approaches can increase the cycle life by incorporating metal alloys as anodes, better polymer gel electrolytes that can improve safety, doping of materials that improves the kinetics of materials, improving coating on cathode materials, adding nano fillers for improving its mechanical strength, etc. [11–15]. Despite all these developments, lithium-ion batteries are still inadequate for better battery performance. In this context, magnesium-ion batteries (MIBs) provide a new opportunity in the battery world. MIB was found to be another alternative because of its attractive properties such as its high volumetric capacity (3833 mAh cc^{-1}) which is higher than lithium (2046 mAh cc^{-1}), bivalency, and its abundance [16–18]. Non-dendrite formation is also an attractive property compared to that of other existing batteries. MIBs are the eighth most abundant in the earth's crust and hence are inexpensive compared to Li-ion batteries ($2700/ton for Mg vs $64,000/ton for Li). Mg can also exhibit twice the capacity owing to the divalent nature compared to monovalent LIBs, if an appropriate cathode and anode have been synthesized. Working principle behind MIB is similar to that of MIB and the schematic illustration of general MIB is shown in Figure 6.1.

DOI: 10.1201/9781003310167-6

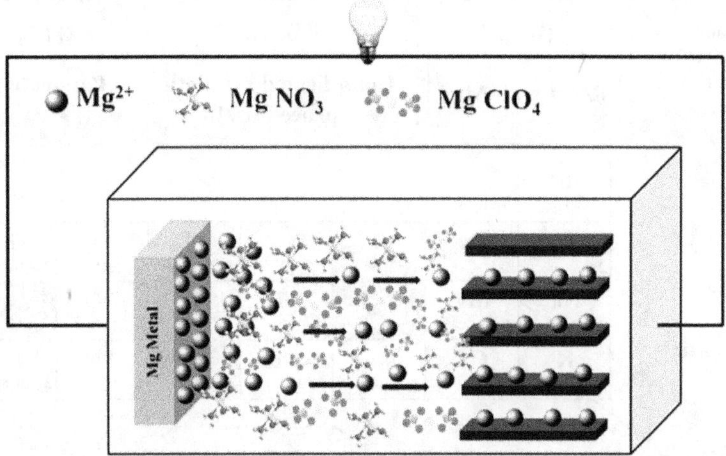

FIGURE 6.1 Schematic illustration of the working principle behind rechargeable magnesium-ion battery.

Research and developments mainly focus on stable anodes and electrolytes that show high cyclability and wide electrochemical potential window. Despite these development, major challenges faced by rechargeable magnesium-ion battery (rMIB) electrolytes are air sensitivity, corrosion, and voltage instability [19]. Development of a better electrolyte is a crucial hurdle to the research field. Fabrication of an efficient electrolyte and boosting the capacity are the prior importance to Pellion Technologies as well as Bar-Ilan University [20,21]. They mainly target electrolyte studies based on the synthesis and characterization of suitable electrolytes for better battery performance. The scope of this chapter is to discuss some of the novel works that have been achieved so far and the future prospects.

6.2 MAJOR ELECTROLYTES IN MG-ION BATTERIES

Electrolytes play a vital role in the advancement of batteries. MIB is at secondary phase of development owing to its advantages such as low cost, abundance in the earth's crust, and safety. Development of a better electrolyte is a crucial hurdle in the research field. They mainly target electrolyte studies based on the synthesis and characterization of suitable electrolytes for better battery performance. Different types of electrolytes such as inorganic halide-based electrolytes, inorganic solid ceramic electrolytes, Grignard reagent, non-nucleophilic electrolytes, non-Grignard reagents, imide-based Mg electrolytes, and magnesium borohydrides, polymer electrolytes, etc. are some of the known electrolytes investigated in MIBs [22,23]. Several liquid electrolytes such as Mg chlorides, inorganic halides, and magnesium perchlorates have been widely used in the scientific community which have certain advantages and disadvantages (Figure 6.2) [24]. This chapter mainly focuses on the liquid electrolytes, Grignard-based electrolytes, and some imide-based electrolytes that are widely explored in the development of rMIB.

6.2.1 INORGANIC SOLID CERAMIC ELECTROLYTES FOR RMIB

Solid ceramic inorganic fillers consist of major division such as NASICON-type structures having three-dimensional framework configurations. The name NASICON derived from the word **Na** **s**uper **i**on **con**ductor is oxide-type electrolytes that are promising candidates for batteries owing to its thermal and chemical stability. NASICON-type electrolytes are widely used in lithium-ion batteries [25] and are now instigated in Mg-ion batteries too. Ikeda et al. [26] were the first to investigate the electrochemical properties of solid ceramic electrolyte in Mg-ion batteries. The crystal structure

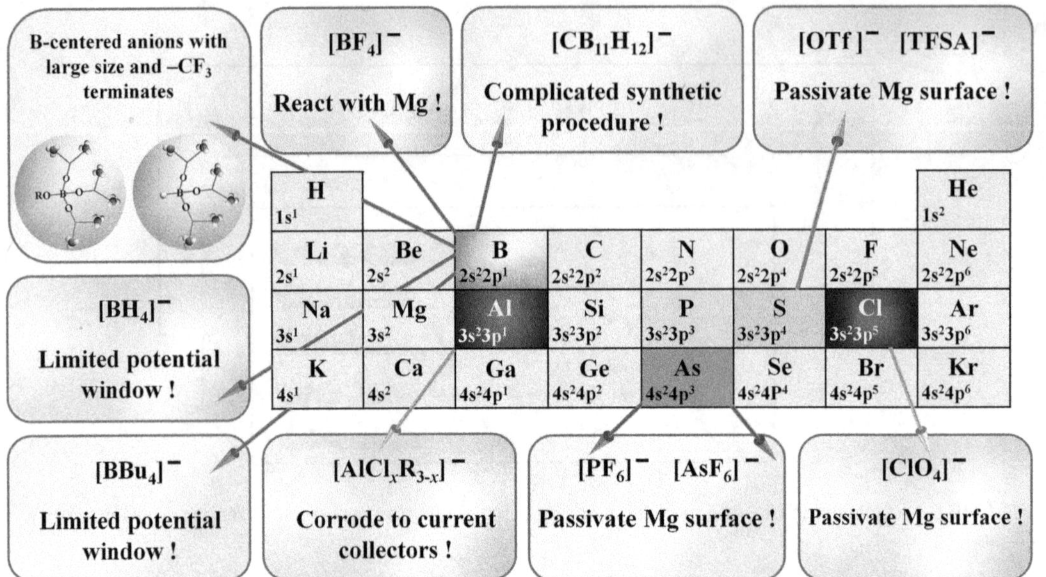

FIGURE 6.2 Schematic diagram of magnesium electrolytes used in rechargeable magnesium-ion battery. Adapted and reproduced from Ref. [24] with permission. Copyright © 2017, Wiley.

of $NaZr_2(PO_4)_2$ was similar to that of $MgZr_2(PO_4)_6$ and the ionic conductivity of $MgZr_2(PO_4)_6$ was found to be 2.9×10^{-5} at 400°C and 6.1×10^{-5} at 800°C. Later, Imanaka et al. [27] investigated phosphate-based $MgZr_2(PO_4)_6$ and the partial substitution of Nb to form the electrolyte $Mg_{1-2x}(Zr_{1-x}Nb_x)_4P_6O_{24}$. The general properties of divalent cations such as Mg, Ni, and Cu, including phase transition, structural elucidation, and transport properties, were reported by Nomura et al. [28] with the general formula $M^{II}Zr_4(PO_4)_6$. Mg in $MgZr_2(PO_4)_6$ showed β-$Fe_2(SO_4)_3$-type framework structure, whereas Ca, Cd, etc. showed NASICON-type structure. $MgZr_2(PO_4)_6$ exists in monoclinic-type crystal structure having an ionic radius of Mg to 0.86 A° and cell volume of 978 (v A^{-3}). The ionic conductivity of $MgZr_2(PO_4)_6$ was 1.6×10^{-6} at 500°C and 1.4×10^{-3} at 800°C. In the same year, Kohler et al. [29] reported the effect of multivalent cations in M^+ $(Zr_2(PO_4)_3)$ (M=metal) possessing an average ionic conductivity ranging from two to three orders at elevated temperatures (700°C–800°C). β-Alumina-type structures and NASICON-type structures possess high ionic conductivity, very less electronic conductivity, and fast diffusion properties. However, this solid polymer electrolyte requires large conduction pathways for the easy migration of cations. Omota et al. [30] reported $MgHf(WO_4)_3$-type structure synthesized by hot press method exhibiting an ionic conductivity of 2.5×10^{-4} at 600°C. One-dimensional alignment in the Hf^+ and Mg ions offers stable 3D structure attributing to the higher ionic conductivity of $MgHf(WO_4)_3$. Similarly, Tamura et al. [31] reported partial substitution of Nb to NASICON-type $(Mg_xHf_{1x})_4/_{4-2x})Nb(PO_4)_3$ electrolyte and enhanced ionic conductivity even at ambient temperature (2.1×10^{-4} S cm^{-1}). The major chemical reaction involved in $(Mg_{0.1}Hf_{0.9})_{4/3.8}Nb\ (PO_4)_3$ is given in equation 6.1:

$$19/2\left(Mg_{0.1}Hf_{0.9}NbPO_4\right)_3 \rightarrow Mg^{2+} + 2e^- + 19/2\left(18/19\ HfO_2\right) + 1/2Nb_2O_5 + 3/2\ P_2O_5 + 1/2\ O_2 \quad (6.1)$$

Lee et al. [32] investigated similar NASICON-type electrolyte containing Hf and Nb with divalent cations in $(M_xHf_{1-x})_{4/(4-2x)}Nb(PO_4)_3$ (M=Mg, Ca, Sr, Ni). Among the divalent cations, Mg and Ni perform good ionic conductivity compared to other transition metals. The ionic conductivity of $(Mg_{0.1}Hf_{0.9})_{4/3}8Nb\ (PO_4)_3$ is 1.24×10^{-4} S cm^{-1} and $(Ni_{0.06}Hf_{0.94})_{4/3.88}Nb(PO_4)_3$ is 2.27×10^{-4} S cm^{-1}. Recently, Nakona et al. [33] investigated the theoretical studies of NASICON-type structure as

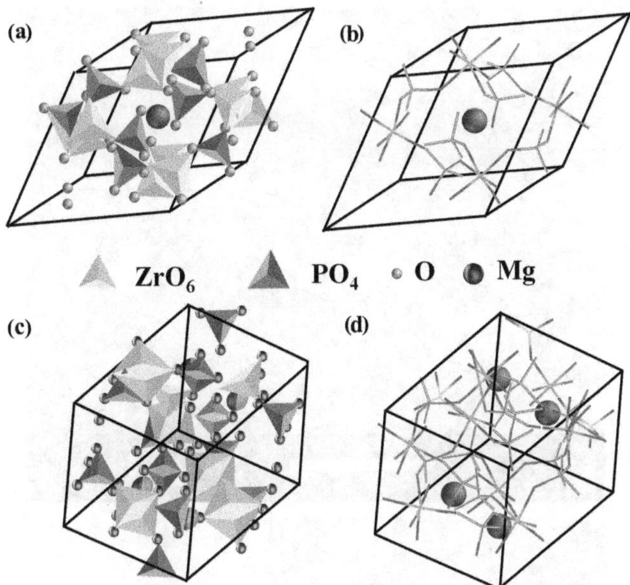

FIGURE 6.3 (a and b) The NASICON-type structure of MZP and (c and d) the β-iron sulfate-type structure of MZP with the ZrO_6 octahedra depicted in green, PO_4 tetrahedra depicted in lavender, O atoms in red, and Mg atoms in orange. This journal trace of Mg atom in FPMD simulation at 1973 K. (a–d) The NASICON-type and β-iron sulfate-type structures. Adapted and reproduced from Ref. [33] with permission. Copyright © 2019, Royal Society of Chemistry.

well as β-$Fe_2(SO_4)_3$-type framework structure and revealed that NASICON-type structure possesses higher migration energy (0.71 eV) compared to β-$Fe_2(SO_4)_3$-type structure (0.63 eV) (Figure 6.3).

6.2.2 Inorganic Halide-Based Electrolytes for rMIB

Inorganic halide groups are a major class of liquid electrolytes that consist of halogen group usually Cl used in rMIBs. The main research in the inorganic halide-based electrolyte system was pristine $MgCl_2$ or $MgCl_2$/Lewis acid combination dissolved in different solvents such as DME< THF. Earlier, Gregory et al. [34] reported inorganic halide-based Mg with Cl, Br, etc. as Lewis base and $AlCl_3$ as Lewis acid. Later, Doe et al. [35] reported that the inorganic halide $MgCl_2$ reacts with Lewis acid $AlCl_3$ to form Magnesium Aluminum Chloride Complex (MACC) electrolyte possessing high anodic stability (3.1 V). The general reaction behind MACC after Lewis acid–base reaction is given in equation 6.2. He et al. [36] reported that the electrochemical performance of this $MgCl_2/AlCl_3$ (1:1 ratio) complex in DME solvent exhibits improved electrochemical stability (3.5 V) than that reported by Doe et al. [35]. The higher electrochemical performance is due to the formation of octahedral structure attributing to the complex formation of $[MgCl(k_2\text{-DME})_2]AlCl_4$. However, the presence of water content from the DME solvent offers the better electrochemical performance in rMIBs. Levi et al. [37] investigated the performance of $MgCl_2/AlCl_3$ complex in 1:1 ratio which is dissolved in THF solvent concluding that the amount of accumulated $MgCl_2$ depends on the ratio of inorganic ligands and Cl molecules. The electrochemical performance of $AlCl_3$/ $MgCl_2$ complex was dissolved in the mixture of THF and DPSO (dipropyl sulfone) in the ratio of 1:1 which was reported by Kang et al. [38]. The surface morphology of $MgCl_2$ DPSO/THF reveals that spherical particle-like deposits of Mg were observed with micrometer size range, followed by the presence of subgrains without any dendrite formation. In addition to this, no such impurities on Mg were observed from the EDX spectrum (Figure 6.4). However, the anodic stability was only 3.3 V offering a specific capacity of 80 mAh g^{-1} stable up to 300 cycles with an ionic conductivity of

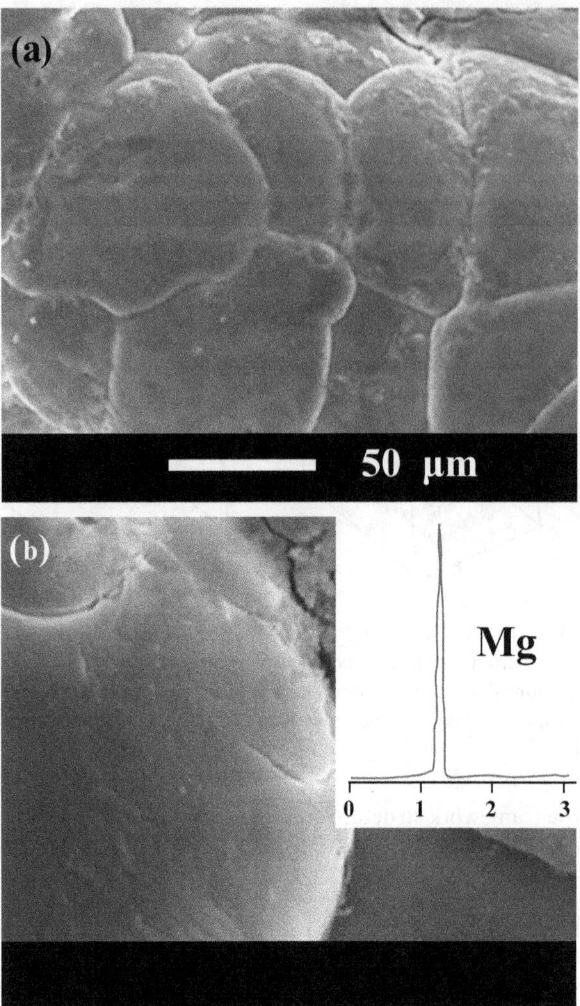

FIGURE 6.4 Scanning electron micrographs on the morphology of (SEM image): (a) Mg deposits on a Cu electrode and (b) magnified SEM image of (a). The inset shows the EDX spectrum of Mg deposits without the signal from the electrolyte decomposition. Adapted and reproduced from Ref. [38] with permission. Copyright © 2016, American Chemical Society.

1.1 mS cm^{-1}. The reason for the incorporation of sulfone compound was to enhance chemical stability and increase the reversible Mg deposition/dissolution. He et al. [39] reported the detailed investigation of general inorganic halide-based system with Lewis acid explored in rechargeable Mg-ion batteries. $MgCl_2/AlCl_3$ complex dissolved in THF solution forms $[(\mu\text{-}Cl)_3Mg_2(THF)_6]^+$ dimer cation via transmetalation mechanism (equation 6.3).

$$mMgCl_2 + nAlCl_3 \rightarrow Mg_mAl_nCl\,[(2*m)+(3*n) \tag{6.2}$$

$$2MgCl_2 + AlCl_3 \rightarrow \left[(\mu\text{-}Cl)_3\,Mg_2\,(THF)_6\right]^+ + AlCl_4 \tag{6.3}$$

6.2.3 Boron Clusters for rMIBs

Though inorganic halide-based electrolytes offer high anodic stability, the chloride-containing electrolytes are highly corrosive. Since corrosion on the current collectors affects the electrochemical performance of rMIBs, non-chloride borohydride class of electrolytes were investigated. Borohydrides used in Mg-ion batteries are generally classified as aromatic borohydrides and aliphatic borohydrides. Aromatic borohydrides include boron cluster-type molecules such as *closo-* and *nido-*boron-based electrolytes, which are rare, noncorrosive, and stable. Omita et al. [40] reported the theoretical and experimental studies of closo-boron cluster synthesized from Mg borohydride offering high anodic stability (~ 4 V) and having hydrogen storage applications. Hence, Carter et al. [41] demonstrated the theoretical studies of boron cluster by choosing closo-boranes rather than the host material. Closo-boranes exhibit a specific capacity of 90 mAh g^{-1} at a current density of 6 mA g^{-1} which was cycled up to 30 cycles. Oscar et al. [42] incorporated carbon to form carborane ($CB_{11}H_{12}^{-}$), which is highly stable and performs reversible Mg deposition/dissolution. Though closo-boranes increase the compatibility toward electrodes, its solubility toward organic solvents was very poor [43]. Aurthur et al. [44] synthesized 10 vertex closo-carboranes formed with less-expensive synthetic procedure to form $[Mg^{2+}][HCB_9H_9^{1-}]_2$. Figure 6.5 shows the synthesis of 12 vertex closo-boranes reported by Mohtadi et al. [43], which suffer from purification difficulties and do not meet expected electrochemical performance. However, Scheme 2 synthesis, also known as cation reduction method developed by Arthur et al. [44], offers a facile, inexpensive synthetic route exhibiting electrochemical stability ~4 V.

In the early 1950s, borohydrides are extensively applicable in hydrogen storage systems owing to the low hydrogen release ability. In addition to this, borohydrides are generally used in organic and inorganic synthesis since these compounds are excellent reducing agents [45]. Electrochemical stability was attained to an extent owing to the reduction property persisting in borohydride system and thus its application was extended to lithium-ion batteries [46]. Mohtadi et al. [47] were the first to introduce Mg borohydride system dissolved in THF and DME for rMIBs since it provides enhanced electrochemical performance and is free from chlorides. Mg borohydride decomposes into Mg^{2+} ion and BH_4^- ion by the formation of decomposition of complex $Mg[(\mu\text{-}H)_2BH_2]_2$ as given in equations 6.4 and 6.5. David et al. [48] investigated the electrode deposition of Ag anodes using $MgBH_4$ electrolytes. $MgBH_4$ in diglyme offers very strong Mg deposits and shows anisotropic behaviors. However, these systems are not safe since organic solvents such as THF and DME have been used.

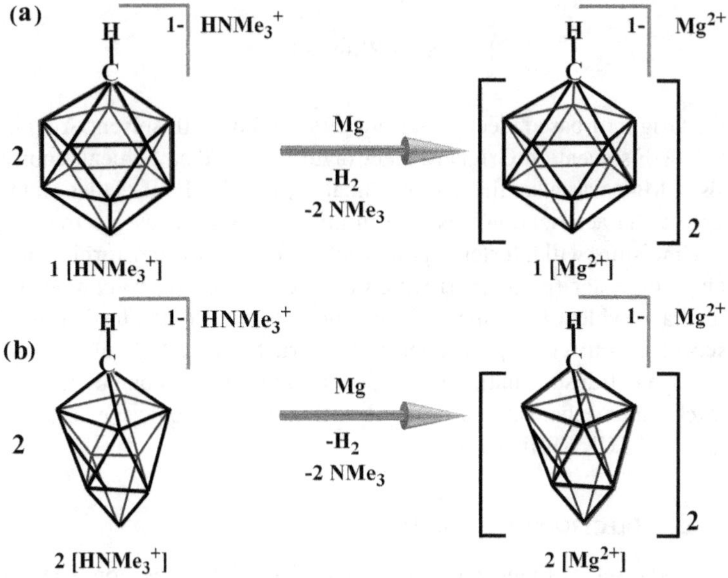

FIGURE 6.5 Schematic illustration of the synthesis of $1[Mg^{2+}]$ and $2[Mg^{2+}]$ via cation reduction. Adapted and reproduced from Ref. [44] with permission. Copyright 2017, Royal Society of Chemistry.

In this context, the incorporation of ionic liquids ensures safety of the battery while also enhancing electrochemical properties. Su et al. [49] incorporated $MgBH_4$ with ionic liquid PP14TFSI that offers an ionic conductivity of 1.33 mScm^{-1}.

$$Mg\left[(\mu\text{-}H)_2\, BH_2\right]_2 \rightarrow Mg\left[(\mu\text{-}H)_2\, BH_2\right]^{+} + BH_4^{-} \tag{6.4}$$

$Mg\left[(\mu\text{-}H)_2\, BH_2\right]^{+}$ again dissociates

$$Mg\left[(\mu\text{-}H)_2\, BH_2\right]^{+} \leftrightarrow Mg^{2+} + BH_4^{-} \tag{6.5}$$

6.2.4 Grignard and Pseudo-Grignard Reagents in MIBs

The general formula for Grignard reagent is RMgX, where R=alkyl group and X=halogen bonded by partial positive group of Mg and partial negative group of Carbon. Organometallic compounds are widely used in Mg-ion batteries since this can act as scavengers, thereby eliminating contaminants. Electrolysis of Grignard compounds and its deposition in electrode was earlier reported by Gaddum et al. [50]. Theoretical explanations and operating mechanisms of Grignard-based electrolytes explored in MIBs were reported by Kim et al. [51]. The solvation energy and the migration energy depend on the R group that was used in Grignard reagent, and the formation of ethereal equilibrium is complex process. Typically, equilibrium exists as Schlenk equilibrium (equation 6.6), which goes through path one by one electron reduction and path 2/3 by two-electron reduction mechanism. Aurbach et al. [52] reported that the effect of Mg deposition and dissolution of Grignard electrolyte RMgX (X=Cl, Br) in THF solvent was undergone not only by a single two-electron transfer mechanism but also by adsorption phenomena. Other than Schlenk equilibrium, ionization equilibrium also existed in this mechanism which is shown in the following equation:

$$2RMgCl \rightarrow 2\, R_2Mg + MgCl_2 \tag{6.6}$$

$$2RMgX \rightarrow RMg^{+} + RMgX_2 \tag{6.7}$$

Since most of the Grignard-based electrolytes are dissolved in THF solvent, Yagi et al. [53] examined the effect of THF solvent on Grignard electrolytes. Other than Schlenk equilibrium and ionization mechanism, Mg^{2+} was also directly coordinated to the THF molecules and Cl groups. Even though the complete characterization was held in inert atmosphere, even a trace amount of water molecule from the moisture will interfere in the reaction to form a white, turbid magnesium hydroxide. Trace amounts of water (ppm level) which exist during cyclic voltammetry will affect the final results which are evident from the morphological difference obtained from FESEM images (Figure 6.6). Recently, Pour et al. [54] reported the structural analysis of aromatic ligands such as PhMgCl in $AlCl_3$ via transmetalation process. Aluminum-based compounds and Mg ions have undergone ion exchange mechanism and form neutral molecules, ionic species, etc. ($MgCl_2{\cdot}4THF$, Ph_yAlCl_{4-y} (y=0–4), $Mg_2Cl_3.6THF$, etc.).

6.2.5 Imide-Based Electrolytes for rMIBs

Tran and his co-workers [55] found an imide-based electrolyte in acetonitrile solution, i.e., Mg $[TFSI]_2$/AN. Mg overpotential was found to be stable while stripping but unstable during plating. Unfortunately, it was corroded after 2.8 V against stainless steel. An advanced non-corrosive-type electrolyte on imide-based electrolyte was synthesized, and it was observed that glymes perform

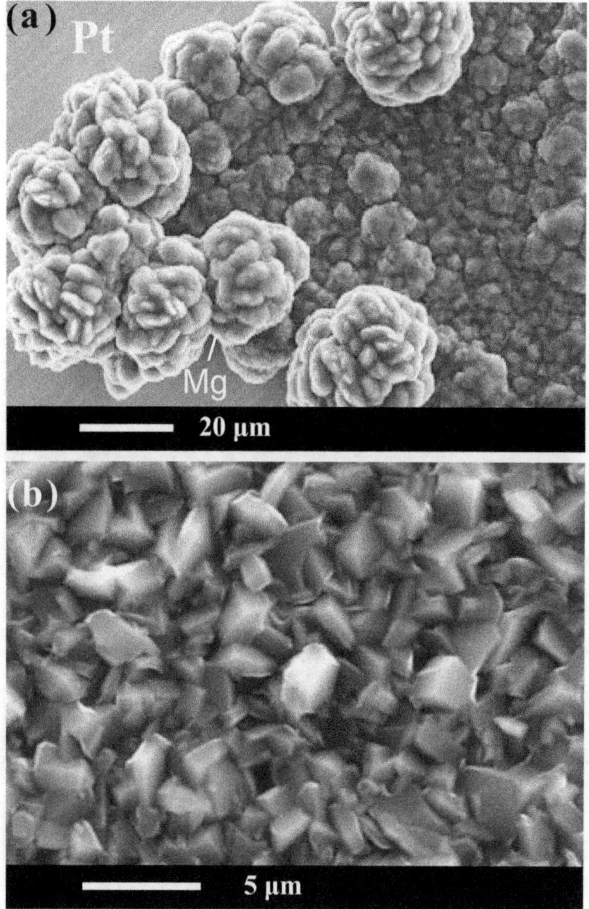

FIGURE 6.6 Scanning electron micrographs (SEM image) on the morphology of Mg deposits on a Pt electrode obtained at −0.1 V vs Mg/Mg^{2+} (electric charge 10 C cm^{-2}) from THF solution containing 0.50 M EtMgCl and 0.25 M $AlCl_3$ with a water concentration of ca. 67 ppm: (a) 0 hour and (b) 24 hours after preparation of the electrolyte. Adapted and reproduced from Ref. [53] with permission. Copyright © 2013, Springer.

as a good solvent showing an anodic stability of 4 V [56]. Surface morphology of Mg electrodes in 0.1 M $Mg(TFSI)_2$ reveals the presence of non-uniform pores after the first cathodic scan (Figure 6.7). In recent years, $MgTFSI_2$-based electrolytes were studied with organic solvents other than glymes with a wide temperature range. While conducting electrochemical studies, it was found that there is a shift in anodic stability resulting in passive layer formation. The major highlights of these electrolytes were its availability, solubility toward organic solvents, commercial availability, and low cost [57]. Sa et al. [58] investigated that the effect of MgTFSI02 on diglyme affects the magnesium plating and stripping in rMIBs. The cyclic voltammetry studies of $Mg(TFSI)_2$/G2 electrolyte conclude that the Mg deposition and dissolution are tedious compared to Grignard-based electrolytes owing to the presence of bulkier sulfone anions. An ~2 V excess potential is required for the deposition of Mg attributing to the formation of surface layers between Mg anode interface and $Mg(TFSI)_2$/G2 electrolyte. Later, Mandai et al. [59] compared the concepts behind Mg imides-based electrolytes and non-Grignard electrolytes. Among $Mg(TFSI)_2$/G2- and $Mg(TFSI)_2$/THF-based electrolytes, higher electrochemical performance was exhibited by glyme-based imides owing to the structural stability and better coordination chemistry behind these electrolytes.

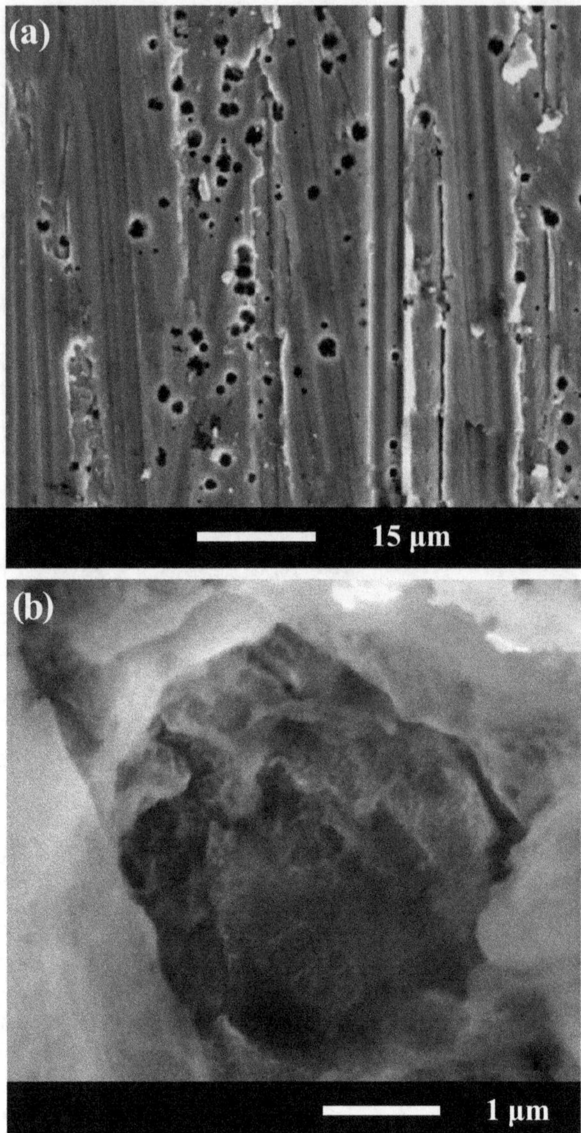

FIGURE 6.7 Scanning electron micrographs on the morphology of (SEM image) of Mg electrodes after first Mg stripping (corresponding capacity = 3.25 mAh cm^{-2}) in 0.1 M $Mg(TFSI)_2$ dissolved in glyme/diglyme (1/1, v/v). (a, b) The SEM images reveal that all Mg electrodes after the first Mg stripping (3.25 mAh cm^{-2}) at a rate of C/40 had non-uniform micropores and their surface morphology was analogous. Adapted and reproduced from Ref. [56] with permission. Copyright © 2014, American Chemical Society.

6.3 CONCLUSION

The primary importance of a battery lies on its durability, cyclability, abundance, conductivity, etc. out of these parameters; MIBs satisfy most of these conditions. There are several hurdles that a MIB must overcome. To improve the performance of a battery, such as voltage, cycle life, shelf life, and charge–discharge rate, research on rechargeable magnesium battery has gained a great impulse. Wiping out the major challenges, a new class of battery materials is emerging with great novelty. MIB is required for the future world since the cost and safety of lithium-ion battery are a primary concern. In a battery, the cathode is the electrode from which current flows through a battery, while electrolytes are the medium through which ionic charge flows through the electrode. Thus, the electrolyte has prime importance in the performance of a battery. In electrolytes, addition of fillers, plasticizers, ionic liquids, etc. enhances the conductivity of the battery system. Polymer electrolytes paved an inevitable role in the technology world. Reports showed that gel polymer electrolytes are adequate and much safer that can work at a wide temperature range.

REFERENCES

1. Yang Z, Zhang J, Kintner-Meyer MC, et al. (2011) Electrochemical energy storage for green grid. *Chem Rev* 111:3577–3613. https://doi.org/10.1021/cr100290v.
2. Dunn B, Kamath H, Tarascon J (2011) Electrical energy storage for the grid: A battery of choices. *Sciences* 334:928–936.
3. Fan Z, Kulkarni P, Gormus S, et al (2012) Smart grid communications: Overview of research challenges, solutions, and standardization activities. *IEEE Commun Surv Tutor* 15:21–38.
4. Cheng F, Liang J, Tao Z, Chen J (2011) Functional materials for rechargeable batteries. *Adv Mater* 23:1695–1715. https://doi.org/10.1002/adma.201003587.
5. Yoshino A (2012) The birth of the lithium-ion battery. *Angew Chemie – Int Ed* 51:5798–5800. https://doi.org/10.1002/anie.201105006.
6. Amon F, Andersson P, Karlson I, Sahlin E (2012) *Fire Risks Associated with Batteries*. SP Technic, Swedan.
7. Tarascon JM (2010) Key challenges in future Li-battery research. *Philos Trans R Soc A Math Phys Eng Sci* 368:3227–3241. https://doi.org/10.1098/rsta.2010.0112.
8. Tarascon J-M and Armand M (1972) Issues and challenges facing rechargeable lithium batteries. *Nature* 359–367. https://doi.org/10.1038/35104644.
9. Pang Q, Liang X, Kwok CY, Nazar LF (2016) Advances in lithium-sulfur batteries based on multifunctional cathodes and electrolytes. *Nat Energy* 1:1–11. https://doi.org/10.1038/nenergy.2016.132.
10. Patel K (2020) Lithium-sulfur battery: Chemistry, challenges, cost, and future. *J. Undergrad Res* 9:39–42.
11. Scrosati B, Garche J (2010) Lithium batteries: Status, prospects and future. *J Power Sources* 195:2419–2430. https://doi.org/10.1016/j.jpowsour.2009.11.048.
12. Lynd NA, Kramer EJ, Hawker CJ, et al (2013) Allyl glycidyl ether-based polymer electrolytes for room temperature lithium batteries. *Macromolecules* 46:8988–8994. https://doi.org/10.1021/ma401267w.
13. Stephan AM (2006) Review on gel polymer electrolytes for lithium batteries. *Eur Polym J* 42:21–42. https://doi.org/10.1016/j.eurpolymj.2005.09.017.
14. Blomgren GE (2017) The development and future of lithium-ion batteries. *J Electrochem Soc* 164:A5019–A5025. https://doi.org/10.1149/2.0251701jes.
15. Scrosati B, Hassoun J, Sun YK (2011) Lithium-ion batteries. A look into the future. *Energy Environ Sci* 4:3287–3295. https://doi.org/10.1039/c1ee01388b.
16. Song J, Sahadeo E, Noked M, Lee SB (2016) Mapping the challenges of magnesium battery. *J Phys Chem Lett* 7:1736–1749. https://doi.org/10.1021/acs.jpclett.6b00384.
17. Besenhard JO, Winter M (2002) Advances in battery technology: Rechargeable magnesium batteries and novel negative-electrode materials for lithium-ion batteries. *ChemPhysChem* 3:155–159. https://doi.org/10.1002/1439-7641(20020215)3:2<155>3.0.CO;2-S.
18. Aurbach D, Suresh GS, Levi E, et al (2007) Progress in rechargeable magnesium battery technology. *Adv Mater* 19:4260–4267. https://doi.org/10.1002/adma.200701495.
19. Muldoon J, Bucur CB, Oliver AG, et al (2012) Electrolyte roadblocks to a magnesium rechargeable battery. *Energy Environ Sci* 5:5941–5950. https://doi.org/10.1039/c2ee03029b.
20. Yoo HD, Shterenberg I, Gofer Y, Gershinsky G, Pour N, Aurbach D (2013) Mg rechargeable batteries: An on-going challenge. *Energy Environ Sci.* https://doi.org/10.1039/C2EE23635D.
21. Van Noorden R (2014) A better battery: Chemists are reinventing rechargeble cells to drive down costs and boost capacity. *Nature* 507:26–28.
22. Sa N, Pan B, Saha-Shah A, et al (2016) Role of chloride for a simple, non-grignard Mg electrolyte in ether-based solvents. *ACS Appl Mater Interfaces* 8:16002–16008. https://doi.org/10.1021/acsami.6b03193.

23. Prabhu SPMR (2018) Development and study of solid polymer electrolytes based on PVdF-HFP/PVAc: $Mg(ClO_4)_2$ for Mg ion batteries. *J Mater Sci Mater Electron* 29:15086–15096. https://doi.org/10.1007/s10854-018-9649-0.
24. Zhang Z, Cui Z, Qiao L, et al (2017) Novel design concepts of efficient Mg-Ion electrolytes toward high-performance magnesium-selenium and magnesium-sulfur batteries. *Adv Energy Mater* 7:1–10. https://doi.org/10.1002/aenm.201602055.
25. Hou M, Liang F, Chen K, Dai Y, Xue D (2020) Challenges and perspectives of NASICON-type solid electrolytes for all solid-state lithium batteries. *Nanotechnology* 31:132003.
26. Ikeda S, Takahashi M, Ishikawa J, Ito K (1987) Solid electrolytes with multivalent cation conduction. 1. Conducting species in $MgZrPO_4$ system. *Solid State Ionics* 23:125–129. https://doi.org/10.1016/0167-2738(87)90091-9.
27. Imanaka N, Okazaki Y, Adachi G (2000) Divalent magnesium ionic conduction in $Mg_{1-2x}(Zr_{1-x}Nb_x)_4P_6O_{24}$ (x=0–0.4) solid solutions. *Electrochem Solid-State Lett* 3:327–329. https://doi.org/10.1149/1.1391138.
28. Nomura K, Ikeda S, Ito K, Einaga H (1992) Framework structure, phase transition, and transport properties in $MIIZr_4(PO_4)_6$ compounds (MII=Mg, Ca, Sr, Ba, Mn, Co, Ni, Zn, Cd, and Pb). *Bull. Chem. Soc. Jpn.* 65:3221–3227.
29. Köhler J, Imanaka N, Adachi GY (1998) Multivalent cationic conduction in crystalline solids. *Chem Mater* 10:3790–3812. https://doi.org/10.1021/cm980473t.
30. Omote A, Yotsuhashi S, Zenitani Y, Yamada Y (2011) High ion conductivity in MgHf(WO4)3 solids with ordered structure: 1-D alignments of Mg^{2+} and Hf^{4+} ions. *J Am Ceram Soc* 94:2285–2288. https://doi.org/10.1111/j.1551-2916.2011.04644.x.
31. Tamura S, Yamane M, Hoshino Y, Imanaka N (2016) Highly conducting divalent Mg^{2+} cation solid electrolytes with well-ordered three-dimensional network structure. *J Solid State Chem* 235:7–11. https://doi.org/10.1016/j.jssc.2015.12.008.
32. Lee W, Tamura S, Imanaka N (2019) Synthesis and characterization of divalent ion conductors with NASICON-type structures. *J Asian Ceram Soc* 7:221–227. https://doi.org/10.1080/21870764.2019.1606141.
33. Nakano K, Noda Y, Tanibata N, et al (2019) Computational investigation of the Mg-ion conductivity and phase stability of $MgZr_4(PO_4)_6$. *RSC Adv* 9:12590–12595. https://doi.org/10.1039/c9ra00513g.
34. Gregory TD, Hoffman RJ, Wwinterton RC (1990) Nonaqueous electrochemistry of magnesium. Applications to energy storage. *J Electrochem Soc* 137:775–780. https://doi.org/10.1002/chin.199022010.
35. Doe RE, Han R, Hwang J, et al (2014) Novel, electrolyte solutions comprising fully inorganic salts with high anodic stability for rechargeable magnesium batteries. *Chem Commun* 50:243–245. https://doi.org/10.1039/c3cc47896c.
36. He S, Luo J, Liu TL (2017) $MgCl_2/AlCl_3$ electrolytes for reversible Mg deposition/stripping: Electrochemical conditioning or not? *J Mater Chem A* 5:12718–12722. https://doi.org/10.1039/c7ta01769c.
37. Viestfrid Y, Levi MD, Gofer Y, Aurbach D (2005) Microelectrode studies of reversible Mg deposition in THF solutions containing complexes of alkylaluminum chlorides and dialkylmagnesium. *J Electroanal Chem* 576:183–195. https://doi.org/10.1016/j.jelechem.2004.09.034.
38. Kang SJ, Lim SC, Kim H, et al (2017) Non-grignard and lewis acid-free sulfone electrolytes for rechargeable magnesium batteries. *Chem Mater* 29:3174–3180. https://doi.org/10.1021/acs.chemmater.7b00248.
39. He S, Nielson KV, Luo J, Liu TL (2017) Recent advances on $MgCl_2$ based electrolytes for rechargeable Mg batteries. *Energy Storage Mater* 8:184–188. https://doi.org/10.1016/j.ensm.2016.12.001.
40. Li HW, Miwa K, Ohba N, et al (2009) Formation of an intermediate compound with a B12H12 cluster: Experimental and theoretical studies on magnesium borohydride $Mg(BH_4)_2$. *Nanotechnology* 20. https://doi.org/10.1088/0957-4484/20/20/204013.
41. Carter TJ, Mohtadi R, Arthur TS, et al (2014) Boron clusters as highly stable magnesium-battery electrolytes. *Angew Chemie – Int Ed* 53:3173–3177. https://doi.org/10.1002/anie.201310317.
42. Tutusaus O, Mohtadi R, Arthur TS, et al (2015) An efficient halogen-free electrolyte for use in rechargeable magnesium batteries. *Angew Chemie – Int Ed* 54:7900–7904. https://doi.org/10.1002/anie.201412202.
43. Tutusaus O, Mohtadi R (2015) Paving the way towards highly stable and practical electrolytes for rechargeable magnesium batteries. *ChemElectroChem* 2:51–57. https://doi.org/10.1002/celc.201402207.
44. McArthur SG, Jay R, Geng L, et al (2017) Below the 12-vertex: 10-vertex carborane anions as non-corrosive, halide free, electrolytes for rechargeable Mg batteries. *Chem Commun* 53:4453–4456. https://doi.org/10.1039/c7cc01570d.

45. Chłopek K, Frommen C, Léon A, et al (2007) Synthesis and properties of magnesium tetrahydroborate, $Mg(BH_4)_2$. *J Mater Chem* 17:3496–3503. https://doi.org/10.1039/b702723k.
46. Singh R, Kumari P, Rathore RK, et al (2018) LiBH4 as solid electrolyte for Li-ion batteries with Bi_2Te_3 nanostructured anode. *Int J Hydrogen Energy* 21709–21714. https://doi.org/10.1016/j.ijhydene.2018.03.068.
47. Mohtadi R, Matsui M, Arthur TS, Hwang SJ (2012) Magnesium borohydride: From hydrogen storage to magnesium battery. *Angew Chemie – Int Ed* 51:9780–9783. https://doi.org/10.1002/anie.201204913.
48. Wetzel DJ, Malone MA, Gewirth AA, Nuzzo RG (2017) Anisotropic Mg electrodeposition and alloying with Ag-based anodes from non-coordinating mixed-metal borohydride electrolytes for Mg hybrid batteries. *Electrochim Acta* 229:112–120. https://doi.org/10.1016/j.electacta.2017.01.077.
49. Su S, NuLi Y, Wang N, et al (2016) Magnesium borohydride-based electrolytes containing 1-butyl-1-methylpiperidinium bis(trifluoromethyl sulfonyl)imide ionic liquid for rechargeable magnesium batteries. *J Electrochem Soc* 163:D682–D688. https://doi.org/10.1149/2.0631613jes.
50. Gaddum LW, French HE (1927) The electrolysis of grignard solutions. *J Am Chem Soc* 49:1295–1299. https://doi.org/10.1021/ja01404a020.
51. Kim DY, Lim Y, Roy B, et al (2014) Operating mechanisms of electrolytes in magnesium ion batteries: Chemical equilibrium, magnesium deposition, and electrolyte oxidation. *Phys Chem Phys* 16:25789–25798. https://doi.org/10.1039/c4cp01259c.
52. Aurbach D (1999) Magnesium deposition and dissolution processes in ethereal grignard salt solutions using simultaneous EQCM-EIS and in situ FTIR spectroscopy. *Electrochem Solid-State Lett* 3:31. https://doi.org/10.1149/1.1390949.
53. Yagi S, Tanaka A, Ichikawa Y, et al (2014) Effects of water content on magnesium deposition from a Grignard reagent-based tetrahydrofuran electrolyte. *Res Chem Intermed* 40:3–9. https://doi.org/10.1007/s11164-013-1449-9.
54. Pour N, Gofer Y, Major DT, Aurbach D (2011) Structural analysis of electrolyte solutions for rechargeable Mg batteries by stereoscopic means and DFT calculations. *J Am Chem Soc* 133:6270–6278. https://doi.org/10.1021/ja1098512.
55. Tran TT, Lamanna WM, Obrovac MN (2012) Evaluation of $Mg[N(SO_2CF_3)_2]_2$/acetonitrile electrolyte for use in Mg-ion cells. *J Electrochem Soc* 159:A2005–A2009. https://doi.org/10.1149/2.012301jes.
56. Ha S, Lee Y, Woo SW, et al (2014) Magnesium (II) bis (tri fl uoromethane sulfonyl) imide-based electrolytes with wide electrochemical windows for rechargeable magnesium batteries. *ACS Appl Mater Interfaces* 6:4063–4073. https://doi.org/10.1021/am405619v.
57. God C, Bitschnau B, Kapper K, et al (2017) Intercalation behaviour of magnesium into natural graphite using organic electrolyte systems. *RSC Adv* 7:14168–14175. https://doi.org/10.1039/c6ra28300d.
58. Sa N, Rajput N, Wang H, et al (2016) Concentration dependent electrochemical properties and structural analysis of a simple magnesium electrolyte: Magnesium bis (tri fluoromethane sulfonyl) imide in diglyme. *RSC Adv* 6:113663–113670. https://doi.org/10.1039/C6RA22816J.
59. Mandai T, Akita Y, Yagi S, Egashira M, Munakata H, Kanamura K (2017) A key concept of utilization of both non-Grignard magnesium chloride and imide salts for rechargeable Mg battery electrolyte. *J Mater Chem A* 5:3152–3156. https://doi.org/10.1039/C6TA10194A.

7 Polymer Electrolytes and Separators for Magnesium-Ion Batteries

Rupali Singh, Mohammed Khalifa, S. Janakiraman, A. Venimadhav, S. Anandhan, and K. Biswas

7.1 INTRODUCTION

The divalent magnesium-ion batteries (MIBs) have the potential to replace commercial lithium-ion batteries (LIBs). MIBs offer a high theoretical specific capacity of ~ 3832 mAh cm^{-1} (volumetric), 2233 mAh g^{-1} (gravimetric), and an energy density of 150–200 Wh kg^{-1}. One of the salient features of MIB is its non-toxic nature, which makes handling and disposal easier compared to the other battery systems. On the other hand, magnesium (Mg) metal is available abundantly and inexpensive [1–5]. Since the MIBs do not have the problem of dendrite formation during the cycling, Mg metal can be used directly as an anode. Although the redox potential of Mg (–2.37 V) is higher than that of Li (–3.04 V), it is sufficiently negative such that with the right cathode, high voltage cells can be envisioned [6]. However, MIBs are still at the early stage of development and their practical implementation is hindered because of unstable electrolyte systems and cathode issues. Figure 7.1 shows the rapid increase in research publications of MIBs and their electrolyte, which confirms a shape growth in MIBs. The MIBs are similar to any electrochemical energy storage system (such as lithium- and sodium-ion batteries), with the main components being an anode, cathode, electrolyte, and separators, as shown in Figure 7.2.

Contrary to lithium metal, in MIBs, pure Mg metal can be used as an anode due to its low reactivity, a smaller amount of dendrite formation, and relatively less sensitivity to moisture, which prevents the internal short circuits. On the other hand, lithium metal is highly sensitive to dendrite formation, and the repetitive formation/obliteration of solid electrolyte interphase (SEI) layer leads to catastrophes and poor performance of battery [7,8]. However, there are some issues with Mg metal such as the formation of an impermeable passivation layer after reacting with most polar organic electrolytes. Thus, researchers focused on finding other anode materials such as Mg-based alloys and alloying type anodes including bismuth, antimony, and tin [9,10].

The electrolytes play an important role in the battery system as they act as a transportation medium for the ion movements from cathode to anode and vice versa. The electrolyte contains salts like magnesium perchlorate ($(MgClO_4)_2$), magnesium chloride ($MgCl_2$), and solvents like propylene carbonate (PC), ethylene carbonate (EC), dimethyl carbonate (DMC), and THF, depending on the requirements [11]. The existing electrolytes that are being used in MIBs have several drawbacks such as: (i) the organic electrolyte forms an impermeable passivation layer on Mg metal, (ii) a low-voltage window, (iii) a low cycling efficiency, (iv) a low reversibility, and (v) a complicated synthesis process.

Electrolytes in different forms such as solid, liquid, and polymer (solid or gel type) electrolytes have been used in the battery. The liquid electrolytes offer high ionic conductivity (~10^{-1} to 10^{-2}), but frequent leakages and fire safety issues are a bottleneck impeding the practical application of this system [12], whereas solid electrolytes have no issues of leakage and thus handling is very easy and highly safe. However, solid electrolytes offer low ionic conductivity, flexibility, and poor

DOI: 10.1201/9781003310167-7

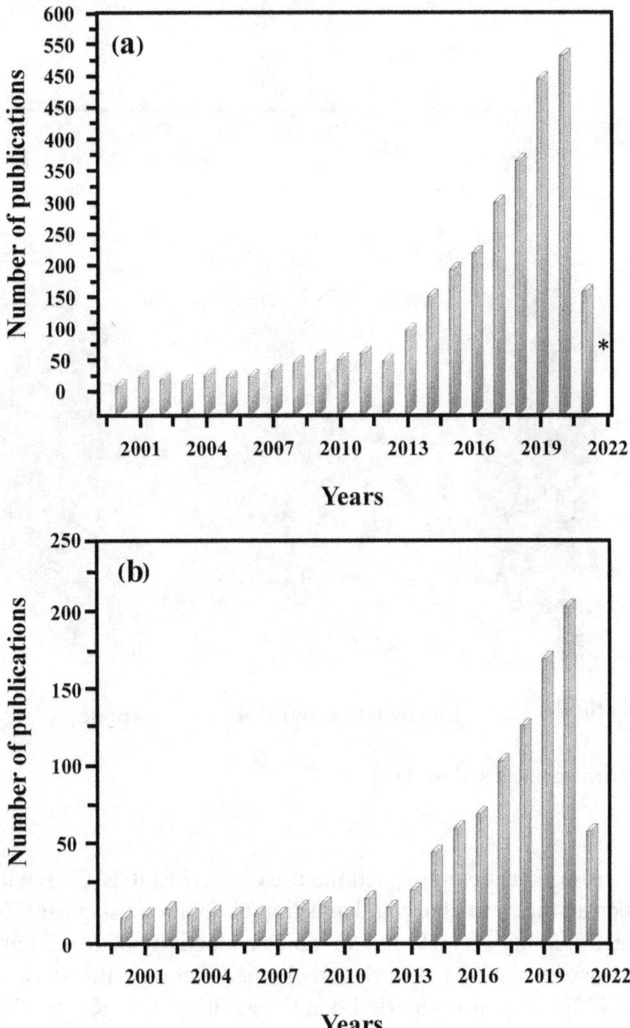

FIGURE 7.1 Number of research publications on MIBs and its electrolyte. From Scopus up to March 2021.

interfacial stability [13–15]. The third type of electrolyte is polymer electrolytes; they have good ionic conductivity, no seepage issues, high flexibility and mechanical properties, and good interfacial contact with electrodes [15].

In LIBs, liquid electrolytes are prepared using salts such as perchlorate (ClO_4^-) and hexafluorophosphate (PF_6^-) dissolved in polar solvents like carbonate and aprotic solvents [11,16]. In LIBs, passivation layer forms called solid electrolyte interface (SEI) occur on the electrode–electrolyte interface, which is permeable through SEI. Conversely, for MIBs, it is not the same because the Mg metal forms a passivation layer that is impermeable hence degrading the properties [11,16]. In LIBs, the SEI layer breaks down and inhibits the possibilities of dendrites formation, thus enhancing the battery life and properties [17]. Similarly, in MIBs, the passivation layer during the cycling must be broken to continue the reversible electrochemical activity. Thus, the electrolyte analogs of Li batteries have been generally considered inappropriate for the Mg-based system [18,19]. Like $LiPF_6$-based liquid electrolytes are an ideal option for LIB because of the high ionic conductivity and voltage stability [20]; however, in MIB, the $MgPF_6$ decomposes to $(PF_6)^{2-}$ and MgF_2, forming a passivation layer on the anode (Mg metal). This makes it incompatible with the MIB system. However,

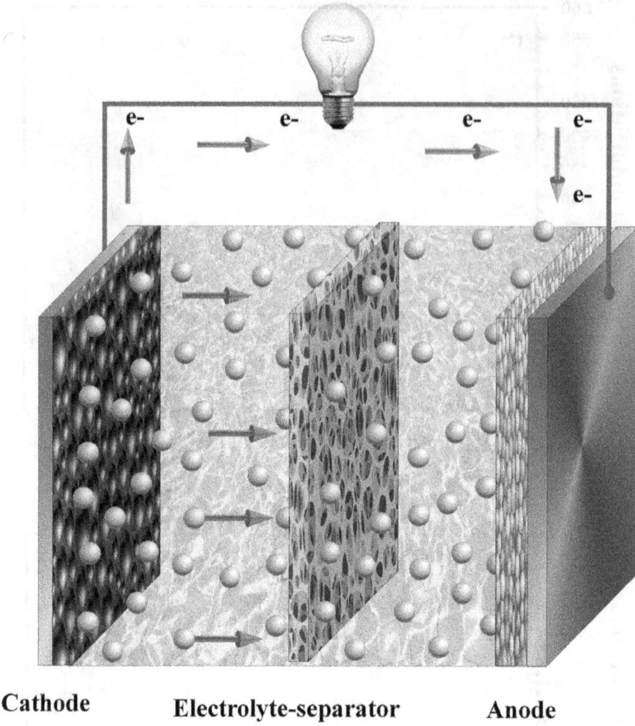

FIGURE 7.2 Battery arrangement and working.

recently Keyzer and co-workers demonstrated the use of $Mg(PF)_6(CH_3CN)_6$ with 1:1 THF/acetonitrile in MIB and reported the reversible Mg deposition/dissolution using Mg/Mo_3S_4 cell [21]. They reported a high voltage stability of 4.0 V and an ionic conductivity of 28 mS cm^{-1}. On the contrary, Shterenberg and co-workers oppose the work and claimed that pure PF_6^--based solution fully passivates the Mg metal [22]. They also reported that the addition of $MgCl_2$ to PF_6^--based electrolyte prevents the formation of a passivation layer on Mg metal. Other salt-based electrolytes, such as $Mg(ClO_4)_2$, are a common choice to demonstrate intercalation into different cathode hosts, although its stability with Mg metal is challenging [11,23–26]. However, the researchers have attempted different routes to activate Mg metal. Xie et al. [27] reported a novel attempt to insert the anion of the salt to activate a fluorinated graphene-based cathode for further cationic intercalation by using $Mg(ClO_4)_2$ with dimethyl sulfoxide (DMSO) as an electrolyte.

The separators are used in the battery to avoid the physical contact of electrodes (anode and cathodes) to prevent the electric short circuits [28,29]. The separators should possess high porosity, chemical resistance, wettability, mechanical, dimensional, and thermal stability [13,29–31]. The separators are generally made of ceramics or polymers; the ceramics (glass fibers) have high wettability and thermal stability, but very poor mechanical properties. In LIBs, generally, polypropylene (PP) Celgard (commercial name) is used as a separator because of its superior mechanical strength and high chemical stability [32]. On the other hand, it has low thermal stability, high shrinkage, poor wettability, and low porosity which make the battery unsafe [33,34]. To improve the battery performance and safety, other polymer films like poly(vinylidene fluoride) (PVDF), poly(vinylidene fluoride-co-hexafluoropropylene) P(VDF-co-HFP), polyethylene oxide (PEO), poly(ethylene carbonate) (PEC), polyacrylonitrile (PAN), and poly(methyl methacrylate) (PMMA) have been investigated as an alternative to PP separators [14,35]. Figure 7.3 shows different types of available cathodes, anodes, and electrolytes for MIB [9].

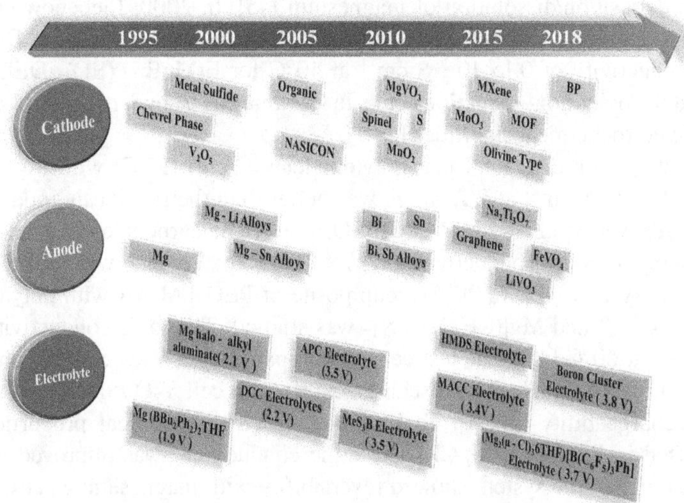

FIGURE 7.3 MIBs reported components (cathodes, anodes, and electrolytes). Adapted and reproduced with permission from Ref. [9]. Copyright © 2019, Elsevier [38].

7.2 POLYMER ELECTROLYTE

The development of polymer electrolytes (PEs) in rechargeable batteries has received considerable interest due to their lightweight, low cost, flexibility, high interfacial contact between the electrodes, avert electrolyte leakage, and better mechanical properties. They also can accommodate the volume change of the electrodes during the charging and discharging processes. The PEs are used in the battery as they provide the advantages of electrolyte and separators. It also reduces the organic liquid evaporation, dendrite formation and corrosion issues, which extends battery life and increases safety [13,29,36]. The PEs are of two types, namely solid polymer electrolytes (SPEs) and gel polymer electrolytes (GPEs), both of which have some advantages and disadvantages [15]. SPEs are solvent-free, resulting in no leakage, lightweight, thin, and easy handling, but have poor ionic conductivity [37,38]. On the other hand, GPEs, which are a combination of liquid electrolyte and polymer matrix, offer ionic conductivity [13,39,40]. In GPEs, an electrolyte is used as an ionic transport medium between electrodes, whereas the polymer membrane acts as a separator and avoids the physical contact between electrodes. Currently, GPE is seen to be attractive in batteries as it improves battery performance. It has high interfacial contact, cycle life, safety, flexibility, interface stability, shelf life, manufacturing integrity, handling, reduce reactivity, and low leakage risk [14,39,41].

7.3 SOLID POLYMER ELECTROLYTE

The SPE for MIBs was first reported in the late 1980s when PEO was combined with $MgCl_2$ and $Mg(ClO_4)_2$. The ionic conductivity of PEO:$MgCl_2$ was 10^{-9} S cm^{-1} and PEO: $Mg(ClO_4)_2$ was 10^{-5} S cm^{-1} at RT [42,43]. Afterward, polyether blends of PEO, PPO, and PPz with $Mg(CF_3SO_3)_2$ were studied and found ionic conductivity of 10^{-5} S cm^{-1} with plasticizer and 10^{-8} S cm^{-1} without plasticizer at 50°C [44]. They confirmed that by using a plasticizer the magnitude of ionic conductivity could be increased up to two orders. Later, PEO with ethyl magnesium bromide (EtMgBr) Grignard reagents was studied. This SPE system showed the highest ionic conductivity 10^{-4} S cm^{-1} at 50°C and activation energy of 50 kJ mol^{-1}. It also showed the reversibility of magnesium ion deposition/dissolution. They concluded that the reversibility of magnesium ion is due to Grignard reagents with THF. This type of electrolyte stops the formation of a passivation layer on the Mg metal surface, which allows the deposition/dissolution of magnesium [45]. In 2000, Liebenow et al. studied the behavior of ethyl magnesium bromide (EtMgBr) in PEO with monomeric ether DEE. They found the maximum conductivity of 0.1×10^{-3} S cm^{-1} at 40°C for EtMgBr-$P(EO)_4$-DEE sample. These SPEs also showed the oxidation–reduction peaks in CV analysis (shown in Figure 7.4), but they have low thermal and electrochemical stabilities [46].

Later, Noto and co-workers found that poly(ethylene glycol) (PEG) with $MgCl_2$ had an ionic conductivity of 1.9×10^{-5} S cm^{-1} at RT, which was better than the previously reported PEO:$MgCl_2$ [47]. PEG-based SPE was studied with $Mg(CF_3SO_3)_2$, ethylene carbonate (EC), and propylene carbonate (PC) showing an ionic conductivity of 1.7×10^{-4} S cm^{-1}, which was higher than the previously reported PEG system [48]. In 2001, a composite of PEO-PMMA with poly(ethylene glycol) dimethyl ether (PEGDE) and $Mg[(CF_3SO_2)_2N]_2$ was studied. The ionic conductivity was found to be 0.4×10^{-3} S cm^{-1} at 60°C [49]. The test cell was prepared by a V_2O_5 cathode and a magnesium metal anode with 1.5 V OCV. The first discharge capacity of cell V_2O_5/SPE/Mg was 100 mAh g^{-1}, but with poor rechargeability [50,51]. To improve the electrochemical properties of PEO, the B_2O_3 filler was used in PEO with $MgCl_2$. The ionic conductivity was improved and found to be 0.7×10^{-5} S cm^{-1}. Further, the system showed reversibility with magnesium electrode, and OCV at 1.9 V was achieved with Mg/SPE/MnO_2 cell [52]. Solid acid polymer electrolyte (SAPE) based on H_2SO_4-doped in poly(vinyl alcohol)-sodium bromide $[(PVA)_{0.7}(NaBr)_{0.3}]$ was prepared by solution cast process. However, the ionic conductivity was low and the standard cell consisting of Mg/SAPE/MnO_2 showed high internal resistance of 900 Ω besides a low capacity of 4.85 mAh g^{-1}, which was significantly low compared to the theoretical capacity of MnO_2 (308 mAh g^{-1}). The cell showed much better 2.73 OCV than the previously reported MnO_2 cell [53]. To overcome the issue, NaBr was replaced with LiBr, along with 5% EC as a plasticizer in SAPE. As a result, the ionic conductivity increased to 1.5×10^{-3} S cm^{-1} and the internal resistance of the cell was decreased to 175 Ω. However, it was found that sulfuric acid was detrimental to Mg anode as it formed $MgSO_4$ during the anodic reaction [54].

Poly(vinylidene fluoride-co-hexafluoropropylene) (PVDF-HFP) and magnesium trifluoromethane sulfonate (Mg(TF))-based SPE were used to improve the ionic conductivity by enhancing the amorphous phase. The highest ionic conductivity of 10^{-3} S cm^{-1} was achieved with 40 wt% MgTF salts in PVDF-HFP [55,56]. Combination of segmented poly(ether urethane) (SPEU) and $Mg(ClO_4)_2$ electrolyte offers excellent voltage stability at 1.9 V. However, the mobility of magnesium Mg ion and transport number was low, i.e., low ionic conductivity of 4.5×10^{-6} S cm^{-1} at RT [57]. SPE

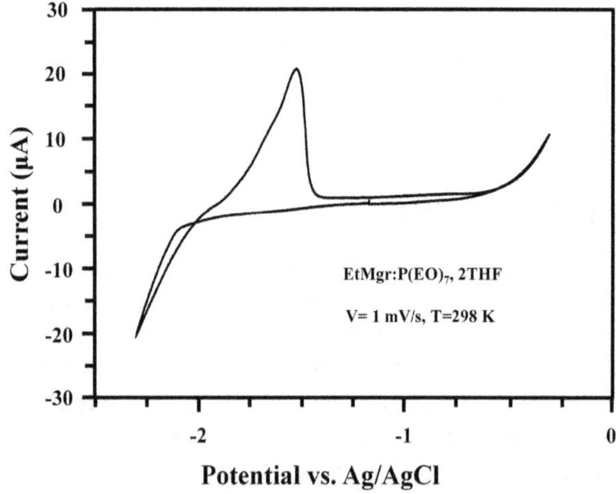

FIGURE 7.4 Cyclic voltammetry (CV) curve of EtMgBr-$P(EO)_7$-2THF with nickel (as a working electrode). Adapted and reproduced with permission from Ref. [46]. Copyright © 2000, Elsevier.

based on solution cast PEO/PVP blend along with $Mg(NO_3)_2$ salt showed an ionic conductivity of 5.8×10^{-4} S cm^{-1} at RT. They found electrochemical stability of 4.0 V, total transference number of 0.99, and Mg^{2+} transference number of 0.33. The CV curve (shown in Figure 7.5) of Mg/SPE/Mg confirmed the reversibility of Mg^{2+}, and Mg/SPE/$MgMn_2O_4$ indicates Mg^{2+} intercalation–deintercalation peaks at 2.24 and 0.79 V; the large difference between voltage is due to slow kinetics reaction and slow diffusion of Mg ion [58].

SPE prepared using PVA/PAN blend electrolyte with $Mg(ClO_4)_2$/DMF showed an ionic conductivity of 2.92×10^{-4} S cm^{-1} in 0.25 m.m.% $Mg(ClO_4)_2$ with a minimum activation energy of 0.21 eV. The total ionic and Mg ion transference numbers were found to be 0.99 and 0.27, respectively, with a voltage stability of 3.65 V [59]. Substituting $Mg(ClO_4)_2$ with $Mg(NO_3)_2$ improved the properties of the SPE. The ionic conductivity of 1.17×10^{-3} S cm^{-1} at RT for 0.3 m.m.% ($Mg(NO_3)_2$ was achieved. Further, the activation energy was found to be 0.36 eV and the transport number of the Mg^{2+} ion was 0.30. The electrochemical voltage stability was found to be 3.4 V suggesting that BPE can be used

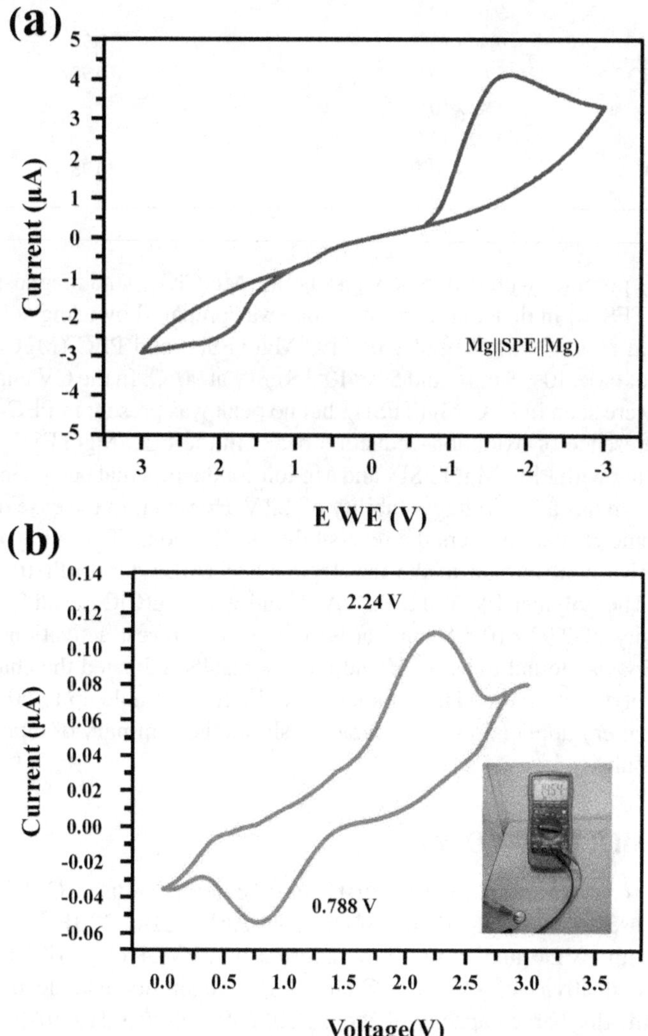

FIGURE 7.5 Cyclic voltammetry (CV) of (a) Mg/SPE/Mg and (b) $MgMn_2O_4$/SPE/Mg; inset image: OCV of ($MgMn_2O_4$/SPE/Mg) coin cell. Adapted and reproduced with permission from Ref. [58]. Copyright © 2017, Elsevier.

TABLE 7.1
Summary of Reported SPE and Their Characterization Results

Composition	Ionic Conductivity ($S\ cm^{-1}$)	Electrochemical Stability (V)	Transference Number	$t_{Mg^{2+}}$	Reference
PEO- $Mg(NO_3)_2$	~10^{-5}	-	0.96	-	[63]
PVA-PVP-$Mg(ClO_4)_2$	~10^{-4}	3.5	0.98	0.31	[64]
PEC-$Mg(TFSI)_2$-LiFSI	~10^{-5} at 80°C	2.0	-	-	[65]
PEO- $Mg(ClO_4)_2$-Al_2O_3	~2.0×10^{-6}				[66]
PEO-$Mg(CF_3SO_3)_2$	2.77×10^{-6}	-	0.98	-	
PEO-$Mg(CF_3SO_3)_2$-SiO_2	5.86×10^{-6}		0.98	0.31	[67]
PEO-$Mg(CF_3SO_3)_2$-TiO_2	1.53×10^{-5}		0.98	0.37	
PEO-$Mg(CF_3SO_3)_2$-MgO	1.67×10^{-5}		0.99	0.38	
PEO-$Mg(CF_3SO_3)_2$-MgO	6.91×10^{-6}		0.98	0.33	
PVA-PEG-$Mg(NO_3)_2$	9.63×10^{-5}		-		[68]
PVA- $Mg(NO_3)_2$	7.36×10^{-7}		0.98	-	[69]
PEG-$Mg(CH_3COO)_2$-TiO_2	5.01×10^{-5}		0.98		[70]
PVA-$Mg(CF_3SO_3)_2$	5.41×10^{-4}		-		[71]
PEO-$Mg(ClO_4)_2$	1.42×10^{-6}				[72]
PVDF-$Mg(NO_3)_2$-MgO	1.04×10^{-4}		0.98		[73]

for MIBs [60]. PEC polymer with different Mg salts like $Mg(ClO_4)_2$ and magnesium trifluoromethanesulfonate ($Mg(TFSI)_2$) in different concentrations was prepared by using solution cast process. They found optimal results with 40 mol% of PEC-$Mg(TFSI)_2$ and PEC-$(MgClO_4)_2$ and reported ionic conductivity of 6.0×10^{-6} S cm^{-1} and 5.2×10^{-4} S cm^{-1} at 90°C. In the CV analysis, the cathodic and anodic peaks were seen in PEC-$Mg(TFSI)_2$, but no peak was present in PEC-$(MgClO_4)_2$, which suggests that the presence of oxidation–reduction peaks in the PEC-$Mg(TFSI)_2$ system confirmed the Mg-metal reacting with PEC-$Mg(TFSI)_2$ and Mg-ion conduction that occurs in the system. While this electrolyte system has a low voltage stability of 2.2 V. However, in the case of PEC-$Mg(ClO_4)_2$, the formation of ionic clusters hinders the reversibility for Mg ions. This study of Mg-ion conductive SPE is a positive contribution to the development of MIBs due to their excellent electrical performance [61]. The polymer PVDF-HFP:PVAc blend with $Mg(ClO_4)_2$ salt THF solvent showed an ionic conductivity of 2.93×10^{-4} S cm^{-1} at 90°C with the lowest activation energy of 0.33 eV. The voltage stability was found to be 3.5 V and the CV results indicated the charge storage nature at electrode–electrolyte interfaces. The samples were thermally stable up to 600°C–650°C, which was sufficient for battery applications [62]. Table 7.1 shows the summary of reported SPE and their characterization results.

7.4 GEL POLYMER ELECTROLYTE

In 2000, Chusid and co-workers modified the first battery prototype by using GPE in the place of liquid electrolytes. They used the two liquid electrolytes ($(Mg(AlCl_2Et\text{-}Bu)_2$/THF and ($Mg(AlCl_2Et\text{-}Bu)_2$/tetraglyme (TG) with PVDF and PEO. They reported that PVDF/($Mg(AlCl_2Et\text{-}Bu)_2$/TG showed the highest ionic conductivity of ~3.7×10^{-3} S cm^{-1}, Mg reversibility, and electrochemical stability of 2.5 V. Further, the discharge capacity of Mo_6S_8/GPE/Mg cell was 110 mAh g^{-1} for 10 cycles at 60°C [74]. In 1999, Kumar and co-workers prepared GPE using polyacrylonitrile (PAN), PC, EC, and $MgTF_2$. They compared the AC impedance of cell-1 (Mg/GPE/Mg) with cell-2 (SS/GPE/SS) and found a semi-circle in cell-1 and a straight line in cell-2. They calculated the specific conductivity for both cells and found 3.5×10^{-3} S cm^{-1} for cell-1 and 1.8×10^{-3} S cm^{-1} for cell-2. The cyclic voltammetry curve showed the anodic and cathodic peaks for cell-1, whereas cell-2 shows a straight line [75,76]. Subsequently, in the extension of their previous work, Mg/GPE/MnO_2 cell was fabricated using the previous reported GPE [75,77]. They found high interfacial resistance (R_i) ~200 KΩ cm^2 of the Mg/GPE/Mg cell, which increased with the aging of the electrochemical cell and decreased with cell temperature. The Mg/GPE/MnO_2 cell showed a maximum of 20 mAh g^{-1} discharge capacity with poor cycling behavior, observed up to 20 cycles. The GPE has poor mechanical strength due to the presence of liquid components (EC, PC) in the composition. Again in 2001, they replaced PAN with PVDF polymer, while other raw materials remained the same ($MgTf_2$, EC, PC). The authors found improvement in results by using PVDF as a polymer. They optimized the composition by changing the EC and PC ratio and found the maximum ionic conductivity of 2.67×10^{-3} S cm^{-1} at 20°C for the mass ratio of 1:2:2:0.8 (PVDF:PC:EC:$MgTf_2$). For charging/discharging results of cell, Mg/GPE/MnO_2 was 50 mAh g^{-1} at C/4 current rate for up to 30 cycles; however, the capacity declined after 30 cycles due to the blocking passive layer formation on the Mg metal surface [78]. Kumar et al. fabricated GPE using polymethylmethacrylate (PMMA) and $MgTf_2$, PC, and EC as a liquid electrolyte. The specific conductivity of 4.2×10^{-4} S cm^{-1} at 20°C with a discharge capacity of 90 mAh g^{-1} using Mg/GPE/MnO_2 cell was observed. However, after few cycles (charge/discharge), the discharge capacity decreased because of passivation layer formation on Mg metal, which hindered the cycle life [79]. Oh and co-workers prepared GPE using 15% PVDF-co-HFP, 73% $Mg(ClO_4)_2$-EC/PC, and 12% SiO_2, respectively. They reported an ionic conductivity of 3.2×10^{-3} S cm^{-1} at RT and voltage stability of 4.3 V. Mg/GPE/V_2O_5 cell that was fabricated showed a low initial discharge capacity of 58 mAh g^{-1} and also poor cyclability, which could be attributed to high interfacial resistance of Mg anode [80].

Yoshimoto et al. [81] prepared GP composed of PEO-PMMA with $Mg[CF_3SO_2)_2N]_2$ and mixed alkyl carbonates by photo-induced radical polymerization. They found ionic conductivity of 2.8 mS cm^{-1} at RT with excellent flexibility, self-standing and mechanical strength. They examine battery performance by using Mg-doped V_2O_5 as the negative electrode and V_2O_5 as the positive electrode. The first discharge capacity was 130 mAh g^{-1} afterward, which gradually decreased with the repetition of the cycle [81]. In 2005, the same group developed a GPE that consisted of PEO-PMMA dissolving ionic liquid mixed with 1-ethyl-3-methylimidazolium bis(trifluoromethyl sulfonyl)imide (EMITFSI) and $Mg(TFSI)_2$ salts. The GPE was self-standing, transparent, and flexible with enough mechanical strength. The highest ionic conductivity of 1.1×10^{-4} S cm^{-1} at RT was obtained for the composition of 50 wt.% of EMITFSI dissolving 20 mol.% $Mg(TFSI)_2$. They also performed polarization experiments that confirmed the mobility of Mg^{2+} ions in the system [82]. Aravindan Vanchiappan and co-workers prepared a GPE using PVDF as a polymer matrix, $Mg(TFSI)_2$ as a salt with various ratios of TG used as a plasticizer, and tetra butyl ammonium chloride (TBACl) as a filler by a solution cast process. The GPE showed an ionic conductivity of 4.42×10^{-4} S cm^{-1} and an electrochemical stability of 1.75 V for a 5% TBACl sample [83].

In 2009, Pandey and co-workers [84] used room temperature ionic liquid (RTIL) for preparing GPE consisting of PVDF-HFP and $Mg(TFSI)_2$, with EMITF as a solvent. They reported a maximum ionic conductivity of 4.8×10^{-3} S cm^{-1} at 20°C. The Mg ion transference number is ~0.26, with a voltage stability of 3.5 V [84]. Further, the effect of different fillers on ionic conductivity and electrochemical voltage stability was analyzed. First, PVDF-HFP dispersed with nano MgO showed a maximum ionic conductivity of 8.0×10^{-3} S cm^{-1} at RT and voltage stability of 3.5 V. Besides, the maximum Mg ion transport number was ~0.44 [41]. Second, micron-sized MgO displayed an ionic conductivity of 6.0×10^{-3} S cm^{-1} at RT and a wider electrochemical voltage window of 3.5 V. Mg-MWCNT/GPE/V_2O_5 cells were fabricated, which showed a discharge capacity of 175 mAh g^{-1} for more than ten cycles and further cycling of cells leads to a decline in capacity, which is attributed to the passivation of a negative electrode [85]. Third, MgO was replaced with fumed silica in the same system and found that 3.0 wt.% fumed silica filler offers a maximum ionic conductivity ~1.1×10^{-2} S cm^{-1} at 25°C. They claimed that the space-charge layers formed between filler particles and gel electrolyte, which are responsible for the enhancement in ionic conductivity. The prototype cell Mg-MWCNT/GPE/MoO_3 was prepared and it showed the discharge capacity of ~175 mAh g^{-1} of MoO_3 for an initial ten charge–discharge cycles. They also reported that the rechargeability of the cell improved when Mg metal is substituted by Mg-MWCCNT composite. The composite gel electrolytes are free-standing and flexible films with enough mechanical strength and excellent thermal and electrochemical stabilities [86]. In 2011. Kumar et al. [87] reported magnesium ion conduction in PEO incorporated with ionic liquid. They found ionic conductivity of 5.6×10^{-4} S cm^{-1} at RT. The maximum Mg ion transference number was 0.45.

In 2012, Tripathi et al. [88] developed GPE via solution cast process using PVDF-HFP as polymer matrix and 0.3M $Mg(ClO_4)_2$ salt as PC solvent. The GPE showed an ionic conductivity of 5.0×10^{-3} S cm^{-1} at RT for 15 wt.% PVDF-HFP system and an activation energy of 0.36 eV. In 2013, Zainol et al. prepared GPE comprised of PMMA with EC, PC, and $Mg(CF_3SO_3)_2$ by solution cast process. At 20 wt.% of $Mg(CF_3SO_3)_2$, GPE showed a maximum ionic conductivity of 1.27×10^{-3} S cm^{-1} at RT, an activation energy of 0.18 eV, and an ion transport number of 0.38 Mg^{2+} [89]. In 2014, they replaced $Mg(CF_3SO_3)_2$ salt with $Mg(ClO_4)_2$ and saw an improvement in the ionic conductivity, while other properties were similar to their previous study [90].

In 2015, Shao et al. [91] have developed nanocomposite PEs consisting of PEO, $Mg(BH_4)_2$, and MgO. They used Mo_6S_8 and Mg metal as electrodes and found high columbic efficiency (98%), cyclability, and efficiency. Further, the theoretical aspects were studied and found that $Mg(BH_4)_2$ glyme electrolyte helps in the enhancement of battery performance [91]. In 2016, Tang et al. [39] studied the PVDF-HFP polymer with $Mg(TF)_2$ (90:10) and EMITF in a different ratio. They found that 40 wt.% of EMITF in (90 PVDF-HFP:10 $Mg(TF)_2$) was showing low relative crystallinity (21.8) and high thermal stability (355°C), melting temperature (112°C), ionic conductivity (4.63×10^{-4} S cm^{-1}), and wide electrochemical stability (4.8 V). They concluded that the high amorphous phase enhances the properties of GPE: PVDF-HFP has a semi-crystalline nature, and as the $Mg(TF)_2$/EMITF amount increases, the amorphicity increases. They also reported that ILS improves the ionic conductivity of GPE [39].

In 2017, Na Wu and co-workers [92] studied the effect of nano MgO on the characteristics of GPE for MIBs applications. Electrospun TPU/PVDF nano fabric was prepared at different concentrations of MgO nanoparticles. They achieved an ionic conductivity of 4.6×10^{-3} S cm^{-1} and an electrochemical stability of 4.7 V for GPE containing 7 wt.% of MgO nanoparticles. Further, CV analysis was studied and found that the maximum current was associated with the reversibility of Mg^{2+} ion. The incorporation of MgO nanoparticles helps in improving the electrochemical activity of GPE [92]. In 2017, Wang et al. [93] have developed PVA/$Mg(Tf)_2$/EMITF-based GPE in different compositions by solution cast process. They found an optimum ratio for PVA:$Mg(Tf)_2$:EMITF, i.e., 85:15:15, which showed a high ionic conductivity of 2.10×10^{-4} S cm^{-1} (2.39×10^{-6} S cm^{-1} without IL) and a voltage stability of 5.0 V, while the activation energy was 0.25 eV with high mechanical properties [93].

Further, they synthesized a biodegradable GPE by replacing PVA with chitosan (CA). They synthesized different compositions varying with the ratio of $Mg(TF)_2$ and EMITF and found the best results for 10 wt.% EMITF-plasticized 90 CA-10 $Mg(TF)_2$. The highest ionic conductivity was 3.57×10^{-5} S cm^{-1} at RT with 0.72 eV of activation energy for a 10 wt.% sample. The ionic transference number was found to be 0.985 and electrochemical voltage stability was 4.15 V, which is sufficient for practical applications. However, the properties declined by replacing PVA with chitosan [94]. In 2019, Du et al. [95] have studied GPE based on polytetrahydrofuran-borate with glass fiber synthesized by in situ crosslinking reaction of [$Mg(BH_4)_2$] and hydroxyl-terminated polytetrahydrofuran. The GPE showed reversible Mg-plating and stripping with outstanding compatibility with Mg metal. They found the ionic conductivity of 4.76×10^{-3} S cm^{-1} at RT and Mg^{2+} ion transference number of 0.73. They also reported the battery performance by using Mo_6S_8 as a cathode and Mg metal as an anode and found good cycling stability and rate capability results as shown in Figures 7.6 and 7.7.

In 2019, Sharma et al. [96] have studied the effect of Al_2O_3 and $MgAl_2O_4$ active fillers in PVDF-HFP-based GPE prepared by solution cast process. The $Mg(TF)_2$, EC, and PC electrolytes

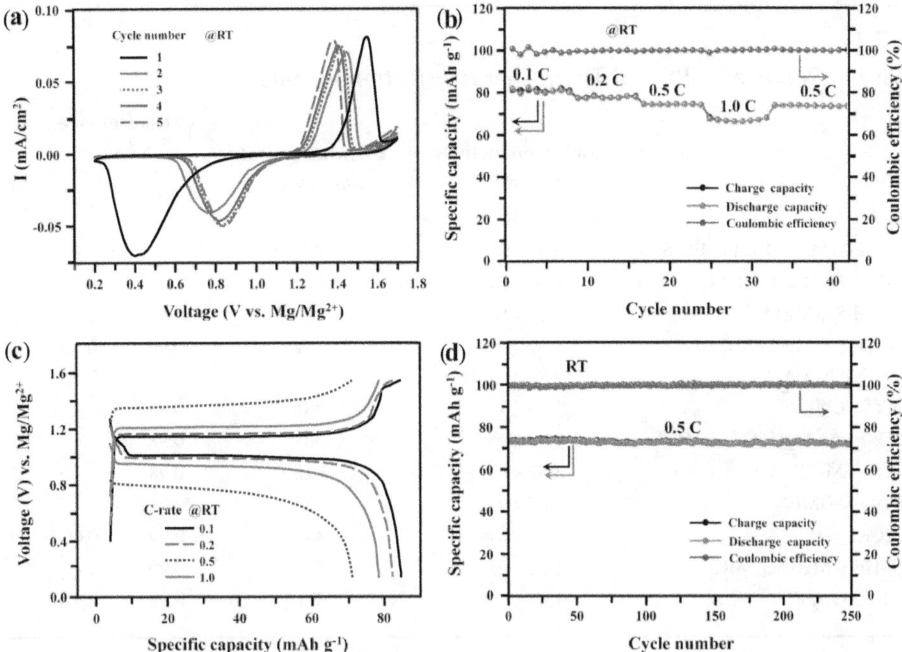

FIGURE 7.6 Electrochemical testing of (Mo_6S_8/PTB@GF-GPE/Mg) cell: (a) Cyclic voltammetry (CV) curves at 0.05 mV s^{-1}, (b) the specific capacity at different C-rate, (c) the galvanostatic charge–discharge profiles at different C-rates, and (d) the cyclic stability at 0.5 C. Adapted and reproduced with permission from Ref. [95]. Copyright © 2019, John Wiley and Sons.

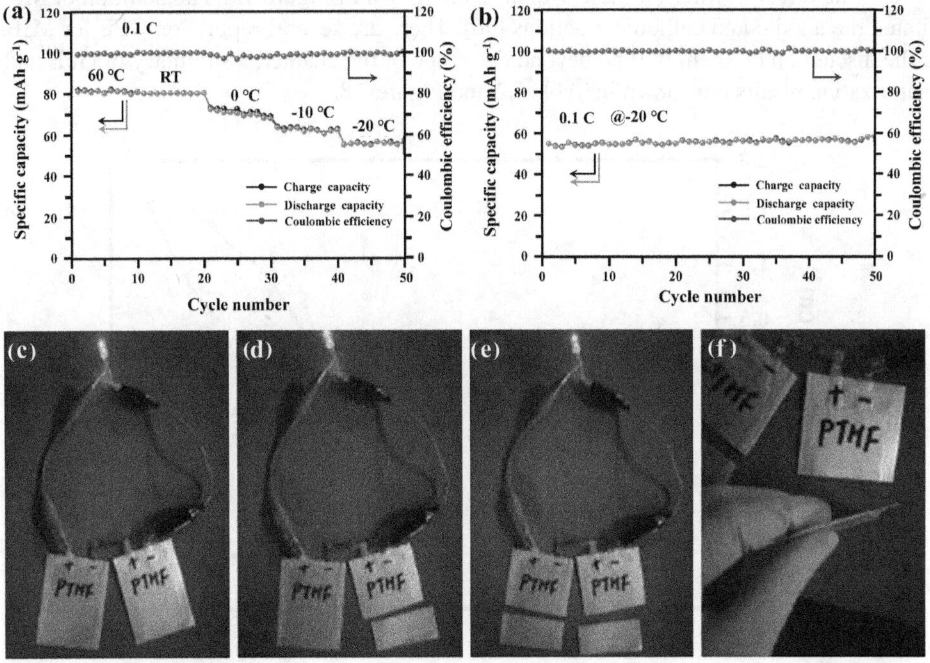

FIGURE 7.7 Electrochemical testing of (Mo_6S_8/PTB@GF-GPE/Mg) cell: (a) the specific capacity at 0.1 C at varied temperatures, (b) the cyclic stability at 0.1 C and –20°C, and (c–f) illustration of soft-package Mo_6S_8/PTB@GF-GPEs/Mg cell undercutting test conditions. Adapted and reproduced with permission from Ref. [95]. Copyright © 2019, John Wiley and Sons.

TABLE 7.2
Summary of Reported GPE and Their Characterization Results

Composition	Ionic Conductivity (S cm^{-1})	Electrochemical Stability (V)	Transference Number t_{ion}	Transference Number t_{Mg}^{2+}	Reference
PAN-$(MgClO_4)_2$-PC-EC	3.2×10^{-3}	-	-	-	[97]
PVDF-co-HFP-$(MgClO_4)_2$-PC-EC-SiO_2	3.2×10^{-3}	4.3	-	-	[98]
PMMA-$Mg(CF_3SO_3)_2$-EC-DEC	5.58×10^{-5}	2.42	-	0.37	[99]
PVdC-co-AN-SN-$MgTf_2$	2.26×10^{-7}	-	0.90	0.33	[100]
PVdC-co-AN-SN-$Mg(TFSI)_2$	1.93×10^{-6}	3.8	0.90	0.59	[101]
PVdC-co-AN-SN-$MgTf_2$	2.82×10^{-7}	3.6		0.56	
PVA-$Mg(Tf)_2$-EMITF	1.2×10^{-3}	5.0	0.99	-	[102]
PVdF-HFP-PVAc-$Mg(ClO_4)_2$-EMITF	9.1×10^{-4}	3.6	0.98	-	[103]
CA-$Mg(Tf)_2$-EMITF	3.57×10^{-5}	4.15	0.98	-	[104]
CA-$Mg(ClO_4)_2$-DMF	4.05×10^{-4}	3.58	-	-	[105]
PAN-$Mg(ClO_4)_2$-PC	3.28×10^{-3}	4.6	0.99	0.60	[106]
PVDF-co-HFP- $Mg(ClO_4)_2$-PC	1.62×10^{-3}	5.5	0.99	-	[107]
PVDF- $Mg(ClO_4)_2$-PC	1.49×10^{-3}	5.0	0.99	0.47	[108]

were mixed in PVDF-HFP solution for 12 hours and then nanofillers were added at different loadings based on the weight of PVDF-HFP (0–30 wt.%). They reported that the highest ionic conductivity was achieved in 30 wt.% of Al_2O_3 and 20 wt.% of $MgAl_2O_4$ samples, 3.3×10^{-3} and 4.0×10^{-3} S cm^{-1}, respectively. By using Al_2O_3 filler, mechanical strength improved, but ion transport was affected. Whereas for $MgAl_2O_4$ nanofillers, the mechanical strength and magnesium ion conduction improved. The effect of filler on the CV analysis is shown in Figure 7.7. The addition of $MgAl_2O_4$ facilitated the anodic and cathodic reactions [96]. There are several reports on GPE for MIBs, but since the discussion on them will be beyond the scope of the chapter, a summary of GPE and their characterization results are shown in Table 7.2 and Figure 7.8.

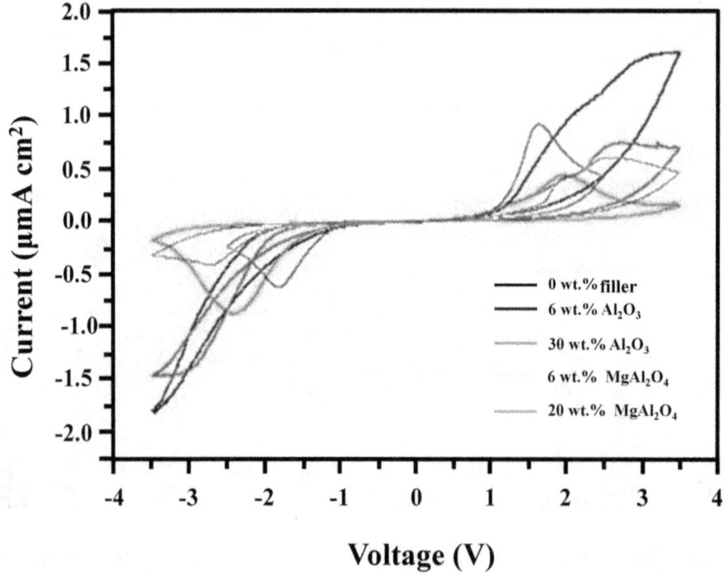

FIGURE 7.8 Cyclic voltammetry results and the effect of different fillers on the PVDF-HFP-based GPE. Adapted and reproduced with permission from Ref. [96]. Copyright © 2018 John Wiley and Sons.

7.5 CONCLUSIONS

The polymer electrolytes are the crucial component in the MIBs and serve as separators and electrolytes. The polymer electrolyte possesses Mg^{+2} ionic conductivity of the order of 10^{-3} S cm^{-1} at ambient temperature and electrochemical voltage stability up to 5.0V vs Mg/Mg^{+2}. The solid polymer electrolyte exhibits relatively low ionic conductivity than the GPE under ambient conditions. The addition of inorganic fillers is a successful way of enhancing ionic conductivity and improving the mechanical and electrochemical properties of polymer electrolytes. Great efforts are still required to enhance the properties of polymer electrolytes like ionic conductivity, better electrolyte/electrode interface, and good thermal, mechanical, and electrochemical properties.

REFERENCES

1. Guo Y, Zhang F, Yang J, et al (2012) Boron-based electrolyte solutions with wide electrochemical windows for rechargeable magnesium batteries. *Energy Environ Sci* 5:9100. https://doi.org/10.1039/c2ee22509c.
2. Doe RE, Han R, Hwang J, et al (2014) Novel, electrolyte solutions comprising fully inorganic salts with high anodic stability for rechargeable magnesium *batteries. Chem Commun* 50:243–245. https://doi.org/10.1039/C3CC47896C.
3. Carter TJ, Mohtadi R, Arthur TS, et al (2014) Boron clusters as highly stable magnesium-battery electrolytes. *Angew Chemie – Int Ed* 53:3173–3177. https://doi.org/10.1002/anie.201310317.
4. Huang ZD, Masese T, Orikasa Y, et al (2014) $MgFePO_4F$ as a feasible cathode material for magnesium batteries. *J Mater Chem A* 2:11578–11582. https://doi.org/10.1039/c4ta01779j.
5. Yoo HD, Shterenberg I, Gofer Y, et al (2013) Mg rechargeable batteries: An on-going challenge. *Energy Environ Sci* 6:2265–2279.
6. Vanysek P (2000) Electrochemical series. *CRC Handbook of Chemistry and Physics* 8:8–33.
7. Aurbach D, Gofer Y, Lu Z, et al (2001) A short review on the comparison between Li battery systems and rechargeable magnesium battery technology. *J Power Sources* 97-98:28–32. https://doi.org/10.1016/S0378-7753(01)00585-7.
8. Zhan Y, Zhang W, Lei B, et al (2020) Recent development of Mg ion solid electrolyte. *Front Chem* 8:1–7. https://doi.org/10.3389/fchem.2020.00125.
9. Zhang Y, Geng H, Wei W, et al (2019) Challenges and recent progress in the design of advanced electrode materials for rechargeable Mg batteries. *Energy Storage Mater* 20:118–138. https://doi.org/10.1016/j.ensm.2018.11.033.
10. Li X, Gao T, Han F, et al (2018) Reducing Mg anode overpotential via ion conductive surface layer formation by iodine additive. *Adv Energy Mater* 8:1701728. https://doi.org/10.1002/aenm.201701728.
11. Deivanayagam R, Ingram BJ, Shahbazian-Yassar R (2019) Progress in development of electrolytes for magnesium batteries. *Energy Storage Mater* 21:136–153. https://doi.org/10.1016/j.ensm.2019.05.028.
12. Tan L, Deng Y, Cao Q, et al (2019) Gel electrolytes based on polyacrylonitrile/thermoplastic polyurethane/polystyrene for lithium-ion batteries. *Ionics (Kiel)* 25:3673–3682. https://doi.org/10.1007/s11581-019-02940-7.
13. Lee H, Yanilmaz M, Toprakci O, et al (2014) A review of recent developments in membrane separators for rechargeable lithium-ion batteries. *Energy Environ Sci* 7:3857–3886. https://doi.org/10.1039/C4EE01432D.
14. Stephan AM (2006) Review on gel polymer electrolytes for lithium batteries. *Eur Polym J* 42:21–42.
15. Jiang Y, Yan X, Ma Z, et al (2018) Development of the PEO based solid polymer electrolytes for all-solid state lithium ion batteries. *Polymers (Basel)* 10:1237. https://doi.org/10.3390/polym10111237
16. Saha P, Datta MK, Velikokhatnyi OI, et al (2014) Rechargeable magnesium battery: Current status and key challenges for the future. *Prog Mater Sci* 66:1–86. https://doi.org/10.1016/j.pmatsci.2014.04.001.
17. An SJ, Li J, Daniel C, et al (2016) The state of understanding of the lithium-ion-battery graphite solid electrolyte interphase (SEI) and its relationship to formation cycling. *Carbon N Y* 105:52–76. https://doi.org/10.1016/j.carbon.2016.04.008.
18. Gregory TD, Hoffman RJ, Winterton RC (1990) Nonaqueous electrochemistry of magnesium: Applications to energy storage. *J Electrochem Soc* 137:775–780. https://doi.org/10.1149/1.2086553.

19. El Kharbachi A, Zavorotynska O, Latroche M, et al (2020) Exploits, advances and challenges benefiting beyond Li-ion battery technologies. *J Alloys Compd* 817:153261. https://doi.org/10.1016/j.jallcom.2019.153261.
20. Xu K (2014) Electrolytes and interphases in Li-ion batteries and beyond. *Chem Rev* 114:11503–11618. https://doi.org/10.1021/cr500003w.
21. Keyzer EN, Glass HFJ, Liu Z, et al (2016) $Mg(PF_6)_2$-based electrolyte systems: understanding electrolyte-electrode interactions for the development of Mg-ion batteries. *J Am Chem Soc* 138:8682–8685. https://doi.org/10.1021/jacs.6b04319.
22. Shterenberg I, Salama M, Gofer Y, Aurbach D (2017) Hexafluorophosphate-based solutions for Mg batteries and the importance of chlorides. *Langmuir* 33:9472–9478. https://doi.org/10.1021/acs.langmuir.7b01609.
23. Huie MM, Bock DC, Takeuchi ES, et al (2015) Cathode materials for magnesium and magnesium-ion based batteries. *Coord Chem Rev* 287:15–27. https://doi.org/10.1016/j.ccr.2014.11.005.
24. Wang R, Chung CC, Liu Y, et al (2017) Electrochemical intercalation of Mg^{2+} into anhydrous and hydrated crystalline tungsten oxides. *Langmuir* 33:9314–9323. https://doi.org/10.1021/acs.langmuir.7b00705.
25. Truong QD, Kempaiah Devaraju M, Tran PD, et al (2017) Unravelling the surface structure of $MgMn_2O_4$ cathode materials for rechargeable magnesium-ion battery. *Chem Mater* 29:6245–6251. https://doi.org/10.1021/acs.chemmater.7b01252.
26. Verrelli R, Black AP, Pattanathummasid C, et al (2018) On the strange case of divalent ions intercalation in V_2O_5. *J Power Sources* 407:162–172. https://doi.org/10.1016/j.jpowsour.2018.08.024.
27. Xie J, Li C, Cui Z, Guo X (2015) Transition-metal-free magnesium-based batteries activated by anionic insertion into fluorinated graphene nanosheets. *Adv Funct Mater* 25:6519–6526. https://doi.org/10.1002/adfm.201503010.
28. Huang X (2011) Separator technologies for lithium-ion batteries. *J Solid State Electrochem* 15:649–662.
29. Costa CM, Silva MM, Lanceros-Méndez S (2013) Battery separators based on vinylidene fluoride (VDF) polymers and copolymers for lithium ion battery applications. *RSC Adv* 3:11404. https://doi.org/10.1039/c3ra40732b.
30. Arora P, Zhang ZJ (2004) Battery separators. *Chem Rev* 104:4419–4462. https://doi.org/10.1021/cr020738u.
31. Seol W-H, Lee YM, Park J-K (2006) Preparation and characterization of new microporous stretched membrane for lithium rechargeable battery. *J Power Sources* 163:247–251. https://doi.org/10.1016/j.jpowsour.2006.02.076.
32. Love CT (2011) Thermomechanical analysis and durability of commercial micro-porous polymer Li-ion battery separators. *J Power Sources* 196:2905–2912. https://doi.org/10.1016/j.jpowsour.2010.10.083.
33. Zhu J, Yanilmaz M, Fu K, et al (2016) Understanding glass fiber membrane used as a novel separator for lithium-sulfur batteries. *J Memb Sci* 504:89–96. https://doi.org/10.1016/j.memsci.2016.01.020.
34. Song J, Ryou MH, Son B, et al (2012) Co-polyimide-coated polyethylene separators for enhanced thermal stability of lithium ion batteries. *Electrochim Acta* 85:524–530. https://doi.org/10.1016/j.electacta.2012.06.078.
35. Ngai KS, Ramesh S, Ramesh K, Juan JC (2016) A review of polymer electrolytes: fundamental, approaches and applications. *Ionics (Kiel)* 22:1259–1279. https://doi.org/10.1007/s11581-016-1756-4.
36. Nestler T, Roedern E, Uvarov NF, et al (2019) Separators and electrolytes for rechargeable batteries: Fundamentals and perspectives. *Phys Sci Rev* 4:1–29. https://doi.org/10.1515/psr-2017-0115.
37. Regu T, Ambika C, Karuppasamy K, et al (2019) Proton transport and dielectric properties of high molecular weight polyvinylpyrrolidone (PVPK90) based solid polymer electrolytes for portable electrochemical devices. *J Mater Sci Mater Electron* 30:11735–11747. https://doi.org/10.1007/s10854-019-01535-2.
38. Ambika C, Karuppasamy K, Vikraman D, et al (2018) Effect of dimethyl carbonate (DMC) on the electrochemical and cycling properties of solid polymer electrolytes (PVP-MSA) and its application for proton batteries. *Solid State Ionics* 321:106–114. https://doi.org/10.1016/j.ssi.2018.04.013.
39. Tang X, Muchakayala R, Song S, et al (2016) A study of structural, electrical and electrochemical properties of PVdF-HFP gel polymer electrolyte films for magnesium ion battery applications. *J Ind Eng Chem* 37:67–74. https://doi.org/10.1016/j.jiec.2016.03.001.
40. Karuppasamy K, Kim H-S, Kim D, et al (2017) An enhanced electrochemical and cycling properties of novel boronic Ionic liquid based ternary gel polymer electrolytes for rechargeable Li/$LiCoO_2$ cells. *Sci Rep* 7:11103. https://doi.org/10.1038/s41598-017-11614-1.
41. Pandey GP, Agrawal RC, Hashmi SA (2009) Magnesium ion-conducting gel polymer electrolytes dispersed with nanosized magnesium oxide. *J Power Sources* 190:563–572. https://doi.org/10.1016/j.jpowsour.2009.01.057.

42. Yang LL (1986) Ionic conductivity in complexes of poly(ethylene oxide) and $MgCl_2$. *J Electrochem Soc* 133:1380. https://doi.org/10.1149/1.2108891
43. Patrick A, Glasse M, Latham R, Linford R (1986) Novel solid state polymer batteries. *Solid State Ion* 18–19:1063–1067. https://doi.org/1037//0033-2909.I26.1.78.
44. Acosta JL, Morales E (1997) Synthesis and characterization of polymeric electrolytes for solid state magnesium batteries. *Electrochim Acta* 43:791–797. https://doi.org/10.1016/S0013-4686(97)00123-0.
45. Liebenow C (1998) A novel type of magnesium ion conducting polymer electrolyte. *Electrochim Acta* 43:1253–1256. https://doi.org/.1037//0033-2909.I26.1.78.
46. Liebenow C (2000) Electrochemical and structural investigations on solutions of organomagnesium bromide in polymeric ether. *Solid State Ionics* 136–137:1211–1214. https://doi.org/10.1016/S0167-2738(00)00586-5.
47. Noto VD, Lavina S, Longo D, Vidali M (1998) A novel electrolytic complex based on δ-$MgCl_2$ and poly(ethylene glycol) 400. *Electrochim Acta* 43:1225–1237. https://doi.org/10.1016/S0013-4686(97)10023-8.
48. Ikeda S, Mori Y, Furuhashi Y, Masuda H (1999) Multivalent cation conductive solid polymer electrolytes using photo-cross-linked polymers. *Solid State Ionics* 121:329–333. https://doi.org/.1037//0033-2909.I26.1.78.
49. Morita M, Yoshimoto N, Yakushiji S, Ishikawa M (2001) Rechargeable magnesium batteries using a novel polymeric solid electrolyte. *Electrochem Solid-State Lett* 4:A177. https://doi.org/10.1149/1.1403195.
50. Yoshimoto N, Tomonaga Y, Ishikawa M, Morita M (2001) Ionic conductance of polymeric electrolytes consisting of magnesium salts dissolved in cross-linked polymer matrix with linear polyether. *Electrochim Acta* 46:1195–1200. https://doi.org/10.1016/S0013-4686(00)00705-2.
51. Nobuko Y, Yakushiji S, Ishikawa M, Morita M (2002) Ionic conductance behavior of polymeric electrolyte containing magnesium salts and their application to rechargeable batteries. *Solid State Ionics* 259–266. https://doi.org/.1037//0033-2909.I26.1.78.
52. Sundar M, Selladurai S (2006) Effect of fillers on magnesium-poly(ethylene oxide) solid polymer electrolyte. *Ionics (Kiel)* 12:281–286. https://doi.org/10.1007/s11581-006-0048-9.
53. Sheha E, El-Mansy MK (2008) A high voltage magnesium battery based on H_2SO_4-doped $(PVA)_{0.7}(NaBr)_{0.3}$ solid polymer electrolyte. *J Power Sources* 185:1509–1513. https://doi.org/10.1016/j.jpowsour.2008.09.046.
54. Sheha E (2009) Ionic conductivity and dielectric properties of plasticized $PVA_{0.7}(LiBr)_{0.3}(H_2SO_4)_{2.7}M$ solid acid membrane and its performance in a magnesium battery. *Solid State Ionics* 180:1575–1579. https://doi.org/10.1016/j.ssi.2009.10.008.
55. Ramesh S, Lu S-C (2010) Structural, morphological, thermal, and conductivity studies of magnesium ion conducting P(VdF-HFP)-based solid polymer electrolytes with good prospects. *J Appl Polym Sci* 117:2050–2058. https://doi.org/10.1002/app.
56. Ramesh S, Lu SC, Morris E (2012) Towards magnesium ion conducting poly(vinylidenefluoride-hexafl uoropropylene)-based solid polymer electrolytes with great prospects: Ionic conductivity and dielectric behaviours. *J Taiwan Inst Chem Eng* 43:806–812. https://doi.org/10.1016/j.jtice.2012.04.004.
57. Jo N-J, Kim M-K, Kang S-W, Ryu K-S (2010) The influence of the cations of salts on the electrochemical stability of a solid polymer electrolyte based on segmented poly(ether urethane). *Phys Scr* T139:014035. https://doi.org/10.1088/0031-8949/2010/T139/014035.
58. Anilkumar KM, Jinisha B, Manoj M, Jayalekshmi S (2017) Poly(ethylene oxide) (PEO) - Poly(vinyl pyrrolidone) (PVP) blend polymer based solid electrolyte membranes for developing solid state magnesium ion cells. *Eur Polym J* 89:249–262. https://doi.org/10.1016/j.eurpolymj.2017.02.004.
59. Manjuladevi R, Thamilselvan M, Selvasekarapandian S, et al (2017) Mg-ion conducting blend polymer electrolyte based on poly(vinyl alcohol)-poly (acrylonitrile) with magnesium perchlorate. *Solid State Ionics* 308:90–100. https://doi.org/10.1016/j.ssi.2017.06.002.
60. Manjuladevi R, Selvasekarapandian S, Thamilselvan M, et al (2018) A study on blend polymer electrolyte based on poly(vinyl alcohol)-poly (acrylonitrile) with magnesium nitrate for magnesium battery. *Ionics (Kiel)* 24:3493–3506. https://doi.org/10.1007/s11581-018-2500-z.
61. Ab Aziz A, Tominaga Y (2018) Magnesium ion-conductive poly(ethylene carbonate) electrolytes. *Ionics (Kiel)* 24:3475–3481. https://doi.org/10.1007/s11581-018-2482-x.
62. Ponmani S, Prabhu MR (2018) Development and study of solid polymer electrolytes based on PVdF-HFP/PVAc: Mg $(ClO_4)_2$ for Mg ion batteries. *J Mater Sci Mater Electron* 29:15086–15096. https://doi.org/10.1007/s10854-018-9649-0.
63. Ramalingaiah S, Reddy DS, Reddy MJ, et al (1996) Conductivity and discharge characteristic studies of novel polymer electrolyte based on PEO complexed with $Mg(NO_3)_2$ salt. *Mater Lett* 29:285–289.

64. Ramaswamy M, Malayandi T, Subramanian S, et al (2017) Magnesium ion conducting polyvinyl alcohol-polyvinyl pyrrolidone-based blend polymer electrolyte. *Ionics (Kiel)* 23:1771–1781. https://doi.org/10.1007/s11581-017-2023-z.
65. Aziz AA, Tominaga Y (2019) Effect of Li salt addition on electrochemical properties of poly(ethylene carbonate)-Mg salt electrolytes. *Polym J* 51:61–67. https://doi.org/10.1038/s41428-018-0113-z.
66. Dissanayake MAKL, Bandara LRAK, Karaliyadda LH, et al (2006) Thermal and electrical properties of solid polymer electrolyte PEO_9 $Mg(ClO_4)_2$ incorporating nano-porous Al_2O_3 filler. *Solid State Ion* 177:343–346. https://doi.org/10.1016/j.ssi.2005.10.031.
67. Agrawal RC, Sahu DK, Mahipal YK, Ashrafi R (2013) Investigations on ion transport properties of hot-press cast magnesium ion conducting Nano-Composite Polymer Electrolyte (NCPE) films: Effect of filler particle dispersal on room temperature conductivity. *Mater Chem Phys* 139:410–415. https://doi.org/10.1016/j.matchemphys.2012.12.056.
68. Polu AR, Kumar R (2011) AC impedance and dielectric spectroscopic studies of Mg^{2+} ion conducting PVA-PEG blended polymer electrolytes. *Bull Mater Sci* 34:1063–1067. https://doi.org/10.1007/s12034-011-0132-2.
69. Polu AR, Kumar R (2013) Preparation and characterization of pva based solid polymer electrolytes for electrochemical cell applications. *Chinese J Polym Sci (English Ed)* 31:641–648. https://doi.org/10.1007/s10118-013-1246-3.
70. Polu AR, Kumar R, Kumar KV, Jyothi NK (2013) Effect of TiO_2 ceramic filler on PEG-based composite polymer electrolytes for magnesium batteries. *AIP Conf Proc* 1512:996–997. https://doi.org/10.1063/1.4791378.
71. Jeong SK, Jo YK, Jo NJ (2006) Decoupled ion conduction mechanism of poly(vinyl alcohol) based Mg-conducting solid polymer electrolyte. *Electrochim Acta* 52:1549–1555. https://doi.org/10.1016/j.electacta.2006.02.061.
72. Reddy MJ, Chu PP (2002) Ion pair formation and its effect in PEO: Mg solid polymer electrolyte system. *J Power Sources* 109:340–346. https://doi.org/10.1016/S0378-7753(02)00084-8.
73. Patel S, Kumar R (2019) Synthesis and characterization of magnesium ion conductivity in PVDF based nanocomposite polymer electrolytes disperse with MgO. *J Alloys Compd* 789:6–14. https://doi.org/10.1016/j.jallcom.2019.03.089.
74. Chusid O, Gofer Y, Gizbar H, et al (2003) Solid-state rechargeable magnesium batteries. *Adv Mater* 15:627–630. https://doi.org/10.1002/adma.200304415.
75. Kumar GG, Munichandraiah N (1999) Reversibility of Mg/Mg^{2+} couple in a gel polymer electrolyte. *Electrochim Acta* 44:2663–2666. https://doi.org/10.1016/S0013-4686(98)00388-0.
76. Kumar G (2000) A gel polymer electrolyte of magnesium triflate. *Solid State Ionics* 128:203–210. https://doi.org/10.1016/S0167-2738(00)00276-9.
77. Kumar GG, Munichandraiah N (2000) Solid-state Mg/MnO_2 cell employing a gel polymer electrolyte of magnesium triflate. *J Power Sources* 91:157–160. https://doi.org/.1037//0033-2909.I26.1.78.
78. Kumar GG, Munichandraiah N (2001) Solid-state rechargeable magnesium cell with poly(vinylidenefluoride)-magnesium triflate gel polymer electrolyte. *J Power Sources* 102:46–54. https://doi.org/10.1016/S0378-7753(01)00772-8.
79. Kumar GG, Munichandraiah N (2002) Poly(methylmethacrylate)-magnesium triflate gel polymer electrolyte for solid state magnesium battery application. *Electrochim Acta* 47:1013–1022. https://doi.org/S0013-4686(01)00832-5.
80. Oh J-S, Ko J-M, Kim D-W (2004) Preparation and characterization of gel polymer electrolytes for solid state magnesium batteries. *Electrochim Acta* 50:903–906. https://doi.org/10.1016/j.electacta.2004.01.099.
81. Yoshimoto N, Yakushiji S, Ishikawa M, Morita M (2003) Rechargeable magnesium batteries with polymeric gel electrolytes containing magnesium salts. *Electrochim Acta* 48:2317–2322. https://doi.org/10.1016/S0013-4686(03)00221-4.
82. Morita M, Shirai T, Yoshimoto N, Ishikawa M (2005) Ionic conductance behavior of polymeric gel electrolyte containing ionic liquid mixed with magnesium salt. *J Power Sources* 139:351–355. https://doi.org/10.1016/j.jpowsour.2004.07.028.
83. Aravindan V, Karthikaselvi G, Vickraman P, Naganandhini SP (2009) Polyvinylidene fluoride-based novel polymer electrolytes for magnesium-rechargeable batteries with $Mg(CF_3SO_3)_2$. *J Appl Polym Sci* 112:3024–3029. https://doi.org/10.1002/app.29877.
84. Pandey GP, Hashmi SA (2009) Experimental investigations of an ionic-liquid-based, magnesium ion conducting, polymer gel electrolyte. *J Power Sources* 187:627–634. https://doi.org/10.1016/j.jpowsour.2008.10.112.

85. Pandey GP, Agrawal RC, Hashmi SA (2011) Performance studies on composite gel polymer electrolytes for rechargeable magnesium battery application. *J Phys Chem Solids* 72:1408–1413. https://doi.org/10.1016/j.jpcs.2011.08.003.
86. Pandey GP, Agrawal RC, Hashmi SA (2011) Magnesium ion-conducting gel polymer electrolytes dispersed with fumed silica for rechargeable magnesium battery application. *J Solid State Electrochem* 15:2253–2264. https://doi.org/10.1007/s10008-010-1240-4.
87. Kumar Y, Hashmi SA, Pandey GP (2011) Ionic liquid mediated magnesium ion conduction in poly(ethylene oxide) based polymer electrolyte. *Electrochim Acta* 56:3864–3873. https://doi.org/10.1016/j.electacta.2011.02.035.
88. Tripathi SK, Jain A, Gupta A, Mishra M (2012) Electrical and electrochemical studies on magnesium ion-based polymer gel electrolytes. *J Solid State Electrochem* 16:1799–1806. https://doi.org/10.1007/s10008-012-1656-0.
89. Zainol NH, Samin SM, Othman L, et al (2013) Magnesium ion-based gel polymer electrolytes: Ionic conduction and infrared spectroscopy studies. *Int J Electrochem Sci* 8:3602–3614.
90. Osman Z, Zainol NH, Samin SM, et al (2014) Electrochemical impedance spectroscopy studies of magnesium-based polymethylmethacrylate gel polymer electroytes. *Electrochim Acta* 131:148–153. https://doi.org/10.1016/j.electacta.2013.11.189.
91. Shao Y, Rajput NN, Hu J, et al (2015) Nanocomposite polymer electrolyte for rechargeable magnesium batteries. *Nano Energy* 12:750–759. https://doi.org/10.1016/j.nanoen.2014.12.028.
92. Wu N, Wang W, Wei Y, Li T (2017) Studies on the effect of nano-sized MgO in magnesium-ion conducting gel polymer electrolyte for rechargeable magnesium batteries. *Energies* 10:1215. https://doi.org/10.3390/en10081215.
93. Wang J, Song S, Muchakayala R, et al (2017) Structural, electrical, and electrochemical properties of PVA-based biodegradable gel polymer electrolyte membranes for Mg-ion battery applications. *Ionics (Kiel)* 23:1759–1769. https://doi.org/10.1007/s11581-017-1988-y.
94. Wang J, Song S, Gao S, et al (2017) Mg-ion conducting gel polymer electrolyte membranes containing biodegradable chitosan: Preparation, structural, electrical and electrochemical properties. *Polym Test* 62:278–286. https://doi.org/10.1016/j.polymertesting.2017.07.016.
95. Du A, Zhang H, Zhang Z, et al (2019) A crosslinked polytetrahydrofuran-borate-based polymer electrolyte enabling wide-working-temperature-range rechargeable magnesium batteries. *Adv Mater* 31:1–7. https://doi.org/10.1002/adma.201805930.
96. Sharma J, Hashmi S (2019) Magnesium ion-conducting gel polymer electrolyte nanocomposites: Effect of active and passive nanofillers. *Polym Compos* 40:1295–1306. https://doi.org/10.1002/pc.24853.
97. Perera K, Dissanayake MAKL, Bandaranayake PWSK (2004) Ionic conductivity of a gel polymer electrolyte based on $Mg(ClO_4)_2$ and polyacrylonitrile (PAN). *Mater Res Bull* 39:1745–1751. https://doi.org/10.1016/j.materresbull.2004.03.027.
98. Oh J-S, Ko J-M, Kim D-W (2004) Preparation and characterization of gel polymer electrolytes for solid state magnesium batteries. *Electrochim Acta* 50:903–906. https://doi.org/10.1016/j.electacta.2004.01.099.
99. Asmara SN, Kufian MZ, Majid SR, Arof AK (2011) Preparation and characterization of magnesium ion gel polymer electrolytes for application in electrical double layer capacitors. *Electrochim Acta* 57:91–97. https://doi.org/10.1016/j.electacta.2011.06.045.
100. Hambali D, Zainuddin Z, Osman Z (2017) Characteristics of novel plastic crystal gel polymer electrolytes based on PVdC-co-AN. *Ionics (Kiel)* 23:285–294. https://doi.org/10.1007/s11581-016-1814-y.
101. Hambali D, Zainol NH, Othman L, et al (2019) Magnesium ion-conducting gel polymer electrolytes based on poly(vinylidene chloride-co-acrylonitrile) (PVdC-co-AN): A comparative study between magnesium trifluoromethanesulfonate ($MgTf_2$) and magnesium bis(trifluoromethanesulfonimide) (Mg(TFSI)2). *Ionics (Kiel)* 25:1187–1198. https://doi.org/10.1007/s11581-018-2666-4.
102. Wang J, Zhao Z, Muchakayala R, Song S (2018) High-performance Mg-ion conducting poly(vinyl alcohol) membranes: Preparation, characterization and application in supercapacitors. *J Memb Sci* 555:280–289. https://doi.org/10.1016/j.memsci.2018.03.068.
103. Ponmani S, Ramesh Prabhu M (2019) Sulfonate based ionic liquid incorporated polymer electrolytes for magnesium secondary battery. *Polym Technol Mater* 58:978–991. https://doi.org/10.1080/03602559.2018.1520259.
104. Wang J, Song S, Gao S, et al (2017) Mg-ion conducting gel polymer electrolyte membranes containing biodegradable chitosan: Preparation, structural, electrical and electrochemical properties. *Polym Test* 62:278–286. https://doi.org/10.1016/j.polymertesting.2017.07.016.

105. Mahalakshmi M, Selvanayagam S, Selvasekarapandian S, et al (2019) Characterization of biopolymer electrolytes based on cellulose acetate with magnesium perchlorate ($Mg(ClO_4)_2$) for energy storage devices. *J Sci Adv Mater Devices* 4:276–284. https://doi.org/10.1016/j.jsamd.2019.04.006.
106. Singh R, Janakiraman S, Khalifa M, et al (2020) A high thermally stable polyacrylonitrile (PAN)-based gel polymer electrolyte for rechargeable Mg-ion battery. *J Mater Sci Mater Electron*. https://doi.org/10.1007/s10854-020-04818-1.
107. Singh R, Janakiraman S, Agrawal A, et al (2020) An amorphous poly(vinylidene fluoride-co-hexafluoropropylene) based gel polymer electrolyte for magnesium ion battery. *J Electroanal Chem* 858:113788. https://doi.org/10.1016/j.jelechem.2019.113788.
108. Singh R, Janakiraman S, Khalifa M, et al (2019) An electroactive β-phase polyvinylidene fluoride as gel polymer electrolyte for magnesium-ion battery application. *J Electroanal Chem* 851. https://doi.org/10.1016/j.jelechem.2019.113417.

8 Electrochemical Mechanisms in Sodium-Ion Batteries

Madhushri Bhar, Udita Bhattacharjee, Shuvajit Ghosh, and Surendra K. Martha

8.1 INTRODUCTION TO SODIUM-ION BATTERIES (SIBs)

The early discovery of sodium-ion batteries (SIBs) began during the 1960s by Kummer and Yao with a new class of material having high Na^+ diffusivity in a non-stoichiometric compound of $5.3Al_2O_3$, Na_2O–$8.5Al_2O_3$, and Na_2O. It is a hexagonal lattice system with the spinel blocks separated by a mirror plane containing one oxygen and one vacancy, where Na-ion delocalizes [1]. In 1966, Ford Company manufactured a 2 V sodium-sulfur (Na-S) battery. In this system, a metallic Na anode was kept inside a β-Al_2O_3 tube, and the whole system was inserted into a separate tube containing liquid sulfur. However, the long-term stability of the dense β-Al_2O_3 phase at a high-temperature reactive condition (i.e., 300°C) was the major obstacle. Though the technology was targeted toward electric vehicle (EV) application, it was maneuvered to stationary grid storage later. At the same time, the ZEBRA battery (ZEolite Battery Research, Africa; 2.6 V) was discovered with β-Al_2O_3 and $NiCl_2$ instead of sulfur in the solid electrolyte. In 1968, Goodenough introduced a covalent network structured NASICON (Na superionic conductor) compound where Na-ion can delocalize via the interconnected cavities with the improvement in conductivity (10^{-3} S cm^{-1}) [2]. Until the mid-1980s, research remained focused on revealing the intercalation chemistry of alkali metal ions into transition metal compound hosts. In 1978, insertion/deinsertion of alkali metal ions (e.g., Li^+ and Na^+) into layered metal oxide phases like Li_xMO_2 (M = Co, Mn, Ni, Fe) was proposed. Delmas and co-workers during 1981 first showed reversible Na^+ intercalation into Na_xCoO_2 system ($0.5 \leq x \leq 1$) between 2 and 3.5 V [3]. Investigation on finding an efficient sodium-ion host was abreast with LIB research. However, the discovery of lithium intercalation phenomena into graphite (specific capacity = 372 mAh g^{-1} corresponding to LiC_6, working voltage < 0.1 V) yielded a paradigm shift in the battery field. LIBs stole all the limelight, while SIB research had hit the bottleneck due to the lack of anode material as sodium intercalation into graphite is not so thermodynamically favorable. LIB surpassed all other analogs, commercialized by SONY in 1991, and has monopolized the portable electronics market (i.e., cell phones, laptops, tablets, etc., along with power tools) for the last three decades. However, SIB research was revitalized during 2000, when Dahn and Stevens invented a disordered carbon material (e.g., hard carbon) [4] via the carbonization of glucose that showed the reversible capacity of 300 mAh g^{-1}. But poor cycling stability restricted its application as a SIB anode. During the past two decades, SIB research is gaining attention on various aspects of electrode architecture design, novel sodium-host framework construction, sodiation–desodiation mechanism elucidation, etc., especially by tuning the LIB electrode materials to improve the capacity, cycling stability, and C-rate performance.

8.2 THE MOTIVATION FOR SIB RESEARCH

It is forecasted by U.S. energy information administration that energy consumption all over the world would grow to 28% between 2015 and 2040. As per the governmental policies of several countries, ~125 million EVs are projected to be produced by various companies within 2030. However, the

DOI: 10.1201/9781003310167-8

uneven distribution of lithium resources in the earth's crust (lithium mines being localized in South America, Australia, Canada, China, etc.), geopolitical conflicts centered around lithium and transition metal mining and supply chain, and the elevation of consumer demand due to globalization and population increase may enhance the LIB manufacturing cost in the near future. Experts forecasted that Li might sustain at most 65 years considering an average growth of 5% in consumption per year [5]. Based on cost, abundant availability of raw materials, and reduction potential (–2.71 V vs SHE), rechargeable SIBs are one of the most promising future batteries. SIBs are having analogous battery design and electrochemical mechanisms to LIBs. However, a larger ionic radius 1.02 Å of Na^+ vs 0.76 Å of Li^+ and higher atomic mass (22.98 g mol^{-1} for Na vs 6.98 g mol^{-1} for Li) for sodium result in poor electrochemical properties and impede Na^+ diffusion inside the host. But, the thermodynamic non-feasibility of sodium-aluminum alloy formation at cell operating conditions brings in the privilege of utilizing Al current collector at both electrodes. Because of their low specific energy density in SIBs, they are preferred for stationary energy storage segments such as UPS, telecom, railroad signaling, switchgear applications, and solar and wind power generation systems.

8.3 OPERATING PRINCIPLE OF SIB

A Na-ion cell consists of a cathode, anode, nonaqueous electrolyte with sodium salt, and a separator placed between the electrodes. A schematic representation of the working principle of SIB is shown in Figure 8.1. Generally, insertion-type layered transition metal oxides (Na_xMO_2, $NaMO_2$), polyanionic type ($NaMPO_4$, $Na_3M_2(PO_4)_3$) materials are used as cathode and hard carbon as the anode [5]. The electrolyte is 1 M $NaPF_6$ or $NaClO_4$ salt dissolved in ester-based organic electrolyte like EC/PC/DEC. It should be highly ionic conductive but electronically nonconductive. During charging, Na^+ is extracted from a high-voltage cathode and inserted at the anode via the electrolyte medium. Opposite phenomena happen during discharging. The movement of electrons through the external circuit produces electricity. The electrochemical cell reaction at both electrodes is shown in Figure 8.1.

8.4 THERMODYNAMICS OF SODIUM-ION BATTERY

A secondary battery is an energy storage device where chemical energy is converted to electrical energy. The maximum voltage and the maximum electrical work can be achieved from a battery when it is operated at a minimum current in reversible condition. This maximum voltage is termed as 'Electrochemical potential' for the electrochemical cell reaction. The thermodynamics says the cell requires an equal amount of energy during the charging and discharging process to maintain reversibility at equilibrium. The reversible electrochemical reaction can be described by the Gibb-Helmholtz equation 8.1:

$$\Delta G = \Delta H - T\Delta S \tag{8.1}$$

where ΔG, i.e., Gibbs free energy, is the energy provided for useful work. ΔH is the enthalpy change for a reaction. ΔS is the entropy and TΔS represents the heat dissipation with the organization and disorganization of materials while T is the absolute temperature.

The maximum electrochemical work is equal to ΔG (equation 8.2):

$$\Delta G = -\, nFE^0{}_{cell} \tag{8.2}$$

where n is the number of electrons transferred during redox process and F is Faraday's constant (1 F = 96,500 coulombs) (equation 8.3).

And,
$$E^0 cell = E\ cathode - E\ anode \tag{8.3}$$

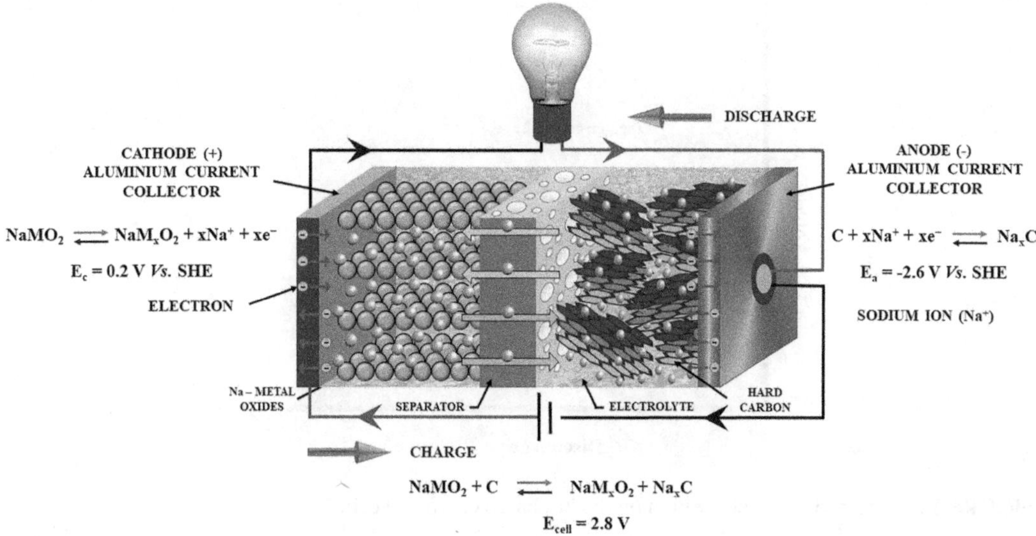

FIGURE 8.1 Pictorial representation of Na-ion cell.

The effect of polarization deviates the electrode potential from equilibrium. There are mainly three types of polarization: (i) activation polarization (or charge-transfer overvoltage is the driving force for an electrochemical reaction to occur at the electrode surface), (ii) concentration polarization (concentration difference arising due to mass transfer on the electrode surface and the bulk), and (iii) ohmic polarization (or internal resistance drop dissipated as waste heat) [6].

The available voltage can be expressed as follows:

$$E = E^0 - [(\eta_{ct})_a + (\eta_c)_a] - [(\eta_{ct})_c + (\eta_c)_c] - IR \tag{8.4}$$

where E^0 is the electromotive force or open-circuit voltage of the cell; $(\eta_{ct})_a$ and $(\eta_{ct})_c$ are activation polarization at anode and cathode, respectively; $(\eta_c)_a$ and $(\eta_c)_c$ are concentration polarization at anode and cathode, respectively; i is the operating current of the cell; and R is the internal resistance of the cell. The cell will operate close to open-circuit potential when the polarization effects and IR drop are negligible, and the maximum theoretically available energy can be obtained. The effect of cell polarization on a cell's discharge curve is represented in Figure 8.2.

8.4.1 Nernst Equation for Electrode Potential

The Nernst equation relates the electrochemical potential of a cell (E_{cell}) to the standard electrode potential ($E^0{}_{cell}$), temperature, and concentration of the chemical species participating in the redox reaction. It can be obtained from Gibb's free energy relation, which is represented in the following equation:

$$\Delta G = \Delta G^0 + RT\ \mathrm{lnk} \tag{8.5}$$

where T is the absolute temperature, R is the gas constant, and equilibrium constant

$K = [C]^C [D]^d / [A]^a [B]^b$ for the reaction: aA + bB = cC+Dd, and $\Delta G°$ is the standard Gibb's free energy.

Putting the value of ΔG from equation 8.3 in equation 8.5 and simplifying, the following equation is obtained:

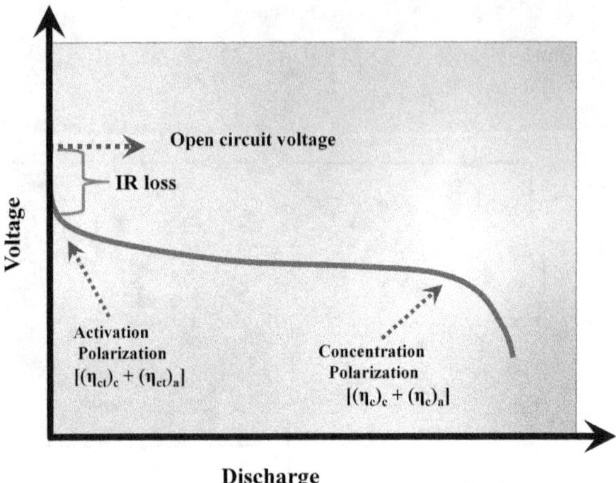

FIGURE 8.2 Effect of cell polarization on the discharge curve of a cell.

$$E_{cell} = E^0{}_{cell} + \left(\frac{RT}{nF}\right) lnK \tag{8.6}$$

This is known as the Nernst equation.

At 298 K, by putting the value of R, F, and T, the following equation becomes

$$E_{cell} = E^0{}_{cell} + \left(\frac{0.0592}{n}\right) lnK \tag{8.7}$$

8.5 CHARGE STORAGE BEHAVIOR IN SIBs

Future energy storage systems need to be a perfect blend of energy density and power density with long cycles and shelf life, especially for electric vehicles and portable electronic applications. It is challenging to design a high-performance SIB system by tailoring both energy density and power density that could be an intermediate between batteries and supercapacitors. Battery-type intercalation materials involve bulk-driven Faradaic redox processes leading to high energy density, whereas capacitor-type materials undergo non-Faradaic electrostatic charge storage at the surface inducing high power density [7]. Pseudocapacitive phenomena overlapping with both the extremes, i.e., battery and capacitor, leverage the advantages of both bulk and surface properties simultaneously as it brings Faradaic redox reactions to the surface. Pristine pseudocapacitive materials (e.g., RuO_2 and MnO_2 in the aqueous system) are scarce in nonaqueous systems. However, nanosizing or exposing covered surfaces may enforce pseudocapacitive nature in the charge storage mechanism [8]. The short diffusion path length facilitates Na-ion migration and electron transport in nanomaterials. Conway classified the Faradaic mechanisms in terms of capacitive charge storage property [9]. Ion (C^+) (e.g., Li^+) intercalation into the layers of electrode material (MA_y) (e.g., TiO_2, T-Nb_2O_5, and MoO_3 in LIB and SIB) and Faradaic charge-transfer reactions that occur without crystallographic phase change result in intercalation pseudocapacitance (equation 8.8):

$$MA_y + xC^+ + xe^- \rightleftharpoons C_xMA_y \tag{8.8}$$

Redox pseudocapacitance (equation 8.9) occurs when electrochemically adsorbed ions (C^+) from electrolyte at the electrode–electrolyte interface experience Faradaic charge-transfer processes in the oxidized species, i.e., RuO_2 and MnO_2, and in the reduced species, i.e., $Ru_{2-z}(OH)_z$ and $Mn_{2-z}(OH)_z$:

$$Ox + xCz+ + xe^- \rightleftarrows RedC_z \tag{8.9}$$

When an adsorption monolayer (e.g., H^+ and Pd^{2+}) is formed on a noble metal (M) (e.g., Pt and Au) surface above its redox potential, it is called underpotential deposition (equation 8.10). The schematic representation of different types of pseudocapacitance mechanisms is represented in Figure 8.3.

$$M + xC^{z-} + xze^- \rightleftarrows C.\ M_{ads} \tag{8.10}$$

Na^+ intercalation can be classified as either pseudocapacitive (surface phenomena) or diffusion-controlled (bulk phenomena) method from the total charge storage (Q_T), the total area under the CV curve at a specific scan rate (v) which is the summation of pseudocapacitive controlled charge (Q_C) and diffusion-controlled charge (Q_D), and can be represented as follows:

$$Q_T = Q_C + Q_D \tag{8.11}$$

The overall charge storage can be determined by using the Trasatti procedure [10–12]. CV curves at different scan rates are used to quantify the pseudocapacitive and diffusion-controlled contribution, according to the following equation:

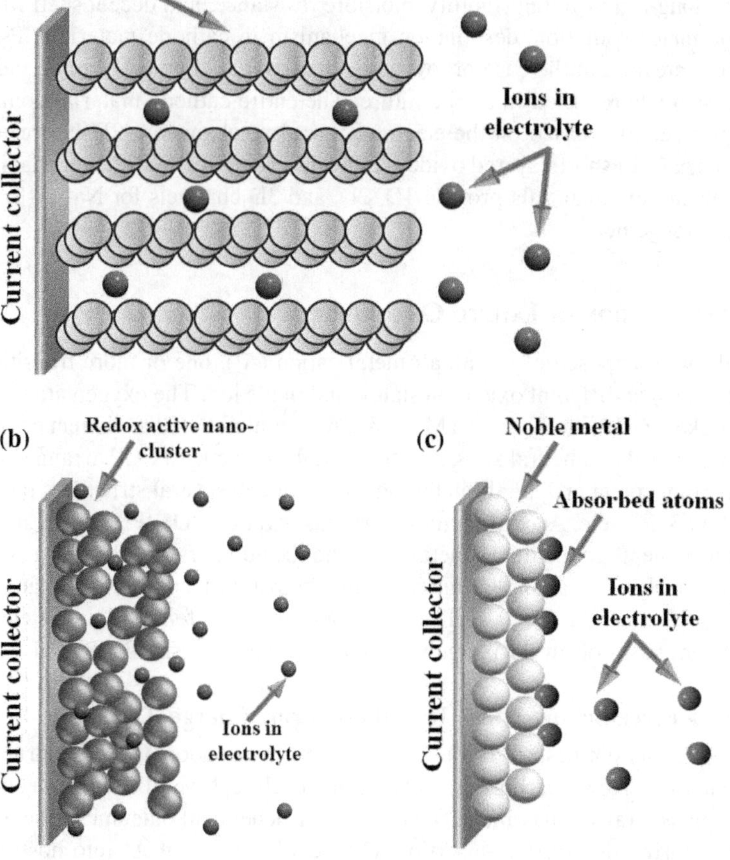

FIGURE 8.3 Types pseudocapacitance mechanisms: (a) Intercalation pseudocapacitance, (b) Redox pseudocapacitance, and (c) underpotential deposition.

$$I\ (v) = k_1 v + k_2 v^{1/2} I(v) = k_1 v + k_2 v^{1/2} \quad (8.12)$$

Equation (8.13) can be rearranged to

$$\frac{I(V)}{V^{1/2}} = K_1 v^{1/2} + K_2 \ldots \frac{I(v)}{v^{\frac{1}{2}}} = k_1 v^{1/2} + k_2 \quad (8.13)$$

where I(v) represents the total current response, which includes current arising from pseudocapacitive surface charge storage ($k_1 v$) and from diffusion-controlled ($k_2 v$) intercalation of Na^+. The capacitive component at different potentials at a specific scan rate can be obtained from the slope at the corresponding potential (equation 8.13).

The charge storage and kinetic behavior of nanomaterial via pseudocapacitive and diffusion-controlled Na-ion intercalation phenomena provide an interesting concept of designing high-performance electrode materials for SIBs. This type of system is highly desired where a huge amount of energy needs to be delivered quickly.

8.6 ELECTROCHEMICAL MECHANISMS FOR VARIOUS CATHODE MATERIALS OF SIBs

Cathode is the limiting factor in terms of overall electrochemical performances of the sodium-ion cell. It is difficult to achieve a suitable cathode material that is a perfect hybrid of good energy-power density balance, long-term cycling stability, moisture resistance, and decent shelf life. Intercalation mechanism dominates sodiation–desodiation mechanism in cathode materials. Pseudocapacitive contribution only prevails in the case of low active material loading and can be induced by nanosizing, but it is scarcely reported in the literature. The entire cathode material domain can be categorized into two sections based on the electrochemical mechanism – 2D-layered oxides and 3D polyanionic materials. In short, layered oxide sandwiches Na^+ ion among hexagonal interlayer pockets, whereas polyanionic materials provide 1D, 2D, and 3D channels for Na^+ diffusion depending upon its lattice arrangements.

8.6.1 Basic Description of Layered Oxide Structure

A_xMO_2 crystals are composed of the alkali metal cation (A^+), one or more transition metal (M^{n+}, where n > 1) cation(s) at different oxidation states, and oxide ion. The oxygen atom is distributed in a cubic close-packed (CCP) array. A^+ and M^{n+} ions are accommodated at distinct octahedral or tetrahedral or prismatic, and octahedral sites, respectively, formed due to CCP arrangement. Depending on the oxygen arrangement and alkali metal ion environment, several structures may arise like O3, P3, P2, O2, T1, O′3, P′3, etc. According to the Delmas notation, 'O', 'P', and 'T' designations refer to the local environment around alkali metal ions, and the numbers '3', '2', and '1' suggest different kinds of oxygen stacking in a single unit cell. Freshly synthesized Na_xMO_2 materials crystallize in either O3 (x = 0.9–1) or P2 ($x \sim 0.7$). T1-type structure occurs for $x = 2$ and is electrochemically irreversible. Other phases occur during galvanostatic cycling for a small range of x [13].

8.6.1.1 Driving Force behind Phase Transition during Charge–Discharge

Fresh A_xMO_2 materials synthesized through ceramic processes adopt O3 or P2 structure depending upon the amount of A. However, during galvanostatic cycling, several subphases occur as an intermediate during phase transition, which affects the electrochemical outcomes. The driving force for phase transition during deinsertion–insertion (charge–discharge) of A^+ into host lattice (A_xMO_2) lies in minimizing all sorts of interionic repulsions (repulsion between A-A, M-O, A-M, M-M, and O-O) to obtain thermodynamically most stable structure according to Pauling's third rule.

The basic structure of all phases is almost the same. The so-called 'transition' only changes the relative stacking sequence of $(MO_2)_n$ sheets and is achievable at room temperature by simple sheet gliding without breaking M-O bonds. This topotactic transformation falls under the category of the first-order phase transition. The stability of phases depends on the size of A, the iono-covalency of M-O bond, the type of transition metal atoms, the temperature of operation, etc. Lithium being the smallest among alkali metal groups in the periodic table is never stable in prismatic coordination due to stronger electrostatic repulsion at short distances and mostly observed in O3 phases. On the other hand, larger sodium is fairly stable at prismatic sites bringing in the possibility of complex phase transitions during cycling. As an example, Li_xCoO_2 undergoes O3-O1 phase transition (O3 at x = 1–0.25 and O1 at x ~ 0), whereas Na_xCoO_2 undergoes O3-P3-O1 phase transition (O3 at x = 1–0.75, P3 at x ~ 0.5, and O1 at x ~ 0) [14].

8.6.1.2 Phase Transitions Observed in SIB

When sodium ions are partially extracted from O3 structure, spacious prismatic sites are formed via gliding of MO_2 sheets one over another at an angle of 60° through $(1/3\,\vec{a} + 2/3\,\vec{b} + 0\,\vec{c})$ gliding vector and give rise to P3 symmetry [15]. Several other distorted structures are also characterized by O3-type oxides. O3-$NaNi_{0.5}Mn_{0.5}O_2$ undergoes O3-O′3-P3-P′3-P3″ transition, whereas O3-$NaFe_{0.5}Mn_{0.5}O_2$ shows O3-P3-OP2 transition [16,17]. On the other hand, desodiation from the P2 structure results in a unique O2 phase, where the stacking of oxygen atoms can be viewed as a hexagonal closed packed (HCP) array (AB-AC). P2-type oxides generally demonstrate O2-P2 transition, except in few cases like $Na_{0.66}Fe_{0.66}Mn_{0.33}O_2$, $Na_{0.66}Mn_{0.67}Ni_{0.26}Zn_{0.07}$, etc., where an additional OP4 phase occurs [18]. However, the transition from O3- or P3-type structure to O2 or P2 form is impossible as it is required to break and reform M-O bonds. The charge–discharge profile of SIB cathodes comprises multiple plateaus, sloping regions, and biphasic mixture, unlike LIB analogs [19].

8.6.1.3 Role of Phase Transition in Electrochemistry

The layered cathode electrochemistry of SIB is controlled by the phase transition. First, the coexistence of multiphase at a particular desodiated state causes kinetic limitations. The slowest diffusive phase in the mixture determines the sodium-ion deinsertion–insertion rate into the host lattice. Phase boundaries impede the motion of Na^+ that consequently slows down charge–discharge rates and increases polarization. Second, each phase possesses its lattice constants. High mechanical stress originating from the difference in lattice parameters (15% unit-cell volume change for O3-P3 transition in $NaNi_{0.5}Mn_{0.3}Ti_{0.2}O_2$) induces surface microcracks [20]. During repeated cycling, it grows in all directions and breaks the stable cathode–electrolyte interphase. As a result, phase transition becomes highly irreversible. Any irreversible changes in the microstructure during cycling are termed as 'Electrochemical creep' [21]. Prevention of creep is one of the main areas of improvement in SIB cathode research for long-term cycling. Exploited tactics include – (i) surface-coated ceramic (Al_2O_3, TiO_2, $NaPO_3$, $AlPO_4$, etc.) or fluoride materials (AlF_3, NaF, etc.) that cannot be directly incorporated into the pristine crystal lattice. They are unable to resist volume changes during cycling and subsequent formation of microcracks is inevitable. They act as a buffer to maintain the integrity of crystallites and inhibit the propagation of microcracks to a certain extent [22]. However, at a very high depth of desodiation, when crystal volume expansion surpasses the elasticity of coated materials, surface coating itself fractures, and cracks spread in all directions to expose new surfaces where the electrolyte decomposes and loses sodium inventory. (ii) Dopants can directly tune the crystal structure to discourage phase alteration and volume changes resulting from it. Electrochemically inactive main group metals (Al^{+3}, Mg^{+2}, Ca^{+2}, Sn^{+4}, etc.) or d^0 transition metals (Ti^{+4}, Zn^{+2}, Cu^{+2}, Nb^{+5}, etc.) do not change their oxidation state during cycling and increase the robustness of lattice. If the substituted atom can form a more ionic M-O bond, then phase transition can be restricted to some extent as O3-structure is favored by more ionic lattice and more covalent lattice has a propensity to adopt P3-type phase. If the substitute can form a shorter M-O bond, it decreases the height of the MO_2 slab, eventually increasing interlayer spacing to favor a higher sodium-ion diffusion rate. Redox inactive substitution also breaks the metal–metal chain in the transition metal layer to induce electron localization. In this way, transition metal dissolution by electrolyte can be blocked in cases of Jahn–Teller active centers [23]. (iii) Restriction of upper charge cutoff voltage helps in avoiding high depth of desodiation and quick capacity fade. The higher the desodiation, the greater the volume change.

8.6.1.4 Practical Performance Comparison of O3 and P2

From theoretical perspectives, P2-type compounds seem much more promising due to less complex phase transition. They have higher capacity retention due to less critical (O2-P2) phase transition and better C-rate performance due to low Na^+ diffusion barrier than O3-type. This holds somewhat true for half cells, especially in cases of P2-$Na_xMn_yFe_{1-y}O_2$, where a higher reversible capacity of >160 mAh g^{-1} is reported to achieve [24,25]. However, in full cell configuration against hard carbon anode, O3-type materials unexpectedly surpass P2-type in terms of power capability (80% capacity retention for P2-$Na_{0.67}Cu_{0.14}Fe_{0.2}Mn_{0.66}$ vs 90% retention for O3-$Na_{0.8}Cu_{0.2}Fe_{0.4}Mn_{0.4}O_2$ at 1 C rate), energy density (33% reduction in energy density in going from half-cell to full cell for P2 vs 20% in O3), and capacity retention. The non-stoichiometric sodium content of P2-type materials is the main reason behind the performance shortcomings in the full cell. Way outs to overcome such issues reported in the literature, such as presodiation and addition of sacrificial salts (Na_2CO_3, NaN_3, etc.), are unscalable at the industry level. Optimization of the cathode to anode mass loading ratio is another challenge for high-energy cells.

8.6.2 Polyanionic Compounds

Polyanionic class includes varieties of crystal structures like sulfates [$Na_2Fe(SO_4)_3$, $Na_2M(SO_4){\cdot}4H_2O$; M = Mg, Fe, Co, Ni], fluorosulfates [$NaFeSO_4F$], phosphates [$NaMPO_4$], fluorophosphates [Na_2MPO_4F; M = Fe, Mn], carbon phosphates [$Na_xM(PO_4)(CO_3)$; M = Fe, Mn], pyrophosphates [$Na_2MP_2O_7$; M = Fe, Mn], mixed anionic [$Na_4M_3(PO_4)_2(P_2O_7)$; M = Fe, Mn, Co, Ni, $NaFe_2(PO_4)(SO_4)_2$], NASICON (Sodium-ion Superionic Conductor) [$Na_xM_2(PO_4)_3$; M = V, Ti; $0 \leq x \leq 3$), etc. MO_6 octahedral and XO_4 (X = P, S, etc.) tetrahedral share corner or edge to form extended M-O-X linkages in three dimensions. More covalent X-O linkage enriches ionic character in M-O bond, which favors $M^{n+}/M^{(n+1)+}$ redox shuttling at lower energy, increasing the operating voltage from 2.5–3.1 V for Na_xMO_2 to 3.3–4.0 V for polyanionic class. Fluorine substitution in the place of oxygen (M-F-X linkage) can increase the voltage further by ~ (250–500) mV like in LIBs. These polyanionic compounds provide multidimensional channels (1D channel in $NaFePO_4$, 2D in $Na_3V_2(PO_4)_2F_3$, and 3D in $Na_3V_2(PO_4)_3$) for sodium diffusion unlike sandwiching phenomena in layered oxides. Therefore, it is robust, accommodates sodium ion reversibly with negligible volume changes (1%–2% in polyanionic vs 15%–25% in layered sodium TM oxides) during charge–discharge (i.e., better long-term cyclability), enables faster sodium-ion migration (i.e., better C-rate performance), and is associated with almost no solid-state phase evolution during cycling. However, most of the polyanionic materials suffer from low electronic conductivity: ~10^{-9} S cm^{-1} for polyanionic compounds vs ~10^{-5} S cm^{-1} for metal oxide and low density resulting in low gravimetric capacity, and complex routes of synthesis. Incorporation of polyanionic materials into carbon matrix enhances electronic conductivity, lowers charge–discharge polarization, and opposes particle agglomeration during high-temperature calcination. For example, 86.3 mAh g^{-1} discharge capacity at 5 C rate was obtained for $Na_3V_2(PO_4)_3$/graphene composite which is doubled compared to bare $Na_3V_2(PO_4)_3$ material. 3D porous nano-architectures facilitate fast electron transport, ease electrolyte penetration, decrease solid-state sodium diffusion path length, and suppress volume change. For instance, 86 mAh g^{-1} discharge capacity at 100 C rate with 64% capacity retention after 10,000 cycles was obtained for $Na_3V_2(PO_4)_3$ particles protected by amorphous carbon layer and rGO nanosheets. On the other hand, elemental doping (Al^{+3}, Ca^{+2}, Ni^{+2}, Ti^{+4}, Ce^{+4}, Mo^{+6}, etc.) modifies intrinsic ionic and electronic conductivity [26].

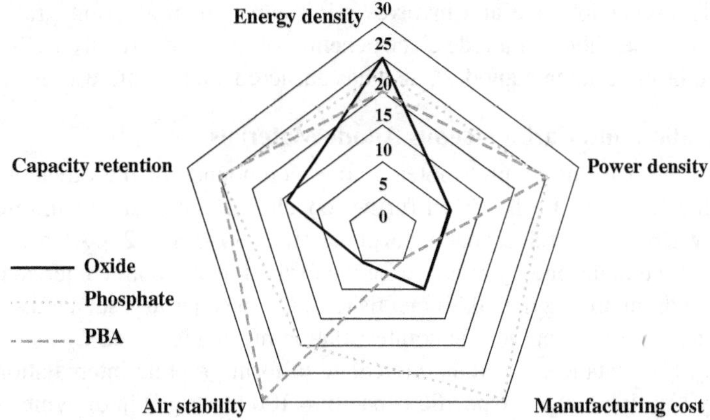

FIGURE 8.4 A comparison radar plot showing practical properties of various SIB cathode types.

8.6.3 Prussian Blue Analogs (PBAs)

Prussian blue analogs (PBAs) [$Na_xMFe(CN)_6$; M = Fe, Ni, Mn, etc.] are another unique type of metal-organic framework (MOF) where octahedral $[Fe(CN)_6]^{4-}$ polymer is crosslinked with Na^+ ions that are attached to a cyanide ligand. This 3D skeleton possesses large interstitial sites (4.6 Å) and open channels (3.2 Å). Weak sodium-cyanide linkage facilitates faster solid-state sodium diffusion. High-quality crystalline materials are prepared by a simple slow-rate precipitation method in an acidic medium. Electrochemical performances depend on phase purity, crystallinity defect, crystalline water content, carbon coating, etc. The average discharge capacity of these materials ranges from 120 to 150 mAh g^{-1} at < 100 mA g^{-1} current density including 3.1–3.5 V nominal voltage with 80%–90% capacity retention after 200 cycles. However, similar kinds of drawbacks like polyanionic compounds along with toxic HCN release at high temperature (>200°C) are still an issue [27]. The radar plot of SIB cathodes is presented in Figure 8.4.

8.7 ELECTROCHEMICAL MECHANISM OF ANODES FOR SODIUM-ION BATTERIES

One of the prominent features of anodes that govern their performance for any electrochemical system is their redox potential of the primary electrochemical reaction. In addition to this, the anodes for sodium-ion batteries are required to possess other essential properties like high sodium storage capacity. Thus, the electrochemical performance of the SIBs essentially depends on the electrochemistry and ion storage mechanism of the anodes. Though sodium metal has a low redox potential (-2.71 V vs SHE) and enormous theoretical capacity (~1165.2 mAh g^{-1}) feasible for being an anode, the intrinsic feature of dendrite formation on electrochemical charge–discharge cycling results in a short circuit and hence limits its potential usage as the anode material. Thus, anode materials for SIBs are chosen while striking a balance between the properties like practically achievable electrochemical capacity, low redox potential, long-term cyclability, safety, etc. A wide range of anode materials have been explored, and they can be broadly classified into three types based on the electrochemical mechanism that these materials follow: intercalation-based, alloy-based, and conversion-based anode materials. These three mechanisms are discussed briefly in the following sub-sections.

8.7.1 Intercalation Mechanism

The intercalation mechanism involves a host matrix that provides the space for sodium ions to intercalate and deintercalate during the electrochemical charge–discharge process. The process of intercalation is a reversible one and involves minimum structural disintegration. Intercalation mechanism, which is the choice of anode electrochemistry for commercially well-established LIBs (e.g., graphite and lithium titanate anodes), also has garnered a lot of interest for SIBs.

8.7.1.1 Intercalation into Carbonaceous Anode Materials

Unlike lithium, sodium forms graphite intercalation compounds of the order of NaC_{64} (specific capacity 35 mAh g^{-1}) [28] and it has been further concluded that sodium can intercalate into the matrix with ease where the interlayer spacing is greater than 3.7–3.8 Å [29]. Also, the theoretical calculations have proved that the energy required for the formation of graphite intercalation compounds is the highest for sodium among the alkali metals [30]. On the contrary, solvated sodium molecules are found to intercalate into graphite via cointercalation mechanism, i.e., $C_n + e^- + A^+ + y$ solvent $\leftrightarrow A^+(\text{solvent})_yC_n$. For instance, Na^+ ions intercalate to form graphite intercalation compounds of the stoichiometry $Na(\text{diglyme})_2C_{20}$ (specific capacity ≈ 100 mAh g^{-1}) in diglyme ether-based electrolyte [31]. Thus, hard carbons being the non-graphitized carbons with the disordered and random assembly of folded graphenic sheets were presented to be an attractive choice for SIB anode material. Also, the prospective of the utility of hard carbons was further validated by the observation of Herold and co-workers during 1970 that sodium vapor gets trapped into non-graphitizable carbon by adsorption and insertion [32]. However, the electrochemical property of hard carbon as anode for SIB is dependent on a lot of factors like the microstructure of the precursor used, calcination temperature implying the disorderness of the carbon formed, the interlayer spacing, defects, presence of surface-active groups, micro- and nanoporosity, etc. Also, there is a largely irreversible capacity loss associated with the hard carbon anodes after the initial charge–discharge cycling due to electrolyte decomposition, the formation of solid electrolyte interface (SEI), and other irreversible surface reactions. Another concern for the hard carbon anodes is that the maximum proportion of the capacity is achieved at low potentials near Na plating potential, and thus, Na dendrites may be formed upon overcharging which may result in short-circuiting of the battery [33]. The mechanism initially proposed for sodium or lithium insertion in hard carbons by Stevens and Dahn [28] described the involvement of two steps: Initially, the intercalation of ions into the graphenic sheets in the higher potential region (>0.1 V) depicted by the sloping potential curve, followed by the insertion into the micropores at lower potential observed as a low potential plateau in the potential curve (Figure 8.5a) [28]. But with the help of theoretical and experimental studies carried out recently, researchers have been able to come up with some other models indicating a probable mechanism that involves the following processes (Figure 8.5b) [34]:

i. Adsorption/chemisorption of sodium ions onto the defect sites: The initial sloping potential capacity at a potential higher than 0.1 V is attributed to the reversible adsorption of sodium ions onto the defects and edges of the graphenic sheets.
ii. Intercalation of sodium ions into the pseudo-graphitic sites: The plateau region in the potential below 0.1 V is attributed to the reversible intercalation of sodium ions into the randomly assembled folded graphenic sheets forming NaC_x.
iii. Filling of the micro/mesoporous regions by sodium ions: The pores formed in between the pseudo-graphitic regions are filled by the deposition of sodium at a lower potential near the reduction potential of sodium, and hence, clusters of sodium are formed which may increase the interlayer distances and in turn give rise to more interstitial space for the accommodation of sodium ions when the electrochemical process is repeated.

The sequence in which this process occurs is usually the one mentioned above but, in some studies, it has been observed that the adsorption/chemisorption process and the interlamellar intercalation processes take place simultaneously [35]. The effect and contribution of individual processes to the practical achievable capacity are dependent on the properties mentioned above. Thus, to achieve the desired capacity with minimum irreversible capacity loss, the properties of the hard

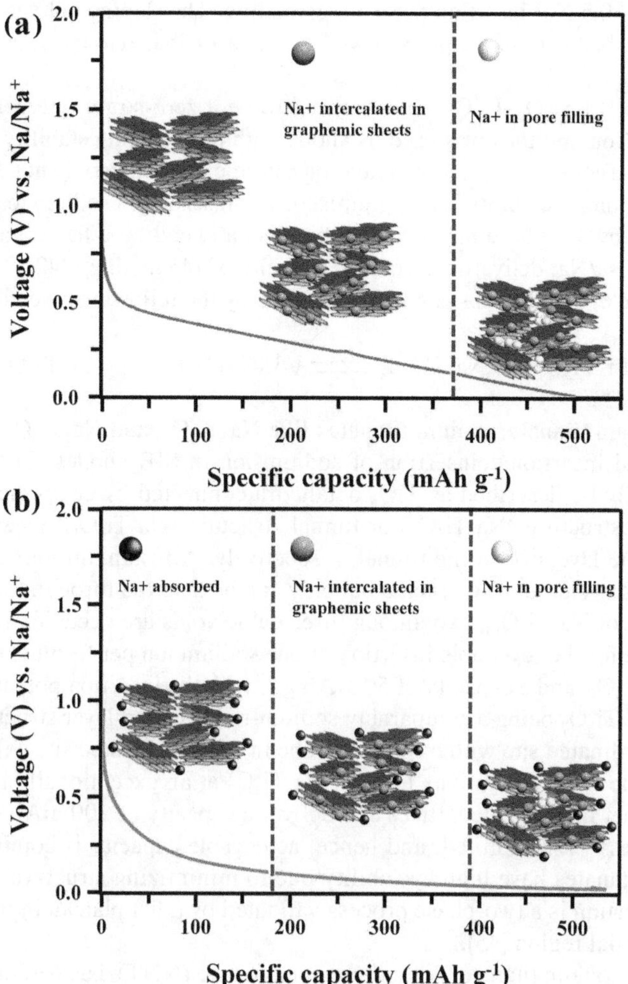

FIGURE 8.5 Schematic mechanism of the Na-ion insertion in hard carbon: (a) according to Stevens and Dahn and (b) according to the latter experimental and theoretical studies.

carbon mentioned above need to be tuned by choosing the appropriate calcination temperature (700°C–2000°C have been reported to produce optimum performing hard carbons), precursor (e.g., sucrose, pitch, and biowastes), and preparatory treatment methods (e.g., chemical treatments using KOH, H_3PO_4, and $ZnCl_2$) [36–39].

8.7.1.2 Intercalation-Based Transition Metal Oxides

Various titanium-based oxides that intercalate sodium ions have been studied as an anode for SIBs. The anatase form of TiO_2 is found to be electrochemically active for Li insertion anodes, but in the case of sodium insertion, only the nanocrystalline forms of anatase TiO_2 are found to be effective. First, the mechanism of sodium-ion intercalation in anatase TiO_2 involves the irreversible formation of an electrochemically active O3-type rhombohedral layered Na_xTiO_2 in which no long-range order exists due to the cation mixing between Na and Ti slabs at ~0.2 V. Upon deintercalation, a disordered three-dimensional structure is formed with short-range anatase like symmetry. The reversible intercalation–deintercalation capacity is derived from this phase transformation process with a sloping voltage profile having an average voltage of 0.8 V and a capacity of 165 mAh g^{-1} was observed owing to ~0.5 Na^+ insertion per TiO_2 crystallite. The sloping voltage profile is indicative of the solid-solution behavior of the process which can be attributed to the amorphous nature of the phases involved [40,41].

Lithium titanate ($Li_4Ti_5O_{12}$-LTO) was observed to be a zero-strain intercalation compound in the case of lithium ion and thus was used as anode with high cycling stability (>3000 cycles). But sodium ion being larger (1.02 Å), its intercalation into the LTO matrix is not a strain less mechanism, and thus, the long cyclability is compromised. The insertion of sodium ions into the LTO host matrix takes place by a 3-phase mechanism which is validated by a flat voltage profile around a potential of 1 V vs Na/Na^+ delivering a specific capacity of 145 mAh g^{-1} [42]. The process of insertion and deinsertion of sodium ions can be represented by the following reversible reaction:

$$Li_4Ti_5O_{12} + 3\,Na^+ + 3\,e^- \rightleftharpoons \tfrac{1}{2}\,LiNa_6Ti_5O_{12} + \tfrac{1}{2}\,Li_7Ti_5O_{12} \quad (8.14)$$

In addition to lithium titanate, sodium titanates like $Na_2Ti_6O_{13}$ and $Na_2Ti_3O_7$ have been used for the electrochemical insertion/deinsertion of sodium ion in SIB anodes. The structure of these titanates can usually be described as TiO_6 octahedra connected by edge- and corner-sharing to give stepped layer structure ($Na_2Ti_3O_7$) or tunnel structure ($Na_2Ti_6O_{13}$) in which sodium ion is located between the layers or in the tunnel, respectively. All titanium ions are already in a +4 state. So, sodium ions cannot be extracted but only can be inserted through the conversion of Ti^{4+} to Ti^{3+}. In the case of $Na_2Ti_6O_{13}$, two among three cubic voids are occupied by sodium ions, and thus, there is space for the reversible insertion of one sodium ion per formula unit resulting in the formation of $Na_3Ti_6O_{13}$ and a capacity of 50 mAh g^{-1}, with the insertion potential being 0.9 V. On the other hand, $Na_2Ti_3O_7$ being a comparably sodium-rich stepped layer structure constitutes one sodium in a 9-coordinated site with oxygen and one in a 7-coordinated site with oxygen. This can additionally take up two sodium ions to form $Na_4Ti_3O_7$ at an exceptionally low potential of 0.3 V compared to other metal oxide lattices and deliver a capacity of 200 mAh g^{-1} [43,44]. Though the insertion process is site limited, and hence, achievable capacity is comparatively low, both of these sodium titanates have high cyclability due to minimizing structural stress and also the mechanism of insertion is a two-phase process validated by a flat plateau in the voltage profile at the insertion potential region [45].

Another type of sodium titanate like sodium nonatitanate (NNT), i.e., $NaTi_3O_6(OH)\cdot 2H_2O$, in its both hydrated and anhydrous form is found to be electrochemically active toward sodium insertion. It consists of interconnected $Ti_4O_{16}^{4-}$ octahedra with hydroxyl groups occupying the steps in the stacking faulted layered structure in which the water molecules occupy the interlayer spaces and the sodium ions occupy the octahedral voids. The anhydrous form is more stable structurally compared to the hydrated form due to the linkage formed and hence delivers a capacity of 125 mAh g^{-1} with a sloping voltage profile having an average voltage of 0.5 V for 20 cycles. The sloping insertion potential for both hydrated and anhydrous form indicates solid-solution behavior for the intercalation of sodium ions and is not site limited, which opens up the prospects for titanates enabling high sodium-ion intercalation capacity if the electrode architecture is improved [45,46].

Inspired by the sodium insertion in layered titanates, Lepidocrocite structure-based titanates (general formula: $A_x[Ti_{2-y}M_y]O_4.zH_2O$, where A is K, Rb, or Cs and M is Mg, Co, Ni, Cu, Zn, Mn, Fe, Li, or a vacancy, where M is placed into the transition metal layers and A is placed between the layers) have also been proposed as SIB anodes. The voltage profile for the intercalation mechanism shows sloping nature similar to sodium nonatitanate due to the solid-solution behavior or multiphase mechanism and the cyclability of these compounds is dependent on factors like voltage cut-offs, compositional variables, etc. [46].

The main issue associated with titanates and titanium dioxide is the low conductivity (~10^{-13} S cm^{-1} for LTO) which is tackled using strategies like nanostructuring, composite formation with carbon or carbon coatings [11].

8.7.2 Alloying Mechanism

The p-block elements for high sodium-rich alloys have a high theoretical capacity of sodium insertion compared to intercalation type anodes mentioned above. For instance, some of the alloys formed along with the deliverable capacities 1108 mAh g^{-1} (Na_3Ge), 847 mAh g^{-1} ($Na_{15}Pb_4$), 2596 mAh g^{-1} (Na_3P), and 660 mAh g^{-1} (NaSb), respectively. The general electrochemical reaction followed by anode alloying mechanism can be represented as follows [47]:

$$M + y\,Na^+ + ze^- \rightleftharpoons Na_yM \tag{8.15}$$

Sn shows a sloping voltage profile, indicating solid-solution topotactic insertion of Na^+ ions, whereas Ge, Sb, and P show a voltage plateau indicating a two-phase insertion mechanism. The main disadvantage associated with alloy type anodes is the high volumetric expansion after charge–discharge cycling (520%, 126%, 300%, and 390% are reported for Sn, Ge, P, and Sb electrodes, respectively) resulting in pulverization. Various strategies, such as modifying the electrode architecture, i.e., nanostructuring, 3D electrode architecture, composite formation with carbon, and choosing a potential window that can alleviate any particular alloying step resulting in high volume expansion, have been applied to utilize the advantage of high theoretical capacity [47].

8.7.2.1 Intermetallic Alloy-Based Anodes

Intermetallic anodes (M-Sn/Si/Sb), where M is another metal, include two types of compounds. One in which M is electrochemically active (represented in equation 8.16) and the other in which M is electrochemically inactive (represented in equation 8.17). The main concept behind the introduction of intermetallic compounds as an anode is that the two alloy forming elements can act as mutual buffers and account for the volume expansion during electrochemical charge–discharge. The first category includes SnSb, SnGe, Zn_4Sb_3, and Sn-Bi-Sb, which have been studied in their nanostructure forms and also in composites with carbon as anode for SIBs. The second category includes Cu_2Sb, NiSb, and Ni_3Sn_2, which were considered because of the presence of an inactive metallic phase to accommodate the volume change [48].

$$M\text{-}A + (x+y)\,Na^+ + ze^- \rightleftharpoons Na_xM + Na_yA \tag{8.16}$$

$$M\text{-}A + y\,Na^+ + ze^- \rightleftharpoons M + Na_yA \tag{8.17}$$

As mentioned earlier, phosphorous in the form of red phosphorous allotrope and its composites are studied as an alloy anode for SIBs, but another allotrope of phosphorous i.e., black phosphorous, also undergoes sodium-ion insertion in a slightly different mechanism involving two steps. First, Na^+ ions are intercalated into the phosphorene layers of black phosphorous along the x-axis channel to give $Na_{0.17}P$, and second, the alloy Na_3P is formed at a voltage below 0.54 V responsible for the maximum share of total capacity delivered. But the application of phosphorous-containing anode is limited due to the formation of toxic PH_3 as a result of hydrolysis during fabrication and large volume expansion [33,48,49].

8.7.3 Conversion Mechanism

The anodes, which follow this mechanism, undergo the electrochemical conversion of the active material completely to form a different compound during the charge–discharge process. The process involves multiple electron transfer. The anodes undergoing conversion following this mechanism can be represented as follows:

$$M_yX_z + nz\,Na^+ + (nz)\,e^- \rightleftharpoons yM + zNa_nX \tag{8.18}$$

where M is the transition/ p-block metal and X is O, S, P, etc. [47].

For metal oxides, there are two types of conversion reaction mechanisms that are followed: The first type in which the M is electrochemically inactive (e.g., Fe, Ni, and Mo oxides), and the capacity is achieved by the formation of metallic clusters encapsulated by Na_2O matrix as per equation 8.18. The second type in which M is electrochemically active (e.g., Sn and Sb oxides) and an alloying reaction takes place between Na and the metal M formed due to the conversion reaction in the first step, thus delivering an additional capacity.

In the case of layered metal sulfides like MoS_2, the sodium ion gets intercalated into Na_xMS_a and finally breaks down to Na_2S and M on further Na^+ insertion following conversion mechanism. Also, if the transition /p-block metal is electrochemically active toward sodium, like in the case of SnS_2, a further alloy formation of Na with Sn takes place after conversion.

In the case of transition/p-block metal-based phosphides, mechanisms similar to oxides and sulfides are followed based on the electrochemical activity between Na and the metal involved. Sodium phosphide is formed during the conversion reaction. The conversion-based oxides and sulfides have higher average voltage (≥1 V) and suffer volume expansion due to the conversion of sulfides and oxides, which overshadows the advantages of high gravimetric capacity as SIB anodes. The comparison of various conversion type anodes is shown in Figure 8.6 [33,48,49].

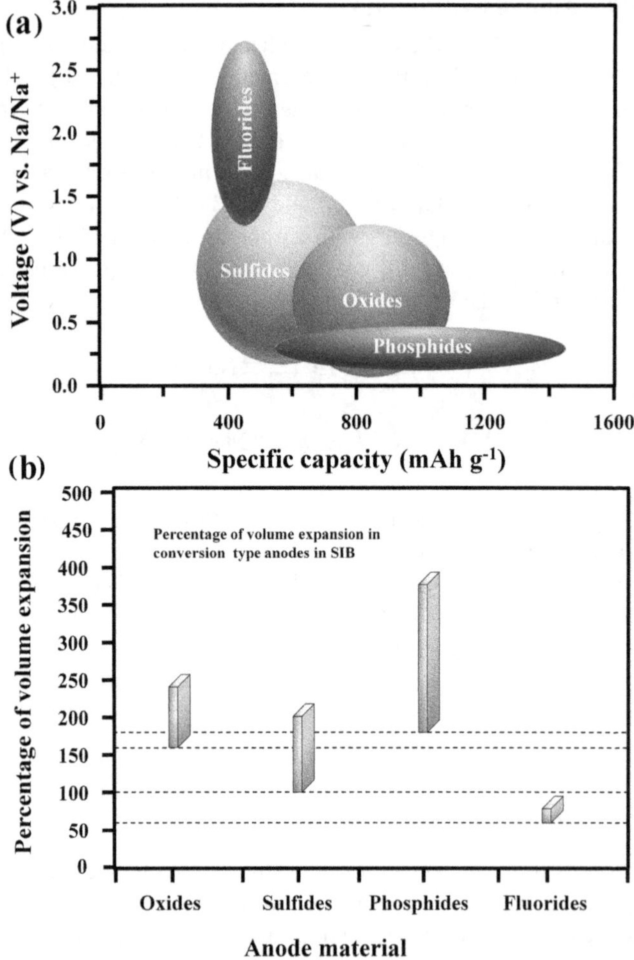

FIGURE 8.6 Comparison charts showing prominent types of conversion anode materials based on (a) specific capacity vs average voltage and (b) volume expansion on intercalation–deintercalation.

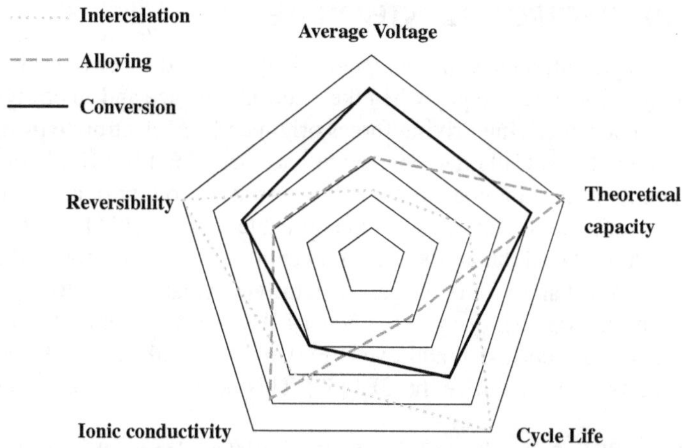

FIGURE 8.7 A radar plot comparing the properties of mechanistically different SIB anodes.

The anodes for SIBs based on the mechanisms briefed above have their own sets of advantages and disadvantages for practical and large-scale applicability. A radar chart showing the comparison in their characteristics is given in Figure 8.7.

8.8 MECHANISM OF SODIUM-ION CONDUCTION THROUGH THE ELECTROLYTE

Electrolytes act as the charge carrier between the electrodes in electrochemical energy storage systems. The main characteristics required for SIB electrolytes are high ionic conductivity, low viscosity, high electrode–electrolyte interface stability, broad operational voltage range, low flammability, etc. The conventional organic liquid-based electrolytes used for SIBs are sodium salts like $NaClO_4$, $NaPF_6$, and NaTFSI in a mixture of carbonate solvents like ethylene carbonate (EC), propylene carbonate (PC), dimethyl carbonate (DMC), diethyl carbonate (DEC), ethyl methyl carbonate (EMC), etc. The mechanism of conduction in these electrolytes is through the transfer of Na^+ ions to the electrode storage sites by the desolvation from the electrolyte. Another mechanism observed in the organic liquid-based electrolytes is the cointercalation mechanism in which the ions are not completely desolvated while insertion into the electrodes and the energy barrier for diffusion is also found to be less for graphite electrodes. The conductivity depends on the coordination of solvation. The reactivity of carbonate solvents at lower anode potentials is one of the prominent disadvantages, and thus, alternatives like ionic liquids (ILs), polymer-based electrolytes, and solid ceramic electrolytes have also been studied. ILs like Na-bis(fluorosulfonyl)amide and Na-[N-methyl-N-propyl pyrrolidinium] [bis(fluorosulfonyl)amide] were observed to follow similar mechanism of Na^+ conduction like organic liquid-based electrolytes. Though the issue of the electrolyte reactivity at lower anode potentials and irreversible trapping of Na^+ is solved, the lower ionic conductivity and high viscosity at room temperature are a major concern for ILs. The organic polymer-based electrolytes (e.g., PVdF, PAN, and PMMA) are of two types: solid polymer electrolytes (SPEs) and gel polymer electrolytes (GPEs). The ion conduction mechanism in the case of GPEs (Na^+ salts liquid solvent filled polymer) is similar to the organic liquid electrolytes and that of SPEs (Na^+ salts dissolved in polymer matrix) is segmental motion of the polymer chain and mobile ion conduction within the polymer matrix. In the case of solid ceramic electrolytes (e.g., Na_3PS_4 and NASICON), the conduction of Na^+ occurs through the vacant/hopping sites and thus depends on the number of mobile ions or vacancies and the hopping energy barrier [50,51].

8.9 ELECTRODE–ELECTROLYTE INTERPHASE

The electrode–electrolyte interphase in nonaqueous batteries is the newly derived substance that bridges the thermodynamic potential gap when the electrode surface and electrolyte come into contact with each other. The underlying driving forces originate from electrolyte instability at the high cathode or low anode voltage. Cathode–electrolyte interface (CEI) in SIBs is much less explored like in LIBs, despite its important contributions in controlling sodium-ion kinetics and hindering transition metal dissolution at the interphase. One probable reason could be difficulty in characterizing nano-thick (5–20 nm) CEI due to the electron beam sensitivity. Although the reducing power of sodium is much weaker than Li (300 mV gap), electrolyte oxidation in SIBs is much more prominent since cathode interphase chemistry at 4.2 V for SIBs is equivalent to 4.5 V in LIBs. The literature reports suggest that fluorinated additive (FEC) and artificial CEI produced by thin ceramic (Al_2O_3) coating can effectively stabilize the CEI [52]. This field of study has many opportunities in the future.

Similar to LIBs, a SEI is also formed during the initial cycles on the anode of SIB as a thin film of a few 10 s of nanometers. This forms as a result of the decomposition of carbonate solvents followed by the reaction with sodium around ~0.25–0.75 V. It is composed of sodium carbonates, semi-carbonates, polymers, and other organic and inorganic sodium salts. This film allows the passage of ions through the structural defects in it [53]. Compared to SEI formed in LIBs, the one formed in SIB is thicker, more homogenous, and is composed of more inorganic salts than organic ones [54]. Though the formation of this layer results in irreversible capacity, this also acts as a protective layer from further anode material dissolution. Modification in electrolyte solvent composition (additives like a fluorinated solvent, vinylene carbonate) helps in optimizing the properties of SEI [51].

8.10 CONCLUSIONS

The research and development on SIBs are directed to produce high energy density (high working voltage and high capacity). The current SIBs have a moderate energy density of 90 Wh kg^{-1} and stable over ~1000 cycles with >95% efficiency. Oxide and polyanionic cathodes having capacities in the range of 120–190 mAh g^{-1} and hard carbon anodes of 200–250 mAh g^{-1} have been achieved due to the feasibility of Na^+ intercalation and deintercalation process. However, the intrinsic challenges of electrode materials like structural deformation, chemical reactivity with electrolyte, slow Na^+ kinetics, and low electronic conductivity limit the electrochemical performance of SIBs. These problems can be mitigated via nanomaterial synthesis: surface coating and doping by eliminating undesirable side reactions and phase transitions. The ongoing progress in SIB research is expected to serve modern civilization as an affordable complementary device in the near future.

ACKNOWLEDGMENTS

MB acknowledges DST-INSPIRE (code: IF180708), UB acknowledges DST-SERB (Sanction Order: CRG/2018/003543), and SG acknowledges CSIR (letter no. 09/1001(0067)/2019-EMR-I), Govt. of India for fellowships. SKM acknowledges DST-SERB (Sanction Order: CRG/2018/003543), Govt. of India and IIT Hyderabad for financial support to this work.

REFERENCES

1. Delmas C (2018) Sodium and sodium-ion batteries: 50 years of research. *Advanced Energy Materials* 8:1703137. https://doi.org/10.1002/aenm.201703137.
2. Goodenough JB, Hong HY- P, Kafalas JA (1976) Fast Na+-ion transport in skeleton structures. *Materials Research Bulletin* 11:203–220. https://doi.org/10.1016/0025-5408 (76)90077-5.

3. Delmas C, Braconnier J-J, Fouassier C, Hagenmuller P (1981) Electrochemical intercalation of sodium in $NaxCoO_2$ bronzes. *Solid State Ionics* 3-4:165–169. https://doi.org/10.1016/0167-2738 (81)90076-X.
4. Stevens DA, Dahn JR (2000) High capacity anode materials for rechargeable sodium-ion batteries. *Journal of the Electrochemical Society* 147:1271. https://doi.org/10.1149/1.1393348.
5. Yabuuchi N, Kubota K, Dahbi M, Komaba S (2014) Research development on sodium-ion batteries. *Chemical Reviews* 114:11636–11682. https://doi.org/10.1021/cr500192f.
6. Reddy T, Linden D (2010) *Linden's Handbook of Batteries* (4th Ed.). McGraw-Hill Professional Publishing, New York, NY
7. Augustyn V, Simon P, Dunn B (2014) Pseudocapacitive oxide materials for high-rate electrochemical energy storage. *Energy & Environmental Science* 7:1597–1614. https://doi.org/10.1039/C3EE44164D.
8. Trasatti S, Buzzanca G (1971) Ruthenium dioxide: A new interesting electrode material. Solid state structure and electrochemical behaviour. *Journal of Electroanalytical Chemistry and Interfacial Electrochemistry* 29:A1–A5. https://doi.org/10.1016/S0022-0728(71)80111-0.
9. Conway BE (2013) *Electrochemical Supercapacitors: Scientific Fundamentals and Technological Applications.* Springer Science & Business Media ISBN: 1475730586, 9781475730586, pages 698.
10. Ardizzone S, Fregonara G, Trasatti S (1990) "Inner" and "outer" active surface of RuO_2 electrodes. *Electrochimica Acta* 35:263–267.
11. Ghosh S, Kumar VK, Kumar SK, Martha SK (2019) An insight of sodium-ion storage, diffusivity into TiO_2 nanoparticles and practical realization to sodium-ion full cell. *Electrochimica Acta* 316:69–78. https://doi.org/10.1016/j.electacta.2019.05.109.
12. Narsimulu D, Ghosh S, Bhar M, Martha SK (2019) Electrochemical studies on kinetics and diffusion of Li-ions in MnO_2 electrodes. *Journal of the Electrochemical Society* 166:A2629. https://doi.org/10.1149/2.1161912jes.
13. Delmas C, Fouassier C, Hagenmuller P (1980) Structural classification and properties of the layered oxides. *Physica B+C* 99:81–85. https://doi.org/10.1016/0378-4363(80)90214-4.
14. Radin MD, Alvarado J, Meng YS, Van der Ven A (2017) Role of crystal symmetry in the reversibility of stacking-sequence changes in layered intercalation electrodes. *Nano Letters* 17:7789–7795. https://doi.org/10.1021/acs.nanolett.7b03989.
15. Wang P-F, You Y, Yin Y-X, Guo Y-G (2018) Layered oxide cathodes for sodium-ion batteries: Phase transition, air stability, and performance. *Advanced Energy Materials* 8:1701912. https://doi.org/10.1002/aenm.201701912.
16. Komaba S, Yabuuchi N, Nakayama T, et al (2012) Study on the reversible electrode reaction of $Na_{1-x}Ni_{0.5}Mn_{0.5}O_2$ for a rechargeable sodium-ion battery. *Inorgonic Chemistry* 51:6211–6220. https://doi.org/10.1021/ic300357d.
17. Yabuuchi N, Kajiyama M, Iwatate J, et al (2012) P2-type $Na_x[Fe_{1/2}Mn_{1/2}]O_2$ made from earth-abundant elements for rechargeable Na batteries. *Nature Materials* 11:512–517. https://doi.org/10.1038/nmat3309.
18. Wu X, Guo J, Wang D, et al (2015) P2-type $Na_{0.66}Ni_{0.33}$-$xZn_xMn_{0.6}7O_2$ as new high-voltage cathode materials for sodium-ion batteries. *Journal of Power Sources* 281:18–26. https://doi.org/10.1016/j.jpowsour.2014.12.083.
19. Wang P-F, Yao H-R, Liu X-Y, et al (2017) Ti-substituted $NaNi_{0.5}Mn_{0.5-x}Ti_xO_2$ cathodes with reversible O3–P3 phase transition for high-performance sodium-ion batteries. *Advanced Materials* 29:1700210. https://doi.org/10.1002/adma.201700210.
20. Radin MD, Van der Ven A (2016) Stability of prismatic and octahedral coordination in layered oxides and sulfides intercalated with alkali and alkaline-earth metals. *Chemical of Materials* 28:7898–7904. https://doi.org/10.1021/acs.chemmater.6b03454.
21. Liu Y, Fang X, Zhang A, et al (2016) Layered P_2-$Na_{2/3}[Ni_{1/3}Mn_{2/3}]O_2$ as high-voltage cathode for sodium-ion batteries: The capacity decay mechanism and Al_2O_3 surface modification. *Nano Energy* 27:27–34. https://doi.org/10.1016/j.nanoen.2016.06.026.
22. Li X, Wang Y, Wu D, et al (2016) Jahn-teller assisted Na diffusion for high performance Na ion batteries. *Chem Mater* 28:6575–6583. https://doi.org/10.1021/acs.chemmater.6b02440.
23. Mariyappan S, Wang Q, Tarascon JM (2018) Will sodium layered oxides ever be competitive for sodium ion battery applications. *Journal of the Electrochemical Society* 165:A3714. https://doi.org/10.1149/2.0201816jes.
24. Talaie E, Kim SY, Chen N, Nazar LF (2017) Structural evolution and redox processes involved in the electrochemical cycling of P2-$Na_{0.67}[Mn_{0.66}Fe_{0.20}Cu_{0.14}]O_2$. *Chemistry of Materials* 29:6684–6697. https://doi.org/10.1021/acs.chemmater.7b01146.

25. Kumar VK, Ghosh S, Biswas S, Martha SK (2020) Practical realization of O3-type $NaNi_{0.5}Mn_{0.3}Co_{0.2}O_2$ cathodes for sodium-ion batteries. *Journal of the Electrochemical Society* 167:080531. https://doi.org/10.1149/1945-7111/ab8ed5.
26. Zhang X, Rui X, Chen D, et al (2019) $Na_3V_2(PO_4)_3$: An advanced cathode for sodium-ion batteries. *Nanoscale* 11:2556–2576. https://doi.org/10.1039/C8NR09391A.
27. Qian J, Wu C, Cao Y, et al (2018) Prussian blue cathode materials for sodium-ion batteries and other ion batteries. *Advanced Energy Materials* 8:1702619. https://doi.org/10.1002/aenm.201702619.
28. Stevens DA, Dahn JR (2001) The mechanisms of lithium and sodium insertion in carbon materials. *Journal of the Electrochemical Society* 148:A803. https://doi.org/10.1149/1.1379565.
29. Tsai P, Chung S-C, Lin S, Yamada A (2015) Ab initio study of sodium intercalation into disordered carbon. *Journal of Materials Chemistry A* 3:9763–9768. https://doi.org/10.1039/C5TA01443C.
30. Nobuhara K, Nakayama H, Nose M, et al (2013) First-principles study of alkali metal-graphite intercalation compounds. *Journal of Power Sources* 243:585–587. https://doi.org/10.1016/j.jpowsour.2013.06.057.
31. Jache B, Adelhelm P (2014) Use of graphite as a highly reversible electrode with superior cycle life for sodium-ion batteries by making use of co-intercalation phenomena. *Angewandte Chemie – International Edition* 53:10169–10173. https://doi.org/10.1002/anie.201403734.
32. Li L, Zheng Y, Zhang S, et al (2018) Recent progress on sodium ion batteries: Potential high-performance anodes. *Energy & Environmental Science* 11:2310–2340. https://doi.org/10.1039/C8EE01023D.
33. Chayambuka K, Mulder G, Danilov DL, Notten PHL (2018) Sodium-ion battery materials and electrochemical properties reviewed. *Advanced Energy Materials* 8:1800079. https://doi.org/10.1002/aenm.201800079.
34. Dou X, Hasa I, Saurel D, et al (2019) Hard carbons for sodium-ion batteries: Structure, analysis, sustainability, and electrochemistry. *Materials Today* 23:87–104. https://doi.org/10.1016/j.mattod.2018.12.040.
35. Morita R, Gotoh K, Fukunishi M, et al (2016) Combination of solid state NMR and DFT calculation to elucidate the state of sodium in hard carbon electrodes. *Journal of Materials Chemistry A* 4:13183–13193. https://doi.org/10.1039/C6TA04273B.
36. Irisarri E, Ponrouch A, Palacin MR (2015) Review-hard carbon negative electrode materials for sodium-ion batteries. *Journal of the Electrochemical Society* 162:A2476. https://doi.org/10.1149/2.0091514jes.
37. El Moctar I, Ni Q, Bai Y, et al (2018) Hard carbon anode materials for sodium-ion batteries. *Functional Materials Letters* 11:1830003. https://doi.org/10.1142/S1793604718300037.
38. Damodar D, Ghosh S, Usha Rani M, et al (2019) Hard carbon derived from sepals of Palmyra palm fruit calyx as an anode for sodium-ion batteries. *Journal of Power Sources* 438:227008. https://doi.org/10.1016/j.jpowsour.2019.227008.
39. Ghosh S, Kumar VK, Kumar SK, et al (2020) Binder less-integrated freestanding carbon film derived from pitch as light weight and high-power anode for sodium-ion battery. *Electrochimica Acta* 353:136566. https://doi.org/10.1016/j.electacta.2020.136566.
40. Wu L, Bresser D, Buchholz D, et al (2015) Unfolding the mechanism of sodium insertion in anatase TiO_2 nanoparticles. *Advanced Energy Materials* 5:1401142. https://doi.org/10.1002/aenm.201401142.
41. Li W, Fukunishi M, Morgan Benjamin J, et al (2017) A reversible phase transition for sodium insertion in anatase TiO_2. *Chemistry of Materials* 29:1836–1844. https://doi.org/10.1021/acs.chemmater.7b00098.
42. Zhao L, Pan H-L, Hu Y-S, et al (2012) Spinel lithium titanate ($Li_4Ti_5O_{12}$) as novel anode material for room-temperature sodium-ion battery. *Chinese Physics B* 21:028201. https://doi.org/10.1088/1674-1056/21/2/028201.
43. Rudola A, Saravanan K, Devaraj S, et al (2013) $Na_2Ti_6O_{13}$: A potential anode for grid-storage sodium-ion batteries. *Chemical Communications* 49:7451–7453. https://doi.org/10.1039/C3CC44381G.
44. Nava-Avendaño J, Morales-García A, Ponrouch A, et al (2015) Taking steps forward in understanding the electrochemical behavior of $Na_2Ti_3O_7$. *Journal of Materials Chemistry A* 3:22280–22286. https://doi.org/10.1039/C5TA05174F.
45. Shirpour M, Cabana J, Doeff M (2013) New materials based on a layered sodium titanate for dual electrochemical Na and Li intercalation systems. *Energy & Environmental Science* 6:2538–2547. https://doi.org/10.1039/C3EE41037D.
46. Doeff MM, Cabana J, Shirpour M (2014) Titanate anodes for sodium ion batteries. *Journal of Inorganic and Organometallic Polymers and Materials* 24:5–14. https://doi.org/10.1007/s10904-013-9977-8.
47. Perveen T, Siddiq M, Shahzad N, et al (2020) Prospects in anode materials for sodium ion batteries - A review. *Renewable and Sustainable Energy Reviews* 119:109549. https://doi.org/10.1016/j.rser.2019.109549.

48. Wang T, Su D, Shanmukaraj D, et al (2018) Electrode materials for sodium-ion batteries: Considerations on crystal structures and sodium storage mechanisms. *Electrochemical Energy Reviews* 1:200–237. https://doi.org/10.1007/s41918-018-0009-9.
49. Skundin AM, Kulova TL, Yaroslavtsev AB (2018) Sodium-ion batteries (a Review). *Russian Journal of Electrochemistry* 54:113–152. https://doi.org/10.1134/S1023193518020076.
50. Wang Y, Song S, Xu C, et al (2019) Development of solid-state electrolytes for sodium-ion battery-A short review. *Nano Materials Science* 1:91–100. https://doi.org/10.1016/j.nanoms.2019.02.007.
51. Bommier C, Ji X (2018) Electrolytes, SEI formation, and binders: A review of nonelectrode factors for sodium-ion battery anodes. *Small* 14:1703576. https://doi.org/10.1002/smll.201703576.
52. Wang E, Niu Y, Yin Y-X, Guo Y-G (2021) Manipulating electrode/electrolyte interphases of sodium-ion batteries: Strategies and perspectives. *ACS Materials Letters* 3:18–41. https://doi.org/10.1021/acsmaterialslett.0c00356.
53. Peled E (1979) The electrochemical behavior of alkali and alkaline earth metals in nonaqueous battery systems-The solid electrolyte interphase model. *Journal of the Electrochemical Society* 126:2047. https://doi.org/10.1149/1.2128859.
54. Komaba S, Murata W, Ishikawa T, et al (2011) Electrochemical Na insertion and solid electrolyte interphase for hard-carbon electrodes and application to Na-ion batteries. *Advanced Functional Materials* 21:3859–3867. https://doi.org/10.1002/adfm.201100854.

9 Cathodes for Sodium-Ion Batteries

Monika Wilamowska-Zawłocka, Anita Cymann-Sachajdak, Zuzanna Zarach, and Magdalena Graczyk-Zajac

9.1 INTRODUCTION

A concept of Na-ion batteries (NIBs) has been studied before commercialization of commonly used today Li-ion batteries (LIBs) [1–3]. The success and progress of Li-ion cells postponed works on Na-ion technology for some time. Nowadays, scientists revisit Na-ion systems to develop energy storage devices from more abundant elements. Lithium abundance in the earth's crust is limited to only 20 ppm. Moreover, cobalt (25 ppm in the earth's crust) plays an integral part in the common LIBs cathode. The price of both elements has steeply increased since the beginning of this century. Another challenging issue to solve is the uneven distribution of these elements, leading to geopolitical problems. In contrast, sodium is one of the most abundant elements in the earth's crust, and infinite sodium resources are also available in oceans. The promising cathode materials for NIBs consist of cheap and available elements such as manganese, vanadium, phosphorous, iron, or titanium [4,5].

Sodium is, next to lithium, the smallest and lightest alkali metal. Furthermore, electrode chemistry for Na ion is analogous to Li ion as many families of compounds exhibit similar insertion properties for Na and Li ions. Thus, recent tremendous growth in the development of electrode materials for sodium-ion cells was possible owing to the broad knowledge gained from the decades of studies of lithium-ion batteries. Taking into account the properties and abundance of sodium, NIBs are the most promising alternative for LIBs. However, there are a couple of issues that have to be solved. First of all, the higher ionic radius and molar mass of Na ions compared to Li ions (1.02 Å and 23.0 g mol^{-1} vs 0.76 Å and 6.94 g mol^{-1} for Na^+ and Li^+, respectively [4,6]) impede the diffusion of ions in the electrode material resulting in a worse rate capability. The larger sodium ions cause higher volume expansion of electrode materials, which may lead to irreversible phase transformation. The size of sodium ions is the reason of their poor intercalation into graphite having 0.335 nm distance between graphene layers [7,8], although there are reports that careful tailoring of electrolyte composition may lead to advantageous solvent co-intercalation, improving performance of graphitic anodes in NIBs [9–11]. Another drawback of Na ions compared to Li is their less negative redox potential (–2.71 V vs standard hydrogen electrode (SHE) for Na/Na^+ couple compared to –3.04 V for Li/Li^+), which reduces the operating voltage of the cell and thus lowers the energy density. These drawbacks hinder finding stable Na host materials with satisfactory electrochemical performance and make the development of high power, high energy density and durable NIBs a tough challenge.

Sodium favors sixfold coordination, which limits the types of structures suitable for Na-intercalation cathodes. Two main classes of materials that can hold Na ions in six-coordinated geometry are layered transition metal oxides (TMOs) and polyanion compounds. Also, Prussian Blue analogs (PBAs) with open interstitial sites exhibit Na-ion insertion properties, which make them appealing candidates for NIBs positive electrodes [12]. Therefore, this chapter will focus on these three classes of compounds.

DOI: 10.1201/9781003310167-9

9.2 LAYERED SODIUM TRANSITION METAL OXIDES

Sodium TMOs with the general formula of Na_xMO_2 (where M is a transition metal, e.g., Ni, Mn, Fe, Cr, V, Cu, and Ti; M can be a single element or mixture of two, three, or more elements) are two-dimensional structures consisting of sodium cations placed between $(MO_2)_n$ sheets, which are formed by edge-sharing MO_6 octahedra. Na ions are coordinated either in an octahedral (O), tetrahedral (T), or trigonal prismatic geometry (P) [13,14]. The ability of sodium ions to occupy both octahedral and prismatic sites gives a huge advantage over lithium ions, which can reside only in octahedral sites. This feature provides richer chemistry of sodium layered oxides compared to lithium analogs. The most common types of sodium TMOs are O3-, P2-, and P3-phases or the mixed one such as P2–O3 intergrowths, where the number reveals the quantity of unique transition metal oxide sheets in the repeating unit cell [13–15], as shown in Figure 9.1 [4]. Additionally, the prime symbol (′) indicates monoclinic distortion, e.g., O′3 signifies monoclinic distortion of the O3 phase.

A huge variety of compounds with numerous combinations of transition metal ions pose ability to reversibly shuttle sodium ions with high capacity values approaching 200 mAh g^{-1} [16–20]. Sodium TMOs are more abundant and less expensive than lithium analogs, but also exhibit problems that have to be overcome like: (i) low operating voltage, (ii) structural instability and possible volume changes upon multiple phase transitions during long-term charge–discharge cycles, (iii) high polarization, or (iv) Na-deficiency in the case of P2 phase oxides, which can deliver high capacity in half cells [20–23].

Oxides with one sodium ion per transition metal ion (full sodium stoichiometry) thermodynamically prefer the O3-type phase. Na_xMO_2 O3-phases are stable when $x > 0.67$ and the average oxidation state of TM is close to 3+. Electrochemical extraction/insertion of Na ions from/to the O3-type

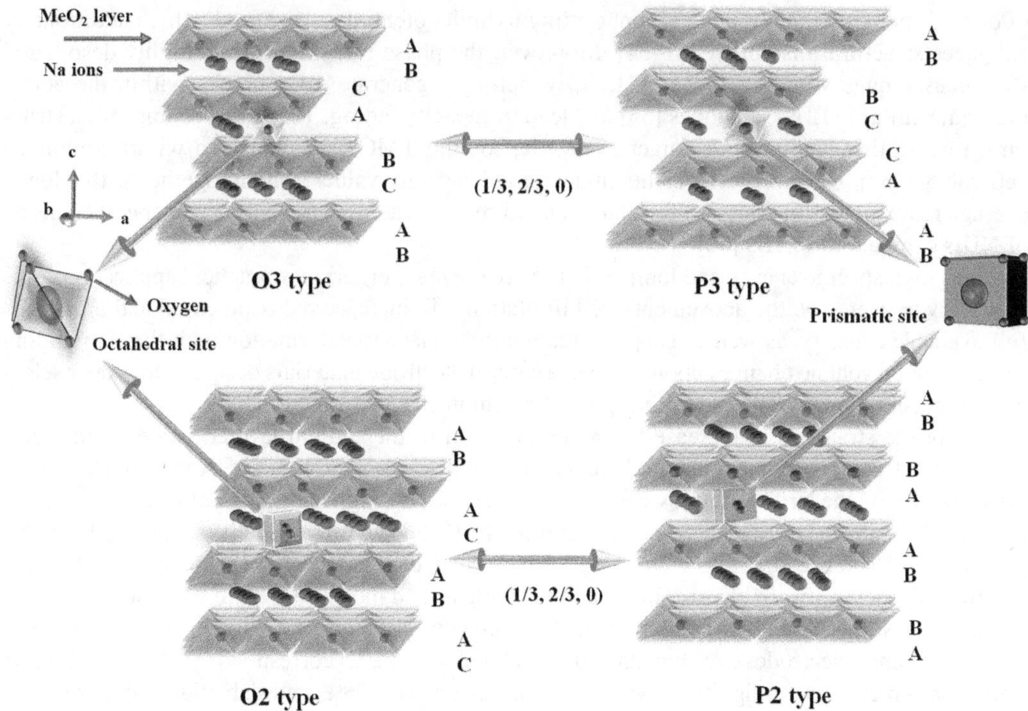

FIGURE 9.1 Various types of packing in transition metal oxides and the process of phase transition induced by extraction of sodium ions. Adapted and reproduced with permission from Ref. [4]. Copyright © 2021 American Chemical Society.

host progresses with reversible structural transformations: O3 ↔ O′3 ↔ P3 ↔ P′3. When sodium ions are partly extracted from the crystal lattice and vacancies are created, Na^+ energetically prefers a prismatic coordination. This induces strong repulsion of O atoms in the Na layers leading to expansion of the interlayer distance. Higher interlayer distance in the P′3 phase results in faster diffusion of Na^+ compared to O3. These transitions cause gliding of the MO_2 slabs without breaking the M-O bonds as first reported by Delmas et al. [1].

P-type structures are solely stabilized with sodium deficiency (typically $Na^+ \leq 0.67$), while the average oxidation state of TM is above 3.3+. The phase transition takes place when the Na^+ content is lower than 0.33. The capacity of P2-type TMO is limited by the sodium content. Fully sodiated P-type compounds can be obtained by electrochemical insertion of sodium, but this process is not feasible in full cells due to lack of extra sodium. Compensation for this non-stoichiometry could be solved by (i) presodiation process, (ii) using sodium metal as an anode (problematic in the full cell due to dendrite formation as in LIBs) [24,25], or (iii) the addition of sacrificial additives (e.g., NaN_3, Na_2CO_3, and Na_3P), which decompose to provide extra sodium [26–29]. All three solutions require additional time-consuming and expensive processes. Uncompensated Na-deficiency in the P2- and P3-type oxides causes too low initial charge capacity to meet the requirement of commercial cells. Therefore, O3-type phases seem to be more suitable for practical applications [30]. On the other hand, the P2-type layered structures, contrary to O3-type frameworks, have no interstitial tetrahedral sites and provide open paths for Na ions, which lowers the diffusion barrier. Thus, the ionic conductivity of P2-type layered oxides is higher than that of O3-type phases. Furthermore, due to the large size of Na^+, multiple phase transitions are inevitable upon the intercalation/deintercalation of sodium ions to/from the O3-type host structure. These structural evolutions result from competing van der Waals and Coulombic forces that induce gliding of the MO_2 layers (i.e., O3 ↔ P3 ↔ O3/O1) upon changes of the Na^+ ions content during charge–discharge cycles. To utilize the full theoretical capacity of the insertion/extraction of one Na^+ per one transition metal ion, cycling to high potentials is required, which additionally enhances structural instability [31–34]. Moreover, the phase transition in the highly desodiated state causes huge volume changes [21,23], which may generate microcracks within the active material similar to LIBs' cathodes [35] and lead to capacity fading. Therefore, to improve a structural reversibility leading to a longer cycle life, sodium TMOs are polarized within a reduced cell voltage [36]. However, this results in decreased capacity values, which together with a lower average redox potential than that for analogous Li oxides lead to noticeably lower specific energy of NIBs compared to LIBs [30].

Since high specific energy and long cycle life are the main criteria for practical applications, it is necessary to overcome the abovementioned limitations. To increase redox potential and utilize the full available capacity, as well as suppress the multiple phase transformation with the slab gliding and associated volume changes upon cycling, a careful electrode materials design with proper selection of transition metal ions and tailoring their stoichiometry is needed.

One of the strategies to increase the redox potential of the sodium layered oxides is to incorporate metal centers that exhibit high redox potential (e.g., Ni^{2+}/Ni^{4+}, Cu^{2+}/Cu^{3+}, or Fe^{3+}/Fe^{4+}). Oxides based on the Ni^{2+}/Ni^{4+} redox couple exhibit a relatively high operating voltage. For instance, P2-$Na_{2/3}[Ni^{2+}{}_{1/3}Mn^{4+}{}_{2/3}]O_2$, which was first reported by Lu and Dahn [37], delivers approx. 160 mAh g^{-1} (the theoretical one equals to 173 mAh g^{-1}) in the voltage range of 2–4.5 V with a Ni^{2+}/Ni^{4+} redox reaction. Moreover, this oxide is stable in contact with air and moisture. Figure 9.2a shows a charge curve of the $Na_{2/3}Ni_{1/3}Mn_{2/3}O_2$ cathode in the half-cell configuration (with metallic sodium as counter and reference electrodes). A long plateau at high voltage (4.2 V) corresponds to the P2-O2 phase transition, which causes a significant volume shrinkage (approx. 23%) and slab gliding (Figure 9.2b), leading to an isolation of the active material from conductive carbon additive and current collector, resulting in a drastic capacity fading upon cycling up to 4.5 V. When the cell voltage is reduced, the cycling stability is significantly improved as the phase transitions and the corresponding volume changes are avoided (Figure 9.2c) [19,36].

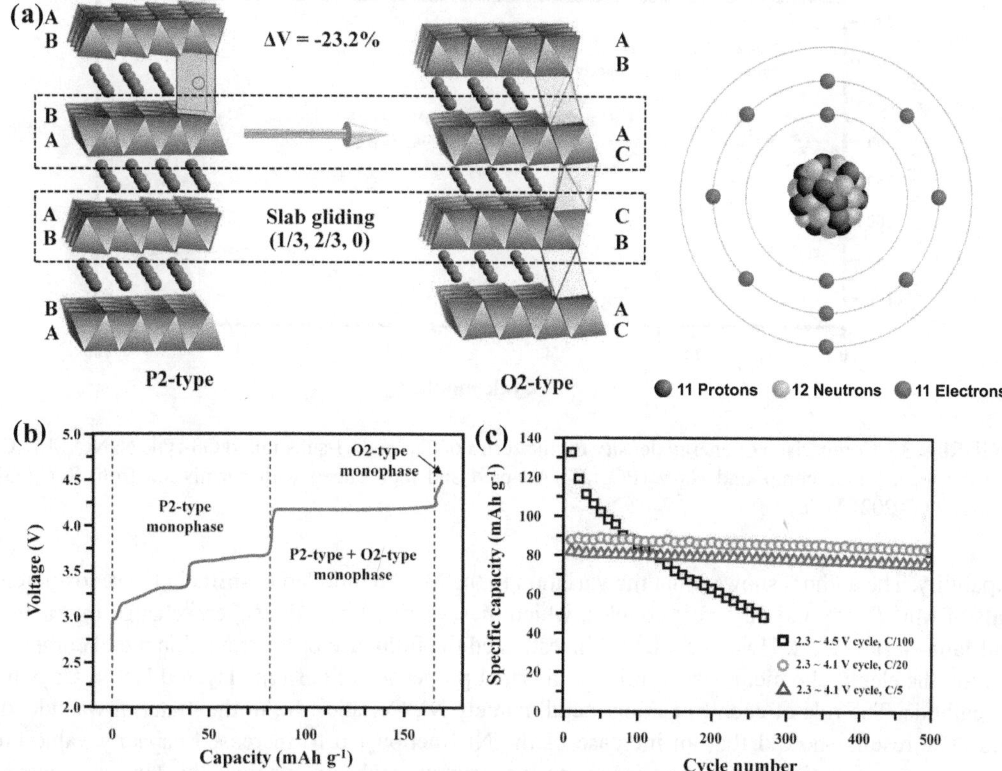

FIGURE 9.2 (a) Charge curve of the $Na_{2/3}Ni_{1/3}Mn_{2/3}O_2$ cathode in the half cell with metallic Na. (b) Schematic illustration of the reversible structural transformation from P2- to O2-type phase with the slab gliding. Adapted and reproduced with permission from Ref. [38]. Copyright © 2015 European Physical Society. (c) Cycling stability of $Na_{2/3}[Ni_{1/3}Mn_{2/3}]O_2$ cathode in the half cell with metallic Na at different voltage ranges (2.3–4.1 V and 2.3–4.5 V) and various scan rates. Adapted and reproduced with permission from Ref. [36]. Copyright © 2013 Royal Society of Chemistry.

On the other hand, a standard example of O3-phase $NaNi_{0.5}Mn_{0.5}O_2$ delivers relatively high capacity (185 mAh g^{-1} in the voltage range of 2.5–4.5 V vs Na/Na^+ and 120 mAh g^{-1} in 2–3.5 V window) and moderate cycling performance (81% of initial capacity after 20 charge/discharge cycles) as reported by Komaba et al. [39]. This oxide was further changed by substitution of Ni and Mn ions by Fe ions. The obtained $NaFeO_2$-$NaNi_{0.5}Mn_{0.5}O_2$ solid solution exhibited a capacity of 130 mAh g^{-1} in a voltage range of 2.0–3.8 V and energy of 400 Wh kg^{-1} in the half cell with metallic Na [18]. Yuan et al. [40] investigated a series of Fe-substituted O3-type $NaNi_{0.5}Mn_{0.5}O_2$. The partial substitution of Ni and Mn with Fe in the O3-phase lattice can greatly improve the electrochemical performance and the structural stability. The same O3-type $NaNi_{0.5}Mn_{0.5}O_2$ was substituted with copper and titanium ions by Wang et al. [33]. $NaNi_{0.4}Cu_{0.1}Mn_{0.4}Ti_{0.1}O_2$ and $NaNi_{0.45}Cu_{0.05}Mn_{0.3}Ti_{0.2}O_2$ delivered energy of 310 and 320 Wh kg^{-1}, respectively, in full cells with significantly improved cycling stability compared to the pristine material. Such values may compete with $Na_3V_2(PO_4)_2F_3$ as shown in Figure 9.3.

It has been found that copper substitution improves air and moisture stability of TMO, which is of high importance from a technological point of view [41–45]. Xu et al. [46] investigated the role of transition metal centers in the Na_x(Cu–Fe–Mn)O_2 structures with various stoichiometry. A strong influence of the pristine valence state of manganese ions on the capacity values was observed. On the other hand, the increase of iron amount resulted in the decreased average redox potential of Na_x(Cu–Fe–Mn)O_2 system, while the incorporation of copper ions improved cyclability and rate

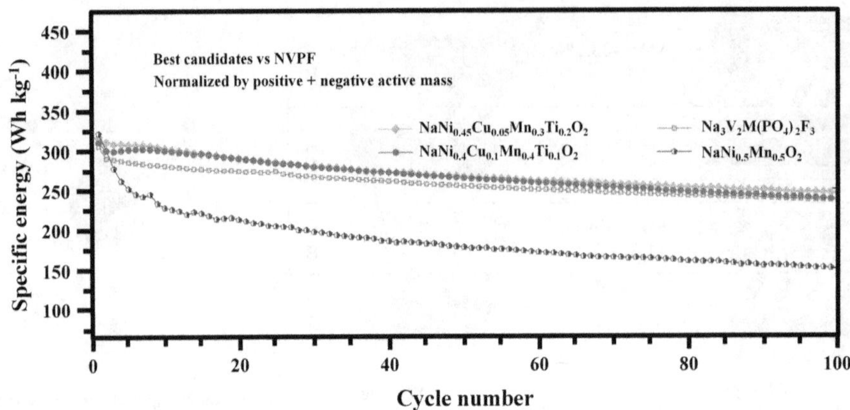

FIGURE 9.3 Comparison of energy density for the series of Cu- and Ti-substituted O3-type $NaNi_{0.5}Mn_{0.5}O_2$ and the top cathode compound $Na_3V_2(PO_4)_2F_3$. Adapted and reproduced with permission from Ref. [33]. Copyright © 2021 Wiley.

capability. The authors showed that the variation in the Mn/Fe ratio led to shifts in the redox potentials of Mn^{3+}/Mn^{4+} and Fe^{3+}/Fe^{4+} couples, which they ascribed it to the superexchange interaction and Jahn–Teller effect. Hwang et al. [47] investigated the influence of the transition metal composition on the electrochemical, structural, and thermal properties of O3-type layered $Na[Ni_xCo_yMn_z]O_2$ cathode. The role of each transition metal, namely Ni, Co, and Mn, in the material was identified. The results showed that an increase of the Ni fraction led to increased capacity value but poor capacity retention. Co centers enhanced the structure stability, whereas the Mn ions contributed to the improved capacity retention and better thermal stability. A quaternary O3-type layered oxide with Ni, Co, Mn, and Fe was studied by Li et al. [48]. The electrochemical performance of $Na(Mn_{0.25}Fe_{0.25}Co_{0.25}Ni_{0.25})O_2$ was greatly improved in terms of capacity value (180 mAh g^{-1}) and cycling stability over the extended voltage range (1.9–4.3 V) compared to the binary systems $Na(Fe_{0.5}Co_{0.5})O_2$ and $Na(Ni_{0.5}Mn_{0.5})O_2$.

Another way to improve the electrochemical performance of TMO is to incorporate non-redox centers. Examples of non-active metals are Mg^{2+}, Zn^{2+}, Ti^{4+}, or Sn^{4+}. Heterometal doping or partial substitution into transition metal sites enhances the structure stability as reported for variety of metals used as substituent or dopants of TM sites [23,41,47,49–55]. The key issue is to optimize the ionicity of transition metal-oxygen bond, which regulates the oxidation/reduction potential and the stability of O3 vs P3 phase. Substitution of Fe or Mn by Ti^{4+} in both P2 and O3 phases has been found beneficial in improving cycling stability and rate capability due to increased ionicity of the crystal lattice, increased redox potential, and reduced number of phase transitions in the broader voltage window [32,56–60]. Wang et al. [51] investigated the influence of partial substitution of Mn^{4+} by Ti^{4+} in the O3-type $NaNi_{0.5}Mn_{0.5}O_2$ cathode. Similar ionic radii of Ti and Mn ions (53 and 60.5 pm for Mn^{4+} and Ti^{4+}, respectively) and the same valence state ensured homogenous distribution of Ti in the TM layers. Substitution of Mn by Ti ions increased interlayer distance, suppressing the irreversible phase transitions in the high voltage range, which in turn led to improved cycle stability and higher capacity values. Introduction of non-transition Sn^{4+} ions in O3-type $NaNi_{0.5}Mn_{0.5}O_2$ phase was presented by Sathiya et al. [55]. $NaNi_{0.5}Sn_{0.5}O_2$ showed an increased redox potential of 3.2 V in comparison to 2.8 V (vs Na/Na^+) for the pristine $NaNi_{0.5}Mn_{0.5}O_2$. Moreover, tin-based oxide delayed O3 $\leftrightarrow$ P3 phase transition, which occurred upon removal of 0.5 Na from the system compared to 0.2 for the pristine phase without tin. This, in turn, led to improved cycling stability (85% of initial capacity value after 200 cycles for Sn-based phase compared to 74% for the $NaNi_{0.5}Mn_{0.5}O_2$). Wang et al. [61] synthesized O3-type $Na_{0.7}Ni_{0.35}Sn_{0.65}O_2$ oxide, which displays the highest reported redox potential for Ni^{2+}/Ni^{3+} redox couple (3.7 V vs Na/Na^+). The presence of Sn^{4+} ions of the same ionic radii as Ni^{2+} (0.69 Å) in the crystal lattice causes cationic disorder and a significant increase of Ni-O bond ionicity.

Improved electrochemical performance of another O3-type Ni-, Fe-, and Mn-based layered oxide was caused by the substitution of Mn ions by Mg^{2+}. The presence of Mg in the $NaMn_{0.48}Ni_{0.2}Fe_{0.3}Mg_{0.02}O_2$ lattice enlarged interlayer spacing leading to enhanced diffusion of Na^+ and mitigation of the lattice strains induced by insertion/extraction of sodium ions. This suppressed the volume changes, which in turn minimized the irreversible phase transitions. Moreover, Mg substituent reduced the TM-O bond length and thickness of MO_6 sheets stabilizing the layered structure [50]. Mariyappan et al. [23] reported substitution of redox-active Ni^{2+} cations by inactive Zn^{2+} in the O3-phase $NaNi_{0.5}Mn_{0.5-z}Ti_zO_2$ (NMT). The resulting $NaNi_{0.45}Zn_{0.05}Mn_{0.4}Ti_{0.1}O_2$ (ZNMT) structure exhibited a high capacity value (170 mAh g^{-1}) with extraction/insertion of approx. 0.8 of Na^+ per one TM center compared to 0.9 Na for unsubstituted $NaNi_{0.5}Mn_{0.4}Ti_{0.1}O_2$ (Figure 9.4a and b).

The fully charged phase (with 0.2 Na^+) displayed a P3-O1 intergrowth structure, with the O1 phase present as nanodomains. Zinc ions substitution suppressed the phase transitions at high voltage, which minimized material degradation upon multiple cycling. The unsubstituted samples heated at 1000°C and 900°C retained approx. 70% and 65% of their initial capacity after 100 cycles, whereas for Zn-substituted samples, the capacitance retention was much higher, namely 85% and 75% for the samples treated at 1000°C and 900°C, respectively (Figure 9.4a and b). The energy density of the NMT and ZNMT upon cycling was compared with the $Na_3V_2(PO_4)_2F_3$ polyanionic compound (Figure 9.4c). There is also a trend to incorporate lithium ions in the TMO structure. It has been found that the Li ions occupy the transition metal sites and that Li substitution mitigates phase transformation at high voltages, improving cyclability of the cathode material [49,62–65]. In the case of P2-type $Na_{0.80}[Li_{0.12}Ni_{0.22}Mn_{0.66}]O_2$, the presence of monovalent lithium ions in the transition metal layers enabled more sodium ions to occupy the prismatic sites, leading to stabilization of the overall charge balance of the oxide [64]. Oh et al. reported high capacity values (180.1 mAh g^{-1} at 0.1C rate) as well as better rate capability and cycle stability for the lithium substituted O3-type $Na[Li_{0.05}(Ni_{0.25}Fe_{0.25}Mn_{0.5})_{0.95}]O_2$ oxide compared to the pristine $Na[Ni_{0.25}Fe_{0.25}Mn_{0.5}]O_2$ [49].

Another approach to combine high capacity values with improved stability (also for LIBs cathodes) is a design of sophisticated structures that improve diffusion of Na^+ (or Li^+), which is often combined with gradually changed concentration of transition metals along the particle in order to take advantage of unique properties of selected metal ions (Figure 9.5a and b) [66–69]. Hwang et al. [66] prepared radially aligned hierarchical columnar (RAHC) structures with an average composition of $Na[Ni_{0.6}Co_{0.05}Mn_{0.35}]O_2$. They obtained spherical particles (10 μm in diameter) made from columns (5–6 μm in length) with gradually changed Ni and Mn concentration along their entire length. Nickel-rich core provided high capacity, whereas manganese-rich shell ensured good cycling stability. Liu et al. [67] reported preparation and electrochemical performance of porous nanofibers composed of nanoparticles (20–90 nm) of P2-phase $Na_{2/3}Ni_{1/3}Mn_{2/3}O_2$ (Figure 9.5c). The $Na_{2/3}Ni_{1/3}Mn_{2/3}O_2$ nanostructures were obtained through electrospinning of nanofibers containing sodium, nickel, and manganese salts and polyvinylpyrrolidone (PVP) and their subsequent heat treatment. The material exhibited high capacity value of 166.7 mAh g^{-1} at 0.1C with significantly improved rate capability (73.4 mAh g^{-1} at 20 C rate) and enhanced cycling stability (approx. 81% of initial capacity after 500 cycles) compared to the bulk $Na_{2/3}Ni_{1/3}Mn_{2/3}O_2$.

As presented, a proper selection of transition metal centers, a variation of their stoichiometry, and substitution or doping with non-transition metal ions enable to achieve reasonable capacity values and energy densities, although the values achieved in the full cell are considerably lower, especially for P2-phases (Figure 9.6). The decrease in energy in full cells is due to the consumption of Na ions in the formation of solid electrolyte interphase (SEI) on anodes and the decrease of voltage range while replacing metallic Na with hard carbon electrode.

Besides the capacity and energy values, essential issues for practical application are the cycle life of cells and easy handling of electrode materials. TMOs suffer from multiple phase transition upon cycling, which leads to capacity fading. This problem can be partially solved by carefully designed

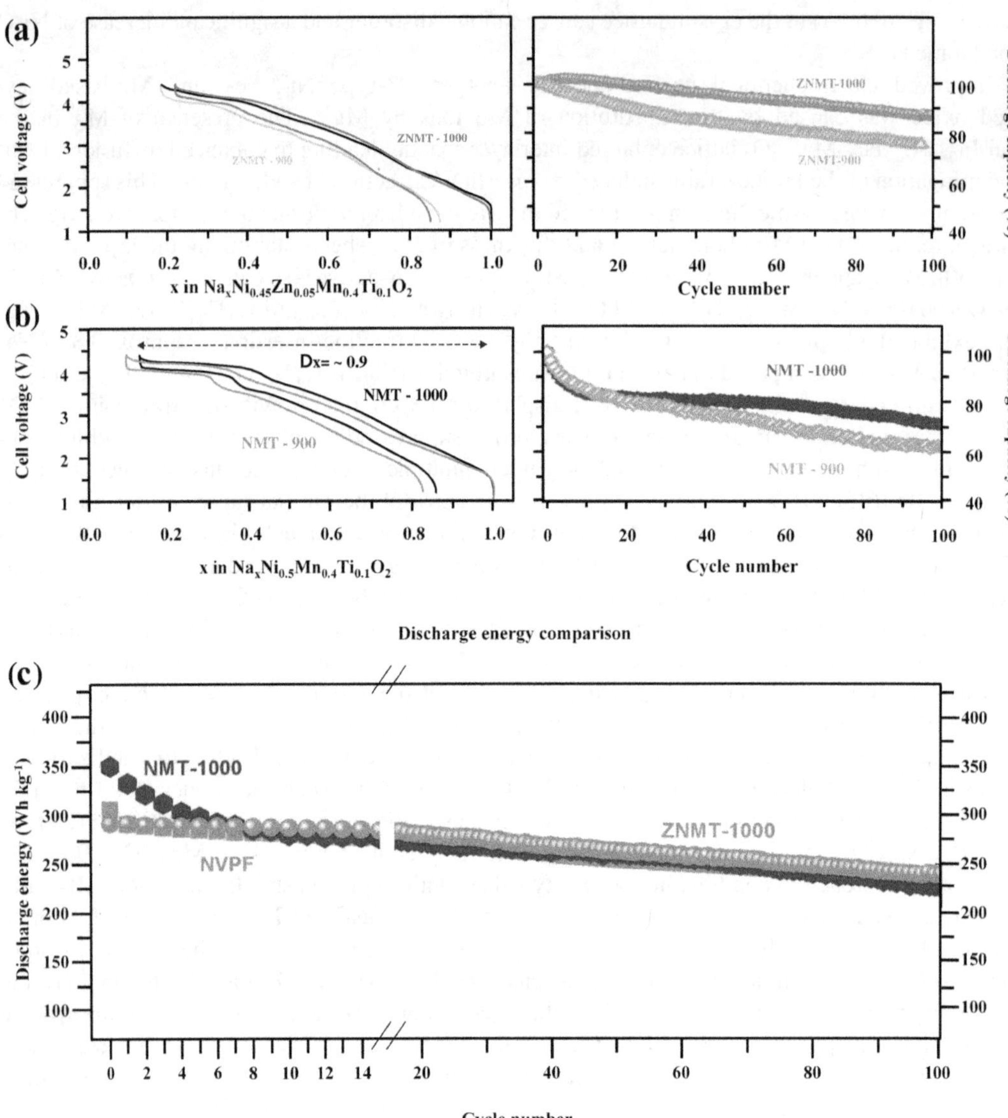

FIGURE 9.4 Electrochemical response of $NaNi_{0.5}Mn_{0.4}Ti_{0.1}O_2$ and Zn-substituted analog $NaNi_{0.45}Zn_{0.05}Mn_{0.4}Ti_{0.1}O$ in full cells with hard carbon anode samples heat-treated at 900 and re-annealed at 1000°C. (a, b) Left side: galvanostatic charge/discharge curves and right side: capacity retention over 100 cycles (cycled with C/10 rate in a voltage range of 1.2–4.4 V) of (a) ZNMT and (b) NMT. (c) Comparison of discharge energy of NMT-1000, ZNMT-1000, and $Na_3V_2(PO_4)_2F_3$ as a benchmark cathode; energy values calculated per mass of both positive and negative (hard carbon) active electrode materials. Adapted and reproduced with permission from Ref. [23] Copyright © 2021 American Chemical Society.

composition and stoichiometry or reduction of cell voltage. However, another disadvantage of sodium layered oxides is their moisture sensitivity, which requires a strictly controlled atmosphere for material storage and processing, thus impeding the commercial application. Highly hygroscopic layered cathodes undergo surface hydration even upon brief exposure to air. This results in the formation of insulating NaOH on their surface according to the following equations [38]:

$$NaMO_2 + H_2O \rightarrow Na_{1-x}H_xMO_2 + xNaOH \quad (9.1)$$

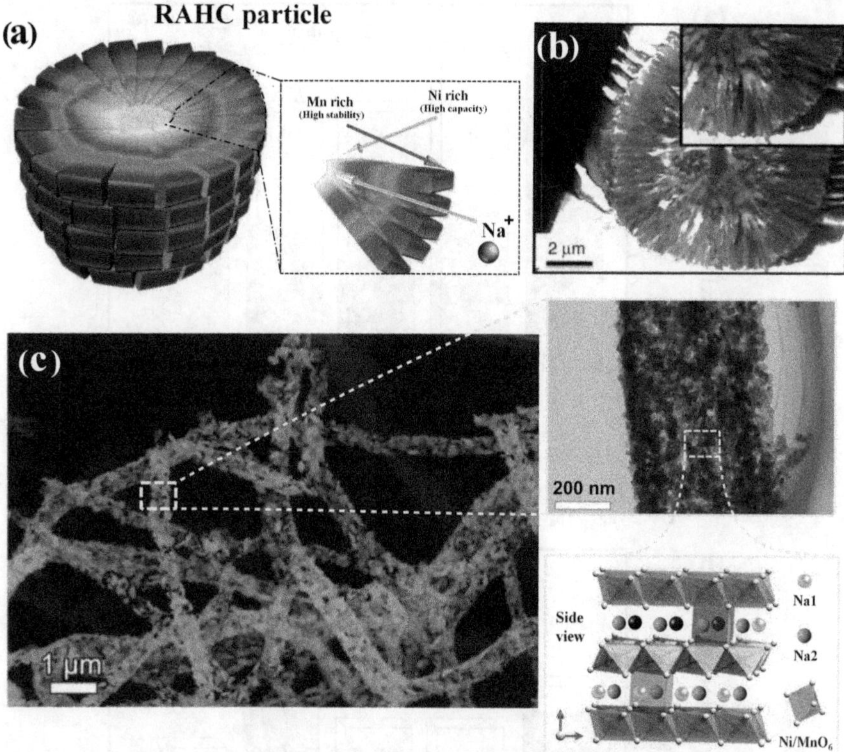

FIGURE 9.5 (a) Schematic illustration of a RAHC particle with Ni-rich core and Mn-rich shell. (b) Cross-sectional TEM image of RAHC $Na[Ni_{0.60}Co_{0.05}Mn_{0.35}]O_2$. Adapted and reproduced with permission from Ref. [66]. Copyright © 2021 Springer Nature. (c) Left side: SEM image of $Na_{2/3}Ni_{1/3}Mn_{2/3}O_2$ nanofibers; right side: TEM of $Na_{2/3}Ni_{1/3}Mn_{2/3}O_2$ nanofiber (top) and illustration of the P2-type $Na_{2/3}Ni_{1/3}Mn_{2/3}O_2$ crystal structure (bottom). Adapted and reproduced with permission from Ref. [67] Copyright © 2021 Wiley.

$$NaMO_2 + xO_2 \rightarrow Na_{1-4x}MO_2 + 2x\ Na_2O \tag{9.2}$$

$$NaMO_2 + xH_2O \rightarrow Na_{1-2x}MO_2 + xN_{a2}O + xH_2 \tag{9.3}$$

$$Na_2O + H_2O \rightarrow 2NaOH \tag{9.4}$$

NaOH produced in equations 9.1 and 9.4 is partially dissolved in N-methyl pyrrolidone solvent used for the preparation of the electrode slurry. NaOH in solution leads to the elimination of hydrogen fluoride (HF) from the poly(vinylidene fluoride) (PVDF) binder resulting in the creation of double carbon–carbon bonds in this polymer and further creation of oxygen-rich functional groups (e.g., C-OH, C-O-C, and C=O) [70,71].

Surface coating decreases moisture/air sensitivity and additionally suppresses side reactions that occur at the material/electrolyte interphase. It has been reported that the coating layers of nanometer thickness can significantly reduce undesirable reactions with the electrolyte [72–76]. For instance, Al_2O_3 nanolayer on the P2-type $Na_{2/3}[Ni_{1/3}Mn_{2/3}]O_2$ was observed to minimize exfoliation of the TMO layers leading to improved cyclability, as presented in Figure 9.7a–c [72]. Similarly, Al_2O_3 nanoparticle coating improved cycling stability of the O3-type $Na[Ni_{0.6}Co_{0.2}Mn_{0.2}]O_2$ cathode due to scavenging of HF coming from the degradation of $NaPF_6$-based electrolyte. Al_2O_3 reacts

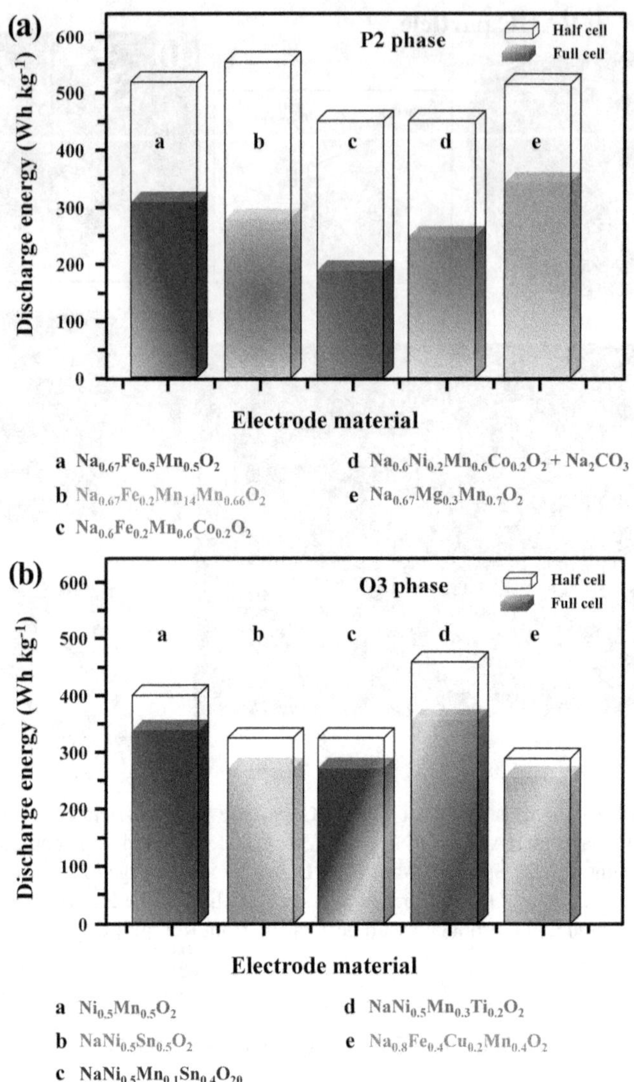

FIGURE 9.6 Energy densities in half-cell (metallic Na) and full-cell (hard carbon) configuration for the selected O3- and P2-type layered transition metal oxides. In both cases (half and full cells), the energy values are calculated per mass of active cathode material only. Adapted and reproduced with permission from Ref. [32]. Copyright © 2021 IOP Publishing.

with HF, forming AlF_3, and in this way, protects active sodium oxides from undesirable reactions with HF extending the cycle life of the cell [73].

AlF_3 can also be used as a protective coating for cathodes as it is more stable than Al_2O_3 and does not react with the electrolyte. Moreover, AlF_3 is more resistive to moisture than oxides, thus AlF_3 coating enables easier handling of the materials. Sun et al. [75] reported suppressed capacity degradation of AlF_3-coated $Na[Ni_{0.65}Co_{0.08}Mn_{0.27}]O_2$ compared to the raw material. It is worth noticing that AlF_3 coating has already been tested in LIBs [75,77–80]. Another example of effective cathode protection is coating with TiO_2, which limited the surface side reactions with the electrolyte and dissolution of Mn from O3-type $NaMn_{0.33}Fe_{0.33}Ni_{0.33}O_2$ [74]. $AlPO_4$ coating on the $Na[Li_{0.05}Mn_{0.50}Ni_{0.30}Cu_{0.10}Mg_{0.05}]O_2$ surface was reported to stabilize the cathode surface, diminishing the dissolution of transition metals, which in turn protected hard carbon anode by suppressing

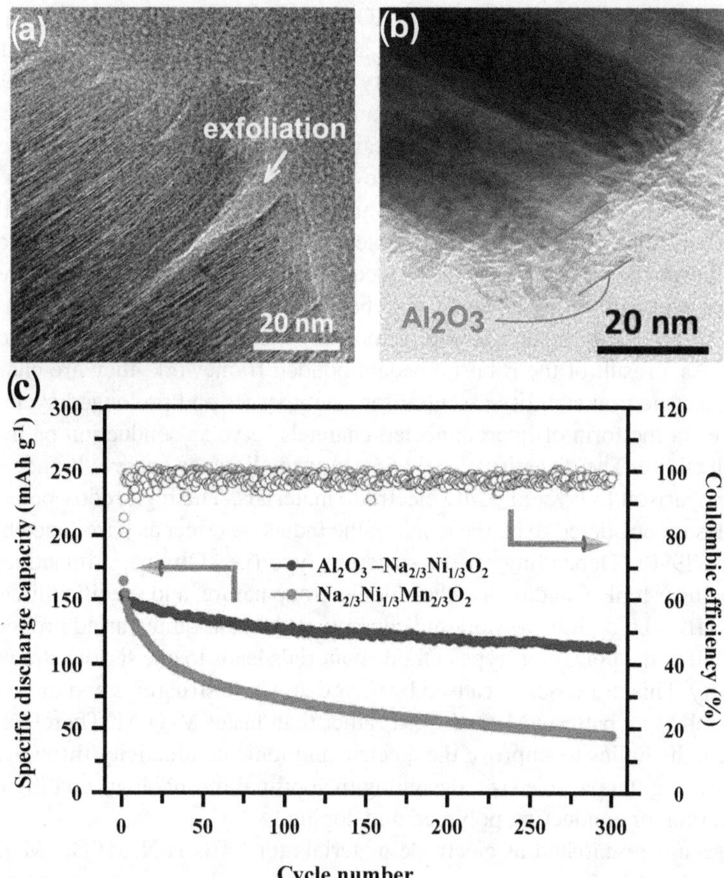

FIGURE 9.7 Protective Al_2O_3 coating on cathode: (a) exfoliated bare P2-type $Na_{2/3}[Ni_{1/3}Mn_{2/3}]O_2$ sample after cycling, (b) Al_2O_3-coated $Na_{2/3}[Ni_{1/3}Mn_{2/3}]O_2$ sample after cycling, and (c) electrochemical performance of bare- and Al_2O_3-coated $Na_{2/3}[Ni_{1/3}Mn_{2/3}]O_2$ sample. Adapted and reproduced with permission from Ref. [72]. Copyright © Elsevier.

the dissolution–migration–deposition process and sodium plating on the anode surface during charge/discharge cycles [30,76].

As shown here, TMOs with optimized composition, stoichiometry, and structure exhibit specific energies comparable to the best performing polyanionic compounds. Their advantage is chemistry based on cheap and non-toxic transition metals and high density, which provides higher volumetric energy density compared to polyanionic compounds (e.g., 250 Wh L^{-1} achieved for $NaNi_{0.4}Cu_{0.1}Mn_{0.4}Ti_{0.1}O_2$ in the full cell compared to 175 Wh L^{-1} for polyanionic $Na_3V_2(PO_4)_2F_3$) [33]. The main disadvantages of TMO are the relatively poor cycling stability resulting from phase transitions and low stability in contact with air/moisture.

9.3 POLYANIONIC COMPOUNDS

Polyanion-type compounds exhibit 3D structures with open channels that facilitate the diffusion of alkali metal ions. A unique framework of these materials enables a minimal structural rearrangement during large sodium ions insertion and extraction, resulting in remarkably small volume changes, leading to outstanding cyclability. They can be described with the following formula: $Na_xM_y(XO_4)_z$, where M is the transition metal and X is either Si, P, S, As, Mo, or W. Such materials are built of corner-shared polyhedron units: $(XO_4)^{z-}$ or their derivatives $(X_nO_{3n+1})^{z-}$ and MO_m in which strong covalent bond exists [81,82]. The advantage of polyanion compounds is the strong bond between X and O, which introduces ionicity in the M-O bonds. This feature leads to a higher redox potential of M^{m+}/M^{n+} as compared to the same metal center in layered metal oxides. Also, the strong covalent bonds between X and O significantly improve the stability of oxygen atoms in the lattice, providing a highly stable framework for reversible Na-ion insertion [83]. Note that structural instability decreases the safety of the materials and is one of the reasons limiting the broader use of Na layered TMOs that suffer from irreversible phase transition processes and low cell voltage (in most cases lower than 3.5 V). The latter factor is extremely limiting for the energy density. Moreover, the presence of strong covalent X-O bonds in polyanion-type materials can inhibit the oxygen evolution [84]. Due to such structure, polyanion-type materials are considered good Na host electrodes. As a result of the robust covalent-bonded framework, they are characterized by a high thermal and structural stability essential for safety issues and prolonged cycle life. The large interstitial spaces in the form of interconnected channels serve as conduction pathways leading to high ionic conductivity. The diversity of structure in turn allows to achieve higher values of redox potentials in comparison to layered TMO electrode materials. The high redox potential of polyanion-type cathodes is considered to be the result of the inductive effect as introduced by Goodenough et al. [85] in the 1990s. Depending on the structure type (i.e., Olivine, Alluaudite, Tavorite, and NASICON (Na Super Ionic Conductor)), the anionic group nature, and specifically the profile of the diffusion paths (1D, 2D, or 3D), polyanionic compounds present quite varied properties. However, the structure of all of the polyanion-type cathode materials leads to one serious drawback: low electrical conductivity. This drawback is caused by their distinctive structures and unfavorable electron transfer, which follows a pattern: M-O-X-O-M rather than faster M-O-M. Therefore, a lot of effort is spent on research studies to improve the electric and ionic conductivity through various strategies, such as reducing the particle size, designing the optimal morphology, coating polyanion-type particles with carbon or conducting polymer, and doping.

One of the groups researched as electrode materials for NIBs is $NaMPO_4$ (M=Fe, Mn) structures that belong to phosphate compounds. Due to the tremendous commercial success of olivine-type $LiFePO_4$ in LIBs, it was reasonable to consider the sodium analog, $NaFePO_4$, for NIBs. $NaFePO_4$ exists in two polymorphs, namely triphylite and maricite. Triphylite phase is isostructural to lithium analog $LiFePO_4$ and exhibits a theoretical capacity of 154 mAh g^{-1}, which is the highest among the iron-based phosphates. However, triphylite phase is thermodynamically metastable (unlike olivine $LiFePO_4$), and in the temperature above 480°C undergoes irreversible phase transition to maricite structure, which is electrochemically significantly less active due to lack of Na-ion diffusion channels [86,87]. Although both structures of $NaFePO_4$ have been reported as potential Na^+ hosts, the specific capacity of pure (uncoated) maricite $NaFePO_4$ (28 mAh g^{-1} at 0.1 C rate [88]) is few times lower than that of triphylite-type (142 mAh g^{-1} at 0.1 C rate [89]). To avoid the creation of the undesirable maricite phase, $NaFePO_4$ cannot be synthesized by high-temperature processes; thus, it is often obtained from the olivine-$LiFePO_4$ through the ion exchange process [89]. Liu et al. [90] synthesized carbon-coated $NaFePO_4$ by milling nanosheets of amorphous iron phosphate with carbon black and further electrochemical sodiation for conductivity improvement. The as-prepared self-nucleated nanocrystals with sizes of 10–15 nm were used as a hybrid cathode material, which showed the exceptionally high reversible capacity of 168.9 mAh g^{-1} at 0.1 C, high rate capability (45.6%) at 10 C, and excellent cycling stability (92.3% of initial capacitance) after 1000 cycles.

$NaMnPO_4$ possesses olivine and maricite phases [91–93]. Contrary to $NaFePO_4$, the olivine-type $NaMnPO_4$ compounds (including substitution with 1/2 Fe and/or Mg) can be prepared directly by the molten salt method at low temperatures (below 100°C) [92]. The olivine $NaMnPO_4$ converts to the thermally stable maricite phase at the temperature of 550°C [92]. Thus, under normal operating conditions of an electrochemical cell with a nonaqueous electrolyte, the metastable olivine phase is fully stable. Similarly, as for the $NaFePO_4$, in the maricite structure of $NaMnPO_4$, sodium-ion mobility is blocked.

The pyrophosphate [P_2O_7 or $(PO_{4-x})_2$] units are another type of polyanionic compounds formed in the temperature range of 500°C–550°C due to the loss of oxygen from the phosphates. Thus, pyrophosphates are thermally more stable than phosphates. They exhibit high intrinsic stability resulting from the presence of stable pyrophosphate anions $P_2O_7^{4-}$. Since the discovery of the first pyrophosphate-class material ($NaFeP_2O_7$) in 1982, several types of pyrophosphate structures have been described, but no application in NIB was proposed until 2003 when Uebou et al. [94] studied the electrochemical insertion of Li^+ and Na^+ into an orthorhombic $(MoO_2)_2P_2O_7$ structure. Later, Barpanda et al. [95] reported carbon-coated $Na_2FeP_2O_7$ as a positive Na^+ insertion host operating at the average voltage of 3 V (vs Na/Na^+) with the capacity of 82 mAh g^{-1} (at C/20 current rate) and a good rate capability. Carbon coating, incorporation in various carbon frameworks, and downsizing the particles improve the electrochemical performance of $Na_2FeP_2O_7$ [96–98].

Another analog, $Na_2MnP_2O_7$, has the same theoretical capacity as $Na_2FeP_2O_7$ (97.5 mAh g^{-1}), but a superior energy density due to the higher redox potential of Mn^{3+}/Mn^{2+} couple. Park et al. [99] reported good electrochemical performance for triclinic $Na_2MnP_2O_7$ in contradiction to its almost inactive lithium counterpart $Li_2MnP_2O_7$. Various structures of $Na_2CoP_2O_7$ were also studied as a cathode material for NIB. Orthorhombic-layered $Na_2CoP_2O_7$ delivered 80 mAh g^{-1} involving Co^{3+}/Co^{2+} redox couple at an average potential of 3 V [100], whereas triclinic $Na_{3.12}Co_{2.44}(P_2O_7)_2$ structure exhibited the impressively high redox potential of 4.3 V (vs Na/Na^+) and high stability during cycling but the low capacity of 40 mAh g^{-1} [101]. These examples illustrate that the electrochemical performance is significantly dependent on the crystal structure [101,102].

Vanadium-based pyrophosphate materials have also been proposed as perspective cathodes for NIB. Monoclinic phase β-$NaVPO_4$ demonstrated a reversible capacity of 104 mAh g^{-1} at an average operating potential of ~3.9 V [103]. Kim et al. [104] proposed $Na_7V_3(P_2O_7)_4$ compound with a moderate theoretical capacity of 80 mAh g^{-1}, low volume change during cycling (~1%), and the highest redox potential of 4.13 V among all vanadium-based polyanionic structures. To date, a variety of polymorphs and analogs of $Na_2MP_2O_7$ have been studied, including $Na_2FeP_2O_7$ [95–98,105], $Na_2MnP_2O_7$ [99,106–108], $Na_2CoP_2O_7$ [100–102], $Na_2CuP_2O_7$ [109,110], and $Na_2ZnP_2O_7$ [111]. To improve the energy density of polyanionic cathodes, the substitution of XO_4 with XO_4F unit has been employed. Thanks to the addition of light fluorine atoms of high electronegativity, the fluorine-based materials usually present a higher theoretical capacity (120–143 mAh g^{-1}) and a higher redox potential (3.4–3.9 V) compared to the fluorine-free analogs. The theoretical specific capacity of $NaVPO_4F$, assuming the reversible cycling of one Na^+ per formula unit, equals to 143 mAh g^{-1}. Zhuo et al. [112] reported $NaVPO_4F$ exhibiting the specific capacity of 87.7 mAh g^{-1} (at 10 mA g^{-1}, in 3.0–4.5 V) with only 73.5% retention after 20 cycles. On the other hand, Ling et al. [113] achieved 135 mAh g^{-1} (in 2.0–4.3 V at 0.2 C) for the carbon-coated nanostructures of $NaVPO_4F$.

Among sulfate-based polyanionic compounds, which are attractive due to high redox potentials, alluaudite-type structures seem most promising. The general formula is $Na_{2+2x}Fe_{2-x}(SO_4)_3$, in which Na^+ ions reside in three distinct sites with different degrees of occupancy. The first alluaudite-type polysulfate, reported by Barpanda et al. [114], delivered a reversible capacity of ~100 mAh g^{-1}. Among the Fe-based cathodes, $Na_2Fe_2(SO_4)_3$ has the highest value of the Fe^{3+}/Fe^{2+} redox potential equal to 3.8 V vs Na^+/Na [115], which is caused by the shortest Fe-Fe distance among Fe-based compounds. To make most of the electrochemical capacity of alluaudite-type polysulfates (theoretical values in the range of 100–120 mAh g^{-1}, depending on stoichiometry), it is demanded to obtain $Na_{2+2x}M_{2-x}(SO_4)_3$ compounds that would be characterized by a Na:M ratio close to 2:2. Unfortunately, a majority of the adopted synthesis methods result in a large non-stoichiometry product with less 3d metals. Plewa et al. [116] showed that an excess of sodium during synthesis may prevent the formation of impurities and enhance the creation of stoichiometric $Na_2FeM(SO_4)_3$ (M=Fe, Mn, Ni) structures. On the other hand, Jungers et al. [117] postulated that obtaining the desired Na:Fe ratio depends on homogeneity of the precursor used. To fully utilize the high redox potential of the sulfate-based polyanionic compounds for high-power applications, it is necessary to enhance their electronic conductivity by the creation of composites with various types of carbonaceous materials [118–121].

Many other SO_4-based materials have been reported, namely $Na_2Co_2(SO_4)_3$ [122], $Na_2Mg_2(SO_4)_3$ [123], or sodium fluorosulfates [124–126]. Sodium fluorosulfates exhibit a very small degree of Na^+ (de)insertion, but the ability might be enhanced for potassium fluorosulfate analogs, which constitute more suitable Na^+ insertion hosts [127], delivering the capacity of 120 mAh g^{-1} and a high voltage (3.5 V).

Due to the two-electron exchange per formula unit, silicates with a general formula of Na_2MSiO_4 (M=Mn, Fe, Co, Ni) may deliver a relatively high specific capacity [128]. Thanks to Si-O bonds, which promote high thermodynamic stability of the structure, volume changes are relatively low [129]. Moreover, considering the abundance of Si and Na resources, they may be competitive when it comes to the cost-effectiveness of cathode materials production. However, their redox potential is incomparably lower due to much lower electronegativity compared to the abovementioned polyanionic compounds. Utilizing Na_2FeSiO_4 as a cathode material may seem to be one of the directions for reducing costs of commercial batteries, and it is gaining more attention as it is characterized by a theoretical capacity of 276 mAh g^{-1}. Na_2FeSiO_4 was first synthesized by Kee et al. [130] with an initial discharge capacity of 126 mAh g^{-1}. However, the material was irreversibly losing its crystalline structure upon cycling. The stable structure was obtained by Li et al. [131], but it delivered a lower capacity of 106 mAh g^{-1}. Wu et al. developed Na_2FeSiO_4/H-N-doped hard carbon nanosphere [132], which enhanced the electron conductivity and provided more sites for sodium-ion storage. Additionally, the mesoporous structure of such composite improved the Na^+ migration rate. Presented cathode material delivered a discharge capacity of 218.4 mAh g^{-1} at 0.1 C, a voltage range of 1.2–4.6 V (vs Na/Na^+), and an impressive cycling life with the capacity retention of 73.8% at 1 C after 3300 cycles. Feng et al. fabricated 3D carbon nanotubes (CNTs)-decorated Na_2FeSiO_4 microspheres, resulting in the increase of the electronic conductivity [133] compared to the bare polysilicate.

Apart from iron-based orthosilicate electrodes, some studies report Na_2CoSiO_4 and $NaMnSiO_4$ materials. Rangasamy et al. combined Na_2CoSiO_4 with functionalized multi-walled CNTs, achieving a capacity of 125 mAh g^{-1} at the C/20 rate, as well as noticeable improvement of the redox kinetics and the conductivity of the material [134]. Zhang et al. reported the outstanding performance of Na_2MnSiO_4/C cathode material (207 mAh g^{-1} at 0.1 C, and 76% of initial capacity after 345 cycles at 2 C), assigned to the honeycomb-like 3D porous structure enabling a short way for sodium-ion diffusion [135].

The first NASICON material, $Na_{1+x}Zr_2P_{3x}Si_xO_{12}$ ($0 \leq x \leq 3$), was reported in the 1960s by Goodenough et al. [85] as a solid state electrolyte. The general formula of NASICON material can be defined as $A_xM_2(XO_4)_3$ (A=Li, Na, K, Mg, Ca; $1 \leq x \leq 4$; M=V, Fe, Ni, Mn, Ti, Cr, Zr, etc.; X=P, S, Si, Se, Mo, etc.) [136–139]. Both cation M and anion X are replaceable in this structure, thus offering a great versatility of NASICON compounds. Furthermore, NASICONs of the same composition may exhibit different crystal structures leading to different electrochemical behaviors. The most representative material of this family is $Na_3V_2(PO_4)_3$ (NVP) (Figure 9.8a) with a theoretical capacity of 117 mAh g^{-1} and a plateau on the charge/discharge curve located at 3.3 V vs Na/Na^+ (Figure 9.8b and c) [140–145].

The thermodynamically stable phase of NVP is the rhombohedral structure [150], which is different from the monoclinic of its lithium counterpart $Li_3V_2(PO_4)_3$ [151]. Electrochemical properties of $Na_3V_2(PO_4)_3$ were first studied by Uebou et al. [140] in 2002. During charging/discharging processes, two Na^+ can be reversibly inserted/extracted into/from NVP material based on the V^{3+}/V^{4+} redox couple [152,153] with a typical two-phase transition reaction from the $Na_3V_2(PO_4)_3$ phase, in the fully charged state, to the $NaV_2(PO_4)_3$ phase, in the discharged state, and a relatively small volume changes (8.26%) [154,155]. However, its low intrinsic electronic conductivity, which results from the separation between the metal polyhedron and electron-rich polyanion in the rhombohedral structure, is the main drawback for commercial use [156]. The capacity obtained for the pure NVP (synthesized via a solution-based carbothermal reduction) was equal to 113 mAh g^{-1} at 0.1 C in the voltage range of 2–4.6 V (vs Na/Na^+) with a corresponding Coulombic efficiency of

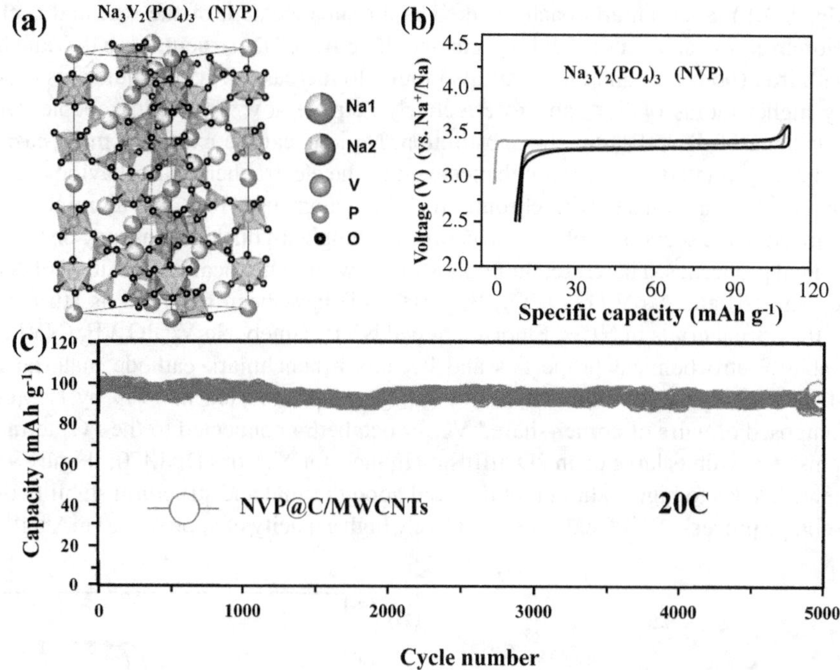

FIGURE 9.8 (a) Unit cell structures of the α-phase (high-temperature) $Na_3V_2(PO_4)_3$ NASICON compound. Crystallographic representation produced with VESTA software [146], based on crystallographic data from [147]. (b) Charge–discharge curves recorded for $Na_3V_2(PO_4)_3$. Adapted and reproduced from Ref. [148], Copyright © 2021 Elsevier. (c) Cycling stability for $Na_3V_2(PO_4)_3$-based composite with carbon nanotubes (CNTs) in half-cell configuration (with Na metal anode) at 20 C rate. Adapted and reproduced with permission from Ref. [149]. Copyright © 2021 Elsevier.

97.4%. After 50 charge/discharge cycles at 0.1 C, the capacity retention was equal to 95% [157]. Carbon coating of NVP combined with the downsizing of its particles is a good method to enhance both electronic and ionic conductivity, and thus improving its electrochemical performance [158]. For instance, Duan et al. [158] synthesized core–shell $Na_3V_2(PO_4)_3$@C nanomaterial (with a particle size of around 40 nm) using a simple hydrothermal method. The nanocomposite exhibited superior electrochemical performance (104.3 mAh g^{-1} at 0.5 C with 99.6% retention after 50 cycles and 96% at 5 C after 700 cycles) compared to the pure NVP. Kang et al. [159] obtained higher value of the capacitance for highly crystalline carbon-coated NVP material synthesized via polyol-mediated pyro-synthetic method. The composite delivered a capacity up to 235 mAh g^{-1}, which corresponds to the theoretical value resulted from insertion/extraction of 4 Na^+ ions per formula when cycled in the voltage range of 3.8–1.2 V. Such high capacity is due to reactions of vanadium V^{3+}/V^{4+} and V^{3+}/V^{2+} redox couples, which occur at 3.4 and 1.62 V potentials (vs Na/Na^+), respectively. Other study [160] showed that carbon-coated nanoparticles embedded in a porous carbon matrix (C@NVP)@pC can exhibit the ultrafast charge–discharge performance and long cycling stability. The delivered specific discharge capacity at current rate of 1 C and 10 C equaled to 104 mAh g^{-1} and 103 mAh g^{-1}, respectively. Moreover, after 1000 cycles at 10 C, the material maintained its original morphology and structure with 80% of the initial capacity value.

Another way to optimize the electrochemical performance of $Na_3V_2(PO_4)_3$ is element doping, which has already been proven to be effective in improving the intrinsic electronic conductivity in Li-ion batteries [161]. For example, potassium ions play an important role in enlarging the Na-ion diffusion pathways with simultaneous increase of the lattice volume [162]. The ionic and electronic conductivities are also significantly improved after Mg doping of the NVP material

($Na_3V_{2-x}Mg_x(PO_4)_3$). Such an approach resulted in an enhancement of the rate and cycle performances. Moreover, the substitution did not change the NASICON structure [163]. Vanadium substitution with iron (both Fe^{2+} and Fe^{3+}) was also found to increase the cell volume, as expected for the slightly higher radius of Fe^{3+}, and to effectively activate a V^{4+}/V^{5+} redox couple. Due to the increasing volume of the cell after the substitution, Na ions can be extracted more easily during electrochemical cycling [164]. Similar enhancement of the electrochemical behavior was found for substitution of the vanadium site with chromium [165], manganese [166], and aluminum [167].

As mentioned in the section on olivine structures, doping with fluoride ions may be beneficial for electrochemical properties. Therefore, fluorophosphates with a chemical composition of Na_2MPO_4F (M=V, Fe, Mn, Ni) and $Na_3V_2O_{2-2x}(PO_4)_2F_{3-x}$ ($0 \leq x \leq 1$) have been regarded as attractive candidates for cathodes materials in NIBs. Fluorine-doped NVP, namely $Na_3V_2(PO_4)_2F_3$ (NVPF), exhibits remarkable electrochemical properties and became a benchmark cathode material for NIBs [148,168–171]. The crystal structure of $Na_3V_2(PO_4)_2F_3$, first reported in 1999 by LeMeins et al. [172], is composed of pairs of corner-shared VO_4F_2 octahedra connected to the PO_4 tetrahedra via oxygen atoms and exhibits large open 2D diffusion tunnels for Na^+ ions [168,170] (Figure 9.9a). Such a 3D framework leads to high sodium mobility and good thermal and structural stability during the charge/discharge process. NVPF exhibits a relatively high capacity of approx. 120 mAh g^{-1} with two

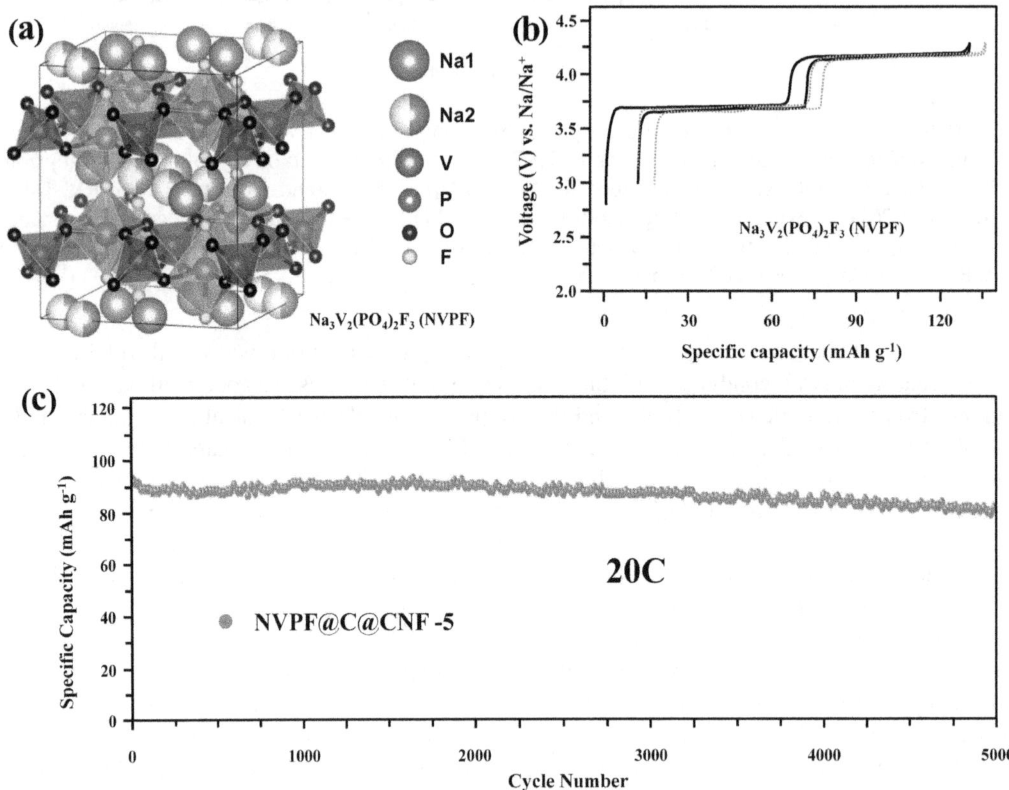

FIGURE 9.9 (a) Unit cell structures of the β-phase (room-temperature) $Na_3V_2(PO_4)_2F_3$ compounds. Crystallographic representation produced with VESTA software. Adapted and reproduced with permission from Ref. [146], based on crystallographic data from [172]. (b) Charge–discharge curves recorded for $Na_3V_2(PO_4)_2F_3$. Adapted and reproduced with permission from Ref. [148] Copyright © 2021 Elsevier. (c) Cycling stability for $Na_3V_2(PO_4)_2F_3$-based composite with carbon nanofibers (CNF) in half-cell configuration (with Na metal anode) at 20C rate. Adapted and reproduced with permission from Ref. [174]. Copyright © 2021 Wiley.

distinct plateaus at approx. 3.6 and 4.0 V (and sometimes small one at 3.3 V) vs Na/Na^+ (Figure 9.9b) [148,173,174], and exceptional cycling stability [174] (Figure 9.9c).

The most frequently used methods for preparation of a well-working cathode material for NIB based on $Na_3V_2(PO_4)_2F_3$ involve a (i) low-temperature synthesis route [175,176], (ii) downsizing the particles sizes to micro- [177,178] or even nanoscale [175,176,178], (iii) uniform carbon coating of the NVPF particles [177], or (iv) embedding them in the conductive carbon matrix [175,176]. Avoiding the high-temperature step not only saves energy and makes the synthesis more cost-effective, but it also enhances the rate performance and cycle stability. The low-temperature synthesis process allows to obtain particles of smaller sizes and with more uniform carbon coating, resulting in higher electronic conductivity. Liu et al. [175] synthesized cathode electrode material that consisted of nanocube-like NVPF (of tetragonal phase) embedded in the matrix of single-walled CNTs via a low-temperature solvothermal method. The material showed a capacity of 116.7 mAh g^{-1} (at 0.5C in the 2.5–4.3 V range) with a voltage plateau at 4.1 V vs Na/Na^+ (theoretical capacity: 1 C = 128 mAh g^{-1}). Recently, Semykina et al. [178] utilized mechanical activation for the improvement of the electrochemical performance of carbon-coated NVPF submicron particles. The capacity of the material after the mechanical activation (MA) raised slightly (from 102 mAh g^{-1} without MA to 108 mAh g^{-1}), and the capacity loss was lowered from 14% to only 2% thanks to the mesoporous structure developed after activation.

From the polyanionic-type compounds, $Na_3V_2(PO_4)_3$ and $Na_3V_2(PO_4)_2F_3$ are considered the most competitive with other potential cathodes for NIBs. This is due to their high energy densities of approx. 400 and 500 Wh kg^{-1} for NVP and NVPF, respectively, and outstanding cycle life. Moreover, both compounds are stable in moist air (contrary to most TMO) and exhibit thermal stability in the charged state of the battery [179]. Their drawback compared to TMO is a relatively low gravimetric capacity.

9.4 METAL HEXACYANOMETALLATE

Metal hexacyanometallates are a group of compounds with the general formula $M_xMe'_y[Me''(CN)_6]{\cdot}zH_2O$, where Me′ and Me″ are divalent and trivalent metal ions, coordinated in structure by nitrogen and carbon from cyanide groups, and M is often monovalent alkali metal cation (e.g., Li^+, Na^+, K^+, and Cs^+), but can be also divalent one such as Ca^{2+}, Zn^{2+}, and Mg^{2+}, which occupies the interstitial sites in the crystal lattice to ensure electroneutrality (see equation 9.5). Water molecules may also be present in the interstitial space (zeolitic water) or coordinate metals at vacancies in the hexacyanometallate complex (fraction of lattice defects is marked as z in equation 9.5). These compounds exhibit a cubic structure. The metals present in the crystal lattice are coordinated by six carbon or six nitrogen atoms of cyanide groups, ensuring octahedral symmetry around the central atom (Figure 9.10) [180–188].

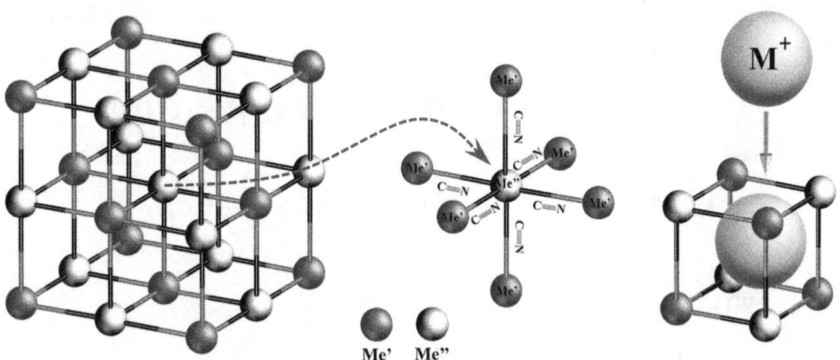

FIGURE 9.10 Schematic illustration of metal hexacyanometallate structure.

$$Me^{x}\left[Me^{y}(CN)_6\right]_{1-z}.w\,H_2O + e^- + M^+ \rightleftarrows MMe^{x-1}\left[Me^{Y}(CN)_6\right]_{1-Z}.\,w\,H_2O \quad (9.5)$$

The parent compound, namely iron hexacyanoferrate, called Prussian Blue (PB), has been known for over 300 years as a dye [189], but its structure and composition were discovered much later [180–182,190,191]. PB is, among all its analogs, the most widely studied and described compound [180–182,184,190–195].

PBAs, due to the possibility of selecting the appropriate metals in the structure, exhibit various properties. The possibility of changing the oxidation state of metals and selecting interstitial cations creates enormous potential for controlling the properties of this group of compounds. Due to their structure, metal hexacyanometallates exhibit electrochromic and thermochromic [183,196,197], ion exchange [198–200], electrocatalytic [201–207], optical [208,209], and magnetic [208–210] properties. Due to the Faradaic reactions and stability upon polarization in various electrolytes, PBAs are attractive for energy storage application [12,188,211–216]. The properties of metal hexacyanometallates may also be influenced by external factors such as temperature [209,217], radiation [209,218–221], or polarization [187,209,222]. These features make metal hexacyanometallates attractive materials for many applications such as (bio)sensors [201,203,205,223–226], batteries [12,188,213–216], electrochromic devices [227,228], molecular sieves [229,230], high-temperature molecular magnets [231,232], or molecular optical switches [209,233]. Metal hexacyanometallates can be obtained chemically by simple precipitation method or electrochemically by electrodeposition directly on the electrode substrate (potentiodynamic, potentiostatic, or galvanostatic methods). The simplicity of preparation can be misleading since relatively small changes in the preparation procedure of the material may result in changes of the materials' structure and properties. As a result, there are many inconsistencies in the literature regarding the same compound, even in the case of the most described PB. The most common differences concern the interstitial spaces, which can be filled with solvent molecules and ions present in the synthesis environment. The electrolyte and the method of synthesis may affect the value of the formal redox potential of the active metals' centers [192,234–237]. Solids may also contain several phases with different chemical surroundings of metal atoms or contain oxide impurities [191]. Cyanide groups can undergo linkage isomerism, binding the metal via carbon or nitrogen interchangeably [187,238–240]. The change in the orientation of the cyanide group in the complex may be induced thermally [238,240]. The degree of this phenomenon depends on the electronic structure of the metals present in the crystal lattice and occurs only in some types of PBAs [187].

Both metal centers in PBAs structure, the one coordinated by carbon and the other coordinated by nitrogen from cyanide groups, may exhibit redox activity. In the case of PB, two Faradaic reactions can be observed – low-spin iron coordinated by carbon (Fe^{LS}) and high spin iron coordinated by nitrogen (Fe^{HS}). The oxidation and reduction reactions of PB in the presence of a potassium salt were proposed by Itaya et al. [184,192] (equations 9.6–9.9):

Reduction of PB to Everett salt:

$$Fe_4{}^{III}\left[Fe^{II}(CN)_6\right]_3 + 4e^- + 4K^+ \rightleftarrows K_4Fe_4{}^{II}\left[Fe^{II}(CN)_6\right]_3 \quad (9.6)$$

$$KFe_4{}^{III}\left[Fe^{II}(CN)_6\right] + e^- + K^+ \rightleftarrows K_2Fe^{II}\left[Fe^{II}(CN)_6\right] \quad (9.7)$$

Oxidation of PB to Berlin green:

$$Fe_4{}^{III}\left[Fe^{II}(CN)_6\right]_3 - 3e^- + 3A^- \rightleftarrows Fe_4{}^{III}\left[Fe^{III}(CN)_6\,A\right]_3 \quad (9.8)$$

(where A is an anion)

$$KFe^{III}\left[Fe^{II}(CN)_6\right]-e^- - K^+ \rightleftarrows Fe_4^{III}\left[Fe^{III}(CN_6)\right] \tag{9.9}$$

As the oxidation and reduction reactions of PBAs are associated with the (de-)insertion of cations, anions, and solvent molecules into interstitial sites, the size of this interstitial space determines the possibility of redox reactions to occur in specific electrolytes.

PB, depending on whether it contains potassium ions in its structure ($KFe^{III}[Fe^{II}(CN)_6]_3$) or not ($Fe^{III}_4[Fe^{II}(CN)_6]_3$), exists in the so-called "soluble" or "insoluble" form. This nomenclature was created for the production of dyes and had nothing to do with the actual solubility in water because both solubility constants (of "soluble" and "insoluble" forms) are very low (in the order of 10^{-40}). The nomenclature only determines the ability of the KFe^{III} $[Fe^{II}(CN)_6]_3$ form to peptize [241].

PBAs are attractive electrode materials for batteries due to several features: (i) a high specific capacity coming from two-electron reactions (when both metal centers, high spin coordinated by N and low spin coordinated by C from -CN groups, are involved), (ii) cubic geometry with wide channels creating an open framework structure for rapid ionic conduction enabling good rate capability, (iii) minimal volume changes during ion insertion, ensuring impressively long cycle lives, (iv) possibility to be prepared via a simple and inexpensive co-precipitation reaction at low temperatures in aqueous solution from non-toxic, abundant elements, (v) stability in moist air, and (vi) highly tunable properties for various applications. The main drawback, however, is their low density (1.96 g cm^{-3} for sodium iron hexacyanoferrate), leading to poor volumetric energy density, especially compared to sodium layered TMOs. Another issue that seems problematic is the presence of cyanide group and the resulting possibility of releasing toxic HCN gas during decomposition. However, the risk is minimal as PB is thermally stable up to 200°C [242].

From the PB family, several compounds may reversibly (de-)insert sodium ions. These are mainly metal hexacyanoferrates (MehcFe), where Me = Fe, Ni, Co, Cu, Mn, Zn [188]. Among these compounds, FehcFe, MnhcFe, and CohcFe are the most interesting ones as both metals (high- and low-spin) undergo redox reaction, contributing to the capacity values [235,236,243,244]. The stoichiometric composition of $Na_2FeFe(CN)_6$ may exchange 2 Na^+, leading to a high capacity of ~170 mAh g^{-1} in the voltage range from 2 to 4 V vs Na/Na^+ (with an average value of 2.7 V) [245]. On the other hand, sodium manganese hexacyanoferrate $Na_xMnFe(CN)_6$ (called Prussian white, due to the white color of the powder) is more promising as it exhibits high capacity (~140 mAh g^{-1}) with redox reactions at higher potentials (average 3.4 V) [235,244]. This compound, obtained by an optimized synthesis procedure resulting in sodium-rich phase (showing a capacity of 160 mAh g^{-1}, which is close to the theoretical value), has been employed in the prototype battery with a hard carbon anode. This prototype pouch cell (Novasis) exhibited good rate capability (82% of capacity at 10 C relatively to 100% recorded at C/3 rate), outstanding cycle stability (98.6% of the initial capacity after 500 cycles at 1 C), and exceptional performance at low temperatures (at –20°C, the pouch cell reached 83% of capacity delivered at room temperature), as shown in Figure 9.11 [246].

Also, MnhcMn [247] was reported to undergo redox reactions of both Mn centers delivering an exceptionally high capacity of 209 mAh g^{-1}, which corresponds to three distinct redox reactions: $Mn^{II}–N≡C–Mn^{III/I}$, $Mn^{II}–N≡C–Mn^{III/II}$, and $Mn^{III/II}–N≡C–Mn^{III}$, accompanied by sodium ions insertion steps, which occur at potentials equal to 1.8, 2.65, and 3.55 V vs Na/Na^+ [247]. It is worth noticing that a huge influence on electrochemical performance has a quality of crystal structure. The presence of vacancies involves water coordination to maintain electroneutrality, which was found to limit the sodium-ion storage capacity [245,248]. As reported by You et al. [245], high-quality PBA crystals with a low amount of defects ($Na_{0.61}Fe[Fe(CN)_6]_{0.94}$) exhibit significantly better electrochemical performance than highly defected material ($Na_{0.13}Fe[Fe(CN)_6]_{0.68}$) (Figure 9.12).

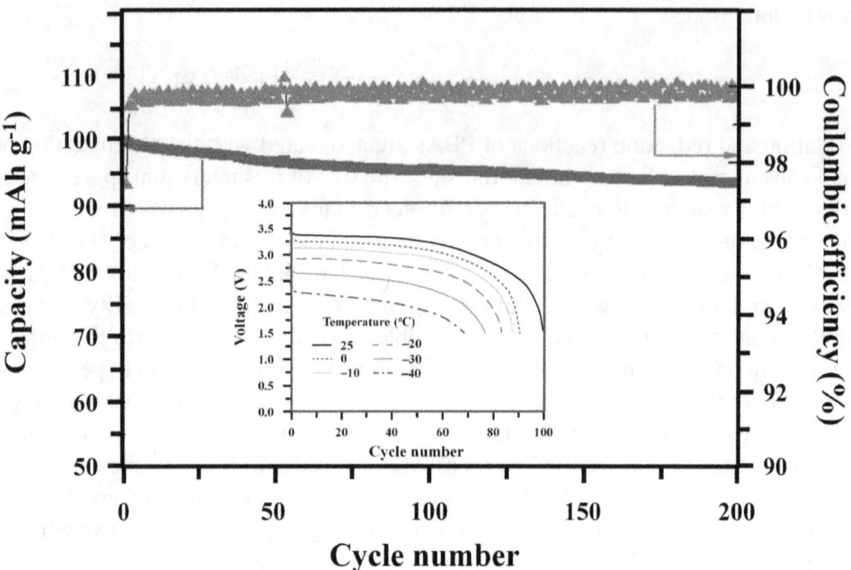

FIGURE 9.11 Cycling stability of the Novasis pouch cell ($Na_xMnFe(CN)_6$ cathode with hard carbon anode) with a charge/discharge rate of 1 C; inset: pouch cell performance at low temperatures with a charge/discharge rate of 1 C. Adapted and reproduced with permission from Ref. [246]. Copyright © 2021 Wiley.

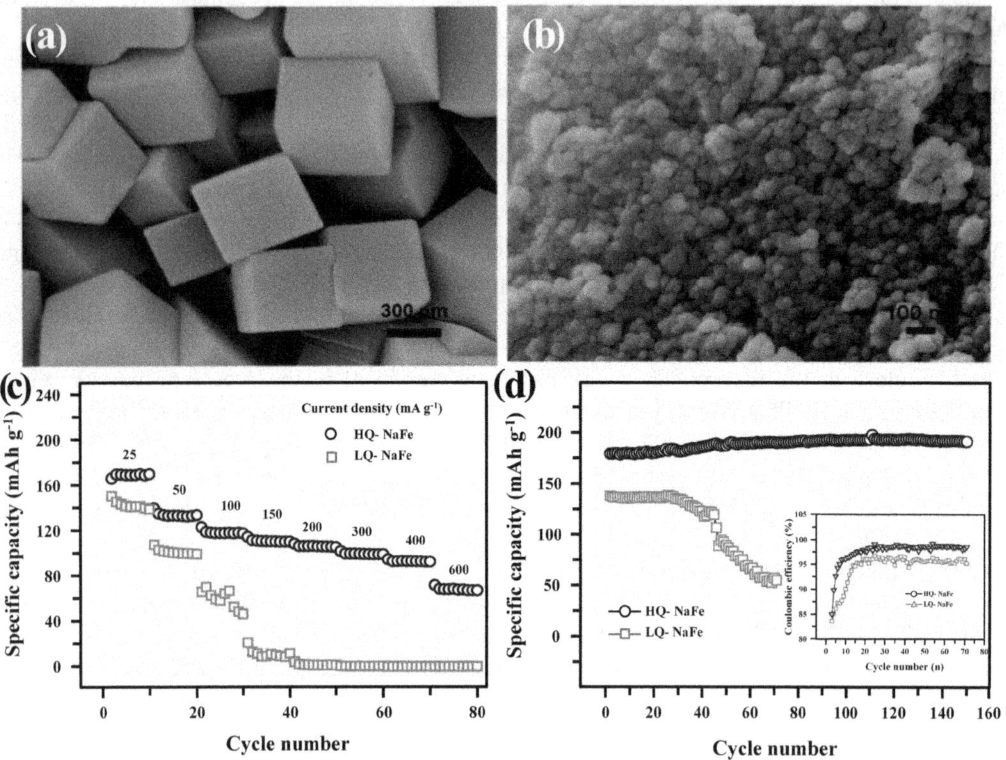

FIGURE 9.12 SEM image of (a) high-quality of PB (HQ-NaFe) with small amount of defects in the structure, (b) low-quality of PB (LQ-NaFe) with a high amount of vacancies (32%); electrochemical performance, (c) rate capability, and (d) cycling stability of high- and low-quality crystals of PB. Adapted and reproduced with permission from Ref. [245]. Copyright © (2014) Royal Society of Chemistry.

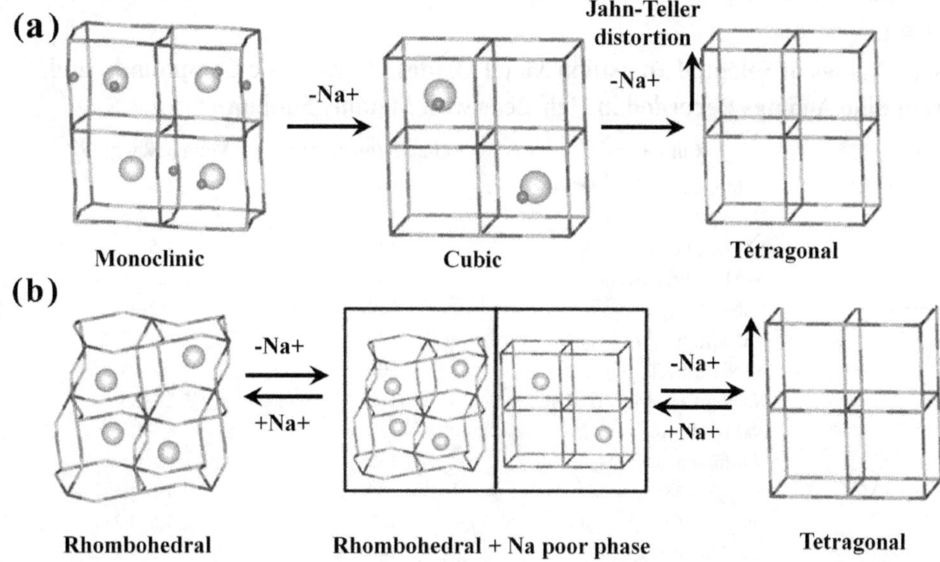

FIGURE 9.13 Schematic illustration of structural evolution of (a) monoclinic $Na_{2-\delta}MnFe(CN)_6$ (hydrated) and (b) rhombohedral $Na_{2-\delta}MnFe(CN)_6$ (dehydrated). Adapted and reproduced with permission from Ref. [237]. Copyright © (2015) American Chemical Society.

Also, removing of interstitial water molecules from $Na_{2-\delta}MnFe(CN)_6$ phase resulted in the structure change from monoclinic (hydrated sample) to rhombohedral (dehydrated sample) (Figure 9.13), which led to an improved electrochemical response in terms of increased capacity (~150 mAh g^{-1}) at the higher potential of 3.5 V [237]. The redox reaction and involved intercalation of metal cation generally cause phase transition, as shown in Figure 9.13 for hydrated and dehydrated Prussian white samples.

As shown, PBAs are promising cathodes for NIBs. Their open framework structure enables fast kinetics, whereas metal redox centers provide competitive capacity values. Other advantages are their highly tunable properties and stability in moist air. The main drawback, however, is the necessity to strictly control synthesis procedures to obtain defect-free crystal phases, as structure significantly affects capacity values and cycle life.

9.5 CONCLUSIONS

The presented electrochemical performance of three types of cathode materials and proposed strategies for solving problematic issues raise expectations that sodium-ion batteries will reach technological maturity in the near future and will partially replace LIBs. By careful material design and optimized synthesis routes, it is possible to achieve high capacity values (as presented in Table 9.1 for the selected materials), which lead to the competitive energy density values to those achieved for LIBs. There are still new materials to be explored for NIB cathodes. Moreover, attention should be paid to reactions that take place on electrode/electrolyte interphases to reduce undesirable side reactions and improve cycling stability. Additionally, electrochemically stable electrolytes need to be recognized, enabling polarization to high voltages, which will further increase energy density. A huge need for sustainable and green energy production is a driving force to produce energy storage systems, which will secure energy supplies.

TABLE 9.1
Capacity Values of Selected Transition Metal Oxides, Polyanionic Compounds, and Prussian Blue Analogs Recorded in Half Cells with Metallic Sodium

Type	Compound	Capacity/mAh g^{-1}	Voltage Range/V	Ref.
O3-phase $NaMO_2$	$NaMn_{0.25}Fe_{0.25}Co_{0.25}Ni_{0.25}O_2$	180	1.9–4.3	[48]
	$NaFe_{0.33}Co_{0.33}Ni_{0.33}O_2$	180	2–4.5	[249]
	$NaMn_{0.33}Fe_{0.33}Co_{0.33}O_2$	180	1.5–4.5	[249]
	$NaMn_{0.33}Fe_{0.33}Ni_{0.33}O_2$	185	2–4.5	[249]
	$NaNi_{0.5}Mn_{0.5}O_2$	140	2–4	[55]
	$NaNi_{0.45}Zn_{0.05}Mn_{0.4}Ti_{0.1}O_2$	170	2–4.5	[23]
	$NaNi_{0.5}Mn_{0.3}Ti_{0.2}O_2$	200	2–4.5	[33]
	$NaNi_{0.45}Cu_{0.05}Mn_{0.3}Ti_{0.2}O_2$	180	2–4.5	[33]
	$NaMn_{0.5}Fe_{0.5}O_2$	153	2–4.5	[249]
	$NaMn_{0.5}Fe_{0.5}O_2$	135	2–4.5	[249]
P2-phase Na_xMO_2	$Na_{0.67}Fe_{0.5}Mn_{0.5}O_2$	190	1.5–4.3	[5]
	$Na_{0.67}Fe_{0.2}Cu_{0.14}Mn_{0.66}O_2$	200	1.5–4.3	[250]
	$Na_{0.6}Ni_{0.2}Mn_{0.6}Co_{0.2}O_2$	140	1.5–4.2	[27]
	$Na_{0.67}[Ni_{0.33}Mn_{0.67}]O_2$	160	2–4.5	[36]
	$Na_{0.67}Mg_{0.28}Mn_{0.72}O_2$	160	2–4.5	[251]
Polyanionic	$NaFePO_4$	168	1.4–3.8	[90]
	$Na_2Fe_2(SO_4)_3$	100	2–4.5	[115]
	Na_2MnSiO_4	207	1.5–4.5	[135]
	$Na_3V_2(PO_4)_3$	113	2–4.6	[157]
	$Na_3V_2(PO_4)_3$	235	1.2–3.8	[159]
	$Na_3V_2(PO_4)_2F_3$	120	2–4.3	[174]
PBAs	$Na_2FeFe(CN)_6$	170	2–4	[245]
	$Na_xMnFe(CN)_6$	160	2–3.7	[246]
	$Na_2Mn[Mn(CN)_6]$	209	1.5–4	[247]

ACKNOWLEDGMENTS

Financial support from BEETHOVEN CLASSIC 3 program of National Science Centre is gratefully acknowledged (Project "Beyond Li-ion batteries: on novel and Efficient electrode materials for Sodium Storage", Grant No. UMO-2018/31/G/ST5/02056).

REFERENCES

1. Delmas C, Braconnier J-J, Fouassier C, Hagenmuller P (1981) Electrochemical intercalation of sodium in Na_xCoO_2 bronzes. *Solid State Ion* 3:165–169.
2. Abraham KM (1982) Intercalation positive electrodes for rechargeable sodium cells. *Solid State Ion* 7:199–212. https://doi.org/10.1016/0167-2738(82)90051-0.
3. Delmas C (1989) Alkali metal intercalation in layered oxides. *Mater Sci Eng B* 3:97–101. https://doi.org/10.1016/0921-5107(89)90185-2.
4. Yabuuchi N, Kubota K, Dahbi M, Komaba S (2014) Research development on sodium-ion batteries. *Chem Rev* 114:11636–11682. https://doi.org/10.1021/cr500192f.
5. Yabuuchi N, Kajiyama M, Iwatate J, et al (2012) P2-type Nax$[Fe_{1/2}Mn_{1/2}]O_2$ made from earth-abundant elements for rechargeable Na batteries. *Nat Mater* 11:512–517. https://doi.org/10.1038/nmat3309.
6. Shannon RD (1976) Revised effective ionic radii and systematic studies of interatomic distances in halides and chalcogenides. *Acta Crystallogr Sect A* 32:751–767. https://doi.org/10.1107/S0567739476001551.

7. Doeff MM, Ma Y, Visco SJ, De Jonghe LC (1993) Electrochemical insertion of sodium into carbon. *J Electrochem Soc* 140:L169–L170. https://doi.org/10.1149/1.2221153.
8. Ge P, Fouletier M (1988) Electrochemical intercalation of sodium in graphite. *Solid State Ion* 28-30:1172–1175. https://doi.org/10.1016/0167-2738(88)90351-7.
9. Jache B, Adelhelm P (2014) Use of graphite as a highly reversible electrode with superior cycle life for sodium-ion batteries by making use of co-intercalation phenomena. *Angew Chemie – Int Ed* 53:10169–10173. https://doi.org/10.1002/anie.201403734.
10. Hasa I, Dou X, Buchholz D, et al (2016) A sodium-ion battery exploiting layered oxide cathode, graphite anode and glyme-based electrolyte. *J Power Sources* 310:26–31. https://doi.org/10.1016/j.jpowsour.2016.01.082.
11. Park J, Xu ZL, Kang K (2020) Solvated ion intercalation in graphite: Sodium and beyond. *Front Chem* 8:1–14. https://doi.org/10.3389/fchem.2020.00432.
12. Hurlbutt K, Wheeler S, Capone I, Pasta M (2018) Prussian blue analogs as battery materials. *Joule* 2:1950–1960. https://doi.org/10.1016/j.joule.2018.07.017.
13. Delmas C, Fouassier C, Hagenmuller P (1980) Structural classification and properties of the layered oxides. *Physica* 99B:81–85.
14. Hasa I, Buchholz D, Passerini S, Hassoun J (2015) A comparative study of layered transition metal oxide cathodes for application in sodium-ion battery. *ACS Appl Mater Interfaces* 7:5206–5212. https://doi.org/10.1021/am5080437.
15. Lee E, Lu J, Ren Y, et al (2014) Layered P2/O3 intergrowth cathode: Toward high power Na-ion batteries. *Adv Energy Mater* 4:1–8. https://doi.org/10.1002/aenm.201400458.
16. Kubota K, Kumakura S, Yoda Y, et al (2018) Electrochemistry and solid-state chemistry of $NaMeO_2$ (Me = 3d Transition Metals). *Adv Energy Mater* 8:1703415. https://doi.org/10.1002/aenm.201703415.
17. Liu Y, Liu X, Wang T, et al (2017) Research and application progress on key materials for sodium-ion batteries. *Sustain Energy Fuels* 1:986–1006. https://doi.org/10.1039/c7se00120g.
18. Yabuuchi N, Yano M, Yoshida H, et al (2013) Synthesis and electrode performance of O3-type $NaFeO_2$-$NaNi_{1/2}Mn_{1/2}O_2$ solid solution for rechargeable sodium batteries. *J Electrochem Soc* 160:A3131–A3137. https://doi.org/10.1149/2.018305jes..
19. You Y, Manthiram A (2018) Progress in high-voltage cathode materials for rechargeable sodium-ion batteries. *Adv Energy Mater* 8:1–11. https://doi.org/10.1002/aenm.201701785.
20. Han MH, Gonzalo E, Singh G, Rojo T (2015) A comprehensive review of sodium layered oxides: Powerful cathodes for Na-ion batteries. *Energy Environ Sci* 8:81–102. https://doi.org/10.1039/c4ee03192j.
21. You Y, Kim SO, Manthiram A (2017) A honeycomb-layered oxide cathode for sodium-ion batteries with suppressed P3-O1 phase transition. *Adv Energy Mater* 7:1–7. https://doi.org/10.1002/aenm.201601698.
22. Zhang X, Jiang K, Guo S, et al (2018) Exploring a high capacity O3-type cathode for sodium-ion batteries and its structural evolution during an electrochemical process. *Chem Commun* 54:12167–12170. https://doi.org/10.1039/C8CC05888A.
23. Mariyappan S, Marchandier T, Rabuel F, et al (2020) The role of divalent (Zn^{2+}/Mg^{2+}/Cu^{2+}) substituents in achieving full capacity of sodium layered oxides for Na-ion battery applications. *Chem Mater* 32:1657–1666. https://doi.org/10.1021/acs.chemmater.9b05205.
24. Lee B, Paek E, Mitlin D, Lee SW (2019) Sodium metal anodes: Emerging solutions to dendrite growth. *Chem Rev*. https://doi.org/10.1021/acs.chemrev.8b00642.
25. Fan L, Li X (2018) Recent advances in effective protection of sodium metal anode. *Nano Energy* 53:630–642.
26. Singh G, Acebedo B, Cabanas MC, et al (2013) An approach to overcome first cycle irreversible capacity in P2-$Na_{2/3}[Fe_{1/2}Mn_{1/2}]O_2$. *Electrochem commun* 37:61–63. https://doi.org/10.1016/j.elecom.2013.10.008.
27. Sathiya M, Thomas J, Batuk D, et al (2017) Dual stabilization and sacrificial effect of Na_2CO_3 for increasing capacities of Na-ion cells based on P2-Na_xMO_2 electrodes. *Chem Mater* 29:5948–5956. https://doi.org/10.1021/acs.chemmater.7b01542.
28. Shanmukaraj D, Kretschmer K, Sahu T, et al (2018) Highly efficient, cost effective, and safe sodiation agent for high-performance sodium-ion batteries. *ChemSusChem* 11:3286–3291. https://doi.org/10.1002/cssc.201801099.
29. Zhang B, Dugas R, Rousse G, et al (2016) Insertion compounds and composites made by ball milling for advanced sodium-ion batteries. *Nat Commun* 7:1–9. https://doi.org/10.1038/ncomms10308.
30. Sun YK (2020) Direction for commercialization of O3-type layered cathodes for sodium-ion batteries. *ACS Energy Lett* 5:1278–1280. https://doi.org/10.1021/acsenergylett.0c00597.

31. Komaba S, Yabuuchi N, Nakayama T, et al (2012) Study on the reversible electrode reaction of $Na_{1-x}Ni_{0.5}Mn_{0.5}O_2$ for a rechargeable sodium-ion battery. *Inorg Chem* 51:6211–6220. https://doi.org/10.1021/ic300357d.
32. Mariyappan S, Wang Q, Tarascon JM (2018) Will sodium layered oxides ever be competitive for sodium ion battery applications? *J Electrochem Soc* 165:A3714–A3722. https://doi.org/10.1149/2.0201816jes.
33. Wang Q, Mariyappan S, Vergnet J, et al (2019) Reaching the energy density limit of layered O3-$NaNi_{0.5}Mn_{0.5}O_2$ electrodes via dual Cu and Ti substitution. *Adv Energy Mater* 9:1–11. https://doi.org/10.1002/aenm.201901785.
34. Silván B, Gonzalo E, Djuandhi L, et al (2018) On the dynamics of transition metal migration and its impact on the performance of layered oxides for sodium-ion batteries: $NaFeO_2$ as a case study. *J Mater Chem A* 6:15132–15146. https://doi.org/10.1039/c8ta02473a.
35. Ryu HH, Park KJ, Yoon CS, Sun YK (2018) Capacity fading of ni-rich $Li[Ni_xCoyMn_{1-x-y}]O_2$ ($0.6 \leq x \leq 0.95$) cathodes for high-energy-density lithium-ion batteries: Bulk or surface degradation? *Chem Mater* 30:1155–1163. https://doi.org/10.1021/acs.chemmater.7b05269.
36. Lee DH, Xu J, Meng YS (2013) An advanced cathode for Na-ion batteries with high rate and excellent structural stability. *Phys Chem Chem Phys* 15:3304–3312. https://doi.org/10.1039/c2cp44467d.
37. Lu Z, Dahn JR (2001) In situ x-ray diffraction study of P2-$Na_{2/3}[Ni_{1/3}Mn_{2/3}]O_2$. *J Electrochem Soc* 148:A1225. https://doi.org/10.1149/1.1407247.
38. Kubota K, Komaba S (2015) Review-practical issues and future perspective for Na-ion batteries. *J Electrochem Soc* 162:A2538–A2550. https://doi.org/10.1149/2.0151514jes.
39. Komaba S, Nakayama T, Ogata A, et al (2009) Electrochemically reversible sodium intercalation of layered $NaNi_{0.5}Mn_{0.5}O_2$ and $NaCrO_2$. *ECS Trans* 16:43–55. https://doi.org/10.1149/1.3112727.
40. Yuan DD, Wang YX, Cao YL, et al (2015) Improved electrochemical performance of Fe-substituted $NaNi_{0.5}Mn_{0.5}O_2$ cathode materials for sodium-ion batteries. *ACS Appl Mater Interfaces* 7:8585–8591. https://doi.org/10.1021/acsami.5b00594.
41. Yao HR, Wang PF, Gong Y, et al (2017) Designing air-stable O3-type cathode materials by combined structure modulation for Na-ion batteries. *J Am Chem Soc* 139:8440–8443. https://doi.org/10.1021/jacs.7b05176.
42. Li Y, Yang Z, Xu S, et al (2015) Air-stable copper-based P2-$Na_{7/9}Cu_{2/9}Fe_{1/9}Mn_{2/3}O_2$ as a new positive electrode material for sodium-ion batteries. *Adv Sci* 2:1–7. https://doi.org/10.1002/advs.201500031.
43. Deng J, Luo W Bin, Lu X, et al (2018) High energy density sodium-ion battery with industrially feasible and air-stable O3-type layered oxide cathode. *Adv Energy Mater* 8:1–9. https://doi.org/10.1002/aenm.201701610.
44. Mu L, Xu S, Li Y, et al (2015) Prototype sodium-ion batteries using an air-stable and Co/Ni-free O3-layered metal oxide cathode. *Adv Mater* 27:6928–6933. https://doi.org/10.1002/adma.201502449.
45. Mu L, Hou Q, Yang Z, et al (2019) Water-processable P2-$Na_{0.67}Ni_{0.22}Cu_{0.11}Mn_{0.56}Ti_{0.11}O_2$ cathode material for sodium ion batteries. *J Electrochem Soc* 166:A251–A257. https://doi.org/10.1149/2.0881902jes.
46. Xu J, Chen J, Zhang K, et al (2020) Nax(Cu-Fe-Mn)O_2 system as cathode materials for Na-ion batteries. *Nano Energy* 78:105142. https://doi.org/10.1016/j.nanoen.2020.105142.
47. Hwang JY, Yoon CS, Belharouak I, Sun YK (2016) A comprehensive study of the role of transition metals in O3-type layered $Na[Ni_xCo_yMn_z]O_2$ (x = 1/3, 0.5, 0.6, and 0.8) cathodes for sodium-ion batteries. *J Mater Chem A* 4:17952–17959. https://doi.org/10.1039/c6ta07392a.
48. Li X, Wu D, Zhou YN, et al (2014) O3-type $Na(Mn_{0.25}Fe_{0.25}Co_{0.25}Ni_{0.25})O_2$: A quaternary layered cathode compound for rechargeable Na ion batteries. *Electrochem commun* 49:51–54. https://doi.org/10.1016/j.elecom.2014.10.003..
49. Oh S-M, Myung S-T, Hwang J-Y, et al (2014) High capacity O3-type $Na[Li0._{05}(Ni_{0.25}Fe_{0.25}Mn_{0.5})_{0\cdot 95}]O_2$ cathode for sodium ion batteries. *Chem Mater* 26:6165–6171. https://doi.org/10.1021/cm502481b.
50. Zhang C, Gao R, Zheng L, et al (2018) New insights into the roles of Mg in improving the rate capability and cycling stability of O_3-$NaMn_{0.48}Ni_{0.2}Fe_{0.3}Mg_{0.02}O_2$ for sodium-ion batteries. *ACS Appl Mater Interfaces* 10:10819–10827. https://doi.org/10.1021/acsami.7b18226.
51. Wang PF, Yao HR, Liu XY, et al (2017) Ti-substituted $NaNi_{0.5}Mn_{0.5-x}Ti_xO_2$ cathodes with reversible O3–P3 phase transition for high-performance sodium-ion batteries. *Adv Mater* 29:1–7. https://doi.org/10.1002/adma.201700210.
52. Luo M, Ortiz AL, Shaw LL (2020) Enhancing the electrochemical performance of $NaCrO_2$ through structural defect control. *ACS Appl Energy Mater* 3:7216–7227. https://doi.org/10.1021/acsaem.0c01302.
53. Wang Y, Cui P, Zhu W, et al (2019) Enhancing the electrochemical performance of an O3-$NaCrO_2$ cathode in sodium-ion batteries by cation substitution. *J Power Sources* 435. https://doi.org/10.1016/j.jpowsour.2019.226760.

54. Zhou C, Yang L, Zhou C, et al (2019) Co-substitution enhances the rate capability and stabilizes the cyclic performance of O3-type cathode $NaNi_{0.45-x}Mn_{0.25}Ti_{0.3}Co_xO_2$ for sodium-ion storage at high voltage. *ACS Appl Mater Interfaces* 11:7906–7913. https://doi.org/10.1021/acsami.8b17945.
55. Sathiya M, Jacquet Q, Doublet ML, et al (2018) A chemical approach to raise cell voltage and suppress phase transition in O3 sodium layered oxide electrodes. *Adv Energy Mater* 8:1–10. https://doi.org/10.1002/aenm.201702599.
56. Park JK, Park GG, Kwak HH, et al (2017) Enhanced rate capability and cycle performance of titanium-substituted P2-type $Na_{0.67}Fe_{0.5}Mn_{0.5}O_2$ as a cathode for sodium-ion batteries. *ACS Omega* 3:361–368. https://doi.org/10.1021/acsomega.7b01481.
57. Zheng L, Obrovac MN (2017) Investigation of O3-type $Na_{0.9}Ni_{0.45}MnxTi_{0.55-x}O_2$ ($0 \leq x \leq 0.55$) as positive electrode materials for sodium-ion batteries. *Electrochim Acta* 233:284–291. https://doi.org/10.1016/j.electacta.2017.03.033.
58. Zhao W, Tanaka A, Momosaki K, et al (2015) Enhanced electrochemical performance of Ti substituted P2-$Na_{2/3}Ni_{1/4}Mn_{3/4}O_2$ cathode material for sodium ion batteries. *Electrochim Acta* 170:171–181. https://doi.org/10.1016/j.electacta.2015.04.125.
59. Yu H, Guo S, Zhu Y, et al (2014) Novel titanium-based O3-type $NaTi_{0.5}Ni_{0.5}O_2$ as a cathode material for sodium ion batteries. *Chem Commun* 50:457–459. https://doi.org/10.1039/c3cc47351a.
60. Yoshida H, Yabuuchi N, Kubota K, et al (2014) P2-type $Na_{2/3}Ni_{1/3}Mn_{2/3-x}TixO_2$ as a new positive electrode for higher energy Na-ion batteries. *Chem Commun* 50:3677–3680. https://doi.org/10.1039/c3cc49856e.
61. Wang PF, Xin H, Zuo TT, et al (2018) An abnormal 3.7 volt O3-type sodium-ion battery cathode. *Angew Chemie – Int Ed* 57:8178–8183. https://doi.org/10.1002/anie.201804130.
62. Xu B, Fell CR, Chi M, Meng YS (2011) Identifying surface structural changes in layered Li-excess nickel manganese oxides in high voltage lithium ion batteries: A joint experimental and theoretical study. *Energy Environ Sci* 4:2223–2233. https://doi.org/10.1039/c1ee01131f.
63. Kim D, Kang SH, Slater M, et al (2011) Enabling sodium batteries using lithium-substituted sodium layered transition metal oxide cathodes. *Adv Energy Mater* 1:333–336. https://doi.org/10.1002/aenm.201000061.
64. Xu X, Lee DH, Clement R, et al (2014) Identifying the critical role of Li substitution in P2-$Na_x(Li_yNi_xMn_{1-y-z})O_2$ ($0<x, y, z<1$) intercalation cathode materials for high energy Na-ion batteries. *Chem Mater* 2:1260–1269. https://doi.org/10.1021/cm403855t.
65. Xu J, Liu H, Meng YS (2015) Exploring Li substituted O3-structured layered oxides $NaLi_xNi_{1/3-x}Mn_{1/3+x}Co_{1/3-x}O_2$ (x=0.07, 0.13, and 0.2) as promising cathode materials for rechargeable Na batteries. *Electrochem commun* 60:13–16. https://doi.org/10.1016/j.elecom.2015.07.023.
66. Hwang JY, Oh SM, Myung ST, et al (2015) Radially aligned hierarchical columnar structure as a cathode material for high energy density sodium-ion batteries. *Nat Commun* 6:1–9. https://doi.org/10.1038/ncomms7865.
67. Liu Y, Shen Q, Zhao X, et al (2020) Hierarchical engineering of porous P2-$Na_{2/3}Ni_{1/3}Mn_{2/3}O_2$ nanofibers assembled by nanoparticles enables superior sodium-ion storage cathodes. *Adv Funct Mater* 30:1–11. https://doi.org/10.1002/adfm.201907837.
68. Sun YK, Chen Z, Noh HJ, et al (2012) Nanostructured high-energy cathode materials for advanced lithium batteries. *Nat Mater* 11:942–947. https://doi.org/10.1038/nmat3435.
69. Noh HJ, Chen Z, Yoon CS, et al (2013) Cathode material with nanorod structure - An application for advanced high-energy and safe lithium batteries. *Chem Mater* 25:2109–2115. https://doi.org/10.1021/cm4006772.
70. Marchand-Brynaert J, Jongen N, Dewez JL (1997) Surface hydroxylation of poly(vinylidene fluoride) (PVDF) film. *J Polym Sci Part A Polym Chem* 35:1227–1235. https://doi.org/10.1002/(SICI)1099-0518(199705)35:7<1227::AID-POLA8>3.0.CO;2-Z.
71. Awanis Hashim N, Liu Y, Li K (2011) Stability of PVDF hollow fibre membranes in sodium hydroxide aqueous solution. *Chem Eng Sci* 66:1565–1575. https://doi.org/10.1016/j.ces.2010.12.019.
72. Liu Y, Fang X, Zhang A, et al (2016) Layered P2-$Na_{2/3}[Ni_{1/3}Mn_{2/3}]O_2$ as high-voltage cathode for sodium-ion batteries: The capacity decay mechanism and Al_2O_3 surface modification. *Nano Energy* 27:27–34. https://doi.org/10.1016/j.nanoen.2016.06.026.
73. Hwang JY, Myung ST, Choi JU, et al (2017) Resolving the degradation pathways of the O3-type layered oxide cathode surface through the nano-scale aluminum oxide coating for high-energy density sodium-ion batteries. *J Mater Chem A* 5:23671–23680. https://doi.org/10.1039/c7ta08443a.
74. Yu Y, Kong W, Li Q, et al (2020) Understanding the multiple effects of TiO_2 coating on $NaMn_{0.33}Fe_{0.33}Ni_{0.33}O_2$ cathode material for Na-ion batteries. *ACS Appl Energy Mater* 3:933–942.

75. Sun HH, Hwang JY, Yoon CS, et al (2018) Capacity degradation mechanism and cycling stability enhancement of AlF_3-coated nanorod gradient $Na[Ni_{0.65}Co_{0.08}Mn_{0.27}]O_2$ cathode for sodium-ion batteries. *ACS Nano* 12:12912–12922. https://doi.org/10.1021/acsnano.8b08266.
76. Zhang Q, Gu Q-F, Li Y, et al (2019) Surface stabilization of O3-type layered oxide cathode to protect the anode of sodium ion batteries for superior lifespan. *iScience* 19:244–254. https://doi.org/10.1016/j.isci.2019.07.029.
77. Zheng J, Gu M, Xiao J, et al (2014) Functioning mechanism of AlF_3 coating on the Li- and Mn-rich cathode materials. *Chem Mater* 26:6320–6327. https://doi.org/10.1021/cm502071h.
78. Lee SH, Yoon CS, Amine K, Sun YK (2013) Improvement of long-term cycling performance of $Li[Ni_{0.8}Co_{0.15}Al_{0.05}]O_2$ by AlF_3 coating. *J Power Sources* 234:201–207. https://doi.org/10.1016/j.jpowsour.2013.01.045.
79. Sun Y-K, Lee M-J, Yoon CS, et al (2012) The role of AlF_3 coatings in improving electrochemical cycling of Li-enriched nickel-manganese oxide electrodes for Li-ion batteries. *Adv Mater* 24:1192–1196. https://doi.org/10.1002/adma.201104106.
80. Sun Y-K, Yoon CS, Myung S-T, et al (2009) Role of AlF_3 coating on $LiCoO_2$ particles during cycling to cutoff voltage above 4.5 V. *J Electrochem Soc* 156:A1005. https://doi.org/10.1149/1.3236501.
81. Padhi AK, Manivannan V, Goodenough JB (1998) Tuning the position of the redox couples in materials with NASICON structure by anionic substitution. *J Mater Process Technol* 145:1518–1520.
82. Padhi AK, Nanjundaswamy KS, Goodenough JB (1997) Phospho-olivines as positive-electrode materials for rechargeable lithium batteries. *J Electrochem Soc* 144:1188.
83. Ni Q, Bai Y, Wu F, Wu C (2017) Polyanion-type electrode materials for sodium-ion batteries. *Adv Sci* 4:1600275. https://doi.org/10.1002/advs.201600275
84. Jin T, Li H, Zhu K, et al (2020) Polyanion-type cathode materials for sodium-ion batteries. *Chem Soc Rev* 49:2342–2377. https://doi.org/10.1039/c9cs00846b.
85. Goodenough JB, Hong HY, Kafalas JA (1976) Fast Na+-ion transport in skeleton structures. *Mat Res Bull* 11:203–220. https://doi.org/10.1016/0025-5408(76)90077-5.
86. Avdeev M, Mohamed Z, Ling CD, et al (2013) Magnetic structures of $NaFePO_4$ maricite and triphylite polymorphs for sodium-ion batteries. *Inorg Chem* 52:8685–8693. https://doi.org/10.1021/ic400870x.
87. Zheng MY, Bai ZY, He YW, et al (2020) Anionic redox processes in maricite- and triphylite-$NaFePO_4$ of sodium-ion batteries. *ACS Omega* 5:5192–5201. https://doi.org/10.1021/acsomega.9b04213.
88. Yadav SN, Rajoba SJ, Kalubarme RS, et al (2021) Solution combustion synthesis of $NaFePO_4$ and its electrochemical performance. *Chinese J Phys* 69:134–142. https://doi.org/10.1016/j.cjph.2020.11.020.
89. Tang W, Song X, Du Y, et al (2016) High-performance $NaFePO_4$ formed by aqueous ion-exchange and its mechanism for advanced sodium ion batteries. *J Mater Chem A* 4:4882–4892. https://doi.org/10.1039/c6ta01111j.
90. Liu T, Duan Y, Zhang G, et al (2016) 2D amorphous iron phosphate nanosheets with high rate capability and ultra-long cycle life for sodium ion batteries. *J Mater Chem A* 4:4479–4484. https://doi.org/10.1039/c6ta00454g.
91. Koleva V, Boyadzhieva T, Zhecheva E, et al (2013) Precursor-based methods for low-temperature synthesis of defectless $NaMnPO_4$ with an olivine- and maricite-type structure. *Cryst Eng Comm* 15:9080–9089. https://doi.org/10.1039/c3ce41545g.
92. Lee KT, Ramesh TN, Nan F, et al (2011) Topochemical synthesis of sodium metal phosphate olivines for sodium-ion batteries. *Chem Mater* 23:3593–3600. https://doi.org/10.1021/cm200450y.
93. Boyadzhieva T, Koleva V, Stoyanova R (2017) Crystal chemistry of Mg substitution in $NaMnPO_4$ olivine: Concentration limit and cation distribution. *Phys Chem Chem Phys* 19:12730–12739. https://doi.org/10.1039/c7cp01947e.
94. Uebou Y, Okada S, Yamaki JI (2003) Electrochemical insertion of lithium and sodium into $(MoO_2)_2P_2O_7$. *J Power Sources* 115:119–124. https://doi.org/10.1016/S0378-7753(02)00648-1.
95. Barpanda P, Ye T, Nishimura SI, et al (2012) Sodium iron pyrophosphate: A novel 3.0 V iron-based cathode for sodium-ion batteries. *Electrochem Commun* 24:116–119. https://doi.org/10.1016/j.elecom.2012.08.028.
96. Chen X, Du K, Lai Y, et al (2017) In-situ carbon-coated $Na_2FeP_2O_7$ anchored in three-dimensional reduced graphene oxide framework as a durable and high-rate sodium-ion battery cathode. *J Power Sources* 357:164–172. https://doi.org/10.1016/j.jpowsour.2017.04.075.
97. Song HJ, Kim DS, Kim JC, et al (2017) An approach to flexible Na-ion batteries with exceptional rate capability and long lifespan using $Na_2FeP_2O_7$ nanoparticles on porous carbon cloth. *J Mater Chem A* 5:5502–5510. https://doi.org/10.1039/c7ta00727b.

98. Kim H, Shakoor RA, Park C, et al (2013) $Na_2FeP_2O_7$ as a promising iron-based pyrophosphate cathode for sodium rechargeable batteries: A combined experimental and theoretical study. *Adv Funct Mater* 23:1147–1155. https://doi.org/10.1002/adfm.201201589.
99. Park CS, Kim H, Shakoor RA, et al (2013) Anomalous manganese activation of a pyrophosphate cathode in sodium ion batteries: A combined experimental and theoretical study. *J Am Chem Soc* 135:2787–2792. https://doi.org/10.1021/ja312044k.
100. Barpanda P, Lu J, Ye T, et al (2013) A layer-structured $Na_2CoP_2O_7$ pyrophosphate cathode for sodium-ion batteries. *RSC Adv* 3:3857–3860. https://doi.org/10.1039/c3ra23026k.
101. Ha KH, Kwon MS, Lee KT (2020) Triclinic $Na_{3.12}CO_{2.44}(P_2o_7)_2$ as a high redox potential cathode material for na-ion batteries. *J Electrochem Sci Technol* 11:187–194. https://doi.org/10.33961/jecst.2019.00633.
102. Kim H, Park CS, Choi JW, Jung Y (2016) Defect-controlled formation of triclinic $Na_2CoP_2O_7$ for 4 V sodium-ion batteries. *Angew Chemie – Int Ed* 55:6662–6666. https://doi.org/10.1002/anie.201601022.
103. Drozhzhin OA, Tertov IV, Alekseeva AM, et al (2019) β-$NaVP_2O_7$ as a superior electrode material for Na-ion batteries. *Chem Mater* 31:7463–7469. https://doi.org/10.1021/acs.chemmater.9b02124.
104. Kim J, Park I, Kim H, et al (2016) Tailoring a new 4 V-class cathode material for Na-ion batteries. *Adv Energy Mater* 6:6–9. https://doi.org/10.1002/aenm.201502147.
105. Honma T, Togashi T, Ito N, Komatsu T (2012) Fabrication of $Na_2FeP_2O_7$ glass-ceramics for sodium ion battery. *J Ceram Soc Japan* 120:344–346. https://doi.org/10.2109/jcersj2.120.344.
106. Li H, Chen X, Jin T, et al (2019) Robust graphene layer modified $Na_2MnP_2O_7$ as a durable high-rate and high energy cathode for Na-ion batteries. *Energy Storage Mater* 16:383–390. https://doi.org/10.1016/j.ensm.2018.06.013.
107. Tanabe M, Honma T, Komatsu T (2017) Unique crystallization behavior of sodium manganese pyrophosphate $Na_2MnP_2O_7$ glass and its electrochemical properties. *J Asian Ceram Soc* 5:209–215. https://doi.org/10.1016/j.jascer.2017.04.009.
108. Kim SW, Seo DH, Kim H, et al (2012) A comparative study on Na_2MnPO_4F and Li_2MnPO_4F for rechargeable battery cathodes. *Phys Chem Chem Phys* 14:3299–3303. https://doi.org/10.1039/c2cp40082k.
109. Erragh F, Boukhari A, Abraham F, Elouadi B (1995) The crystal structure of α- and β-$Na_2CuP_2O_7$. *J Solid State Chem* 120:23–31.
110. Etheredge KMS, Hwu SJ (1995) Synthesis of a new layered sodium copper(II) pyrophosphate, $Na_2CuP_2O_7$, via an eutectic halide flux. *Inorg Chem* 34:1495–1499. https://doi.org/10.1021/ic00110a030.
111. Belharouak I, Gravereau P, Parent C, et al (2000) Crystal structure of $Na_2ZnP_2O_7$: Reinvestigation. *J Solid State Chem* 152:466–473. https://doi.org/10.1006/jssc.2000.8714.
112. Zhuo H, Wang X, Tang A, et al (2006) The preparation of $NaV_{1-x}Cr_xPO_4F$ cathode materials for sodium-ion battery. *J Power Sources* 160:698–703. https://doi.org/10.1016/j.jpowsour.2005.12.079.
113. Ling M, Li F, Yi H, et al (2018) Superior Na-storage performance of molten-state-blending-synthesized monoclinic $NaVPO_4F$ nanoplates for Na-ion batteries. *J Mater Chem A* 6:24201–24209. https://doi.org/10.1039/c8ta08842j.
114. Barpanda P, Oyama G, Nishimura SI, et al (2014) A 3.8-V earth-abundant sodium battery electrode. *Nat Commun* 5:1–8. https://doi.org/10.1038/ncomms5358.
115. Oyama G, Pecher O, Griffith KJ, et al (2016) Sodium intercalation mechanism of 3.8 v class alluaudite sodium iron sulfate. *Chem Mater* 28:5321–5328. https://doi.org/10.1021/acs.chemmater.6b01091.
116. Plewa A, Kulka A, Hanc E, et al (2020) Facile aqueous synthesis of high performance $Na_2FeM(SO_4)_3$ (M=Fe, Mn, Ni) alluaudites for low cost Na-ion batteries. *J Mater Chem A* 8:2728–2740. https://doi.org/10.1039/c9ta11565j.
117. Jungers T, Mahmoud A, Malherbe C, et al (2019) Sodium iron sulfate alluaudite solid solution for Na-ion batteries: Moving towards stoichiometric $Na_2Fe_2(SO_4)_3$. *J Mater Chem A* 7:8226–8233. https://doi.org/10.1039/c9ta00116f.
118. Goñi A, Iturrondobeitia A, Gil de Muro I, et al (2017) $Na_{2.5}Fe_{1.75}(SO_4)_3$/Ketjen/rGO: An advanced cathode composite for sodium ion batteries. *J Power Sources* 369:95–102. https://doi.org/10.1016/j.jpowsour.2017.09.087.
119. Chen M, Cortie D, Hu Z, et al (2018) A novel graphene oxide wrapped Na_2Fe_2 $(SO_4)_3$ /C cathode composite for long life and high energy density sodium-ion batteries. *Adv Energy Mater* 8:1800944. https://doi.org/10.1002/aenm.201800944.
120. Yao G, Zhang X, Yan Y, et al (2020) Facile synthesis of hierarchical $Na_2Fe(SO_4)_2$@rGO/C as high-voltage cathode for energy density-enhanced sodium-ion batteries. *J Energy Chem* 50:387–394. https://doi.org/10.1016/j.jechem.2020.03.047.

121. Wang W, Liu X, Xu Q, et al (2018) A high voltage cathode of $Na^{2+}2$: $XFe_{2-x}(SO_4)_3$ intensively protected by nitrogen-doped graphene with improved electrochemical performance of sodium storage. *J Mater Chem A* 6:4354–4364. https://doi.org/10.1039/c7ta11110j.
122. Dwibedi D, Gond R, Dayamani A, et al (2017) $Na_{2.32}Co_{1.84}(SO_4)_3$ as a new member of the alluaudite family of high-voltage sodium battery cathodes. *Dalt Trans* 46:55–63. https://doi.org/10.1039/c6dt03767d.
123. Trussov IA, Male LL, Sanjuan ML, et al (2019) Understanding the complex structural features and phase changes in $Na_2Mg_2(SO_4)_3$: A combined single crystal and variable temperature powder diffraction and Raman spectroscopy study. *J Solid State Chem* 272:157–165. https://doi.org/10.1016/j.jssc.2019.02.014.
124. Barpanda P, Chotard JN, Recham N, et al (2010) Structural, transport, and electrochemical investigation of novel $AMSO_4F$(A=Na, Li; M=Fe, Co, Ni, Mn) metal fluorosulphates prepared using low temperature synthesis routes. *Inorg Chem* 49:7401–7413. https://doi.org/10.1021/ic100583f.
125. Liang J, Li Y, Hou X, et al (2017) Insight into electrochemical and elastic properties in $AFe_{1-x}M_xSO_4F$ (A=Li, Na; M=Co, Ni, Mg) cathode materials: A first principle study. *Electrochim Acta* 251:316–323. https://doi.org/10.1016/j.electacta.2017.08.123.
126. Momida H, Kitajou A, Okada S, Oguchi T (2019) First-principles study of X-ray absorption spectra in $NaFeSo_4F$ for exploring Na-ion battery reactions. *J Phys Soc Japan* 88. https://doi.org/10.7566/JPSJ.88.124709.
127. Recham N, Rousse G, Sougrati MT, et al (2012) Preparation and characterization of a stable $FeSO_4F$-based framework for alkali ion insertion electrodes. *Chem Mater* 24:4363–4370. https://doi.org/10.1021/cm302428w.
128. Bianchini F, Fjellvåg H, Vajeeston P (2017) First-principles study of the structural stability and electrochemical properties of Na_2MSiO_4 (M=Mn, Fe, Co and Ni) polymorphs. *Phys Chem Chem Phys* 19:14462–14470. https://doi.org/10.1039/c7cp01395g.
129. Zhu L, Zeng YR, Wen J, et al (2018) Structural and electrochemical properties of Na_2FeSiO_4 polymorphs for sodium-ion batteries. *Electrochim Acta* 292:190–198. https://doi.org/10.1016/j.electacta.2018.09.170.
130. Kee Y, Dimov N, Staykov A, Okada S (2016) Investigation of metastable Na_2FeSiO_4 as a cathode material for Na-ion secondary battery. *Mater Chem Phys* 171:45–49. https://doi.org/10.1016/j.matchemphys.2016.01.033.
131. Li S, Guo J, Ye Z, et al (2016) Zero-strain Na_2FeSiO_4 as novel cathode material for sodium-ion batteries. *ACS Appl Mater Interfaces* 8:17233–17238. https://doi.org/10.1021/acsami.6b03969.
132. Wu H, Zhang Y, Zhang X, et al (2020) Low cost Na_2FeSiO_4/H-N-doped hard carbon nanosphere hybrid cathodes for high energy and power sodium-ion supercapacitors. *J Alloys Compd* 842:155797. https://doi.org/10.1016/j.jallcom.2020.155797.
133. Feng Z, Tang M, Yan Z (2018) 3D conductive CNTs anchored with Na_2FeSiO_4 nanocrystals as a novel cathode material for electrochemical sodium storage. *Ceram Int* 44:22019–22022. https://doi.org/10.1016/j.ceramint.2018.08.186
134. Rangasamy VS, Thayumanasundaram S, Locquet JP (2018) Solvothermal synthesis and electrochemical properties of Na_2CoSiO_4 and Na_2CoSiO_4/carbon nanotube cathode materials for sodium-ion batteries. *Electrochim Acta* 276:102–110. https://doi.org/10.1016/j.electacta.2018.04.166.
135. Zhang D, Ding Z, Yang Y, et al (2018) Fabricating 3D ordered macroporous Na_2MnSiO_4/C with hierarchical pores for fast sodium storage. *Electrochim Acta* 269:694–699. https://doi.org/10.1016/j.electacta.2018.03.045.
136. Jian Z, Hu YS, Ji X, Chen W (2017) NASICON-structured materials for energy storage. *Adv Mater* 29:1–16. https://doi.org/10.1002/adma.201601925.
137. Delmas C (2018) Sodium and sodium-ion batteries: 50 years of research. *Adv Energy Mater* 8:1–9. https://doi.org/10.1002/aenm.201703137.
138. Chen S, Wu C, Shen L, et al (2017) Challenges and perspectives for NASICON-type electrode materials for advanced sodium-ion batteries. *Adv Mater* 29:1–21. https://doi.org/10.1002/adma.201700431.
139. Rajagopalan R, Zhang Z, Tang Y, et al (2021) Understanding crystal structures, ion diffusion mechanisms and sodium storage behaviors of NASICON materials. *Energy Storage Mater* 34:171–193.
140. Uebou Y, Kiyabu T, Okada S, et al (2002) Electrochemical sodium insertion into the 3D-framework of $Na_3M_2(PO_4)_3$ (M=Fe, V) reports. *Inst Adv Mater Study Kyushu Univ* 16:1–5.
141. Luo S, Li J, Bao S, et al (2018) $Na_3V_2(PO_4)_3$/C composite prepared by sol-gel method as cathode for sodium ion batteries. *J Electrochem Soc* 165:A1460–A1465. https://doi.org/10.1149/2.0961807jes.
142. Jing M, Zhang J, Han C, et al (2018) A flexible $Na_3V_2(PO_4)_3C$ composite fiber membrane cathode for Na-ion and Na-Li hybrid-ion batteries. *J Electrochem Soc* 165:A1761–A1769. https://doi.org/10.1149/2.0801809jes.

143. Li S, Ge P, Zhang C, et al (2017) The electrochemical exploration of double carbon-wrapped $Na_3V_2(PO_4)_3$: Towards long-time cycling and superior rate sodium-ion battery cathode. *J Power Sources* 366:249–258. https://doi.org/10.1016/j.jpowsour.2017.09.032.
144. Zheng LL, Xue Y, Hao SE, Wang Z (2018) Porous $Na_3V_2(PO_4)_3$ prepared by freeze-drying method as high performance cathode for sodium-ion batteries. *Ceram Int* 44:9880–9886. https://doi.org/10.1016/j.ceramint.2018.03.001.
145. Zhao Y, Cao X, Fang G, et al (2018) Hierarchically carbon-coated $Na_3V_2(PO_4)_3$ nanoflakes for high-rate capability and ultralong cycle-life sodium ion batteries. *Chem Eng J* 339:162–169. https://doi.org/10.1016/j.cej.2018.01.088.
146. Momma K, Izumi F (2011) VESTA 3 for three-dimensional visualization of crystal, volumetric and morphology data. *J Appl Crystallogr* 44:1272–1276. https://doi.org/10.1107/S0021889811038970.
147. Chotard JN, Rousse G, David R, et al (2015) Discovery of a sodium-ordered form of $Na_3V_2(PO_4)_3$ below ambient temperature. *Chem Mater* 27:5982–5987. https://doi.org/10.1021/acs.chemmater.5b02092.
148. Hasa I, Mariyappan S, Saurel D, et al (2021) Challenges of today for Na-based batteries of the future: From materials to cell metrics. *J Power Sources* 482:228872. https://doi.org/10.1016/j.jpowsour.2020.228872.
149. Chen L, Zhong Z, Ren S, Han DM (2020) Carbon-Coated $Na_3V_2(PO_4)_3$ supported on multiwalled carbon nanotubes for half-/full-cell sodium-ion batteries. *Energy Technol* 8:1–10. https://doi.org/10.1002/ente.201901080.
150. Gopalakrishnan J, Rangan KK (1992) $V_2(PO_4)_3$: A novel NASICON-type vanadium phosphate synthesized by oxidative deintercalation of sodium from $Na_3V_2(PO_4)_3$. *Chem Mater* 4:745–747. https://doi.org/10.1021/cm00022a001.
151. Fu Q, Liu S, Sarapulova A, et al (2019) Electrochemical and structural investigation of calcium substituted monoclinic $Li_3V_2(PO_4)_3$ anode materials for Li-ion batteries. *Adv Energy Mater* 9. https://doi.org/10.1002/aenm.201901864.
152. Saravanan K, Mason CW, Rudola A, et al (2013) The first report on excellent cycling stability and superior rate capability of $Na_3V_2(PO_4)_3$ for sodium ion batteries. *Adv Energy Mater* 3:444–450. https://doi.org/10.1002/aenm.201200803.
153. Pivko M, Arcon I, Bele M, et al (2012) A $3V_2(PO_4)_3$ (A=Na or Li) probed by in situ X-ray absorption spectroscopy. *J Power Sources* 216:145–151. https://doi.org/10.1016/j.jpowsour.2012.05.037.
154. Jian Z, Han W, Lu X, et al (2013) Superior electrochemical performance and storage mechanism of $Na_3V_2(PO_4)_3$ cathode for room-temperature sodium-ion batteries. *Adv Energy Mater* 3:156–160. https://doi.org/10.1002/aenm.201200558.
155. Jian Z, Yuan C, Han W, et al (2014) Atomic structure and kinetics of NASICON $Na_xV_2(PO_4)_3$ cathode for sodium-ion batteries. *Adv Funct Mater* 24:4265–4272. https://doi.org/10.1002/adfm.201400173.
156. Wang Z, Liu J, Du Z, et al (2020) Enhancing Na-ion storage in $Na_3V_2(PO_4)_3$/C cathodes for sodium ion batteries through Br and N co-doping. *Inorg Chem Front* 7:1289–1297. https://doi.org/10.1039/c9qi01690b.
157. Song W, Ji X, Wu Z, et al (2014) First exploration of Na-ion migration pathways in the NASICON structure $Na_3V_2(PO_4)_3$. *J Mater Chem A* 2:5358–5362. https://doi.org/10.1039/c4ta00230j.
158. Duan W, Zhu Z, Li H, et al (2014) $Na_3V_2(PO_4)_3$@C core-shell nanocomposites for rechargeable sodium-ion batteries. *J Mater Chem A* 2:8668–8675. https://doi.org/10.1039/c4ta00106k.
159. Kang J, Baek S, Mathew V, et al (2012) High rate performance of a $Na_3V_2(PO_4)_3$/C cathode prepared by pyro-synthesis for sodium-ion batteries. *J Mater Chem* 22:20857–20860. https://doi.org/10.1039/c2jm34451c.
160. Zhu C, Song K, Van Aken PA, et al (2014) Carbon-coated $Na_3V_2(PO_4)_3$ embedded in porous carbon matrix: An ultrafast Na-storage cathode with the potential of outperforming Li cathodes. *Nano Lett* 14:2175–2180. https://doi.org/10.1021/nl500548a.
161. Chekannikov A, Novikova S, Kulova T, et al (2016) Electrochemical study of doped $LiFePO_4$ as a cathode material for lithium-ion battery. *J Electrochem Sci Eng* 6:1. https://doi.org/10.5599/jese.234.
162. Lim SJ, Han DW, Nam DH, et al (2014) Structural enhancement of $Na_3V_2(PO_4)_3$/C composite cathode materials by pillar ion doping for high power and long cycle life sodium-ion batteries. *J Mater Chem A* 2:19623–19632. https://doi.org/10.1039/c4ta03948c.
163. Li H, Yu X, Bai Y, et al (2015) Effects of Mg doping on the remarkably enhanced electrochemical performance of $Na_3V_2(PO_4)_3$ cathode materials for sodium ion batteries. *J Mater Chem A* 3:9578–9586. https://doi.org/10.1039/c5ta00277j.
164. Aragón MJ, Lavela P, Ortiz GF, Tirado JL (2015) Effect of iron substitution in the electrochemical performance of $Na_3V_2(PO_4)_3$ as cathode for Na-ion batteries. *J Electrochem Soc* 162:A3077–A3083. https://doi.org/10.1149/2.0151502jes.

165. Aragón MJ, Lavela P, Ortiz GF, Tirado JL (2015) Benefits of chromium substitution in $Na_3V_2(PO_4)_3$ as a potential candidate for sodium-ion batteries. *ChemElectroChem* 2:995–1002. https://doi.org/10.1002/celc.201500052.
166. Klee R, Lavela P, Aragón MJ, et al (2016) Enhanced high-rate performance of manganese substituted $Na_3V_2(PO_4)_3$/C as cathode for sodium-ion batteries. *J Power Sources* 313:73–80. https://doi.org/10.1016/j.jpowsour.2016.02.066.
167. Aragón MJ, Lavela P, Alcántara R, Tirado JL (2015) Effect of aluminum doping on carbon loaded $Na_3V_2(PO_4)_3$ as cathode material for sodium-ion batteries. *Electrochim Acta* 180:824–830. https://doi.org/10.1016/j.electacta.2015.09.044.
168. Bianchini M, Fauth F, Brisset N, et al (2015) Comprehensive investigation of the $Na_3V_2(PO_4)_2F_3$-$NaV_2(PO_4)_2F_3$ system by operando high resolution synchrotron X-ray diffraction. *Chem Mater* 27:3009–3020. https://doi.org/10.1021/acs.chemmater.5b00361.
169. Song W, Cao X, Wu Z, et al (2014) Investigation of the sodium ion pathway and cathode behavior in $Na_3V_2(PO_4)_2F_3$ combined via a first principles calculation. *Langmuir* 30:12438–12446. https://doi.org/10.1021/la5025444.
170. Liu Q, Wang D, Yang X, et al (2015) Carbon-coated $Na_3V_2(PO_4)_2F_3$ nanoparticles embedded in a mesoporous carbon matrix as a potential cathode material for sodium-ion batteries with superior rate capability and long-term cycle life. *J Mater Chem A* 3:21478–21485. https://doi.org/10.1039/c5ta05939a.
171. Broux T, Fauth F, Hall N, et al (2019) High rate performance for carbon-coated $Na_3V_2(PO_4)_2F_3$ in Na-ion batteries. *Small Methods* 3:1–12. https://doi.org/10.1002/smtd.201800215.
172. Le Meins JM, Crosnier-Lopez MP, Hemon-Ribaud A, Courbion G (1999) Phase transitions in the $Na_3M_2(PO_4)_2F_3$ family ($M=Al^{3+}$, V^{3+}, Cr^{3+}, Fe^{3+}, Ga^{3+}): Synthesis, thermal, structural, and magnetic studies. *J Solid State Chem* 148:260–277. https://doi.org/10.1006/jssc.1999.8447.
173. Song W, Ji X, Wu Z, et al (2014) Exploration of ion migration mechanism and diffusion capability for $Na_3V_2(PO_4)_2F_3$ cathode utilized in rechargeable sodium-ion batteries. *J Power Sources* 256:258–263. https://doi.org/10.1016/j.jpowsour.2014.01.025.
174. Zhao J, Gao Y, Liu Q, et al (2018) High rate capability and enhanced cyclability of $Na_3V_2(PO_4)_2F_3$ cathode by in situ coating of carbon nanofibers for sodium-ion battery applications. *Chem Eur J* 24:2913–2919. https://doi.org/10.1002/chem.201704131.
175. Liu S, Wang L, Liu J, et al (2019) $Na_3V_2(PO_4)_2F_3$-SWCNT: A high voltage cathode for non-aqueous and aqueous sodium-ion batteries. *J Mater Chem A* 7:248–256. https://doi.org/10.1039/c8ta09194c.
176. Zhu C, Wu C, Chen CC, et al (2017) A high power-high energy $Na_3V_2(PO_4)_2F_3$ sodium cathode: Investigation of transport parameters, rational design and realization. *Chem Mater* 29:5207–5215. https://doi.org/10.1021/acs.chemmater.7b00927.
177. Shakoor RA, Seo DH, Kim H, et al (2012) A combined first principles and experimental study on $Na_3V_2(PO_4)_2F_3$ for rechargeable Na batteries. *J Mater Chem* 22:20535–20541. https://doi.org/10.1039/c2jm33862a.
178. Semykina DO, Kirsanova MA, Volfkovich YM, et al (2021) Porosity, microstructure and electrochemistry of $Na_3V_2(PO_4)_2F_3$/C prepared by mechanical activation. *J Solid State Chem* 297:122041. https://doi.org/10.1016/j.jssc.2021.122041.
179. Nguyen LHB, Broux T, Camacho PS, et al (2019) Stability in water and electrochemical properties of the $Na_3V_2(PO_4)_2F_3$ - $Na_3(VO)_2(PO_4)_2F$ solid solution. *Energy Storage Mater* 20:324–334. https://doi.org/10.1016/j.ensm.2019.04.010.
180. Robin MB (1962) The color and electronic configurations of prussian blue. *Inorg Chem* 1:337–342. https://doi.org/10.1021/ic50002a028.
181. Buser HJ, Ludi A, Schwarzenbach D, Petter W (1977) The crystal structure of prussian blue: $Fe_4[Fe(CN)_6]_3.xH_2O$. *Inorg Chem* 16:2704–2710. https://doi.org/10.1021/ic50177a008.
182. Herren F, Ludi A, Fischer P, Halg W (1980) Neutron diffraction study of prussian blue, $Fe_4[Fe(CN)_6]_3.xH_2O$. Location of water molecules and long-range magnetic order. *Inorg Chem* 19:956–959. https://doi.org/10.1021/ic50206a032.
183. Giorgetti M, Berrettoni M, Zamponi S, et al (2005) Cobalt hexacyanoferrate in PAMAM doped silica matrix. 2. Structural and electronic characterization. *Electrochim Acta* 51:511–516. https://doi.org/10.1016/j.electacta.2005.05.009.
184. Itaya K, Uchida I, Neff VD (1986) Electrochemistry of polynuclear transition metal cyanides: Prussian blue and its analogues. *Acc Chem Res* 19:162–168. https://doi.org/10.1021/ar00126a001.
185. Widmann A, Kahlert H, Petrovic-Prelevic I, et al (2002) Structure, insertion electrochemistry, and magnetic properties of a new type of substitutional solid solutions of copper, nickel, and iron hexacyanoferrates/hexacyanocobaltates. *Inorg Chem* 41:5706–5715. https://doi.org/10.1021/ic0201654.

186. Kumar A, Yusuf SM, Keller L (2005) Structural and magnetic properties of $Fe[Fe(CN)_6]_4H_2O$. *Phys Rev B - Condens Matter Mater Phys* 71:1–7. https://doi.org/10.1103/PhysRevB.71.054414.
187. Hillman AR, Skopek MA, Gurman SJ (2010) EXAFS structural studies of electrodeposited Co and Ni hexacyanoferrate films. *J Solid State Electrochem* 14: 1997–2010.
188. Lu Y, Wang L, Cheng J, Goodenough JB (2012) Prussian blue: A new framework of electrode materials for sodium batteries. *Chem Commun* 48:6544–6546. https://doi.org/10.1039/c2cc31777j.
189. Miscellanea Berolinensia ad incrementum scientiarum ex scriptis Societati... – Königliche Akademie der Wissenschaften (Berlin) – Google Książki. https://books.google.pl/books?hl=pl&lr=&id=xldFAAAAcAAJ&oi=fnd&pg=PA1&dq=Miscellanea+Berolinensia+ad+incrementum+scientiarum,+Berlin,+(1710)+377.+&ots=wjLn6btAAG&sig=4hf8yA_HvOJJOZUnlWnqXbIIp7g&redir_esc=y#v=onepage&q&f=false. Accessed 15 Feb 2021.
190. Keggin JF, Miles FD (1936) Structures and formulæ of the Prussian blues and related compounds [4]. *Nature* 137:577–578.
191. Wilde RE, Ghosh SN, Marshall BJ (1970) The Prussian blues. *Inorg Chem* 9:2512–2516. https://doi.org/10.1021/ic50093a027.
192. Itaya K, Ataka T, Toshima S (1982) Spectroelectrochemistry and electrochemical preparation method of Prussian blue modified electrodes. *J Am Chem Soc* 104:4767–4772. https://doi.org/10.1021/ja00382a006.
193. Kulesza PJ (1990) Solid-state electrochemistry of iron hexacyanoferrate (Prussian blue type) powders. Evidence for redox transitions in mixed-valence ionically conducting microstructures. *J Electroanal Chem* 289:103–116. https://doi.org/10.1016/0022-0728(90)87209-3.
194. Kulesza PJ, Zamponi S, Berrettoni M, et al (1995) Preparation, spectroscopic characterization and electrochemical charging of the sodium-containing analogue of Prussian Blue. *Electrochim Acta* 40:681–688. https://doi.org/10.1016/0013-4686(94)00348-5.
195. Karyakin AA (2001) Prussian blue and its analogues: Electrochemistry and analytical applications. *Electroanalysis* 13:813–819. https://doi.org/10.1002/1521-4109(200106)13:10<813::AID-ELAN813>3.0.CO;2-Z.
196. Kulesza PJ, Malik MA, Berrettoni M, et al (1998) Electrochemical charging, countercation accommodation, and spectrochemical identity of microcrystalline solid cobalt hexacyanoferrate. *J Phys Chem B* 102:1870–1876. https://doi.org/10.1021/jp9726495.
197. Baioni AP, Vidotti M, Fiorito PA, et al (2007) Synthesis and characterization of copper hexacyanoferrate nanoparticles for building up long-term stability electrochromic electrodes. *Langmuir* 23:6796–6800. https://doi.org/10.1021/la070161h.
198. Faustino PJ, Yang Y, Progar JJ, et al (2008) Quantitative determination of cesium binding to ferric hexacyanoferrate: Prussian blue. *J Pharm Biomed Anal* 47:114–125. https://doi.org/10.1016/j.jpba.2007.11.049.
199. Beheir SG, Benyamin K, Mekhail FM (1998) Chemical precipitation of cesium from waste solutions with iron(II)hexacyanocobaltate(III) and triphenylcyanoborate. *J Radioanal Nucl Chem* 232:147–150. https://doi.org/10.1007/BF02383731.
200. Hao X, Li Y, Pritzker M (2008) Pulsed electrodeposition of nickel hexacyanoferrate films for electrochemically switched ion exchange. *Sep Purif Technol* 63:407–414. https://doi.org/10.1016/j.seppur.2008.06.001.
201. Ricci F, Palleschi G (2005) Sensor and biosensor preparation, optimisation and applications of Prussian blue modified electrodes. *Biosens Bioelectron* 21:389–407.
202. Eftekhari A (2004) Electrochemical properties of lanthanum hexacyanoferrate particles immobilized onto electrode surface by au-codeposition method. *Electroanalysis* 16:1324–1329. https://doi.org/10.1002/elan.200302942.
203. Zhao H, Yuan Y, Adeloju S, Wallace GG (2002) Study on the formation of the Prussian blue films on the polypyrrole surface as a potential mediator system for biosensing applications. *Anal Chim Acta* 472:113–121. https://doi.org/10.1016/S0003-2670(02)00937-6.
204. Gaitán M, Gonçales VR, Soler-Illia GJAA, et al (2010) Structure effects of self-assembled Prussian blue confined in highly organized mesoporous TiO_2 on the electrocatalytic properties towards H_2O_2 detection. *Biosens Bioelectron* 26:890–893. https://doi.org/10.1016/j.bios.2010.07.026.
205. Liu Y, Chu Z, Jin W (2009) A sensitivity-controlled hydrogen peroxide sensor based on self-assembled Prussian blue modified electrode. *Electrochem commun* 11:484–487. https://doi.org/10.1016/j.elecom.2008.12.029.
206. de Lara González GL, Kahlert H, Scholz F (2007) Catalytic reduction of hydrogen peroxide at metal hexacyanoferrate composite electrodes and applications in enzymatic analysis. *Electrochim Acta* 52:1968–1974. https://doi.org/10.1016/j.electacta.2006.08.006.

207. Malinauskas A, Araminaite R, Mickevičiute G, Garjonyte R (2004) Evaluation of operational stability of Prussian blue- and cobalt hexacyanoferrate-based amperometric hydrogen peroxide sensors for biosensing application. *Mater Sci Eng C* 24:513–519. https://doi.org/10.1016/j.msec.2004.01.002.
208. Yokoyama T, Ohta T, Sato O, Hashimoto K (1998) Characterization of magnetic CoFe cyanides by x-ray-absorption fine-structure spectroscopy. *Phys Rev B - Condens Matter Mater Phys* 58:8257–8266. https://doi.org/10.1103/PhysRevB.58.8257.
209. Sato O, Hayami S, Einaga Y, Gu ZZ (2003) Control of the magnetic and optical properties in molecular compounds by electrochemical, photochemical and chemical methods. *Bull Chem Soc Jpn* 76:443–470. https://doi.org/10.1246/bcsj.76.443.
210. Keene TD, Komm T, Hauser J, Krämer KW (2011) Two-dimensional coordination compounds based on Fe(II) and Co(III) hexacyanometallates with Cu(II)(dien) groups: Structures and magnetic properties. *Inorganica Chim Acta* 373:100–106. https://doi.org/10.1016/j.ica.2011.03.067.
211. Makowski O, Kowalewska B, Szymanska D, et al (2007) Controlled fabrication of multilayered 4-(pyrrole-1-yl) benzoate supported poly(3,4-ethylenedioxythiophene) linked hybrid films of Prussian blue type nickel hexacyanoferrate. *Electrochim Acta* 53:1235–1243. https://doi.org/10.1016/j.electacta.2007.02.083.
212. Jayalakshmi M, Radhika P, Mohan Rao M (2006) A new electrode consisting of Prussian blue/Dibenzo-18-crown-6 ion-pair complex for electrochemical capacitor applications. *J Power Sources* 158:801–805. https://doi.org/10.1016/j.jpowsour.2005.09.034.
213. Eftekhari A (2003) A high-voltage solid-state secondary cell based on chromium hexacyanometallates. *J Power Sources* 117:249–254. https://doi.org/10.1016/S0378-7753(03)00019-3.
214. Gómez-Romero P, Torres-Gómez G (2000) Molecular batteries: Harnessing $Fe(CN)_{63}$-electroactivity in hybrid polyaniline-hexacyanoferrate electrodes. *Adv Mater* 12:1454–1456. https://doi.org/10.1002/1521-4095(200010)12:19<1454::AID-ADMA1454>3.0.CO;2-H.
215. Tung TS, Chen LC, Ho KC (2003) An indium hexacyanoferrate-tungsten oxide electrochromic battery with a hybrid K+/H+-conducting polymer electrolyte. *Solid State Ion* 165:257–267.
216. Moritomo Y, Takachi M, Kurihara Y, Matsuda T (2012) Thin film electrodes of Prussian blue analogues with rapid Li+intercalation. *Appl Phys Express* 5:2–5. https://doi.org/10.1143/APEX.5.041801.
217. Margadonna S, Prassides K, Fitch AN (2004) Zero thermal expansion in a Prussian blue analogue. *J Am Chem Soc* 126:15390–15391. https://doi.org/10.1021/ja044959o.
218. Lezna RO, Romagnoli R, De Tacconi NR, Rajeshwar K (2002) Cobalt hexacyanoferrate: Compound stoichiometry, infrared spectroelectrochemistry, and photoinduced electron transfer. *J Phys Chem B* 106:3612–3621. https://doi.org/10.1021/jp013991r.
219. Manfrin MF, Setti L, Moggi L (1992) Supramolecular adducts between the hexacyanocobaltate(III) anion and poly(Ethyleneimines): Influence on the photoaquation efficiency of the complex. *Inorg Chem* 31:2768–2771. https://doi.org/10.1021/ic00039a020.
220. Milder SJ, Gray HB, Miskowski VM (1984) Photochemistry of hexacyanocobaltate(III) in haloalkanes. *J Am Chem Soc* 106:3764–3767. https://doi.org/10.1021/ja00325a009.
221. Reguera E, Marín E, Calderón A, Rodríguez-Hernández J (2007) Photo-induced charge transfer in Prussian blue analogues as detected by photoacoustic spectroscopy. *Spectrochim Acta - Part A Mol Biomol Spectrosc* 68:191–197. https://doi.org/10.1016/j.saa.2006.11.013.
222. Sauter S, Wittstock G, Szargan R (2001) Localisation of electrochemical oxidation processes in nickel and cobalt hexacyanoferrates investigated by analysis of the multiplet patterns in X-ray photoelectron spectra. *Phys Chem Chem Phys* 3:562–569. https://doi.org/10.1039/b008430l.
223. Florescu M, Brett CMA (2004) Development and characterization of cobalt hexacyanoferrate modified carbon electrodes for electrochemical enzyme biosensors. *Anal Lett* 37:871–886. https://doi.org/10.1081/AL-120030284.
224. Florescu M, Barsan M, Pauliukaite R, Brett CMA (2007) Development and application of oxysilane sol-gel electrochemical glucose biosensors based on cobalt hexacyanoferrate modified carbon film electrodes. *Electroanalysis* 19:220–226. https://doi.org/10.1002/elan.200603714.
225. Vidal JC, Espuelas J, Garcia-Ruiz E, Castillo JR (2004) Amperometric cholesterol biosensors based on the electropolymerization of pyrrole and the electrocatalytic effect of Prussian-Blue layers helped with self-assembled monolayers. *Talanta* 64:655–664. https://doi.org/10.1016/j.talanta.2004.03.038.
226. De Mattos IL, Gorton L, Laurell T, et al (2000) Development of biosensors based on hexacyanoferrates. *Talanta* 52:791–799. https://doi.org/10.1016/S0039-9140(00)00409-4.
227. Somani P, Radhakrishnan S (1998) Electrochromic response in polypyrrole sensitized by Prussian blue. *Chem Phys Lett* 292:218–222. https://doi.org/10.1016/S0009-2614(98)00646-0.

228. Kulesza PJ, Miecznikowski K, Chojak M, et al (2001) Electrochromic features of hybrid films composed of polyaniline and metal hexacyanoferrate. *Electrochim Acta* 46:4371–4378. https://doi.org/10.1016/S0013-4686(01)00681-8.
229. Boxhoorn G, Moolhuysen J, Coolegem JGF, Van Santen RA (1985) Cyanometallates: An underestimated class of molecular sieves. *J Chem Soc Chem Commun* 0:1305–1307. https://doi.org/10.1039/c39850001305.
230. Mardan A, Ajaz R, Mehmood A, et al (1999) Preparation of silica potassium cobalt hexacyanoferrate composite ion exchanger and its uptake behavior for cesium. *Sep Purif Technol* 16:147–158. https://doi.org/10.1016/S1383-5866(98)00121-X.
231. Martinez-Garcia R, Knobel M, Reguera E (2006) Modification of the magnetic properties in molecular magnets based on Prussian blue analogues through adsorbed species. *J Phys Condens Matter* 18:11243–11254. https://doi.org/10.1088/0953-8984/18/49/016.
232. Ng CW, Ding J, Gan LM (2001) Microstructural changes induced by thermal treatment of cobalt(II) hexacyanoferrate(III) compound. *J Solid State Chem* 156:400–407. https://doi.org/10.1006/jssc.2000.9013.
233. Sato O, Einaga Y, Iyoda T, et al (1997) Reversible photoinduced magnetization. *J Electrochem Soc* 144:L11–L13. https://doi.org/10.1149/1.1837356.
234. Chen SM, Chan CM (2003) Preparation, characterization, and electrocatalytic properties of copper hexacyanoferrate film and bilayer film modified electrodes. *J Electroanal Chem* 543:161–173. https://doi.org/10.1016/S0022-0728(03)00017-2.
235. Wang L, Lu Y, Liu J, et al (2013) A superior low-cost cathode for a Na-ion battery. *Angew Chemie – Int Ed* 52:1964–1967. https://doi.org/10.1002/anie.201206854.
236. Wu X, Wu C, Wei C, et al (2016) Highly crystallized $Na_2CoFe(CN)_6$ with suppressed lattice defects as superior cathode material for sodium-ion batteries. *ACS Appl Mater Interfaces* 8:5393–5399. https://doi.org/10.1021/acsami.5b12620.
237. Song J, Wang L, Lu Y, et al (2015) Removal of interstitial H_2O in hexacyanometallates for a superior cathode of a sodium-ion battery. *J Am Chem Soc* 137:2658–2664. https://doi.org/10.1021/ja512383b.
238. Harris TD, Long JR (2007) Linkage isomerism in a face-centered cubic $Cu_6Cr_8(CN)_{24}$ cluster with an S = 15 ground state. *Chem Commun* 0:1360–1362. https://doi.org/10.1039/b615141h.
239. Baggio Saitovitch E, Danon J (1976) Kinetic studies of the cyanide linkage isomerism in silver hexacyano ferrates by Mössbauer spectroscopy. *Chem Phys Lett* 39:296–299. https://doi.org/10.1016/0009-2614(76)80079-6.
240. Gadet V, Mallah T, Castro I, et al (1992) High-Tc Molecular-based magnets: A ferromagnetic bimetallic chromium(III)-nickel(II) cyanide with Tc=90 K. *J Am Chem Soc* 114:9213–9214. https://doi.org/10.1021/ja00049a078.
241. Ellis D, Eckhoff M, Neff VD (1981) Electrochromism in the mixed-valence hexacyanides. 1. Voltammetric and spectral studies of the oxidation and reduction of thin films of Prussian blue. *J Phys Chem* 85:1225–1231. https://doi.org/10.1021/j150609a026.
242. Aparicio C, Machala L, Marusak Z (2012) Thermal decomposition of Prussian blue under inert atmosphere. *J Therm Anal Calorim* 110:661–669. https://doi.org/10.1007/s10973-011-1890-1.
243. Wang L, Song J, Qiao R, et al (2015) Rhombohedral Prussian white as cathode for rechargeable sodium-ion batteries. *J Am Chem Soc* 137:2548–2554. https://doi.org/10.1021/ja510347s.
244. Shen Z, Guo S, Liu C, et al (2018) Na-rich Prussian white cathodes for long-life sodium-ion batteries. *ACS Sustain Chem Eng* 6:16121–16129. https://doi.org/10.1021/acssuschemeng.8b02758.
245. You Y, Wu XL, Yin YX, Guo YG (2014) High-quality Prussian blue crystals as superior cathode materials for room-temperature sodium-ion batteries. *Energy Environ Sci* 7:1643–1647. https://doi.org/10.1039/c3ee44004d.
246. Bauer A, Song J, Vail S, et al (2018) The scale-up and commercialization of nonaqueous Na-ion battery technologies. *Adv Energy Mater* 8:1–13. https://doi.org/10.1002/aenm.201702869.
247. Lee HW, Wang RY, Pasta M, et al (2014) Manganese hexacyanomanganate open framework as a high-capacity positive electrode material for sodium-ion batteries. *Nat Commun* 5:1–6. https://doi.org/10.1038/ncomms6280.
248. Bueno PR, Giménez-Romero D, Gabrielli C, et al (2006) Changeover during in situ compositional modulation of hexacyanoferrate (Prussian blue) material. *J Am Chem Soc* 128:17146–17152. https://doi.org/10.1021/ja066982a.
249. Li X, Wang Y, Wu D, et al (2016) Jahn-Teller assisted Na diffusion for high performance Na ion batteries. *Chem Mater* 28:6575–6583. https://doi.org/10.1021/acs.chemmater.6b02440.

250. Talaie E, Kim SY, Chen N, Nazar LF (2017) Structural evolution and redox processes involved in the electrochemical cycling of P2-$Na_{0.67}[Mn_{0.66}Fe_{0.20}Cu_{0.14}]O_2$. *Chem Mater* 29:6684–6697. https://doi.org/10.1021/acs.chemmater.7b01146.
251. Maitra U, House RA, Somerville JW, et al (2018) Oxygen redox chemistry without excess alkali-metal ions in $Na_{2/3}[Mg_{0.28}Mn_{0.72}]O_2$. *Nat Chem* 10:288–295. https://doi.org/10.1038/nchem.2923.

10 Anode Materials for Sodium-Ion Battery

Magdalena Graczyk-Zajac, Alexander Kempf, Monika Wilamowska-Zawlocka, and Magdalena Graczyk-Zajac

10.1 INTRODUCTION

Lithium-ion batteries (LIBs) have dominated portable electronic devices and have also been regarded as a possible solution to the electrical grid concerns. The fundamental components of the present lithium-ion batteries are lithium and cobalt. Analysis by researchers of the Karlsruhe Institute of Technology (KIT) shows that the availability of both elements could become seriously critical [2]. Lithium is a non-abundant and unevenly distributed element. The increasing lithium demand in mobile electronic equipment, hybrid electrical vehicles (HEVs), and electrical vehicles (EVs) will ultimately push up the price of lithium compounds, leading to a prohibitively expensive large-scale storage. Cobalt plays an integral part in the common LIB cathode, and as battery-powered applications such as electric vehicles become ubiquitous, cobalt mining will need to grow proportionally to avoid supply bottlenecks. The foremost risk, and perhaps the most challenging to solve, is geopolitical. Sixty-two percent of the world's cobalt comes from the Democratic Republic of Congo, and combined with production from Zambia, Madagascar, Republic of South Africa and Zimbabwe, the five countries mine more than 71% of the world's cobalt. Companies process ore locally and export more than 80% of the total to China for further processing and refining to produce commercial cobalt compounds used in batteries. This exclusive trade between African countries and China exposes the world market to Chinese regulatory volatility and export restrictions, a recent example being that of the rare earths market, which saw extreme shortages after the Chinese enacted export restriction in 2010. According to this, the economical focus changed to alternate sources, materials and stockpiling. Cobalt-free battery technologies, including post-lithium technologies based on non-critical elements such as sodium, represent the possibility to avoid this criticality situation on the market in a long-term view (compare Figure 10.2). Sodium is one of the most abundant elements in the earth's crust; meanwhile, infinite sodium resources are also available in the ocean. Due to the high abundance and low cost of battery components, combined with low sodium redox potential of –2.71 V (Na+/Na) vs standard hydrogen electrode (SHE), which is only 0.3 V above the one of Li+/Li, sodium-ion batteries (NIBs) are very promising for large-scale energy storage application. Large-scale stationary applications demand lower energy and power densities than mobile applications, as they are not constrained by volume or weight. Instead, stationary batteries must demonstrate a longer battery lifetime and lower costs (Figure 10.1). Perspective potential application fields of Na-ion batteries are schematically depicted in Figure 10.1.

Development of a stable, high capacity, safe negative electrode for NIB remains a challenge. Graphite, which is a commercial choice for LIB, is not suitable for sodium-ion batteries as intercalation of Na into graphite is thermodynamically unfavorable [3,4]. Despite this fact, there have been numerous works realized to adopt the solid-state electrochemistry of graphite to facilitate/allow Na intercalation/insertion. The works of Kim et al. and Jache et al. [5,6] present a promising approach in which a co-intercalation of a solvent makes the intercalation of Na into graphite

DOI: 10.1201/9781003310167-10

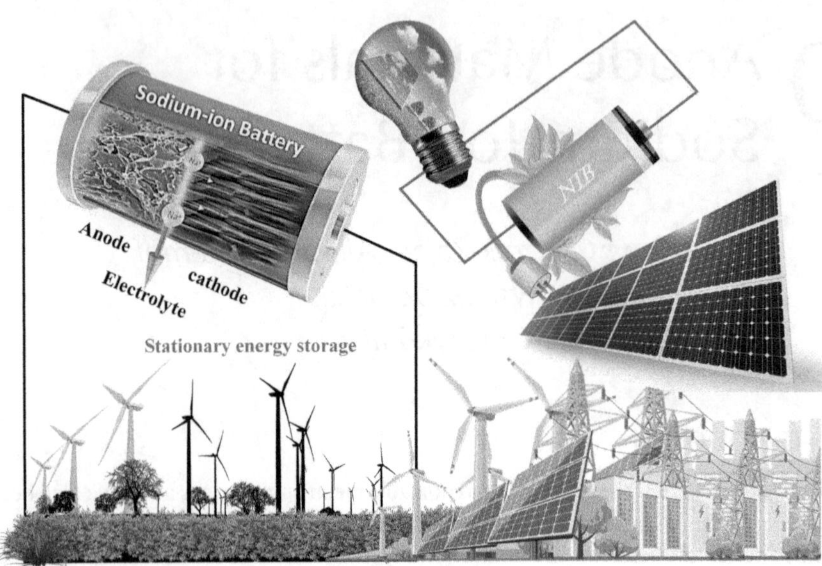

FIGURE 10.1 Perspective potential application of Na-ion batteries. Adapted and reproduced from ref. [1]. Copyright © 2021 Elsevier.

possible. Although this solution leads to an interesting electrochemical performance, several drawbacks hinder its further development, namely: (i) only a part of graphite capacity is exploited for energy storage purposes (sodium intercalation), which leads to lower gravimetric energy density, and (ii) ether-based electrolyte is used for co-intercalation, which means losing the advantage of well-known and well-explored carbonate-based electrolytes. It has also been reported that in terms of energy efficiency, the electrochemical performance of graphite for sodium storage is better than for lithium [7] due to much smaller hysteresis between the charge and discharge profiles. Nonetheless, the status of graphite anodes does not satisfy the practical requirements [4]. According to Parveen et al. [8], the cost of NIB anode accounts for 14% of the total cell cost. For the commercialization of NIBs, anode materials with large interstitial spaces that can easily store Na ions (with ionic radii of almost 34% greater than Li ions; 1.02 vs 0.76 Å) are required. Thus, the ideal anode material should fulfill the following requirements:

- High gravimetric and volumetric capacity, cycling stability and efficiency
- High electronic and ionic conductivity
- A potential close to that of pure sodium metal, stable with different Na content
- No reaction/no dissolution in the solvent of the electrolyte
- Cost efficient and environmentally friendly

Dependent on the mechanism of the electrochemical reaction, the negative electrode materials can be classified into three groups: insertion-based electrodes, alloy-based materials and conversion-type materials. Insertion compounds include soft and hard carbons, as well as titanium-based oxides. The alloy-forming materials are represented by individual elements from group 14 and group 15, with Sn, Sb and P being the most prominent examples. The reaction within conversion-type materials involves the formation of new phases during sodiation that are usually structurally very different from the starting material. Transition metal oxides (Fe, Cu, and Ni) have been considered in this context due to their intrinsic stability and relative abundancy (Figure 10.2).

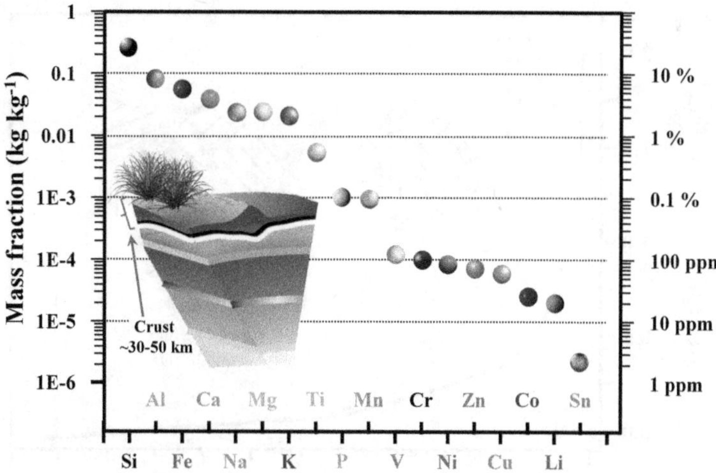

FIGURE 10.2 Elemental abundance in the earth's crust. Adapted and reproduced from ref. [3]. Copyright © 2014 American Chemical Society.

10.2 INSERTION MATERIALS

Historically, the first reversible electrochemical sodium insertion at room temperature was demonstrated using TiS_2 in 1980 [9]. Insertion or intercalation mechanism is defined as the acceptance of the guest species inside the crystal structure without a significant disturbing of crystalline parameters such as bond distances, unit cell volume, crystal phase and spacing between crystal planes. Thus, the amount of the intercalated species is determined by thermodynamic equilibrium attained at the electrode/electrolyte interface. When the charge specie is inserted into the structure, it will force some of the electrons to move in/out of the structure. This means that the structure should possess a readily oxidizable and reducible atom, as well as moderate electronic conductivity. For this, the presence of large interstitial spaces to accommodate the fast track ions is required [10].

Nowadays, carbonaceous and titanium-based oxides have been extensively studied as insertion anodes for NIBs [11–15]. In the group of carbon-based materials, one can distinguish non-graphitic anodes based on "soft" graphitizable carbons such as carbon black [16] and pitch-based carbon-fibers and non-graphitizable "hard carbons" [17]. Hard carbons are synthesized at high temperatures from carbon-based precursors and have been comprehensively modeled [18,19], characterized [20] and thermally tested [21] in Na cells [11]. In particular, hard-disordered carbon materials are widely studied because of their ability to accommodate Na^+ ions into their structure with reasonable capacities up to ~300 mAh g^{-1} and low operating potential (almost zero, ~0 V vs Na^+/Na) [17,22]. Although extensively studied, the Na^+ storage mechanism in a disordered carbon structure is still controversial [17,23–26]. Titanium-based oxide compounds have been widely studied because of their low operating voltage and cost [27]. Similar to LIBs, titanium-based oxide anodes, including various polymorphs of titanium dioxide (TiO_2), spinel-lithium titanate ($Li_4Ti_5O_{12}$) and sodium titanate ($Na_xTi_yO_z$), were reported [28–32]. The capacities of such materials are limited by compound stability and stoichiometry, and specific capacities higher than 200–250 mAh g^{-1} are hardly achieved. On the other hand, these materials have less volume expansion and are suitable for high current charge–discharge cycles [33,34].

10.2.1 Hard Carbons

A typical discharge curve of HC vs sodium metal is composed of a slope region (region 1) and a plateau region (region 2/2+3; Figure 10.3). There is an ongoing debate on the assignment of these regions to different storage mechanisms [35]. Figure 10.3a and b shows an illustration of two possible

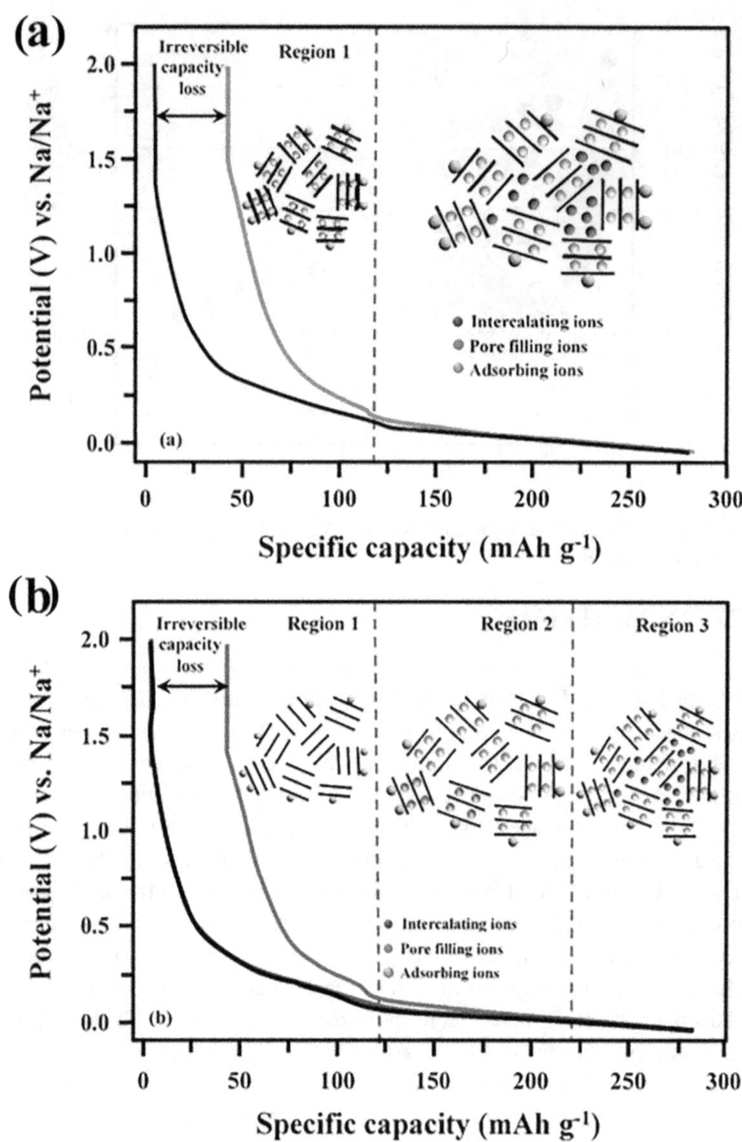

FIGURE 10.3 (a) Illustration of HC with an intercalation–adsorption mechanism. (b) Schematic illustration of HC with an adsorption–intercalation–pore filling mechanism. Adapted and reproduced from ref. [11]. Copyright © 2020 Elsevier.

mechanisms, namely the intercalation–pore filling mechanism and the adsorption–intercalation–pore filling mechanism, respectively. In the early model proposed by Dahn and Stevens [17] (Figure 10.3a), region 1 is attributed to the intercalation of sodium ions into the parallel graphitic layers and region 2 is assigned to the filling of free space (often called "nanoporosity") between randomly stacked layers [17]. The intercalation of sodium ions follows a sloping trend because the potential for further insertion of sodium ions can be altered if there are more sodium ions in the graphene layer. There have been many studies which supported this mechanism. A negative shift has been identified by using ex situ XRD for HC discharged to 0.1 V, indicating the expansion of the interlayer spacing due to Na-ion insertion [36]. Small-angle X-ray scattering (SAXS) analysis on the sample discharged below 0.2 V showed a reversible decrease in intensity, corresponding to pores with sizes of about 14 Å. These results indicated that, within this voltage range, sodium ions inserted into free spaces between randomly stacked layers. Furthermore, Bai et al. [37] also associated the low voltage plateau with the pore filling mechanism as it disappeared after filling the HC micropores with sulfur. The first mechanism was commonly accepted to describe the insertion of lithium ions in the disordered carbons; however, the debate has been initiated once the studies of highly porous HC for Na-ion storage started. It has been found that the low voltage plateau capacity was not positively proportional to the porosity of the HC material [38,39]. The d spacing of the graphene layers expanded from 3.96 to 4.16 Å, within the voltage window of 0.1–0.2 V, indicating that Na intercalation took place. This leads to the doubts that the higher-voltage part of the region 2 is due to pore filling and the early Dahn model may be subject to amendments.

Many other studies confirm the "amending" of the first sodiation mechanism. Figure 10.3b shows a competitive model of the sodiation sequence of HC. Bommier et al. [25] tuned the microstructure of HC through the pyrolysis of sucrose at 1100°C, 1400°C and 1600°C, so that the resultant HCs displayed decreasing d spacing between the graphene sheets and an increasing size of turbostratic nanodomains with increasing temperature. They identified a decreased capacity from the slope region, whereas an increased capacity from the plateau region was observed with the increase of size of the turbostratic nanodomains in contrast to the trend observed in the conventional intercalation–pore filling model. Because larger turbostratic nanodomains are unambiguously more favorable for Na-ion intercalation, this indicates that the plateau region, instead of the sloping region, is associated with Na-ion intercalation. Inspired by theoretical calculations on the effect of defects on improving the sodiation potential, speculation was made that the slope region is more likely to be linked with defects in the HC [40–42]. Following this work, experimental studies on the effects of defects on HC performance have been conducted, including introducing defects into the HC structure through heteroatom doping. Li et al. [40] used both boron and phosphorus to dope HC and found that the interlayer spacing was expanded after doping, which in turn contributed to higher capacity of the plateau region. By increasing the doping level of boron and phosphorus, the slope region capacity also increased. The wide range of binding energy explains why adsorption proceeds in the slope region. The study revealed that the slope region was related to binding of Na ions to defects and the plateau region was due to Na intercalation [43]. The balance between these mechanisms is highly dependent on the precursor and the synthetic conditions, and thus, its assignment to the various voltage regions of HC is often contradictive between reports [25,35,40,44–47]. In general, it should be underlined that the electrochemical properties of HC are determined by its microstructure. The microstructure of HC in turn is determined by processing temperature and atmosphere as well as the source of carbon. The role of those factors has been intensively studied [35]. The common value of the optimum temperature is around 1400°C–1500°C [48,49]. Recently, Ji et al. [50] reported that the rate capability of HC had been underestimated in the literature. The reason is the high overpotential of the counter sodium electrode in a three-electrode cell. Therefore, when using an HC anode in a full cell, the rate capability is much higher than the fundamental studies. This suggests that the practical potential of HC anodes might be more than what has already been reported in the literature.

10.2.2 Titanium Oxide-Based Materials

Analogous to LIBs, metal oxide compounds have been studied as Na^+-ion insertion host materials. Titanium-based oxides are particularly interesting as anodes due to their reasonable operating voltage, cost and nontoxicity [28,51]. In general, the low operating potential of anode material is of advantage as it provides a higher cell voltage. It can, however, cause serious safety issues for practical applications such as metallic sodium plating and sodium dendrite formation on the surfaces of anodes [24,25,52]. The most representative examples of titanium-based electrode materials are titanium dioxides [53–64] and sodium titanate compounds [29,30,65–73]. The electrochemical reactions of these compounds are driven by a $Ti^{4+/3+}$ redox couple in Na cells. Recent works have focused on finding the sodiation/desodiation mechanism and improving the electrochemical performance of such materials.

Several TiO_2 polymorphs, including anatase-TiO_2, rutile-TiO_2, brookite-TiO_2 and bronze-TiO_2, have been investigated as anode materials for SIBs [11]. Among them, most research results have been reported for anatase TiO_2. This titania polymorph possesses the activation barrier for Na^+ insertion into the anatase lattice comparable to that of lithium, which is a remarkable feature considering its significantly larger ionic radius. On the other hand, Mattsson et al. [74] claimed that high crystalline and/or micronized TiO_2 cannot easily support Na^+ insertion because of the ionic size of Na^+ and a much higher sodium diffusion barrier compared with Li. Recently, high electrochemical activity of TiO_2 with Na^+ ions was achieved by reducing the particles to the nanometer size for shortening of the diffusion paths for Na^+ insertion. In general, sodium storage in TiO_2 suffers from the sluggish sodium kinetics due to the larger ionic size of Na^+ ions. To overcome this drawback, strategies such as the nano-architecture modified by metallic Ti and high conductivity carbon additives are introduced [62].

Sodium titanates ($Na_2O{\cdot}nTiO_2$) have been extensively investigated for the application as anode material for LIBs and NIBs. They are characterized by a peculiar structure, in which building block is given from edge- and corner-sharing TiO_6 octahedra forming corrugated sheets responsible for the outstanding chemical stability and high ion conductivity [11]. Moreover, being a zero-strain material, they guarantee structural stability [75]. Among various titanates, $Na_2Ti_3O_7$ has been investigated due to the lowest operating potential for SIBs. Typical voltage profiles of a composite of $Na_2Ti_3O_7$ with carbon black indicated an irreversible electrochemical process at ca. 0.7 V vs Na/Na^+, which corresponds to the reaction of carbon black and a reversible plateau around 0.3 V vs Na/Na^+ with concomitant intercalation of additional $2Na^+$ ions in the structure. According to the calculations of the electrostatic interaction in the crystal structure, 2 mol of Na^+ ions are intercalated into the $Na_2Ti_3O_7$ structure to form $Na_4Ti_3O_7$. It leads to a strong electrostatic repulsion, which in turn leads to structural instability and low operating voltage. It has been stated that this strong electrostatic repulsion in the fully sodiated state induces the self-relaxation phenomena.

Other types of sodium titanates have also been widely studied as potential anodes for SIBs. A reversible electrochemical activity of $Na_2Ti_6O_{13}$ in sodium cells with less than half a mol of sodium per formula unit has been reported [76]. Rudola et al. [66], based on the ex situ XRD measurements in the voltage range of 0.5–2.5 V, proposed the sodium insertion/extraction mechanism of $Na_2Ti_6O_{13}$ as follows: $Na_2Ti_6O_{13}+xNa^++xe^- \rightleftarrows Na_{2+x}Ti_6O_{13}$ with $x=0.85$.

Later, Shen et al. [72] show that the capacity of the $Na_2Ti_6O_{13}$ anode material can be enhanced from 49.5 mAh g^{-1} ($Na_{2+1}Ti_6O_{13}$) to 196 mAh g^{-1} ($Na_{2+4}Ti_6O_{13}$) by lowering the cutoff voltage from 0.3 to 0 V. Simultaneously with experimental works, Shen et al. [77] predicted the structure and average voltage change of reduced phases at various compositions of $Na_{2+x}Ti_6O_{13}$ ($x=0$–4) by using density functional theory (DFT) calculations. In our recent work, we report the molten salt synthesis of $Na_2Ti_6O_{13}$ nanorods, which is a facile and up-scalable synthesis approach providing a high purity material in a cost-effective way, with no need of additional ad hoc processing. Figure 10.4a presents the structure of the material, namely three edge-shared TiO_6 octahedra (highlighted with the brighter blue) representing the repeating unit composed of the tunnels within two Na ions are placed, whereas the first cycle charge–discharge transients of the material prepared at various temperatures are presented in Figure 10.4b.

10.3 ALLOY-BASED MATERIALS

Sodium can form alloy with elements from groups 14 and 15 (Sn, Pb, Ge, P, Sb, Bi and Si). Most of the materials studied for this purpose are represented by rather benign and abundant elements, thus making them potentially promising. Their common feature is a high theoretical capacity, since they can uptake multiple Na^+ ions per single atom with an average voltage of less than 1 V. Table 10.1 presents the overview of alloying elements in line with sodiation products and theoretical capacities.

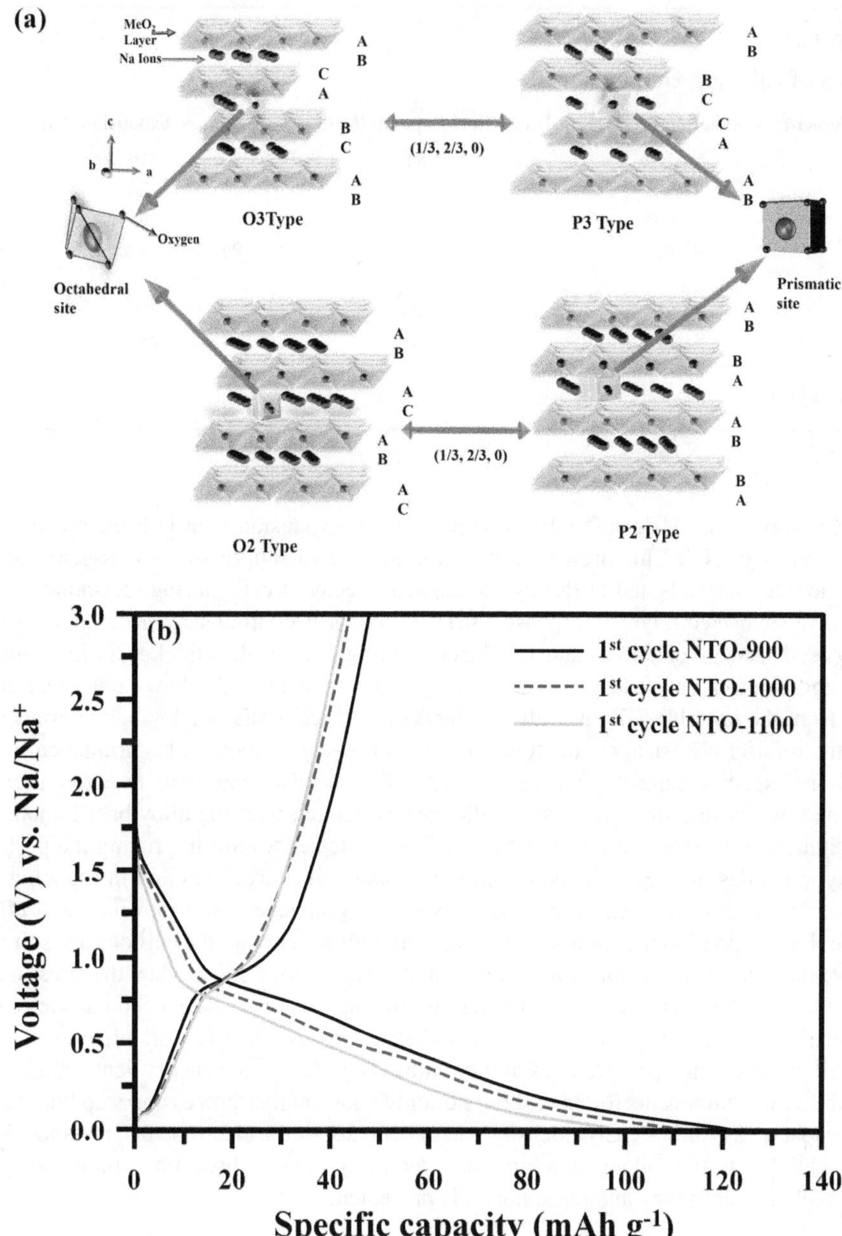

FIGURE 10.4 (a) Structure of $Na_2Ti_6O_{13}$ with the Na^+ ions marked in yellow, filling the tunnel structure built up by the TiO_6 framework. The red balls represent the oxygen atoms and the blue ones the Ti atoms. The structural model was realized using the 3D visualization software VESTA. (b) First cycle's charge and discharge profiles of NTO-900, NTO-1000 and NTO-1100. Adapted and reproduced from Ref. [77]. Copyright © 2021 John Wiley and Sons.

However, depending on the host materials and electrochemical sodiation levels, the reaction causes huge volume changes during the alloying–dealloying (see Table 10.1). Volume expansion is a key obstacle that hinders the commercialization of these alloy-based anodes. This leads to a severe capacity fading upon cycling, which is ascribed to a couple of reasons. One commonly recognized reason is the aggregation (slower kinetics due to loss of nanoscale diffusion distance) and pulverization (loss of electrical contact) of active materials induced by severe volume variation [83,84].

TABLE 10.1
Properties of Alloying Elements

Alloying Element	Sodiation Product	Theoretical Capacity(MAh g^{-1})	Volume Expansion (%)	Ref
Sn	$Na_{3.75}Sn$	847	410	[78]
Ge	NaGe	369	126	[79]
Pb	$Na_{3.75}Pb$	485	365	[80]
P	Na_3P	2596	Red 440;Black 499	Red [81]; Black [82]
Sb	Na_3Sb	660	290	[80]
Bi	Na_3Bi	385	250	[80]

Source: From [11].

Meanwhile, anodes for SIBs suffer from larger volume expansion than LIB electrodes owing to the larger radius of Na^+. This presents a bigger challenge in suppressing stress generated within the active anode materials and buffering the capacity decay of cells during (de)sodiation process. Additionally, solid electrolyte interphase (SEI) layers form continuously accompanied with the newly exposed surfaces of active materials during cycling. The resulted thick SEI films significantly block charge transfer and lead to quick capacity fading. Besides, the low Coulombic efficiency resulting from the unstable SEI layer also presents one critical issue of alloying materials in NIBs. Further, the inferior electrical conductivity of some alloy-based anodes also dramatically restricts their delivered specific capacity and the rate capability. To date, enormous research attention has been devoted to address these intrinsic challenges associated with the alloy-based anode materials, and significant progress has been achieved. The strategies toward improving the performance of alloy-type anodes mainly focus on designing efficient nanostructures and introducing conductive carbon host/substrate (e.g., carbon nanofibers and graphene), both of which can effectively accelerate the reaction kinetics and mitigate capacity fading. Though the carbons are soft and well conductive matrices, they do not exhibit sufficient robustness to accommodate the stress developed during the large volume changes [83] Although promising properties have been reported in numerous publications, the main question remains: Is there a way to effectively stabilize these materials so they will provide long-term cycling stability without compromising energy density? The addition of the stabilizing components (often carbon) not only adds another processing step but also affects practical capacity as well as energy density due to decrease of the electrode density and addition of "dead weight" [1]. In the following, a more detailed discussion of three most studied alloying elements, namely tin, antimony and phosphorus, is presented.

10.3.1 Tin

The theoretical capacity of Sn alloying up to $Na_{15}Sn_4$ is 847 mAh g^{-1} [78,84–87]. The first study by Komaba et al. [88] revealed capacities of 500 mAh g^{-1} over more than 20 cycles in the presence of FEC as an additive in electrolyte which promote the formation of stable SEI layer. A two-step reaction mechanism was observed starting from Na-poor amorphous phase Na_xSn followed by several rich amorphous phases which at the end transferred to crystalline Na_{15}Sn4 without any cracking [85]. Ellis et al. [78] studied the mechanism revealing different stages including $NaSn_3$, NaSn, Na_9Sn_4 and crystalline $Na_{15}Sn_4$ which were in accordance to that observed by DFT without realizing some of the amorphous phases as understood by in situ TEM analysis. The huge volume expansion is a hindrance in long-term cycling stability of the Sn-based anode and is usually addressed by making composites with carbon materials or other alloying elements [10,89,90]

To stabilize the electrochemical response of sodiation and desodiation reaction, various techniques have been applied. In 2015, Luo et al. [91] synthesized a novel anode where tin nanoparticles were encapsulated into graphene-backboned carbon foam to compare the electrochemical results in Figure 10.5. Graphene and the outermost carbon coating serve as a physical boundary to prevent the aggregation of well-distributed tin nanoparticles and alleviate the huge volume changes of tin particles. The unique structure is prepared by uniformly growing SnO_2 on the surface of graphene oxide and coating with porous carbon through a hydrothermal process, finally calcinating in a reducing atmosphere. The resulting composite shows excellent cycle stability and exceptional rate performance in LIBs as well as in SIBs. A reversible specific capacity of 506 mAh g^{-1} can be achieved at a current density of 400 mAh g^{-1} and retained at 270 mAh g^{-1}, and even at 3200 mA g^{-1} after 500 cycles (Figure 10.5).

Chen and Deng prepared Sn particles encapsulated in carbon nanospheres under CVD conditions forming unique deflated Sn@C nanoparticles firmly attached on the surface of the 3D carbon derived from walnut shell membranes [92]. The first insertion capacity was 260 mAh g^{-1}, and the first extraction was 163 mAh g^{-1} at 10 mA g^{-1}, but significant capacity fading has been found. Better results have been obtained with C/Sn/Ni/TMV1cys, binder-free composite electrode, where TMV1cys stands for a novel mutant of tobacco mosaic virus (TMV) created via genetic engineering, wherein a cysteine codon is expressed within the N-terminus of coat proteins. These tin-coated viral nanoforests as anodes retained a capacity of 405 mAh g^{-1} after 150 deep cycles at a current

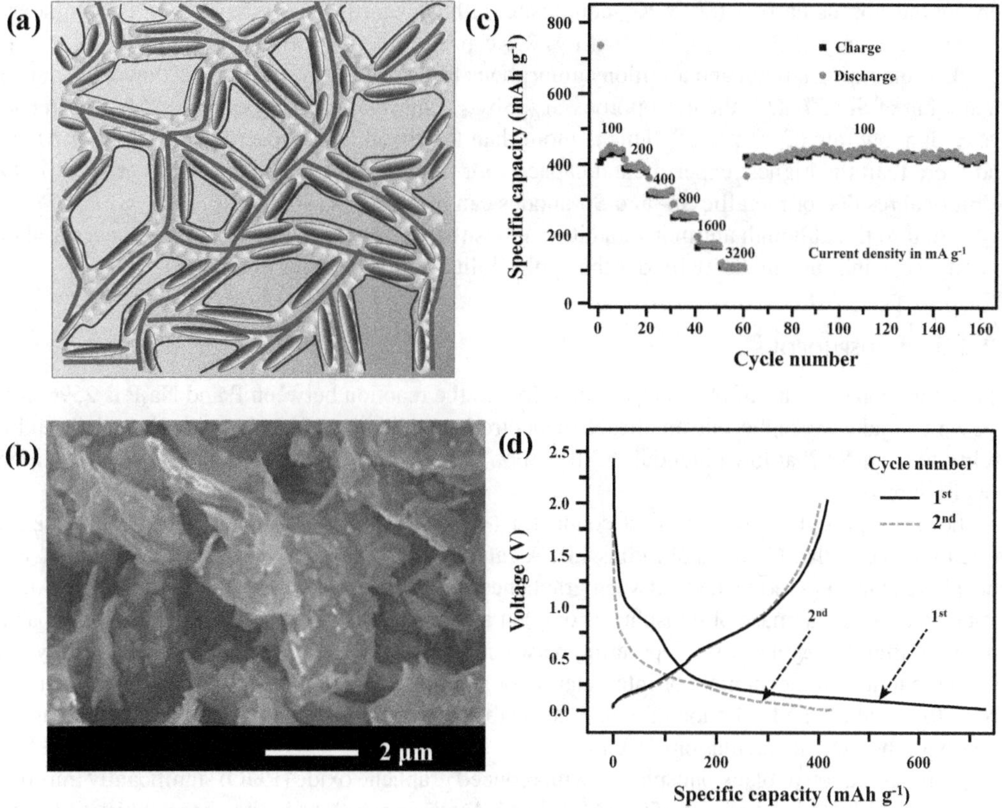

FIGURE 10.5 Schematic illustration (a) and SEM image (b) of tin nanoplates encapsulated in foam-like graphene-backboned carbonaceous carbon matrix (F-G/Sn@C). (c) Galvanostatic discharge–charge voltage profiles of the F-G/Sn@C electrode between 10 mV and 2.0 V vs Na/Na^+ at a current density of 100 mA g^{-1}. (d) Rate capability and cycling performance of the F-G/Sn@C electrodes in the voltage range of 10 mV to 2.0 V for Na-ion storage. Adapted and Reproduced from Ref. [91]. Copyright © 2016 Elsevier.

density of 50 mA g^{-1} [86]. Similar results were obtained with a Sn-Cu nanocomposite that delivered a capacity of 420 mAh g^{-1} at 0.2 C rate, retaining 97% of their maximum observed capacity after 100 cycles [93]. The best results, however, were obtained with a Sn–C composite (58 wt.% Sn and 42 wt.% N-doped carbon) in which 5–50 nm spherical Sn particles are on nitrogen-doped graphite nanoplatelets [87]. This anode delivered 429 mA h $(g_{Sn+C})^{-1}$ at 0.2 C and most of all maintained 290 mA h $(g_{Sn+C})^{-1}$ after 1000 cycles at 1 A g^{-1} (i.e., 82.6% retention in capacity referring to the second cycle (350 mA h $(g_{Sn+C})^{-1}$). This exceptional cycle life for a Sn-based anode was attributed to the effective electrode expansion reduced to 14% during sodiation, compared with 420% expected for Sn. Note, however, that the large amount of carbon reduces the effective capacity of the electrode. Although tin-based anode materials are widely studied for the application in NIB, one has to keep in mind that tin availability in the earth's crust is low (compare Figure 10.2), and in a view of the amount of electrode materials necessary for stationary storage system, the interest (and research progress) in tin-based anodes will rather remain purely academic.

10.3.2 Antimony

The theoretical capacity of Na_3Sb is 660 mAh g^{-1} [90,94,95]. In its carbon nanocomposite form, Sb/C can reversibly alloy up to 3 Na, with satisfactory rate capability and a long-term cycling stability with 94% capacity retention over 100 cycles [96]. Employing porous carbon allows to achieve a capacity of 385 mAh g^{-1} (capacity retention of 88.5%) after 500 cycles [97]. Antimony/nitrogen-doping porous carbon (Sb/NPC) composite with polyaniline nanosheets as a carbon source delivers a capacity of 529.6 mA hg^{-1}, with 97.2% capacity retention after 100 cycles at 100 mA g^{-1} [98]. By combining silicon and antimony amorphous films with bilayer thickness down to 2 nm and an amount of Si of 7 at.%, the mesoporous $Si_{0.07}Sb_{0.93}$ achieves a capacity of 663 mAh g^{-1} after 140 cycles at a low rate of 20 mA g^{-1}. This is more than the theoretical capacity for Sb (660 mAh g^{-1}) and more than the highest experimental capacity for pure Si reported so far (~600 mAh g^{-1}) [99]. Additional results for metallic Sn- and Sb-anodes can be found in [84,89,100].

Similar to tin, although antimony and its composites are widely investigated, the concerns about toxicity, cost and sustainability hinder the applicability of this element in NIBs.

10.3.3 Phosphorous

The theoretical capacity of phosphorus according to the reaction between P and Na_3P is 2596 mAh g^{-1}, which makes phosphorous an attractive electrode candidate with its low atomic weight and its ability to form Na_3P at low potential [82,101]. For SIBs, the red- and black-phosphorous allotropes are of interest.

The black phosphorous is a good conductor (~300 S m^{-1}), and the interlayer channel size is large (3.08 Å), so that Na^+ ions of radius 1.02 Å can be stored between the phosphorene layers. Few phosphorene layers sandwiched between graphene layers show a specific capacity of 2440 mAh g^{-1} (calculated using the mass of phosphorus only) at a current density of 0.05 A g^{-1} and 83% capacity retention after 100 cycles while operating between 0 and 1.5 V [101]. This very high capacity was attributed to a dual mechanism of intercalation of sodium ions along the x axis of the phosphorene layers followed by the formation of a Na_3P alloy that accompanies the P-P bond breaking in agreement with theoretical calculations [102].

The modification of black phosphorus with reduced graphene oxide (RGO) significantly improves its rate capability. A capacity of 650 mAh g^{-1} at 1 A g^{-1} over 200 cycles has been achieved for this composite [103]. The drawback, however, was a capacity at low rate (1400 mAh g^{-1} at 0.1 A g^{-1}) smaller than the best results that can exceed 2000 mAh g^{-1}: 2060 mAh g^{-1} at 0.2 C, with capacity retention of 75.3% after 200 cycles for a composite of black phosphorus and multiwall carbon nanotubes (BP–CNT) prepared via a surface oxidation-assisted chemical bonding procedure [104].

Red phosphorous has also a high sodium storage theoretical capacity (2595 mAh g^{-1}), but it shows much lower electronic conductivity (≈ 10^{-14} S cm^{-1}), so a conductive material is required as additive to fabricate performing anodes for SIBs [105]. In addition, the volume expansion upon sodiation is large, so that hollow and porous structures have been fabricated to increase the cycle life. For instance, wet-chemical synthesis of hollow red-phosphorus nanospheres with porous shells used as an anode delivered 1364 and 1100 mAh g_{em}^{-1} (g_{em}=gram of electrode materials) at 0.2C. The corresponding areal capacities was 2.3 and 1.8 mA h cm^{-2} at 0.52 and 1.3 mA cm^{-2}, respectively. At 1C, a stable capacity of 969.8 mAh g^{-1} was demonstrated over 600 cycles [106]. Combining electroless deposition with chemical dealloying to control the shell thickness and composition of a red-phosphorus (RP)@Ni–P core@shell nanostructure, Liu et al. [107] obtained an anode with remarkable properties: 1256 mAh g^{-1} after 200 cycles at 260 mA g^{-1}, while at the high current density of 5.2 A g^{-1}, the capacity was 491 mAh g^{-1} retained at 409 mAh g^{-1} after 2000 cycles (the data are per gram of the composite).

The drawback of both red and black phosphorus anodes, which hinders the application of this material, is large volume expansion–contraction changes during the sodium alloying–dealloying electrochemical process (compare Table 10.1). However, unlike the isotropic volume swelling of red phosphorus, the volume expansion of black phosphorus during the reaction is anisotropic, which mainly occurs in the second step of the alloying reaction. Recently, various nanostructured phosphorus-based anodes, which efficiently restrained the pulverization and supplied faster reaction kinetics, have been developed to solve these issues [101,108,109].

10.4 CONVERSION-BASED MATERIALS

A conversion reaction with sodium can be generalized by the following reaction (equation 10.1):

$$M_aX_b+(bc)\times N_a \leftrightarrow aM + bNa_cX \qquad (10.1)$$

with M being a transition metal (Cu, Fe, Ni, …) and X being an anion (O, S, P, …) [110,111]. The majority of transition metals, such as Fe, Co, Nb, and Cu, are electrochemically inactive in the oxide, so metal oxides react with Na^+ through a one-step conversion reaction. This reaction involves the formation of new phases during sodiation that are usually structurally very different from the starting material. In most cases, sodiation leads to the formation of an amorphous matrix of Na_cX in which nanoparticles of M are dispersed. In some cases, such as M=P, additional sodium can be stored through alloy formation [1,14]. Nevertheless, the ability to store more than one Na atom leads to huge volume changes, which are more significant than in the case of the reaction with lithium due to a bigger radius of Na^+ (Figure 10.6). This challenge is usually tackled by engineering morphology, incorporating carbon materials or modifying the electrolyte to obtain stable SEI layer [10].

Transition metal oxides have been considered as potential negative electrode material due to their intrinsic stability, intermediate voltages against Na/Na^+, volume expansion between sulfides and phosphides and relative abundancy [11,110,111]. While iron, tin, cobalt and copper oxides represent the largest studied area, others such as antimony, nickel and molybdenum oxides have also been studied [112]. To increase electrode/electrolyte contact, promote kinetics of Na^+ and buffer volume changes, nano-structuring of anode materials is usually preferred. Hariharan et al. [113] synthesized Fe_3O_4 nanoparticles which delivered the initial capacity of 643 mAh g^{-1}. The investigation of the mechanism showed the presence of crystalline Na_2O embedded in a matrix of metallic Fe. After desodiation, it showed incomplete oxidation leading to a low initial Columbic efficiency of 57%. Porous morphology also accommodates volume changes by providing more free volume for Na^+ storage and increases electrode/electrolyte contact due to larger surface area. Porous nanofibers and nanotubes of SnO_2 were synthesized by Duran et al. [114], and better performance was observed in terms of capacity retention and cyclability.

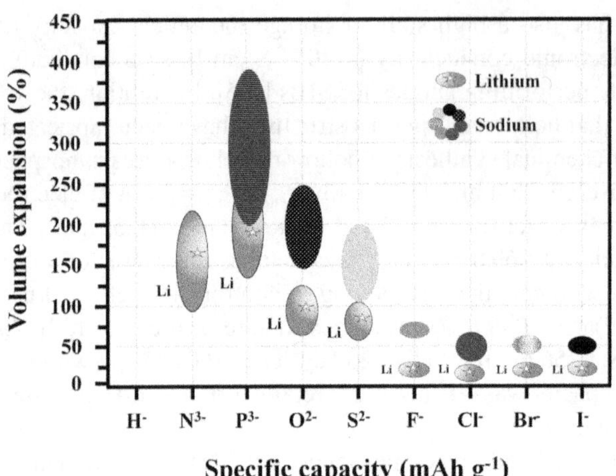

FIGURE 10.6 Calculated volume expansions for lithium- and sodium-based conversion reactions. Values are calculated as follows: volume expansion (%) = $100\cdot([V(bA_cX) + V(aM)]/V(M_aX_b)) - 100$. Adapted and reproduced from Ref. [110]. Copyright © 1999 Royal Society of Chemistry.

Due to better reversibility of the sodiation–desodiation process resulting in higher Coulombic efficiency and better stability, many transition metal sulfides have been studied in view of their performance as negative electrodes in SIBs [115,116]. Among them, iron- and tin-based sulfides are the most promising [111,117]. Iron chalcogenides (FeS_2) show better rate performance and cycling stability than iron oxides, but their high operating voltage (above 0.6 V vs Na^+/Na) hinders their applications. Sulfides of tin (II) and tin (IV) rely on the conversion reaction followed by the alloying with tin and therefore deliver high capacities of around 600 mAh g^{-1} [118]; however, their operating voltage is still quite high.

In addition to oxides and chalcogenides, metal phosphides represent one of the promising candidates in the present group [119]. They combine the conversion reaction resulting in the formation of phosphorus within the network of metal atoms which provides electronic conductivity. The formed phosphorus undergoes the alloying reaction described above delivering high capacity. It should be noted that the pulverization problems associated with the use of phosphorus are partially mitigated in these compounds [120].

An overview of the specific capacities and cell potentials vs Na/Na^+ for conversion reactions of different classes of materials is presented in Figure 10.7. The high redox potentials of conversion materials with a halogen anion make this group not suitable for negative electrodes. In contrary, sulfides, oxides and phosphides are potentially promising due to lower potential and high capacities, but the large volume changes during the redox process slow down the possible application [110].

10.5 2D MATERIALS

The tremendous interest in 2D materials gave rise to the whole subfamily of the layered chalcogenides as potential negative electrode materials for SIBs. Among those, MoS_2 with S–Mo–S motifs stacked together by van der Waals forces is the most common representative; however, a variety of similar chalcogenides have been reported in the literature and covered in several reviews [121–124]. As a typical conversion-type material, MoS_2 allows to achieve a theoretical capacity of ~670 mAh g^{-1} for complete sodiation [125]. However, its layered structure provides a variety of opportunities for material engineering aiming the mitigation of the problems of volumetric changes, low electrical conductivity and improvement of ion diffusion. Nevertheless, the complexity of these materials in terms of their mass production in a controlled fashion raises a question of their applicability in real batteries in the nearest future [1].

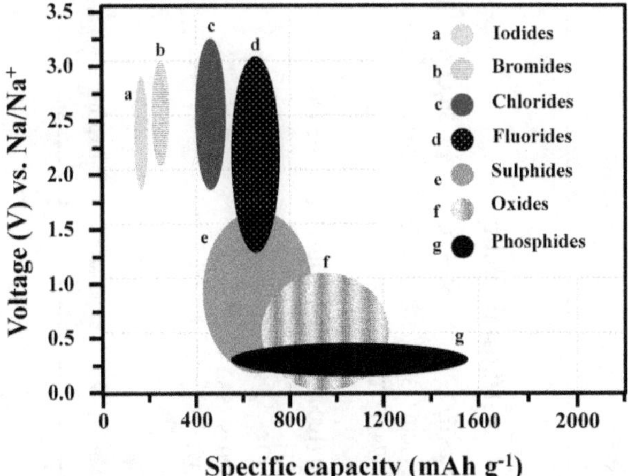

FIGURE 10.7 Specific capacities and cell potentials vs Na/Na+ for conversion reactions of different classes of materials with sodium. Adapted and reproduced from Ref. [110]. Copyright © 1999 Royal Society of Chemistry.

10.6 SUMMARY AND OUTLOOK

Table 10.2 presents the summary of properties of the negative electrode materials mostly applied in NIB in terms of energy, rate stability and cost/sustainability

Due to the low operating voltage and extraordinary cycling stability, hard carbon is nowadays considered as the better candidate than any other electrode material for anode materials. Since the first LIBs adopted hard carbon as an anode material, it can be the first commercial anode materials even in SIBs as well. The most urgent issue is to hinder the sodium metal deposition onto the surface of hard carbon at low voltage, in which safety issues should be considered because of the high reactivity of sodium metal. In addition, capacity below 300 mAh g^{-1}, which is lower than graphite anodes for LIBs, is not advantageous to improve the energy density of SIBs. Thus, conversion and alloying reaction-based materials are intensively being studied to find alternative anode materials that deliver high capacity and do not have sodium metal deposition on discharge. It is evident that the conversion and alloying materials have superiority in capacity, while volume expansion of electrodes in the sodiated state arising from the large ionic size of Na^+ is the intrinsic problem for these electrodes. So far, various anode materials are under investigation, and it is believed that hard carbon is thought to be the best candidate in practical applications due to its low operating voltage, cycling stability and high Coulombic efficiency at the first cycle.

The constant progress experienced these last 5 years evidenced that the NIBs will find an increasing market. They will not compete with the lithium-ion chemistry in terms of energy density and rate capability, and the lithium chemistry will keep the market of the electric cars for instance. Energy produced from renewable sources is, however, increasingly integrated into our electricity grid. These sources are in principle inexhaustible but operate intermittently, thus interest in energy storage technologies for grid stabilization is growing. Electrochemical energy storage systems represent the most promising solution to fill the gap between energy production and utilization securing energy supply. Among the battery systems available, the sodium-ion technology has the potential to represent the next generation of low-cost and environmentally friendly electrochemical energy storage systems, especially for stationary energy storage applications.

TABLE 10.2
Summary of Properties of the Mostly Studied Anode Materials in Terms of Capacity, Rate, Stability and Cost/Sustainability

Material	Hard Carbon	Sn/C	P/C	Sb/C	TiO_2	$NaxTi_yO_z$	MoS_2/C	Sb_2O_3/C
Capacity (mAh g^{-1}) (current density (mA g^{-1}))	310 (20) [126] 225 (50) [127] 286 (20) [128]	494 (200) [129] 750 (84.7) [130]	1682 (100) [131] 1764 (250) [132]	440 (100) [133] 628 (200) [134]	252 (100) [135] 239 (50) [136]	172 (100) [137] 40 (74) [77])	702 (20) [138]	503 (100) [112]
Rate performance Capacity mAh g^{-1} (current density in mA g^{-1})	142 (500) [126] 110 (5000) [127] 117 (100) [128]	349 (4000) [129] 190 (4235) [130]	606 (2000) [131] 640 (4000) [132]	88 (6000) [133] 302 (3000) [134]	99 (2000) [135] 113 (5000) [136]	119 (1000) [137]	352 (640) [138]	315 (2000) [112]
Long-term cyclability Cycle number (capacity retention)	2000 (98%) [126] 800 (87%) [127]	500 (≈100%) [129] 1000 (≈100%) [130]	140 (≈67%) [132] 1000 (76%) [131]	100 (93%) [134] 300 (≈82%) [133]	500 (≈65%) [135]	2800 (90.8%) [137]		
Cost/Sustainability	Cheap/available	Cheap/rare	Cheap/available	Expensive/rare	Cheap/available	Cheap	Expensive	Expensive

ACKNOWLEDGMENTS

MGZ and AK acknowledge the financial support of the German Research Foundation (DFG) within the project with Grant no. 4440/4-1.

REFERENCES

1. Hasa I, Mariyappan S, Saurel D et al. (2021) Challenges of today for Na-based batteries of the future: From materials to cell metrics. *Journal of Power Sources* 482:228872. https://doi.org/10.1016/j.jpowsour.2020.228872.
2. Vaalma C, Buchholz D, Weil M et al. (2018) A cost and resource analysis of sodium-ion batteries. *Nature Reviews Materials* 3:18013. https://doi.org/10.1038/natrevmats.2018.13.
3. Sangster J (2007) C-Na (carbon-sodium) system. *Journal of Phase Equilibria and Diffusion* 28:571–579.
4. Eftekhari A, Kim D-W (2018) Sodium-ion batteries: New opportunities beyond energy storage by lithium. *Journal of Power Sources* 395:336–348. https://doi.org/10.1016/j.jpowsour.2018.05.089.
5. Kim S-W, Seo D-H, Ma X et al. (2012) Electrode materials for rechargeable sodium-ion batteries: Potential alternatives to current lithium-ion batteries. *Advanced Energy Materials* 2:710–721.
6. Jache B, Adelhelm P (2014) Use of graphite as a highly reversible electrode with superior cycle life for sodium-ion batteries by making use of co-intercalation phenomena. *Angewandte Chemie – International Edition in English* 53:10169–10173. https://doi.org/10.1002/anie.201403734.
7. Eftekhari A (2017) Energy efficiency: A critically important but neglected factor in battery research. *Sustainable Energy Fuels* 1:2053–2060. https://doi.org/10.1039/c7se00350a.
8. Wu Y (2015) *Lithium-Ion Batteries: Fundamentals and Applications.* CRC Press, Boca Rato.
9. Newman GH, Klemann LP (1980) Ambient temperature cycling of an Na-TiS_2 cell. *Journal of the Electrochemical Society* 127:2097–2099. https://doi.org/10.1149/1.2129353.
10. Perveen T, Siddiq M, Shahzad N et al. (2020) Prospects in anode materials for sodium ion batteries - A review. *Renewable and Sustainable Energy Reviews* 119:109549. https://doi.org/10.1016/j.rser.2019.109549.
11. Hwang J-Y, Myung S-T, Sun Y-K (2017) Sodium-ion batteries: Present and future. *Chemical Society Reviews* 46:3529–3614. https://doi.org/10.1039/C6CS00776G.
12. Mauger A, Julien CM (2020) State-of-the-art electrode materials for sodium-ion batteries. *Materials* 13. https://doi.org/10.3390/ma13163453.
13. (2018) Energy union and climate. https://ec.europa.eu/commission/priorities/energy-union-and-climate_en.
14. Li L, Zheng Y, Zhang S et al. (2018) Recent progress on sodium ion batteries: Potential high-performance anodes. *Energy and Environmental Science* 11:2310–2340. https://doi.org/10.1039/c8ee01023d.
15. Skundin AM, Kulova TL, Yaroslavtsev AB (2018) Sodium-ion batteries (a review). *Russian Journal of Electrochemistry* 54:113–152. https://doi.org/10.1134/S1023193518020076.
16. Thomas P, Ghanbaja J, Billaud D (1999) Electrochemical insertion of sodium in pitch-based carbon fibres in comparison with graphite in $NaClO_4$-ethylene carbonate electrolyte. *Electrochimica Acta* 45:423–430. https://doi.org/10.1016/S0013-4686(99)00276-5.
17. Stevens DA, Dahn JR (2000) An in situ small-angle x-ray scattering study of sodium insertion into a nanoporous carbon anode material within an operating electrochemical cell. *Journal of the Electrochemical Society* 147:4428. https://doi.org/10.1149/1.1394081.
18. Stevens DA, Dahn JR (2001) The mechanisms of lithium and sodium insertion in carbon materials. *Journal of the Electrochemical Society* 148:A803. https://doi.org/10.1149/1.1379565.
19. Joncourt L, Mermoux M, Touzain P et al. (1996) Sodium reactivity with carbons. *Journal of Physics and Chemistry of Solids* 57:877–882. https://doi.org/10.1016/0022-3697(95)00366-5.
20. Xia X, Dahn JR (2012) Study of the reactivity of Na/hard carbon with different solvents and electrolytes. *Journal of the Electrochemical Society* 159:A515–A519. https://doi.org/10.1149/2.jes111637.
21. Palomares V, Serras P, Villaluenga I et al. (2012) Na-ion batteries, recent advances and present challenges to become low cost energy storage systems. *Energy & Environmental Science* 5:5884–5901.
22. Stevens DA, Dahn JR (2000) High capacity anode materials for rechargeable sodium ion batteries. *Journal of the Electrochemical Society* 147:1271–1273.
23. Lee L-L, Tsai D-S (1999) A hydrogen-permselective silicon oxycarbide membrane derived from polydimethylsilane. *Journal of the American Ceramic Society* 82:2796–2800. https://doi.org/10.1111/j.1151-2916.1999.tb02158.x.

24. Irisarri E, Ponrouch A, Palacin MR (2015) Review hard carbon negative electrode materials for sodium-ion batteries. *Journal of the Electrochemical Society* 162:A2476–A2482
25. Bommier C, Surta TW, Dolgos M et al. (2015) New mechanistic insights on Na-ion storage in non-graphitizable carbon. *Nano Letters* 15:5888–5892. https://doi.org/10.1021/acs.nanolett.5b01969.
26. Doeff MM, Ma Y, Visco SJ et al. (1993) Electrochemical insertion of sodium into carbon. *Journal of the Electrochemical Society* 140:L169–L170. https://doi.org/10.1149/1.2221153.
27. Mei Y, Huang Y, Hu X (2016) Nanostructured Ti-based anode materials for Na-ion batteries. *Journal of Materials Chemistry A* 4:12001–12013. https://doi.org/10.1039/C6TA04611H.
28. Guo S, Yi J, Sun Y et al. (2016) Recent advances in titanium-based electrode materials for stationary sodium-ion batteries. *Energy & Environmental Science* 9:2978–3006. https://doi.org/10.1039/C6EE0.1807F.
29. Libich J, Máca J, Chekannikov A et al. (2019) Sodium titanate for sodium-ion batteries. *Surface Engineering and Applied Electrochemistry* 55:109–113. https://doi.org/10.3103/S1068375519010125.
30. Doeff MM, Cabana J, Shirpour M (2014) Titanate anodes for sodium ion batteries. *Journal of Inorganic and Organometallic Polymers and Materials* 24:5–14. https://doi.org/10.1007/s10904-013-9977-8.
31. Candelaria SL, Shao Y, Zhou W et al. (2012) Nanostructured carbon for energy storage and conversion. *Nano Energy* 1:195–220.
32. Wang J, Bi J, Wang W et al. (2020) $Na_2Ti_6O1_3$ coated with carbon produced by citric acid as an anode material in sodium ion batteries. *Journal of the Electrochemical Society* 167:90539. https://doi.org/10.1149/1945-7111/ab8fd6.
33. Palacin MR (2009) Recent advances in rechargeable battery materials: A chemist's perspective. *Chemical Society Reviews Journal* 38:2565–2575.
34. Yabuuchi N, Kubota K, Dahbi M et al. (2014) Research development on sodium-ion batteries. *Chemical Reviews* 114:11636–11682. https://doi.org/10.1021/cr500192f.
35. Saurel D, Orayech B, Xiao B et al. (2018) From charge storage mechanism to performance: A roadmap toward high specific energy sodium-ion batteries through carbon anode optimization. *Advanced Energy Materials* 8:1703268. https://doi.org/10.1002/aenm.201703268.
36. Komaba S, Murata W, Ishikawa T et al. (2011) Electrochemical Na insertion and solid electrolyte interphase for hard-carbon electrodes and application to Na-ion batteries. *Advanced Functional Materials* 21:3859–3867.
37. Bai P, He Y, Zou X et al. (2018) Elucidation of the sodium-storage mechanism in hard carbons. *Advanced Energy Materials* 8:1703217. https://doi.org/10.1002/aenm.201703217.
38. Ding J, Wang H, Li Z et al. (2013) Carbon nanosheet frameworks derived from peat moss as high performance sodium ion battery anodes. *ACS Nano* 7:11004–11015.
39. Lotfabad EM, Ding J, Cui K et al. (2014) High-density sodium and lithium ion battery anodes from banana peels. *ACS Nano* 8:7115–7129. https://doi.org/10.1021/nn502045y.
40. Li Z, Bommier C, Chong ZS et al. (2017) Mechanism of Na-ion storage in hard carbon anodes revealed by heteroatom doping. *Advanced Energy Materials* 7:1602894. https://doi.org/10.1002/aenm.201602894.
41. Bommier C, Surta TW, Dolgos M et al. (2015) New mechanistic insights on Na-ion storage in non-graphitizable carbon. *Nano Letters* 15:5888–5892. https://doi.org/10.1021/acs.nanolett.5b01969.
42. Luo W, Shen F, Bommier C et al. (2016) Na-ion battery anodes: Materials and electrochemistry. *Accounts of Chemical Research* 49, 231–240.
43. Xiao B, Rojo T, Li X (2019) Hard carbon as sodium-ion battery anodes: Progress and challenges. *ChemSusChem* 12:133–144. https://doi.org/10.1002/cssc.201801879.
44. Dou X, Hasa I, Saurel D et al. (2019) Hard carbons for sodium-ion batteries: Structure, analysis, sustainability, and electrochemistry. *Materials Today* 23:87–104. https://doi.org/10.1016/j.mattod.2018.12.040.
45. Hou H, Qiu X, Wei W et al. (2017) Carbon anode materials for advanced sodium-ion batteries. *Advanced Energy Materials* 7:1602898. https://doi.org/10.1002/aenm.201602898.
46. Stratford JM, Allan PK, Pecher O et al. (2016) Mechanistic insights into sodium storage in hard carbon anodes using local structure probes. *Chemical Communications* 52:12430–12433. https://doi.org/10.1039/C6CC06990H.
47. Anji Reddy M, Helen M, Groß A et al. (2018) Insight into sodium insertion and the storage mechanism in hard carbon. *ACS Energy Letters* 3:2851–2857. https://doi.org/10.1021/acsenergylett.8b01761.
48. Dahbi M, Kiso M, Kubota K et al. (2017) Synthesis of hard carbon from argan shells for Na-ion batteries. *Journal of Materials Chemistry A* 5:9917–9928. https://doi.org/10.1039/C7TA01394A.
49. Jin Y, Zhou G, Shi F et al. (2017) Reactivation of dead sulfide species in lithium polysulfide flow battery for grid scale energy storage. *Nature Communications* 8:462. https://doi.org/10.1038/s41467-017-00537-0.

50. Jiang T, Zhang R, Yin Q et al. (2017) Morphology, composition and electrochemistry of a nano-porous silicon versus bulk silicon anode for lithium-ion batteries. *Journal of Materials Science* 52:3670–3677. https://doi.org/10.1007/s10853-016-0599-8.
51. Aravindan V, Lee Y-S, Yazami R et al. (2015) TiO_2 polymorphs in 'rocking-chair' Li-ion batteries. *Materials Today* 18:345–351. https://doi.org/10.1016/j.mattod.2015.02.015.
52. Balogun M-S, Luo Y, Qiu W et al. (2016) A review of carbon materials and their composites with alloy metals for sodium ion battery anodes. *Carbon* 98:162–178. https://doi.org/10.1016/j.carbon.2015.09.091.
53. Legrain F, Malyi O, Manzhos S (2015) Insertion energetics of lithium, sodium, and magnesium in crystalline and amorphous titanium dioxide: A comparative first-principles study. *Journal of Power Sources* 278:197–202. https://doi.org/10.1016/j.jpowsour.2014.12.058.
54. Agostini M, Scrosati B, Hassoun J (2015) An advanced lithium-ion sulfur battery for high energy storage. *Advanced Energy Materials* 5:1500481. https://doi.org/10.1002/aenm.201500481.
55. Su D, Dou S, Wang G (2015) Anatase TiO_2: Better anode material than amorphous and rutile phases of TiO_2 for Na-ion batteries. *Chemistry of Materials* 27:6022–6029. https://doi.org/10.1021/acs.chemmater.5b02348.
56. Lunell S, Stashans A, Ojamäe L et al. (1997) Li and Na Diffusion in TiO_2 from quantum chemical theory versus electrochemical experiment. *Journal of the American Chemical Society* 119:7374–7380. https://doi.org/10.1021/ja9708629.
57. Xiong H, Slater MD, Balasubramanian M et al. (2011) Amorphous TiO_2 nanotube anode for rechargeable sodium ion batteries. *The Journal of Physical Chemistry Letters* 2:2560–2565. https://doi.org/10.1021/jz2012066.
58. Wu L, Buchholz D, Bresser D et al. (2014) Anatase TiO_2 nanoparticles for high power sodium-ion anodes. *Journal of Power Sources* 251:379–385. https://doi.org/10.1016/j.jpowsour.2013.11.083.
59. Wu L, Bresser D, Buchholz D et al. (2014) Nanocrystalline TiO_2(B) as anode material for sodium-ion batteries. *Journal of the Electrochemical Society* 162:A3052–A3058. https://doi.org/10.1149/2.0091502jes.
60. Wu L, Bresser D, Buchholz D et al. (2014) Unfolding the mechanism of sodium insertion in anatase TiO_2 nanoparticles. *Advanced Energy Materials* 5:1401142. https://doi.org/10.1002/aenm.201401142.
61. Abel PR, Lin YM, de Souza T et al. (2013) Nanocolumnar germanium thin films as a high-rate sodium-ion battery anode material. *The Journal of Physical Chemistry C* 117:18885–18890. https://doi.org/10.1021/jp407322k.
62. Hwang J-Y, Myung S-T, Lee J-H et al. (2015) Ultrafast sodium storage in anatase TiO_2 nanoparticles embedded on carbon nanotubes. *Nano Energy* 16:218–226. https://doi.org/10.1016/j.nanoen.2015.06.017.
63. Usui H, Yoshioka S, Wasada K et al. (2015) Nb-doped rutile TiO_2: A potential anode material for Na-ion battery. *ACS Applied Materials & Interfaces* 7:6567–6573. https://doi.org/10.1021/am508670z.
64. Søndergaard M, Dalgaard KJ, Bøjesen ED et al. (2015) In situ monitoring of TiO_2(B)/anatase nanoparticle formation and application in Li-ion and Na-ion batteries. *Journal of Materials Chemistry A* 3:18667–18674. https://doi.org/10.1039/C5TA04110D.
65. Rudola A, Saravanan K, Devaraj S et al. (2013) $Na_2Ti_6O_{13}$: A potential anode for grid-storage sodium-ion batteries. *Chemical Communications* 49:7451. https://doi.org/10.1039/c3cc44381g.
66. Rudola A, Saravanan K, Mason CW et al. (2013) $Na_2Ti_3O_7$: An intercalation based anode for sodium-ion battery applications. *Journal of Materials Chemistry A* 1:2653. https://doi.org/10.1039/c2ta01057g.
67. Rudola A, Sharma N, Balaya P (2015) Introducing a 0.2 V sodium-ion battery anode: The $Na_2Ti_3O_7$ to $Na_{3-x}Ti_3O_7$ pathway. *Electrochemistry Communications* 61:10–13. https://doi.org/10.1016/j.elecom.2015.09.016.
68. Saravanan K, Mason CW, Rudola A et al. (2012) The first report on excellent cycling stability and superior rate capability of $Na_3V_2(PO_4)_3$ for sodium ion batteries. *Advanced Energy Materials* 3:444–450. https://doi.org/10.1002/aenm.201200803.
69. Cabana J, Monconduit L, Larcher D et al. (2010) Beyond intercalation-based Li-ion batteries: The state of the art and challenges of electrode materials reacting through conversion reactions. *Advanced Materials* 22:E170–E192.
70. Senguttuvan P, Rousse G, Seznec V et al. (2011) $Na_2Ti_3O_7$: Lowest voltage ever reported oxide insertion electrode for sodium ion batteries. *Chemistry of Materials* 23:4109–4111. https://doi.org/10.1021/cm202076g.
71. Ponrouch A, Dedryvere R, Monti D et al. (2013) Towards high energy density sodium ion batteries through electrolyte optimization. *Energy & Environmental Science* 6:2361–2369.
72. Shen K, Wagemaker M (2014) $Na_{2+x}Ti_6O1_3$ as potential negative electrode material for Na-ion batteries. *Inorganic Chemistry* 53:8250–8256. https://doi.org/10.1021/ic5004269.

73. Wang Y, Yu X, Xu S et al. (2013) A zero-strain layered metal oxide as the negative electrode for long-life sodium-ion batteries. *Nature Communications* 4. https://doi.org/10.1038/ncomms3365.
74. Mattsson MS, Veszelei G, Niklasson A, Granqvist C.-G, Stashan A, and Lunell, S (1997) Cation diffusion in electrochromic fluorinated Ti dioxide. In *Electrochromic Materials and Their Applications III*, eds. Ho KC, Greenberg CB, MacArthur DM, The Electrochemical Society, Pennington, NJ.
75. Pérez-Flores JC, Kuhn A, García-Alvarado F (2011) Synthesis, structure and electrochemical Li insertion behaviour of $Li_2Ti_6O_{13}$ with the $Na_2Ti_6O_{13}$ tunnel-structure. *Journal of Power Sources* 196:1378–1385. https://doi.org/10.1016/j.jpowsour.2010.08.106.
76. Soraru GD, Campostrini R, Maurina S et al. (1997) Gel precursor to silicon oxycarbide glasses with ultrahigh ceramic yield. *Journal of the American Ceramic Society* 80:999–1004.
77. de Carolis DM, Vrankovic D, Kiefer SA et al. (2021) Towards a greener and scalable synthesis of $Na_2Ti_6O_{13}$ nanorods and their application as anodes in batteries for grid-level energy storage. *Energy Technology (Weinheim)* 9:2000856. https://doi.org/10.1002/ente.202000856.
78. Ellis LD, Hatchard TD, Obrovac MN (2012) Reversible insertion of sodium in tin. *Journal of the Electrochemical Society* 159:A1801–A1805. https://doi.org/10.1149/2.037211jes.
79. Baggetto L, Keum JK, Browning JF et al. (2013) Germanium as negative electrode material for sodium-ion batteries. *Electrochemistry Communications* 34:41–44. https://doi.org/10.1016/j.elecom.2013.05.025.
80. Ellis LD, Wilkes BN, Hatchard TD et al. (2014) In situ XRD study of silicon, lead and bismuth negative electrodes in nonaqueous sodium cells. *Journal of the Electrochemical Society* 161:A416–A421. https://doi.org/10.1149/2.080403jes.
81. Sun J, Lee H-W, Pasta M et al. (2016) Carbothermic reduction synthesis of red phosphorus-filled 3D carbon material as a high-capacity anode for sodium ion batteries. *Energy Storage Materials* 4:130–136. https://doi.org/10.1016/j.ensm.2016.04.003.
82. Sun J, Lee HW, Pasta M et al. (2015) A phosphorene-graphene hybrid material as a high-capacity anode for sodium-ion batteries. *Nature Nanotechnology* 10:980–985. https://doi.org/10.1038/nnano.2015.194.
83. Lao M, Zhang Y, Luo W et al. (2017) Alloy-based anode materials toward advanced sodium-ion batteries. *Advanced Materials* 29:1700622. https://doi.org/10.1002/adma.201700622.
84. Jing WT, Yang CC, Jiang Q (2020) Recent progress on metallic Sn- and Sb-based anodes for sodium-ion batteries. *Journal of Materials Chemistry A* 8:2913–2933. https://doi.org/10.1039/C9TA11782B.
85. Wang JW, Liu XH, Mao SX et al. (2012) Microstructural evolution of tin nanoparticles during in situ sodium insertion and extraction. *Nano Letters* 12:5897–5902. https://doi.org/10.1021/nl303305c.
86. Liu Y, Xu Y, Zhu Y et al. (2013) Tin-coated viral nanoforests as sodium-ion battery anodes. *ACS Nano* 7:3627–3634. https://doi.org/10.1021/nn400601y.
87. Palaniselvam T, Goktas M, Anothumakkool B et al. (2019) Sodium storage and electrode dynamics of tin-carbon composite electrodes from bulk precursors for sodium-ion batteries. *Advanced Functional Materials* 29:1900790. https://doi.org/10.1002/adfm.201900790.
88. Komaba S, Matsuura Y, Ishikawa T et al. (2012) Redox reaction of Sn-polyacrylate electrodes in aprotic Na cell. *Electrochemistry Communications* 21:65–68. https://doi.org/10.1016/j.elecom.2012.05.017.
89. Sayed SY, Kalisvaart WP, Luber EJ et al. (2020) Stabilizing tin anodes in sodium-ion batteries by alloying with silicon. *ACS Applied Energy Materials* 3:9950–9962. https://doi.org/10.1021/acsaem.0c01641.
90. Mou H, Xiao W, Miao C et al. (2020) Tin and tin compound materials as anodes in lithium-ion and sodium-ion batteries: A review. *Frontiers in Chemistry* 8:141. https://doi.org/10.3389/fchem.2020.00141.
91. Luo B, Qiu T, Ye D et al. (2016) Tin nanoparticles encapsulated in graphene backboned carbonaceous foams as high-performance anodes for lithium-ion and sodium-ion storage. *Nano Energy* 22:232–240. https://doi.org/10.1016/j.nanoen.2016.02.024.
92. Chen W, Deng D (2015) Deflated carbon nanospheres encapsulating tin cores decorated on layered 3-D carbon structures for low-cost sodium ion batteries. *ACS Sustainable Chemistry and Engineering* 3:63–70. https://doi.org/10.1021/sc500543u.
93. Lin Y-M, Abel PR, Gupta A et al. (2013) Sn-cu nanocomposite anodes for rechargeable sodium-ion batteries. *ACS Applied Materials & Interfaces* 5:8273–8277. https://doi.org/10.1021/am4023994.
94. Aurbach D, Markovsky B, Salitra G et al. (2007) Review on electrode-electrolyte solution interactions, related to cathode materials for Li-ion batteries. *Journal of Power Sources* 165:491–499.
95. Darwiche A, Marino C, Sougrati MT et al. (2012) Better cycling performances of bulk sb in na-ion batteries compared to li-ion systems: An unexpected electrochemical mechanism. *Journal of the American Chemical Society* 134:20805–20811. https://doi.org/10.1021/ja310347x.
96. Qian J, Chen Y, Wu L et al. (2012) High capacity Na-storage and superior cyclability of nanocomposite Sb/C anode for Na-ion batteries. *Chemical Communications* 48:7070. https://doi.org/10.1039/c2cc32730a.

97. Zhang N, Liu Y, Lu Y et al. (2015) Spherical nano-Sb@C composite as a high-rate and ultra-stable anode material for sodium-ion batteries. *Nano Research* 8:3384–3393. https://doi.org/10.1007/s12274-015-0838-3.
98. Wu T, Hou H, Zhang C et al. (2017) Antimony anchored with nitrogen-doping porous carbon as a high-performance anode material for Na-ion batteries. *ACS Applied Materials and Interfaces* 9:26118–26125. https://doi.org/10.1021/acsami.7b07964.
99. Kalisvaart WP, Olsen BC, Luber EJ et al. (2019) Sb-Si alloys and multilayers for sodium-ion battery anodes. *ACS Applied Energy Materials* 2:2205–2213. https://doi.org/10.1021/acsaem.8b02231.
100. Xie H, Kalisvaart WP, Olsen BC et al. (2017) Sn-Bi-Sb alloys as anode materials for sodium ion batteries. *Journal of Materials Chemistry A* 5:9661–9670. https://doi.org/10.1039/c7ta01443k.
101. Liu W, Zhi H, Yu X (2019) Recent progress in phosphorus based anode materials for lithium/sodium ion batteries. *Energy Storage Materials* 16:290–322. https://doi.org/10.1016/j.ensm.2018.05.020.
102. Hembram KPSS, Jung H, Yeo BC et al. (2015) Unraveling the atomistic sodiation mechanism of black phosphorus for sodium ion batteries by first-principles calculations. *The Journal of Physical Chemistry C* 119:15041–15046. https://doi.org/10.1021/acs.jpcc.5b05482.
103. Liu H, Tao L, Zhang Y et al. (2017) Bridging covalently functionalized black phosphorus on graphene for high-performance sodium-ion battery. *ACS Applied Materials and Interfaces* 9:36849–36856. https://doi.org/10.1021/acsami.7b11599.
104. Xu G-L, Chen Z, Zhong G-M et al. (2016) Nanostructured black phosphorus/ketjenblack-multiwalled carbon nanotubes composite as high performance anode material for sodium-ion batteries. *Nano Letters* 16:3955–3965. https://doi.org/10.1021/acs.nanolett.6b01777.
105. Capone I, Hurlbutt K, Naylor AJ et al. (2019) Effect of the particle-size distribution on the electrochemical performance of a red phosphorus-carbon composite anode for sodium-ion batteries. *Energy & Fuels* 33:4651–4658. https://doi.org/10.1021/acs.energyfuels.9b00385.
106. Zhou J, Liu X, Cai W et al. (2017) Wet-chemical synthesis of hollow red-phosphorus nanospheres with porous shells as anodes for high-performance lithium-ion and sodium-ion batteries. *Advanced Materials* 29. https://doi.org/10.1002/adma.201700214.
107. Liu S, Feng J, Bian X et al. (2017) A controlled red phosphorus@Ni-P core@shell nanostructure as an ultralong cycle-life and superior high-rate anode for sodium-ion batteries. *Energy and Environmental Science* 10:1222–1233. https://doi.org/10.1039/C7EE00102A.
108. Liu C, Wang Y, Sun J et al. (2020) A review on applications of layered phosphorus in energy storage. *Transactions of Tianjin University* 26:104–126. https://doi.org/10.1007/s12209-019-00230-x.
109. Sui Y, Zhou J, Wang X et al. (2020) Recent advances in black-phosphorus-based materials for electrochemical energy storage. *Materials Today*. https://doi.org/10.1016/j.mattod.2020.09.005.
110. Klein F, Jache B, Bhide A et al. (2013) Conversion reactions for sodium-ion batteries. *Physical Chemistry Chemical Physics* 15:15876. https://doi.org/10.1039/c3cp52125g.
111. Qi S, Xu B, Tiong VT et al. (2020) Progress on iron oxides and chalcogenides as anodes for sodium-ion batteries. *Chemical Engineering Journal* 379:122261. https://doi.org/10.1016/j.cej.2019.122261.
112. Li D, Yan D, Ma J et al. (2016) One-step microwave-assisted synthesis of Sb_2O_3/reduced graphene oxide composites as advanced anode materials for sodium-ion batteries. *Ceramics International* 42:15634–15642. https://doi.org/10.1016/j.ceramint.2016.07.017.
113. Hariharan S, Saravanan K, Ramar V et al. (2013) A rationally designed dual role anode material for lithium-ion and sodium-ion batteries: Case study of eco-friendly Fe_3O_4. *Physical Chemistry Chemical Physics* 15:2945. https://doi.org/10.1039/c2cp44572g.
114. Duran EC, Kizil H (2017) Porous structured SnO_2 nanofibers and nanotubes as anode materials for sodium-ion batteries. *ECS Transactions* 77:353–363. https://doi.org/10.1149/07711.0353ecst.
115. Hu Z, Liu Q, Chou S-L et al. (2017) Advances and challenges in metal sulfides/selenides for next-generation rechargeable sodium-ion batteries. *Advanced Materials* 29. https://doi.org/10.1002/adma.201700606.
116. Xie X, Ao Z, Su D et al. (2015) MoS_2/graphene composite anodes with enhanced performance for sodium-ion batteries: The role of the two-dimensional heterointerface. *Advanced Functional Materials* 25:1393–1403. https://doi.org/10.1002/adfm.201404078.
117. Li Z, Ding J, Mitlin D (2015) Tin and tin compounds for sodium ion battery anodes: Phase transformations and performance. *Accounts of Chemical Research* 48:1657–1665. https://doi.org/10.1021/acs.accounts.5b00114.
118. Liu Y, Kang H, Jiao L et al. (2015) Exfoliated-SnS_2 restacked on graphene as a high-capacity, high-rate, and long-cycle life anode for sodium ion batteries. *Nanoscale* 7:1325–1332. https://doi.org/10.1039/C4NR05106H.

119. Xia Q, Li W, Miao Z et al. (2017) Phosphorus and phosphide nanomaterials for sodium-ion batteries. *Nano Research* 10:4055–4081. https://doi.org/10.1007/s12274-017-1671-7.
120. Kim Y, Kim Y, Choi A et al. (2014) Tin phosphide as a promising anode material for Na-ion batteries. *Advanced Materials* 26:4139–4144. https://doi.org/10.1002/adma.201305638.
121. Chang Y-M, Lin H-W, Li L-J et al. (2020) Two-dimensional materials as anodes for sodium-ion batteries. *Materials Today Advances* 6:100054. https://doi.org/10.1016/j.mtadv.2020.100054.
122. Wu Y, Yu Y (2019) 2D material as anode for sodium ion batteries: Recent progress and perspectives. *Energy Storage Materials* 16:323–343. https://doi.org/10.1016/j.ensm.2018.05.026.
123. Shi H, Yuan A, Xu J (2017) Tailored synthesis of monodispersed nano/submicron porous silicon oxycarbide (SiOC) spheres with improved Li-storage performance as an anode material for Li-ion batteries. *Journal of Power Sources* 364:288–298. https://doi.org/10.1016/j.jpowsour.2017.08.051.
124. Mao J, Zhou T, Zheng Y et al. (2018) Two-dimensional nanostructures for sodium-ion battery anodes. *Journal of Materials Chemistry A* 6:3284–3303. https://doi.org/10.1039/C7TA10500B.
125. Wu J, Ciucci F, Kim J-K (2020) Molybdenum disulfide based nanomaterials for rechargeable batteries. *Chemistry – A European Journal* 26:6296–6319. https://doi.org/10.1002/chem.201905524.
126. Zhao X, Ding Y, Xu Q et al. (2019) Low-temperature growth of hard carbon with graphite crystal for sodium-ion storage with high initial coulombic efficiency: A general method. *Advanced Energy Materials* 9:1–10. https://doi.org/10.1002/aenm.201803648.
127. Zhang F, Yao Y, Wan J et al. (2017) High temperature carbonized grass as a high performance sodium ion battery anode. *ACS Applied Materials and Interfaces* 9:391–397. https://doi.org/10.1021/acsami.6b12542.
128. Zheng P, Liu T, Guo S (2016) Micro-nano structure hard carbon as a high performance anode material for sodium-ion batteries. *Scientific Reports* 6:35620. https://doi.org/10.1038/srep35620.
129. Liu Y, Zhang N, Jiao L et al. (2015) Ultrasmall Sn nanoparticles embedded in carbon as high-performance anode for sodium-ion batteries. *Advanced Functional Materials* 25:214–220. https://doi.org/10.1002/adfm.201402943.
130. Sha M, Zhang H, Nie Y et al. (2017) Sn nanoparticles@nitrogen-doped carbon nanofiber composites as high-performance anodes for sodium-ion batteries. *Journal of Materials Chemistry A* 5:6277–6283. https://doi.org/10.1039/c7ta00690j.
131. Yao S, Cui J, Huang J et al. (2018) Rational assembly of hollow microporous carbon spheres as P hosts for long-life sodium-ion batteries. *Advanced Energy Materials* 8:1–13. https://doi.org/10.1002/aenm.201702267.
132. Qian J, Wu X, Cao Y et al. (2013) High capacity and rate capability of amorphous phosphorus for sodium ion batteries. *Angewandte Chemie – International Edition* 52:4633–4636. https://doi.org/10.1002/anie.201209689.
133. Zhu Y, Han X, Xu Y et al. (2013) Electrospun Sb/C fibers for a stable and fast sodium-ion battery anode. *ACS Nano* 7:6378–6386. https://doi.org/10.1021/nn4025674.
134. Wu L, Lu H, Xiao L et al. (2015) Electrochemical properties and morphological evolution of pitaya-like Sb@C microspheres as high-performance anode for sodium ion batteries. *Journal of Materials Chemistry A* 3:5708–5713. https://doi.org/10.1039/c4ta06086e.
135. Li K, Zhang J, Lin D et al. (2019) Evolution of the electrochemical interface in sodium ion batteries with ether electrolytes. *Nature Communications* 10:725. https://doi.org/10.1038/s41467-019-08506-5.
136. Zhou M, Xu Y, Wang C et al. (2017) Amorphous TiO_2 inverse opal anode for high-rate sodium ion batteries. *Nano Energy* 31:514–524. https://doi.org/10.1016/j.nanoen.2016.12.005.
137. Cao K, Jiao L, Pang WK et al. (2016) $Na_2Ti_6O_{13}$ nanorods with dominant large interlayer spacing exposed facet for high-performance Na-ion batteries. *Small* 12:2991–2997. https://doi.org/10.1002/smll.201600845.
138. Xie X, Ao Z, Su D et al. (2015) MoS_2/graphene composite anodes with enhanced performance for sodium-ion batteries: The role of the two-dimensional heterointerface. *Advanced Functional Materials* 25:1393–1403. https://doi.org/10.1002/adfm.201404078.

11 Electrolytes for Sodium-Ion Batteries

Akhila Das, Gurdial Blugan, and Pradeep Vallachira Warriam Sasikumar

11.1 INTRODUCTION

Non-renewable energy sources such as fossil fuels are at high risk, and there is a strong and ever-growing demand for renewable energy storage devices. Hence, a switching from non-renewable energy storage devices to renewable energy storage devices is inevitable, and hence, rechargeable batteries, supercapacitors, fuel cells, etc. have been developed. Rechargeable batteries are the most successful technology which repeatedly produce electricity from the stored materials. Among rechargeable batteries, lithium-ion batteries (LIBs) are very prominent role, and industries begin demonstration test for electrical energy storage devices for skyrocketing sales [1,2]. However, LIBs are hurdled by its cost, regional scarcity, safety issues, dendrite formation, etc., and the large-scale development is a matter of discussion [3].

Researchers developed an innovative and efficient post-lithium batteries, mainly sodium-ion batteries (SIBs), which are found to be an alternative to the battery technology [4,5]. SIBs are abundant in the earth's crust and can be worked at ambient temperature which paves a new way for large grid-scale application [4,6,7]. The intercalation chemistry of Na ion is similar to that of Li ion and considered as almost similar [8]. In the periodic table, sodium is just below lithium metal within same group, and hence, the chemical properties are similar. The mechanism and fundamental principles are similar to those of LIB [9]. Polyanion cathode materials and sodium metal oxide materials can be synthesized by naturally abundant elements such as Fe and Mn [10]. The cost of extraction of SIBs is very less compared to LIBs, and hence, research focuses on the development of future battery technology [11]. Sustainability and free carbon-based materials are striving and are applicable to short-range electric vehicle application. The development of SIBs is still at infancy. The major barriers for the development of high performance batteries include its big cationic size of Na (1.02 Å) compared to lithium metal (0.76 Å). Table 11.1 shows the physical properties of sodium and lithium metals [12]. The ionic conductivity, hardness, first ionization energy, etc. decrease down the group and anodic electrode potential of LIB is lower than SIB. Among electrodes and electrolytes,

TABLE 11.1
Comparison of Different Properties of Na and Li Metals

Property	Li	Na
Atomic radius	**6.94**	**22.9**
Electronic configuration	**[He] $2s^1$**	**[Ne] $3s^1$**
Cationic radius [Å]	**0.76**	**1.02**
Standard electrode potential [V]	**−3.04**	**−2.71**
Melting point [°C]	**180.5**	**97.7**
Density [g cm^{-3}]	**0.971**	**0.534**
Theoretical gravimetric capacity [mAh g^{-1}]	**3861**	**1165**
Theoretical volumetric capacity [mAh cm^{-3}]	**2062**	**1131**

DOI: 10.1201/9781003310167-11

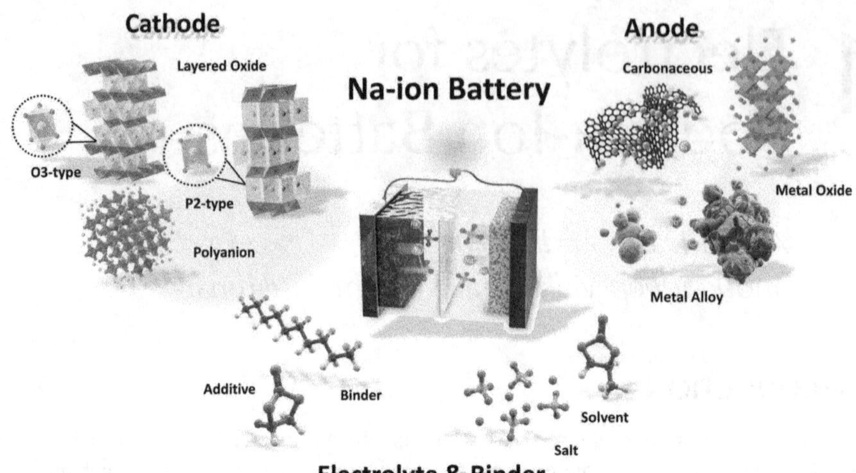

FIGURE 11.1 Schematic illustration of sodium-ion battery and its different anodes, cathode, electrolyte and binder materials. Adapted and reproduced with permission from Ref. [19]. Copyright © 2017 Chemical Society Reviews.

electrolytes play a pivotal role in the development of Na-ion battery [6,13]. Generally, electrolytes are classified as protic solvent consisting of labile H^+ ion and aprotic solvents without labile H^+ ion. Interface between electrode and electrolyte determines the electrochemical performance of the SIB, and hence, the type of electrolyte chosen is very crucial. There are different types of electrolytes employed in the fabrication of SIB such as carbonate-based solvents [14,15], ionic liquid-based electrolytes [16] and polymer electrolytes [17,18]; and schematic illustration of different cathodes, anodes and electrolytes employed in SIB is given in Figure 11.1 [19]. Table 11.1 shows the physical properties of sodium and lithium metals. The general requirement of electrolytes is given in the next session.

11.2 GENERAL REQUIREMENT OF SOLVENT AND ELECTROLYTES

The major criteria for suitable electrolytes dissolvent in different solvents are: (i) environmentally benign material, (ii) economically reliable, (iii) wide operational temperature, (iv) better electrochemical stability, (v) high dielectric permittivity and (vi) low viscosity favoring high ionic conductivity. The voltage range between anodic and cathodic potential determines the chemical stability of the material. The major requirement for high thermal stability is attributed to the higher boiling point and lower melting point. Organic electrolytes have high vapor pressure and are not appropriate since these are flammable at high temperature, and polymer electrolytes and ionic liquid-based electrolytes find a solution for it. Stokes Einstein law reveals the indirect proportionality of ionic conductivity with viscosity, and hence, very less viscous electrolytes favor higher ionic conductivity since the mobility of Na^+ ions could be faster. The major electrolytes explored in Na-ion batteries are discussed in the upcoming section.

11.3 OPTIMIZATION OF SUITABLE ELECTROLYTES

Different electrolytes have different physical, chemical and electrochemical properties. Hence, optimization of electrolytes is a greater challenge since antagonistic behavior such as electrochemical performance, liquid range, chemical and thermal stability should be instantaneously restructured. Pristine carbonate solvents and mixture of different solvents were optimized and reveal that the chosen binary mixtures should have comparable molecular structures and melting point. Ether-based electrolytes provide a new paradigm shift to form graphite sheets as anode materials, but they have less voltage stability range. This dispute can be eradicated by the choice of ionic liquid performing in wide electrochemical voltage stability. Similar to polymer-based lithium electrolytes, polymer electrolytes were also incorporated in SIBs.

11.4 ELECTROLYTES IN SODIUM-ION BATTERIES

Electrolytes play a key role in governing the electrochemical properties of the batteries such as cycling stability, operating temperature range and safety of the battery [20]. Among all these properties, ionic conductivity of the battery is crucial, and choosing the suitable electrolytes is essential. Chemical and physical stability of electrolytes leads to the working of batteries in wide operating voltage range. Transport properties rely on compatibility of electrolytes, and electronically insulating with ionic conducting material provides a better battery for the future application. The major classification of batteries based on the physical state are liquid and solid electrolytes which have been widely used in lithium, sodium and magnesium batteries [21]. The major electrolytes in SIB are $NaClO_4$, $NaPF_6$, etc. dissolved in EC, DMC and so on which have high ion conductivity [22]. However, these inorganic salts are not safe and require porous separators for full cell fabrication. Electrolytes such as ionic liquid-based compounds, NASICON type and polymer electrolytes were then widely monitored for the advancement of rechargeable SIBs [23].

11.4.1 Carbonate Salt-Based Electrolytes in SIBs

Major carbonate solvents employed in SIBs are EC, PC, DMC, etc., which provide solid electrolyte interface (SEI) layer assisting the movement of easy passage of Na ions as well as inhibiting the decomposition of electrolytes. Lithium-based intercalation mechanism was investigated earlier with different carbonate solvents [24,25], and hence, Thomas et al. [26] investigated the influence of ethylene carbonate on Na-ion batteries having $NaClO_4$ as doping salt. Intercalation of Na on EC/$NaClO_4$ with UF4 natural graphite and pitch-based fibers reveals that easy interaction is possible with pitch-based fibers than graphite. Later, intercalation studies of $NaVPO_4F$ were examined by $NaClO_4$ in EC/DMC with hard carbon as anode exhibiting a specific capacity of 80 mAh g^{-1}. Insertion of Na ions in boron-functionalized reduced graphene oxide (BF-rGO) was determined by $NaClO_4$ doping salt in carbonate salts such as EC, diethyl carbonate (DEC) and 1,2-dimethoxyethane (DME) [27]. Electrochemical performance was optimized by choosing different combination of carbonate esters and EC: DEC shows a discharge capacity of ~ 180 mAh g^{-1}, whereas DME delivers a specific capacity of ~175 mAh g^{-1}. Hence, fluorinated ethylene carbonate was incorporated and optimized as EC:DEC:FEC showing a specific capacity ~200 mAhg^{-1} at 100 A g^{-1}. Carbonated esters play a vital role in the electrochemical properties of a battery. Ponrouch et al. [28,29] investigated different electrolytes such as NiClO4, $NiPF_6$ and NiTFSI, and different carbonate solvent mixtures such as EC:DMC, PC:DMC, EC:PC and EC:Triglyme. EC:DMC offers higher conductivity than EC:PC:DMC, and $NiPF_6$ has higher ionic conductivity than $NiClO_4$. Ionic conductivity depends upon the nature of solvents and its viscosity since lower viscous solvents offer ionic mobility and hence transport of Na ions (Figure 11.2). Recently, different combinations of carbonate solvents were optimized by $NiPF_6$ and compared with $LiPF_6$ doping salt and its intercalation behavior through DFT calculations was studied [30].

11.4.2 Ionic Liquid-Based Electrolytes in Sodium-Ion Batteries

Ionic liquid-based electrolytes are fabricated for clean battery technology and green chemistry owing to the advantages such as tunable solvating power, low vapor pressure, non-volatile, high thermal stability and wide electrochemical stability. Ionic liquid-based electrolytes were reported in LIBs [31,32], and later, its synthesis procedure switched to SIBs too [33,34]. Studies on ionic liquid

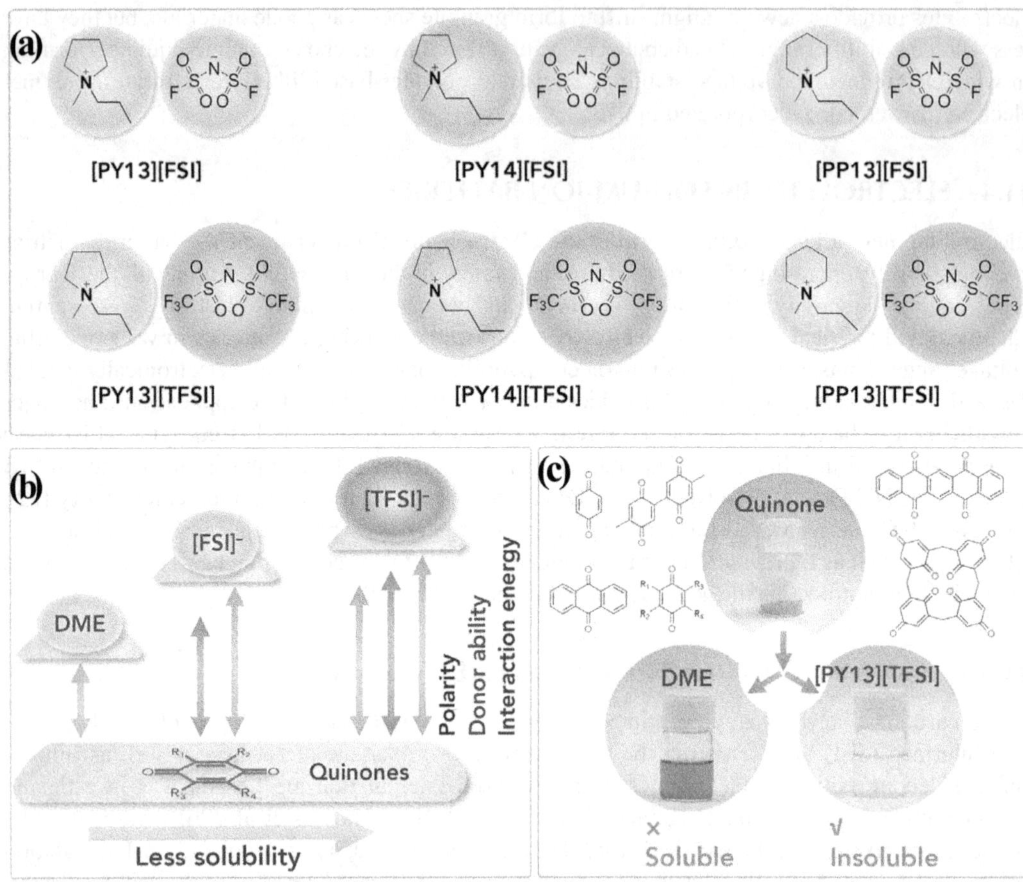

FIGURE 11.2 Electrochemical potential window stability and thermal range (lower y axis) values of (a) PC-based electrolytes with 1 M of various Na salts. Discharge capacity vs cycle number for tape-cast hard carbon electrodes cycled in 1 M $NaClO_4$ in PC alone and EC:PC at C/10 rate up to 110 cycles and further at C/30 Conductivity (left-hand side y axis) and viscosity (right-hand side y axis) values of (c) PC-based electrolytes with 1 M of various Na salts and (b) electrolytes based on 1 M $NaClO_4$ dissolved in various solvents and solvent mixtures. Adapted and reproduced with permission from Ref. [28]. Copyright © 2012 The Royal Society of Chemistry.

incorporated electrolytes in SIB are at infant stage. Monti et al. [34] investigated the electrochemical performance of two ionic liquids such as 1-ethyl-3-methylimidazolium bis(trifluoromethylsulfonyl) imide (EMImTFSI) and 1-butyl-3-methylimidazolium TFSI (BMImTFSI) along with sodium salt (NATFSI). Na salt of EMImTFSI shows better electrochemical performance compared to sodium salt of BMImTFSI. In the same year, Chagas et al. [35] investigated the performance of cathode material in ionic liquid-based polymer electrolytes. The battery performance was determined by 10 mol% of NaTFSI in PYR14FSI-based polymer electrolyte with or without ethylene carbonate; the cathode material used was $Na_{0.45}Ni_{0.22}Co_{0.11}Mn_{0.66}O_2$. Columbic efficiency and specific capacity of NaTFSI/PYR14FSI were 230 mAh g^{-1} and 90% Columbic efficiency, whereas Columbic efficiency and specific capacity of $NaPF_6$/PC were 210 mAh g^{-1} and 88%, respectively. The same ionic liquid [33] was performed on $Na_{0.6}Ni_{0.22}Fe_{0.11}Mn_{0.66}O_2$ cathode electrodes delivering ionic conductivity in the range of 10^{-3} S cm^{-1} and having thermal stability up to 400°C. The specific capacity was less (120 mAh g^{-1}) compared to the above-reported IL-based polymer system [35]. Na[TFSI]/[PY13]-based ionic liquid was reported by Wang et al. [36]. 0.3 M Na[TFSI]/

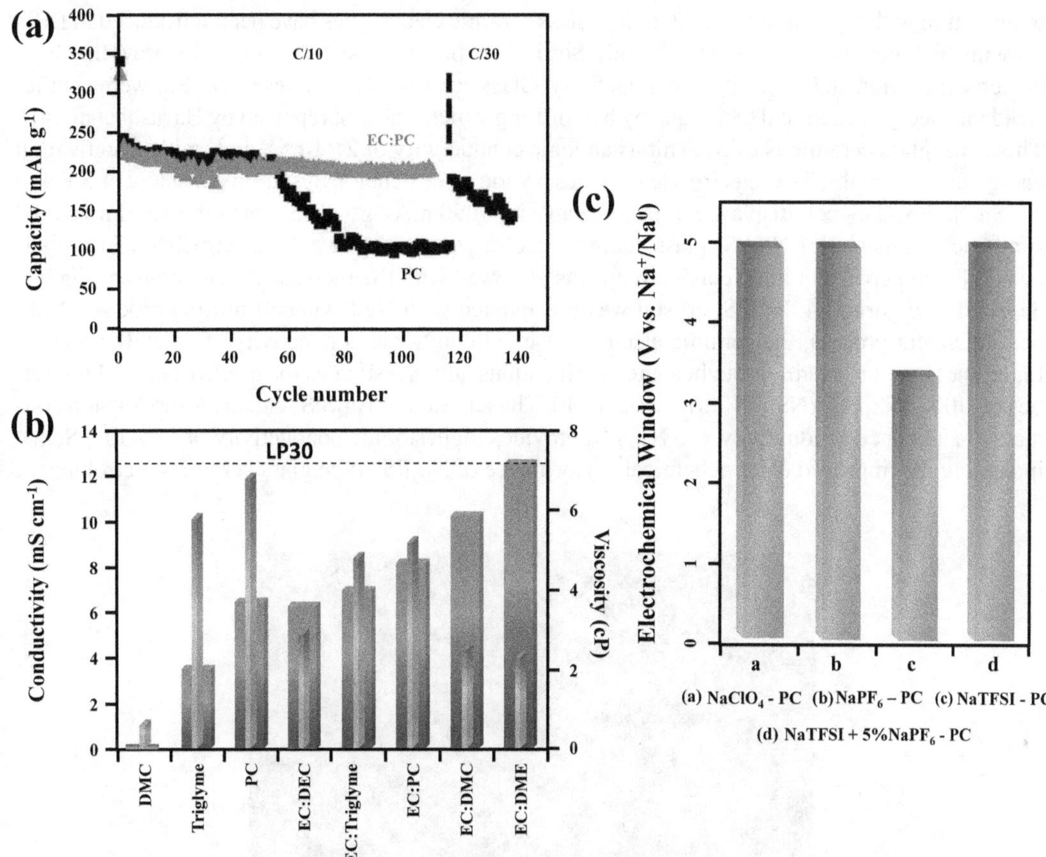

FIGURE 11.3 (a) Chemical structures of six ionic liquids in this study. (b) Schematic comparison of polarity, donor ability and quinone–solvent interactive energy for DME and ionic liquids. (c) Photographs of soluble quinone in DME and insoluble quinone in IL. Adapted and reproduced with permission from Ref. [36]. Copyright © 2018 Elsevier.

[PY13]-based electrolyte with quinone cathode exhibits high specific capacity (> 400 mAh g^{-1}) owing to the advantage of large polarity and weak donor ability. Capacity retention of 99.7% was attained at a specific capacity of 130 mAh g^{-1} with cycling stability up to 300 cycles. Compared to DME and $[FSI]^-$, quinone is molar polar to $[TFSI]^-$ which was evident from Figure 11.3. In the same year, Hagiwara [37] employed binary system with tertiary ammonium salt of 1-ethyl-3-methylimidazolium and N-methyl-N-propylpyrrolidinium bis(fluorosulfonyl)imide exhibiting cycling stability of 500 cycles with a capacity retention of 87%. Recently, Chagas et al. [16] investigated different combination of ionic liquids such as N-butyl-N-methylpyrrolidinium bis(fluorosulfonyl)imide (Pyr14FSI) and N-butyl-N-methylpyrrolidinium bis (trifluoromethanesulfonyl) imide (Pyr13FSI) N-methyl-N-propylpyrrolidinium bis-(fluorosulfonyl)imide (Pyr13FSI) with doping salt such as NaTFSI or NaFSI. Pyr14FSI has better electrochemical performance among all the blends delivering a specific capacity of 140 mAh g^{-1} with cycling stability up to 200 cycles.

11.4.3 Glass Ceramic Electrolytes in Sodium-Ion Batteries

Various electrolytes were scrutinized for the high performance application predominantly Na-ion conduction in rechargeable SIB including polymer electrolytes, inorganic electrolytes and organic electrolytes. Glass ceramic electrolytes are a type of solid electrolytes working under ambient temperature with high Na-ion conductivity. Glass ceramic electrolytes have been introduced in LIBs showing high battery performance [38–40]. Similar studies were performed in SIBs providing high Na ion conduction and hence ionic conductivity. Glass ceramic electrolytes of Na_3PS_4 were synthesized via mechanochemical technique by ball milling which was first reported by Havashi et al. [41]. This cubic glass ceramic Na_3PS_4 exhibits an ionic conductivity of 2×10^{-4} S cm^{-1} with an activation energy of 27 kJ mol^{-1}. This electrolyte provides a wide electrochemical stability window of 5 V and Na-Sn/Na_3PS_4/TiS_2 cell displayed a specific capacity of 90 mAh g^{-1}. The same group optimized the synthetic parameters of Na_3PS_4 glass ceramic electrolytes to improve the electrochemical properties [42]. An increase in ionic conductivity was observed with the increase in temperature which is depicted in Figure 11.4. Na_3PS_4 crystal was precipitated with Na_2S via ball milling process. After heat treatment process, the ceramic electrolytes exhibit an ionic conductivity of 4.6×10^{-4} S cm^{-1}. Later, the same group [43] brought some modifications and investigates the electrochemical properties of (100-x)Na_3PS_4.xNa_4SiS_4, where $0\leq x\leq 10$. The addition of Na_4SiS_4 enhances the Na^+-ion conduction and hence conductivity. 6% Na_4SiS_4 provides highest ionic conductivity of 7.4×10^{-4} Scm^{-1} indicating the improved electrochemical performance of Na_3PS_4-type glass ceramic electrolytes.

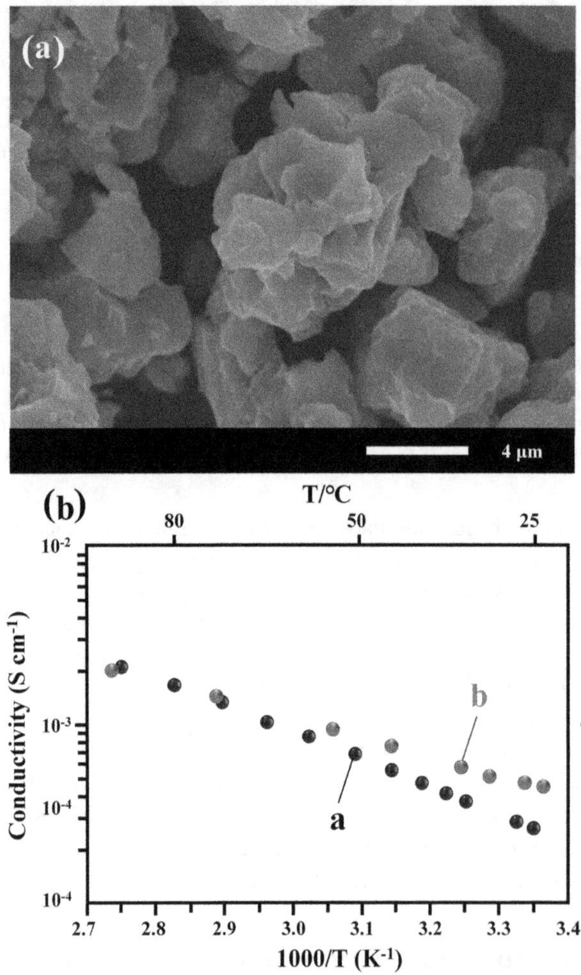

FIGURE 11.4 (a) SEM image for particles of the Na_3PS_4 glass ceramic. (b) Temperature dependence of conductivity for the Na_3PS_4 glass ceramics. Adapted and reproduced with permission from Ref. [42]. Copyright © 2014 Elsevier.

11.4.4 Polymer Electrolytes in Sodium-Ion Batteries

Electrolytes play a vital role in the performance of the battery and must be safer. The safety issues with liquid electrolytes were addressed in Na-ion batteries too. Hence, polymer electrolytes resolved the durability and safety problems. Different types of polymer electrolytes (PEO, PAN and PVdF) are explored in Na-ion batteries with/without the addition of plasticizers, ionic liquids, ceramic fillers, etc. In addition to its safety, polymer electrolytes offer good ionic conductivity, high mechanical strength, high thermal and chemical stability, compatibility with electrode materials and wide electrochemical stability window. Studies on polymer electrolytes were started even in the 1980s by Munshi et al. [44] and Cherng et al. [45]. Later on, modification on polymer electrolytes with blending different polymers, addition of fillers, etc. were made to improve the electrochemical performance of SIBs. Based on the physical state and composition, polymer electrolytes are broadly classified as solid, gel and composite polymer electrolytes. Detailed investigation of different polymer electrolytes is discussed in the next section.

11.4.4.1 Solid Polymer Electrolytes for Rechargeable Sodium-Ion Batteries

Polyethylene oxide (PEO)-based solid polymer electrolytes were investigated at the infant stage of SIBs owing to the electrochemical stability compared to other polyether and its copolymers. Munshi et al. [44] investigated PEO-based solid polymer electrolyte in SIBs with $NaCF_3SO_3$:PEO as electrolyte. The electrochemical behavior of Na/PEO_8.$NaCF_3SO_3$/V_6O_{13} cell was determined and an open circuit voltage of 3.2 V was obtained. Trifluoro sulfonyl imide-based electrolytes show better electrochemical performance which is evident from LIBs. Qi et al. [46] reported PEO/NaFSI (sodium bis (fluorosulfonyl)imide)-based polymer electrolyte owing to the advantages of flame retardant and flexible solid polymer electrolyte. Thermogravimetric analysis shows that PEO/NaFSI membrane is thermally stable up to 263°C and an ionic conductivity reaches up to 4.1×10^{-4} S cm^{-1}. In the same year, Konduru and his co-workers [17] investigated PEO-based solid polymer electrolytes in advanced SIBs. In order to improve battery performance, PEO was blended with PVP (Polyvinylpyrrolidone) with $NaIO_4$ as a doping salt. Blending of PEO with PVP improves the amorphous nature of the polymer since a strong intermolecular interaction arises among the polymer chains. Along with this, the addition of graphene oxide (GO – 0.4 wt. %) to the above polymer matrix increases the ionic conductivity, which is attributed to the increase in the mobility of the mobile charge carriers in polymer electrolytes. PEO/PVP blend polymer electrolyte has an ionic conductivity of 2.24×10^{-9} S cm^{-1}, whereas GO/PEO/PVP polymer matrix exhibits an ionic conductivity of 1.57×10^{-7} S cm^{-1}. Addition of inorganic fillers to PEO-based solid polymer electrolytes boosts the ionic conductivity of the matrix by reducing the crystallinity of the PEO. Incorporation of TiO_2 which was synthesized by hydrothermal method increases the segmental motion of the polymer chain followed by the increment in the ionic mobility and hence conductivity [47]. PEO-based SPE exhibits an ionic conductivity of 1.35×10^{-4} S cm^{-1}, whereas TiO_2-incorporated PEO exhibits an increased ionic conductivity of 2.62×10^{-4} S cm^{-1}. The flexibility of PEO/$NaClO_4$ is evident from its physical appearance itself (Figure 11.5).

11.4.4.2 Gel Polymer Electrolytes and Composite Polymer Electrolytes for SIB

Gel polymer electrolytes (GPEs) are attractive polymer materials in SIBs owing to its high ionic conductivity and higher amorphous nature compared to solid polymer electrolytes. Reports on GPEs in LIBs [48] show an enhancement in ionic mobility of Li salts and consequently ionic conduction pathways; hence, these types of GPEs were also fabricated in Na-ion batteries. Yang et al. [49] introduced PVdF-based polymer electrolytes due to its strong electron-withdrawing fluorine ion and high dielectric constant value (~ 8.4) with $NaClO_4$ as the doping salt. The GPE with Na/GPE/Na cell shows an ionic conductivity of 0.60 mS cm^{-1} and a transference number of 0.30. The same polymer was fabricated with sodium trifluoromethanesulfonate ($NaCF_3SO_3$) as a doping salt along with EC and PC plasticizing agent [50]. Ionic conductivity was highest for 20 wt. % $NaCF_3SO_3$ that

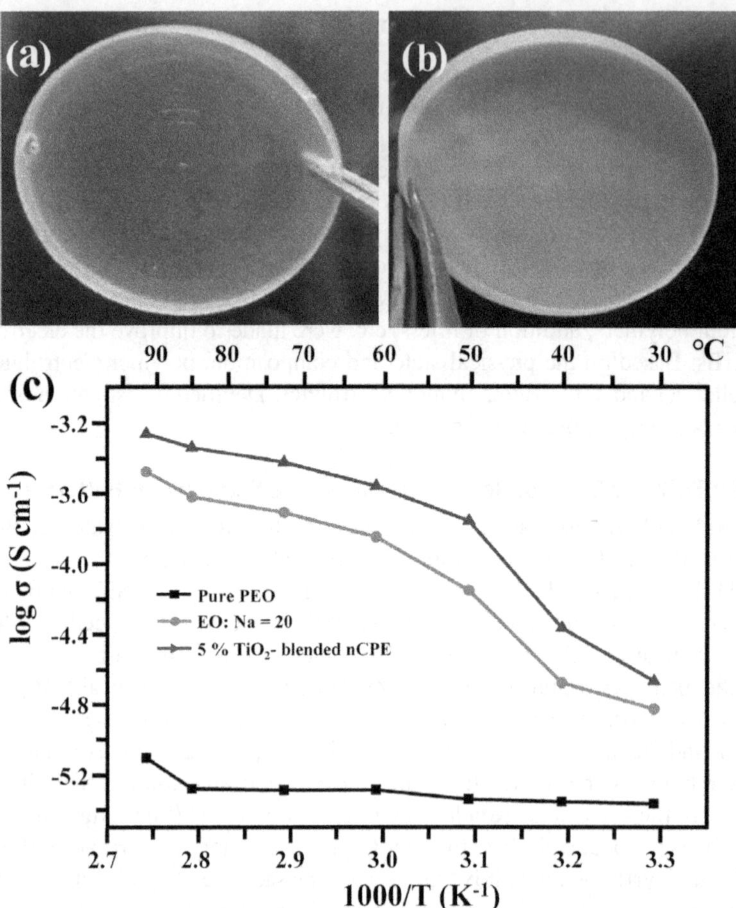

FIGURE 11.5 Photograph of a typical PEO/$NaClO_4$ solid polymer electrolyte film: (a) EO:Na=20 and (b) 5% TiO_2-blended nCPE (EO:Na=20). (c) Temperature-dependent conductivity plots for PEO, PEO-$NaClO_4$ and TiO_2-blended nCPE. Adapted and reproduced with permission from Ref. [47]. Copyright © 2014 Elsevier.

delivers 2.5 mS cm^{-1} which is attributed to the increase in the number of free ions present followed by diffusion of ions and electrochemical stability was attained up to 3.4 V. Kim et al. [51] synthesized PVdF-*co*-HFP-based stretchable solid gel polymer electrolyte (SGPE) in which glass fiber (GF) positioned by phase separation method with water as non-solvent. This SGPE/GF exhibits a specific capacity of 295 Ah g^{-1} at 0.2 C rate and an ionic conductivity of 3.3 mS cm^{-1}. A detailed investigation on the organization of uniform pores in GF was reported by the same group [52]. Surface morphological studies of SGPE/GF with deionized water reveal the porous structure pattern as well as Figure 11.6 shows that porosity was directly proportional to the amount of deionized water. By controlling the solubility parameters and non-solvent properties, morphology was tuned to improve the electrochemical properties of the polymer membrane. Other than PVdF-co-HFP, PAN-based GPEs were also reported with a combination of 11PAN-12$NaClO_4$_40EC-37PC (wt.%) delivering an ionic conductivity of 4.5 mScm^{-1} [53].

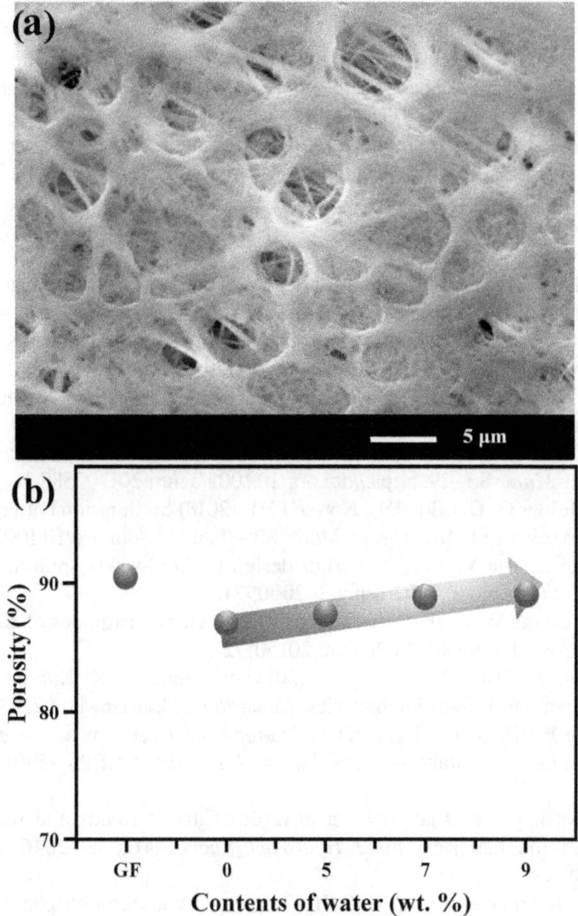

FIGURE 11.6 Scanning electron microscope (SEM) image on the surface morphology of (a) GF/SPGE (9 wt% DI water). (b) Porosity of GF and GF/SPGE with different DI water contents. Adapted and reproduced with permission from Ref. [51] Copyright © 2017 John Wiley and Sons.

11.5 CONCLUSION

SIBs are promising battery technology which could be a new horizon to the modern technological era. Owing to wide geographical contribution of Na metal, SIBs are considered as cost-effective material and can be used as a substitute for LIBs. Research on different types of electrode materials were explored and were less on the electrolytes field. Similar to LIBs, certain types of electrolytes such as ionic liquid-based electrolytes, solid polymer electrolytes and polymer electrolytes were investigated. Different organic electrolytes are examined in SIB with electrochemical voltage stability >3.5 V, and successful operation is achieved by solid electrolyte interface, stability and electrochemical performance of the electrolyte. Safety and energy density are contradictory to each other, and chemically stable matrix such as ionic liquid-based electrolyte and polymer electrolytes is important for the fabrication of SIBs. Despite considerable achievement, SIB research is at still at infancy. Hence, a great deal of novel electrolytes and electrode materials should be fabricated to enhance the electrochemical performance of SIB.

REFERENCES

1. Chau KT, Wu KC, Chan CC (2004) A new battery capacity indicator for lithium-ion battery powered electric vehicles using adaptive neuro-fuzzy inference system. *Energy Convers Manag* 45:1681–1692. https://doi.org/10.1016/j.enconman.2003.09.031.
2. Tran TH, Harmand S, Desmet B, Filangi S (2014) Experimental investigation on the feasibility of heat pipe cooling for HEV/EV lithium-ion battery. *Appl Therm Eng* 63:551–558. https://doi.org/10.1016/j.applthermaleng.2013.11.048.
3. Amon F, Andersson P, Karlson I, Sahlin E (2012) *Fire Risks Associated with Batteries, SP Technic.* Science partner, Sweden.
4. Sawicki M, Shaw LL (2015) Advances and challenges of sodium ion batteries as post lithium-ion batteries. *RSC Adv* 5:53129–53154. https://doi.org/10.1039/c5ra08321d.
5. Wang F, Wu X, Li C, et al. (2016) Nanostructured positive electrode materials for post-lithium-ion batteries. *Energy Environ Sci* 9:3570–3611. https://doi.org/10.1039/c6ee02070d.
6. Wang Y, Song S, Xu C, et al (2021) Development of solid-state electrolytes for sodium-ion battery-A short review. *Elements* 1:91–100. https://doi.org/10.1016/j.nanoms.2019.02.007.
7. Shadike Z, Zhao E, Zhou YN, et al. (2018) Advanced characterization techniques for sodium-ion battery studies. *Adv Energy Mater* 8:1–29. https://doi.org/10.1002/aenm.201702588.
8. Chayambuka K, Mulder G, Danilov DL, Notten PHL (2018) Sodium-ion battery materials and electrochemical properties reviewed. *Adv Energy Mater* 8:1–49. https://doi.org/10.1002/aenm.201800079.
9. Yang C, Xin S, Mai L, You Y (2021) Materials design for high-safety sodium-ion battery. *Adv Energy Mater* 11:1–17. https://doi.org/10.1002/aenm.202000974.
10. Fang C, Huang Y, Zhang W, et al. (2016) Routes to high energy cathodes of sodium-ion batteries. *Adv Energy Mater* 6. https://doi.org/10.1002/aenm.201501727.
11. Yabuuchi N, Kubota K, Dahbi M, Komaba S (2014) N. Yabuuchi, K. Kubota, M. Dahbi, S. Komaba, Research development on sodium-ion batteries, *Chem Rev* 114:11636–11682.
12. Palomares V, Serras P, Villaluenga I, et al (2012) Na-ion batteries, recent advances and present challenges to become low cost energy storage systems. *Energy Environ Sci* 5:5884–5901. https://doi.org/10.1039/c2ee02781j.
13. Vignarooban K, Kushagra R, Elango A, et al (2016) Current trends and future challenges of electrolytes for sodium-ion batteries. *Int J Hydrogen Energy* 41:2829–2846. https://doi.org/10.1016/j.ijhydene.2015.12.090.
14. Dahbi M, Nakano T, Yabuuchi N, et al (2016) Effect of hexafluorophosphate and fluoroethylene carbonate on electrochemical performance and the surface layer of hard carbon for sodium-ion batteries. *ChemElectroChem* 3:1856–1867. https://doi.org/10.1002/celc.201600365.
15. Shakourian-Fard M, Kamath G, Smith K, et al (2015) Trends in Na-ion solvation with alkyl-carbonate electrolytes for sodium-ion batteries: Insights from first-principles calculations. *J Phys Chem C* 119:22747–22759. https://doi.org/10.1021/acs.jpcc.5b04706.
16. Chagas LG, Jeong S, Hasa I, Passerini S (2019) Ionic liquid-based electrolytes for sodium-ion batteries: Tuning properties to enhance the electrochemical performance of manganese-based layered oxide cathode. *ACS Appl Mater Interfaces* 11:22278–22289. https://doi.org/10.1021/acsami.9b03813.
17. Koduru HK, Iliev MT, Kondamareddy KK, et al (2016) Investigations on poly (ethylene oxide) (PEO) - Blend based solid polymer electrolytes for sodium ion batteries. *J Phys Conf Ser* 764. https://doi.org/10.1088/1742-6596/764/1/012006.
18. Gebert F, Knott J, Gorkin R, et al (2021) Polymer electrolytes for sodium-ion batteries. *Energy Storage Mater* 36:10–30. https://doi.org/10.1016/j.ensm.2020.11.030.
19. Hwang JY, Myung ST, Sun YK (2017) Sodium-ion batteries: Present and future. *Chem Soc Rev* 46:3529–3614. https://doi.org/10.1039/c6cs00776g.
20. Xu K (2004) Nonaqueous liquid electrolytes for lithium-based rechargeable batteries. *Chem. Rev* 104, 10:4303–4418.
21. Eftekhari A, Kim DW (2018) Sodium-ion batteries: New opportunities beyond energy storage by lithium. *J Power Sources* 395:336–348. https://doi.org/10.1016/j.jpowsour.2018.05.089.
22. Environ E, Ponrouch A, Marchante E, et al (2012) In search of an optimized electrolyte for Na-ion batteries. *Environ Sci* 8572–8583. https://doi.org/10.1039/c2ee22258b.
23. Monti D, Demet AE, Ponrouch A, et al (2013) Towards high energy density sodium ion batteries through electrolyte optimization. *Environ Sci* 6:2361–2369. https://doi.org/10.1039/c3ee41379a.
24. Yang Z, Wu H (2001) Electrochemical intercalation of lithium into fullerene soot. *Mater Lett* 50:108–114. https://doi.org/10.1016/S0167-577X(00)00425-0.

25. Aurbach D, Ein-Eli Y, Chusid (Youngman) O, et al (1994) The correlation between the surface chemistry and the performance of Li-carbon intercalation anodes for rechargeable 'rocking-chair' type batteries. *J Electrochem Soc* 141:603–611. https://doi.org/10.1149/1.2054777.
26. Thomas P, Ghanbaja J, Billaud D (1999) Electrochemical insertion of sodium in pitch-based carbon fibres in comparison with graphite in $NaClO_4$-ethylene carbonate electrolyte. *Electrochim Acta* 45:423–430. https://doi.org/10.1016/S0013-4686(99)00276-5.
27. Wang Y, Wang C, Wang Y, et al (2016) Boric acid assisted reduction of graphene oxide: A promising material for sodium-ion batteries. *ACS Appl Mater Interfaces* 8:18860–18866. https://doi.org/10.1021/acsami.6b04774.
28. Ponrouch A, Marchante E, Courty M, et al (2012) In search of an optimized electrolyte for Na-ion batteries. *Energy Environ Sci* 5:8572–8583. https://doi.org/10.1039/c2ee22258b.
29. Ponroucha A, Dedryvèreb R, Montia P, Johanssonc P, Croguennece L, Masquelierd C, Palacin MR (2013) High energy density Na-ion batteries through electrolyte optimization. *J Chem Inf Model* 53:1689–1699.
30. Cresce AV, Russell SM, Borodin O, Allen JA, Schroeder MA, Dai M, Peng J, Gobet M, Greenbaum SG, Rogers RE, Xu K (2016) Solvation behavior of carbonate-based electrolytes in sodium ion batteries. *Phys Chem Phys* 19:574–586. https://doi.org/10.1039/C6CP07215A.
31. Mahant YP, Kondawar SB, Nandanwar DV, Koinkar P (2018) Poly(methyl methacrylate) reinforced poly(vinylidene fluoride) composites electrospun nanofibrous polymer electrolytes as potential separator for lithium-ion batteries. *Mater Renew Sustain Energy* 7:1–9. https://doi.org/10.1007/s40243-018-0115-y.
32. von CzarneckiTransactions P (2017) Ionic liquid based electrolytes for electrical storage. *Electrochem Soc* 77:79–87.
33. Hasa I, Passerini S, Hassoun J (2016) Characteristics of an ionic liquid electrolyte for sodium-ion batteries. *J Power Sources* 303:203–207. https://doi.org/10.1016/j.jpowsour.2015.10.100.
34. Monti D, Jónsson E, Palacín MR, Johansson P (2014) Ionic liquid based electrolytes for sodium-ion batteries : Na þ solvation and ionic conductivity. *J Power Sources* 245:630–636. https://doi.org/10.1016/j.jpowsour.2013.06.153.
35. Chagas LG, Buchholz D, Wu L, et al (2014) Unexpected performance of layered sodium-ion cathode material in ionic liquid-based electrolyte. *J Power Sources* 247:377–383. https://doi.org/10.1016/j.jpowsour.2013.08.118.
36. Wang X, Shang Z, Yang A, et al (2019) Combining quinone cathode and ionic liquid electrolyte for organic sodium-ion batteries. *Chem* 5:364–375. https://doi.org/10.1016/j.chempr.2018.10.018.
37. Hagiwara R, Matsumoto K, Hwang J, Nohira T (2019) Sodium ion batteries using ionic liquids as electrolytes. *Chem Rec* 19:758–770. https://doi.org/10.1002/tcr.201800119.
38. Dudney NJ (2003) *Glass and Ceramic Electrolytes for Li and Li-Ion Batteries*. Springer, pp. 624–642.
39. Kimura T, Kato A, Hotehama C, et al (2019) Preparation and characterization of lithium-ion conductive Li_3SbS_4 glass and glass-ceramic electrolytes. *Solid State Ion* 333:45–49. https://doi.org/10.1016/j.ssi.2019.01.017.
40. Huang B, Yao X, Huang Z, et al (2015) Li_3PO_4-doped $Li_7P_3S_{11}$ glass-ceramic electrolytes with enhanced lithium-ion conductivities and application in all-solid-state batteries. *J Power Sources* 284:206–211. https://doi.org/10.1016/j.jpowsour.2015.02.160.
41. Hayashi A, Noi K, Sakuda A, Tatsumisago M (2012) Superionic glass-ceramic electrolytes for room-temperature rechargeable sodium batteries. *Nat Commun* 3:2–6. https://doi.org/10.1038/ncomms1843.
42. Hayashi A, Noi K, Tanibata N, et al (2014) High sodium ion conductivity of glass-ceramic electrolytes with cubic Na_3PS_4. *J Power Sources* 258:420–423. https://doi.org/10.1016/j.jpowsour.2014.02.054.
43. Tanibata N, Noi K, Hayashi A, Tatsumisago M (2014) Preparation and characterization of highly sodium ion conducting Na_3PS_4-Na_4SiS_4 solid electrolytes. *RSC Adv* 4:17120–17123. https://doi.org/10.1039/c4ra00996g.
44. Munshi MZA, Gilmour A, Smyrl WH, Owens BB (1989) Sodium/V_6O_{13} polymer electrolyte cells. *J Electrochem Soc* 136:1847.
45. Cherng JY, Munshi MZA, Owens BB, Smyrl WH (1987) Applications of multivalent ionic conductors to polymeric elecrolyte batteries. *Solid State Ion* 28:857–561.
46. Qi X, Ma Q, Liu L, et al (2016) Sodium bis(fluorosulfonyl)imide/poly(ethylene oxide) polymer electrolytes for sodium-ion batteries. *ChemElectroChem* 3:1741–1745. https://doi.org/10.1002/celc.201600221.
47. Ni'Mah YL, Cheng MY, Cheng JH, et al (2015) Solid-state polymer nanocomposite electrolyte of TiO_2/PEO/NaC_1O_4 for sodium ion batteries. *J Power Sources* 278:375–381. https://doi.org/10.1016/j.jpowsour.2014.11.047.

48. Wu CG, Lu MI, Chuang HJ (2005) PVdF-HFP/P123 hybrid with mesopores: A new matrix for high-conducting, low-leakage porous polymer electrolyte. *Polymer (Guildf)* 46:5929–5938. https://doi.org/10.1016/j.polymer.2005.05.077.
49. Yang YQ, Chang Z, Li MX, et al (2015) A sodium ion conducting gel polymer electrolyte. *Solid State Ionics* 269:1–7. https://doi.org/10.1016/j.ssi.2014.11.015.
50. Isa KB, Othman L, Hambali D, Osman Z (2017) Electrical and electrochemical studies on sodium ion-based gel polymer electrolytes. *AIP Conf Proc* 1877:040001. https://doi.org/10.1063/1.4999867.
51. Kim JI, Choi Y, Chung KY, Park JH (2017) A structurable gel-polymer electrolyte for sodium ion batteries. *Adv Funct Mater* 27:1–7. https://doi.org/10.1002/adfm.201701768.
52. Kim JI, Chung KY, Park JH (2018) Design of a porous gel polymer electrolyte for sodium ion batteries. *J Memb Sci* 566:122–128. https://doi.org/10.1016/j.memsci.2018.08.066.
53. Vignarooban K, Badami P, Dissanayake MAKL, et al (2017) Poly-acrylonitrile-based gel-polymer electrolytes for sodium-ion batteries. *Ionics (Kiel)* 23:2817–2822. https://doi.org/10.1007/s11581-017-2002-4.

12 Redox Flow Battery
An Overview

Irshad U Khan, Rajmohan K. S, and Murali Mohan Seepana

12.1 INTRODUCTION TO REDOX FLOW BATTERY

The advancement of energy technologies has transformed human life throughout history. The rapid development of socially productive powers, the tremendous improvement in the quality of human life and the great improvements in the human environment follow any breakthrough in the field of energy. Global energy demand is increasing rapidly and the rising problem of energy demand is always a burden on the ecosystem and economy of any country. Energy demand is increasing by the rate of 1.52% each year, while the population increasing rate is 1.14% per year. [1]. Energy can also be subcategorized in two more groups: first from non-renewable energy sources and second from renewable energy sources with reference to their self-sustainability and regenerative characteristics. Fossil fuels and nuclear energy are non-renewable sources of energy. Non-renewable energy cannot be reproduced. Renewable energy sources, such as solar, wind, hydroelectric, geothermal, thermal, wave and tidal energy, can be stored repeatedly. The proven global oil and gas reserves are expected to be consumed in next 60 years at present usage rates [2]. In comparison, significant quantities of sulfides, nitrides, CO_2 and greenhouse gases are released through the use of fossil fuels, leading to a degradation of the global atmosphere and increasing the occurrence of unwanted weather conditions. The severe sense of urgency for energy security impels an impending transition from hydrocarbon fuels to renewable and environment friendly energy source. As sun and wind are intermittent, these conditions cannot be synchronized with customer demand. The power generated during peak period can be utilized very well, but during off peak period, generated power is wasted because of the lack of the demand.

There is a requirement for large-scale electrical energy storage (EES) to balance energy generation and demand. Renewable energy generation technologies are emerging, but in order to meet energy demand, there is a need for promising, efficient and low-cost energy storage technologies. The storage system will work as a bridge between generated power and energy demand. Solar and wind power technology can be utilized fully by the addition of energy storage. Among all the storage systems, electrochemical energy storage system is having some advantages. Unlike hydro-pumping and compressed air energy storage system, no any geographical requirements are needed in electrochemical energy storage.

In the 1970s, modern technology for redox flow batteries (RFBs) was invented. 30 Years after its invention, there was sparse and lukewarm research and development in this field recently. Rising demand to move from hydrocarbon to renewable energy like solar and wind energies has stimulated the advancement of large-scale electricity storage technology to regulate intermittent renewable energies into stable, usable electricity. As a result, for recent growth, it has become a prime focus for different uses, such as solar integration and load leveling, as well as upgraded energy storage systems for stationary modern electric grids.

RFBs have come out as potential candidates in the production of large-scale storage system for grid-connected energies. From 2005, there is a swift in the amount of research in the field of flow batteries, because they provide significantly long-life cycle, higher energy efficiency (EE), as well as low costs for applications requiring high energy to power ratios. Flow batteries have a number

DOI: 10.1201/9781003310167-12

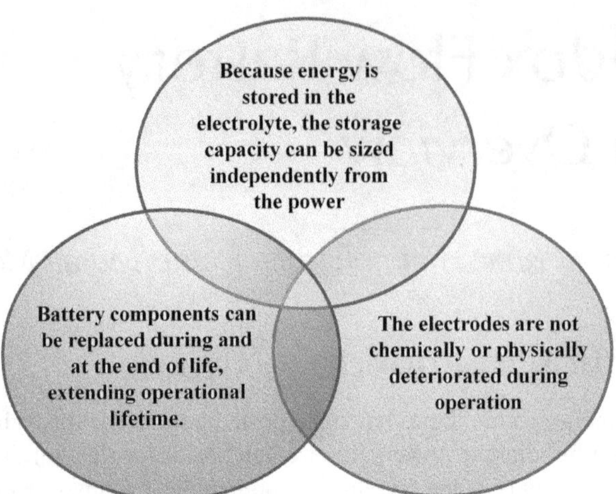

FIGURE 12.1 Features of redox flow battery.

of attractive features as shown in Figure 12.1. Fast response and high efficiency are also the other advantage of electrochemical energy storage. So as per the advantages, RFB is one of the potential candidates for renewable energy storage. The RFB is an electrochemical energy storage device capable of storing electrical energy as chemical energy and returning it when needed. RFB can provide reversible conversion between chemical energy and electrical energy.

12.1.1 Brief History

The zinc-halogen systems which surfaced in the nineteenth century can be traced to early RFB concepts. However, in the 1970s, the development of modern RFBs was demonstrated for the first time by National Aeronautics and Space Administration (NASA) that used the positive and negative sides of Fe^{3+}/Fe^{2+} and Cr^{3+}/Cr^{2+} redox pairs [2]. The machine was, however, subject to crossover and therefore performance was reduced, and the further development of mixed electrolytes was essential to mitigate the problem, as well as catholytes and negolytes. The use of the same element on both sides of the RFB with different oxidation states was another method for alleviating the problem of cross-contamination. This idea was originally introduced in 1949. The University of New South Wales, Australia, initiated this concept for the development of the all-vanadium redox flow battery (VRB) in 1980s. A flow battery of four states of vanadium oxidation which has been more widely explored at current times: on the cathode side V^{5+} and V^{4+} and on the anode side V^{3+} and V^{2+}. VRBs research and development in RFB technology with a focus on improved performance by modifications in electrolytes, membrane and electrode development, as well as customized flow field designs are continuously going on. The electrolytes discussed above for all RFB systems consist of metal salts in aqueous solutions. In the case of aqueous electrolytes, the associated cell voltage of RFBs is restricted by water electrolysis problems and observed inherently low (i.e., < 1.7 V). Research efforts recently have begun to move toward creating new chemistries for redox flow [3]. Because its electrochemical windows in these systems are much broader, the first strategy is to use organic solvents and replace water in order to create nonaqueous RFBs; another approach is to use redox-active organic materials instead of just metal compounds, their diverse chemistry, structural diversity and substantially lower costs that are intended to be used [4]. The fundamental concept and mechanisms of RFB technology in recent years, particularly after 2010, namely the use of expandable storage vessels and flowable redox-active materials, have drawn great attention to the creation of a wide range of new, hybrid energy storage and conversion systems. Flow batteries based on lithium metal, for example, use an anode of lithium metal when an aqueous catholyte is used and a membrane or a ceramic separator is isolated from the catholyte by this anode [5]. In either nonaqueous or aqueous form, the catholyte uses redox pairs, which include organic active materials, iodine, polysulfide and metal oxide as well [6].

The developments of membrane-less and pumpless RFBs based on several working procedure, for example magnetic field mediated transfer, were also stated [7] as well as gravity-induced properties [8] or electrochemical laminar flow cells [9]. More recently, extensive photochemical flow battery analysis has been performed, in which photocatalysts based on titanium or iron are used primarily to store solar power [10]. Metal-air batteries are regarded to be energy storage devices that are promising and environmentally friendly, and it is also possible to combine flow systems with this [11,12]. Timeline of key development of RFBs is presented in Figure 12.2. Different types of flow batteries and their characteristics are tabulated in Table 12.1. A brief description on the flow battery chemistry, along with current technical obstacles in systems, is given in the following sections.

12.1.2 Working Principle and Characteristics

RFBs are secondary batteries in which two redox pairs have reversible electrochemical reactions (A^{2+}/A^{+} and C^{3+}/C^{2+}) which are based on energy conversions, as presented in equation 12.1. In electrolyte solutions, the redox pairs are usually dissolved [13–16].

$$A^{2+} + C^{2+} \underset{\text{Discharge}}{\overset{\text{Charge}}{\rightleftarrows}} A^{+} + C^{3+} \quad (12.1)$$

RFBs are composed of two elements, as shown in Figure 12.3, battery stack, where electrochemical reactions occur, and outer tanks that store electrolytes where these two components are connected by pumps. The battery stacks are typically made of two electrode sets, two bipolar plates and two current collectors, and one membrane is sandwiched between two electrodes. When preventing the mixing of the electrolytes on the two sides, the membranes act as charge-carrier conductors unlike traditional batteries that consist of redox-active materials within electrodes.

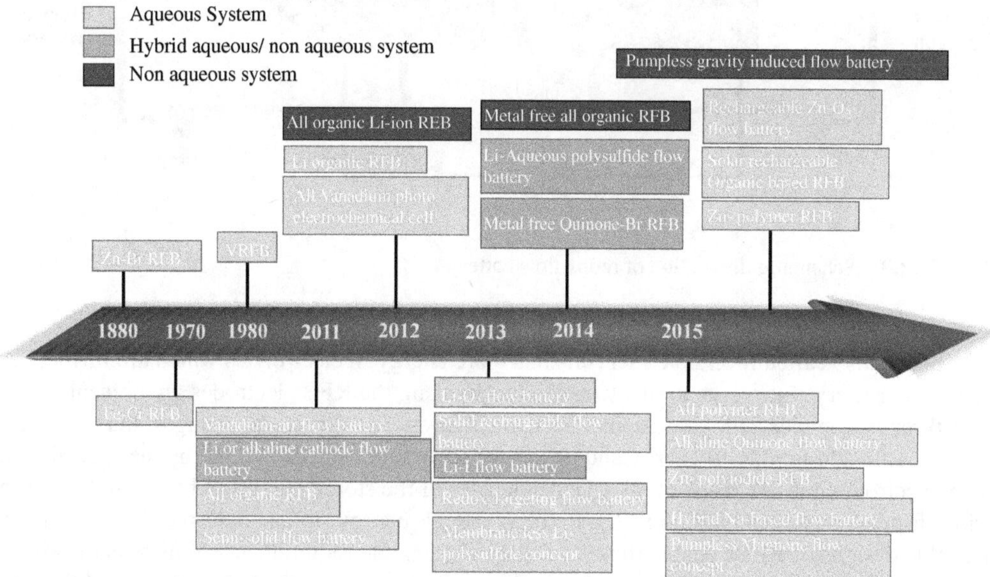

FIGURE 12.2 Timeline of key development of RFBs.

TABLE 12.1
Different Flow Batteries and Their Characteristics [65]

RFB	Demonstrated Charge Concentration (M)	Cell Voltage (V)	Stability (Cycles)	Key Drawbacks	Element Prices ($ per 0.1 kg)
VRB	3	1.25	>1000	High chemical cost	V: 2.7
ICB	1.25	1.18	30–100	Slow kinetics; H_2 evolution	Fe: 0.02
IVB	1.5	1.02	>100	High chemical cost	Cr: 0.28
PSB	2.6	1.5	50	Br_2 crossover; S precipitation; slow kinetics	Br: 0.15 Zn: 0.18
ZBB	4	1.76	300	Zinc dendrite; low utilization zinc	Pb: 0.02
ZIB	6.67	1.3	50	Zinc dendrite; high cost	Ce: 1.2 Cu: 0.66
Zn/Ce	0.8	2.4	50	Gas evolution	S: 0.01
SLFB	1	1.59	2000	Pb dendrite; PbO_2 polymorph	Cl: 0.15
H_2/Br_2	1	1.09	100	Costly Pt catalyst; Br_2 crossover	I: 8.3

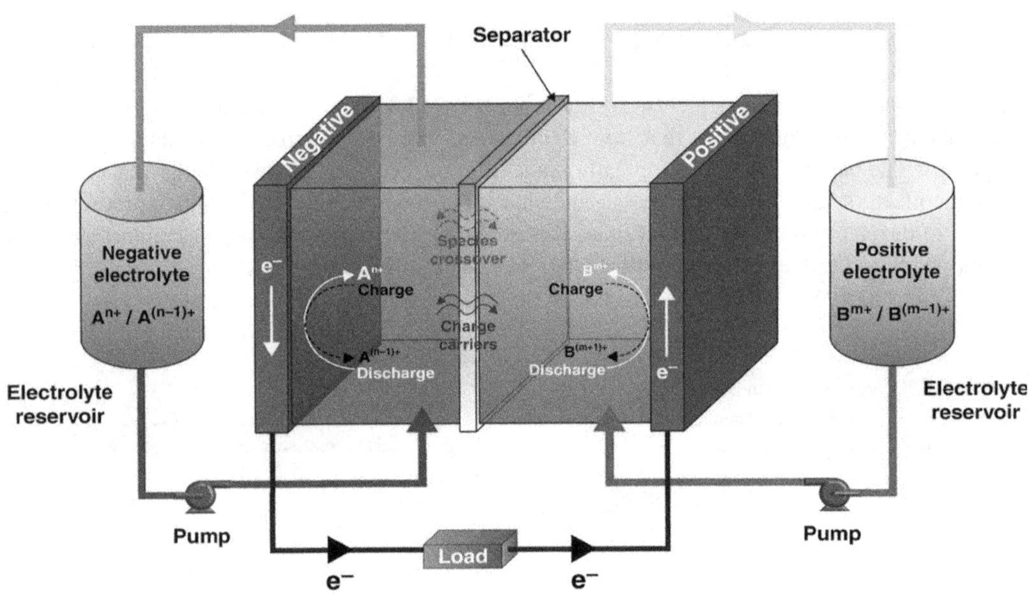

FIGURE 12.3 Schematic illustration of redox flow battery.

RFBs are also called reversible fuel cells that store energy in electrolytes which are further supplied to the battery stack. Due to this working mechanism, the RFB electrodes are prevented from going through complex redox reactions, mechanical stress and structural changes, thus extending service life. In addition, RFB energy and power are separated from each other by the special isolation of electrodes and electrolyte tanks. Dependencies of the stored energy and delivered power on various factors are shown in Figure 12.4. In the grid-scale energy storage industry, the versatile and adjustable designs of RFBs enable those to satisfy the flexible requirements of different energy to power ratios [16]. Energy to power ratios actually play a key role in the cost and optimal flow batteries operating parameters [16].

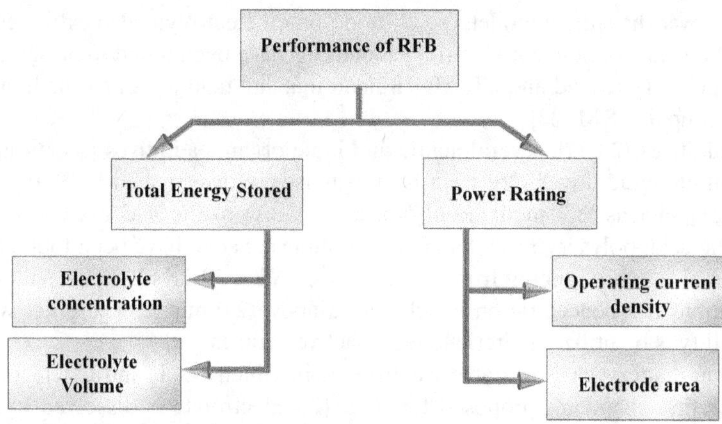

FIGURE 12.4 Energy and power dependency in redox flow battery.

12.2 COMPONENTS OF REDOX FLOW BATTERY

12.2.1 RFB Electrolytes

The RFB electrolytes are independent solutions that contain the solubilized electroactive species within an ion-conductive liquid that is forced along each side of the cell during charging and discharging. C^{2+} is from the catholyte tank and A^{2+} is from anolyte tank, which are fed through the electrodes during charging and discharging, where they gone through reduction and oxidization, resulting in A^{+} and C^{3+}, respectively [17]. This phenomenon is reversed while discharging. The most significant element in the RFB device is the electrolyte. The properties of electrolytes largely decide the performance of the RFB. The solubility of the dissolved redox-active materials in the electrolyte solution is a significant parameter influencing the energy density of an RFB. Furthermore, for future commercial viability, the temperature range, durability, conductivity, costing and stabilization of the electrolyte are also significant. There is nevertheless a large number of variations in RFB active materials and systems that combine redox-active species and electrolytes, each with their advantages and disadvantages. A brief discussion with several different electrolyte systems is presented.

12.2.1.1 Aqueous Electrolytes

The VRFB system is by far the advanced RFB system till date. The highly investigated electrolyte is also vanadium, and historically, in sulfuric acid solvent electroactive vanadium, sulfate species are dissolved. The significant issues are the lower efficiency and temperature stability of the solutions, particularly with pure sulfate vanadium electrolytes. The vanadium (V) (i.e., V^{5+}) species, in particular, demonstrated poor stability at higher concentrations of vanadium (i.e., >2 M) and high temperatures (i.e., >35°C). During battery activity, this thermochemical instability is observed as the accumulation of vanadium pentoxide (V_2O_5) in the tank bottom, resulting in energy loss and subsequent battery failure. Vijayakumar and colleagues at PNNL have evaluated the chemical reaction behind the low VO_2^+ stability in the concentrated sulfuric acid electrolyte (pure) [13]. The two key ways to address this problem are to add a second acid (H_2SO_4 or HCl as supporting electrolyte) [18–21] or an electrolyte stabilizer [21].

Without precipitation, the maximum concentration obtained in a mixed-acid solution of sulfuric acid and hydrochloric acid is 3 M for a large temperature window from –10°C to –50°C. For a wide temperature range in the chloride solution, an improved thermal stability is ascribed over the system using vanadium dinuclear or dinuclear chlorine complex in chloride solution [19,22]. In comparison with sulfate solution, the chloride solution was 30%–40% low viscous, which improved overall efficiency by decreasing adverse pumping losses due to a low pressure drop. With an energy density increase of 30% over the sulfate model, the chloride-based electrolyte also exhibited high reaction kinetics [18]. The viable options for electrolyte solutions have been found to be several other acids, including such polyacrylic acid and CH_3SO_3H blend that can hold all four vanadium ions stable at concentrations of up to 1.8 M [22].

Regarding additives [23,24], several acidic and ionic organic additives [25,26] and amine- and ammonium-containing additives [26] have been extensively investigated by Wang et al. [27] and additive acids [28], such as HCl, methanesulfonic acid (MSA), oxalic acid, boric acid, trifluoroacetic acid, methacrylic acid, polyacrylic acid and phosphotungstic acid, have been found to increase V^{5+} electrolyte thermal stability ranging from –5°C to 45°C. After 30 days [28], the solution of V^{5+} with these acid additions has concentration which stayed above 2.0 mol/L. The other way to enhance electrolyte stability is by utilizing alternate redox-active couples.

Iron-chromium battery has the stable electrolytes in which $FeCl_2$ and $CrCl_3$ (or hydrate) salts were among the first chemicals proposed for RFB [2]. Electrolyte is dissolved in HCl. Fe^{2+}/Fe^{3+} redox reaction kinetics are increasingly proceeding to carbon electrode materials being adopted. However, to improve the weak electrochemical kinetics [29], the Cr^{2+}/Cr^{3+} redox reaction requires a catalyst. Catalysts like Bi [30] when applied to 0.005 M Bi^{3+} (Bi_2O_3) [31] electrolyte lead to deposition on the electrode surface during charging, improving the electrochemical kinetics of the redox Cr^{2+}/Cr^{3+} reaction.

Wang et al. reported an iron-vanadium RFB due to the weak electrochemical reaction kinetics of the Cr^{2+}/Cr^{3+} reaction [32]. At room temperature, $VOSO_4$ and $FeCl_2$ are dissolved in concentrated HCl, and the aboriginal electrolyte of a combined mixture comprising Fe^{2+}, V^{4+}, SO_2^{4-}1.5 M each and Cl^-6.8 M was prepared. The iron-vanadium sulfate-chloride mixed-acid electrolyte's temperature stability at state of charge (SOC) of 0% and 100% for each half-cell was measured at 50°C. There was no any measurable precipitation visible for over 10 days in electrolyte, suggesting the electrolyte's excellent thermal stability. Most reliable composition with the highest conductivity in the temperature window –5°C to 50°C consisted of 1.5 M of Iron (II)Chloride ($FeCl_2$) and Vanadyl sulfate ($VOSO_4$) each with 3 M HCl; this flow battery demonstrated steady cycle efficiency for the broad temperature window –5°C to 40°C. In addition, the cycling capacity for charge–discharge is steady at –5°C and at room temperature [33].

A V/Cl battery has been suggested by the Skyllas-Kazacos, in the negative half-cell using VCl_2/VCl_3 or VBr_2/VBr_3 pairs and at the positive half of the cell Br/$ClBr_2^-$ or Cl^-/$BrCl_2^-$ pairs [34]. This pair of system V/Br possesses all of the benefits of the VFB. Due to the high vanadium(II/III) bromide solubility (up to 4 M) and potential surplus of Br^- at positive half-cell, the vanadium bromide efficiently results in two times increase in the energy densities [35]. A V/Ce RFB [36] was developed by Xia et al. [37,38] for higher open-circuit voltage (OCV) from 1.54 to 1.98 V (based on the supporting electrolyte [38]). Initially, the V/Ce RFB was examined in a single-acid medium [35,39–42], which had a low solubility of the Ce(III) ion (<0.3 mol dm^{-3}) only in sulfuric acid. This was then improved in MSA or mixed-acid solutions [41] more than 0.8 mol dm^{-3} [43,44] for cerium and more than 2.0 mol dm^{-3} [44] for vanadium.

Zn/Zn (E = –0.76 V) can be combined with the high potential of Ce^{3+}/Ce^{4+} of Eeq = 1.28 to 1.67 vs SHE, which results in an OCV of 2.43 V [45]. Since its MSA solubility is around 10 times higher than that of sulfuric acid, Ce(III) carbonate is used [46]. As the organic acids are less detrimental to the atmosphere and possess low toxicological risk, this organic acid can be employed as a supporting electrolyte in flow battery [47,48]. However, strongly precipitating cerium species have been recorded due to the formation of cerous-ceric hydroxysulfate ($Ce(SO_4)_2 \cdot 2H_2O$) or ceric sulfate ($Ce(SO_4)_2$) above 50°C [49,50]. Furthermore, the deposition of zinc on the negative electrode makes the system prone to the mechanical fracture of the system, membrane and electrode design [51,52].

While zinc has been deposited, its redox reaction was combined with other different chemistry such as bromine [52] and iodine [53]. A bromine sequester agent like organic quaternary ammonium bromide [54] usually requires for the zinc-bromine electrolyte. One of that quaternaries is 1-ethyl-1-methylpyrrolidinium bromide ($C_7H_{16}BrN$), which makes big complexes with bromine

TABLE 12.2
Different Organic Flow Battery Chemistry and their Performances [61,62]

S.No.	Negolyte	Posolyte	Supporting Electrolyte	Solubility	Cell Voltage (V)
1	1,4 p-Benzoquinone	hydroquinone	H_2SO_4	0.1 M in neutral pH aqueous solution	0.693
2	1,4-dihydroxybenzene-2-sulfonic acid	$PbSO_4$	H_2SO_4	0.2 M in acidic pH	0.72
3	2,5-dihydroxy-1,4-benzoquinone	$K_4Fe(CN)_6$	KOH	>8 M	1.2

molecule appear on the positively charged electrode and makes a heterogeneous phase with the aqueous electrolyte.

Due to the formation of ligands in between the alcohol group's oxygen and the zinc ions [53], it was noticed in the ZN/I flow battery that the alcohols addition into electrolytes at the cathode side stabilized the catholyte at slightly lower temperatures, while at the anode, it improved the growth of zinc dendrite. This absence of the strongly oxidative VO_{2+} and Br_2 [55], which enables cheaper electrode and membrane materials to be used, attributes the zinc-iodide batteries over the all-Vanadium battery. The benefits of nonaqueous RFB electrolytes with improved solubility of redox-active materials can be used by an all-organic aqueous RFB while also preventing the complex preparation of complexes of metals [56]. Wang et al. [57] demonstrated a customized anthraquinone (AQ) and Aziz et al. [58,59] utilized 1moldm^{-3} anthraquinone like 2,6-dihydroxyanthraquinone (DHAQ) and 9,10-anthraquinone-2,7-disulfonic acid (AQDS) added to ferro/ferricyanide in sulfuric acid (H_2SO_4) electrolyte or DHAQ and AQDS combined with bromine in hydrogen bromine (HBr) electrolyte. An all-organic RFB was developed by Liu et al. [58] using an economic and benign NaCl electrolyte [60], 4-hydroxy-2,2,6,6-tetramethyl-piperidin-1-oxyl (4-HO-TEMPO) with methyl viologen (MV). Few more organic electrolyte pairs for redox flow battery and the cell performances for the respective single cell are tabulated in Table 12.2.

12.2.1.2 Nonaqueous Electrolytes

For some redox species, nonaqueous solvents can serve as a supporting electrolyte, allowing over a broader potential window over aqueous solvents restricted to 1.23 V [63,64]. Organic (nonaqueous) solvents have quite lower freezing points and boiling points, such as tetrahydrofuran and 1-propanol. Singh suggested this notion in 1984 [63]. Through the production of numerous nonaqueous metal-based and metal-free RFBs, nonaqueous RFB technology is still evolving. Solvents like lithium hexafluorophosphate ($LiPF_6$, PF_6^-), sodium perchlorate ($NaClO_4$, ClO_4^-) and tetraethyl ammonium tetrafluoroborate (ET_4NBF_4, BF_4^-) are the most common supporting electrolytes as they have the higher ionic conductivities.

12.2.2 RFB Electrodes

The RFB's role is responsible for providing surface for an electrolyte redox reaction in comparison to solid-state batteries. Electrode redox reactions in a traditional solid battery usually include challenging, tedious problems like phase transformation, alteration of crystallographic structure and modification of morphology; the electrodes in RFB are prevented from these complex changes that provide fundamentally to the longevity of operational life.

The power associated with RFB system is based on the size of the electrode (area), the operating current density and the size of the stack [64]. RFB efficiency is significantly affected by electrode material's electrical resistance and its catalytic impact over the redox reaction. The same reaction can require some physical or chemical changes during cycling to the electrode [65], and if the catalytic effect is optimal, the power output for a given region of the electrode will increase significantly [64].

To allow electrolyte flow, the electrode must have a three-dimensional porous structure [66]. Additionally, the electrode must preferably possess low electrical resistance, a significantly higher surface area and high electrochemical activity against flow battery reactions. The electrode should be chemically and mechanically inert toward the supporting electrolyte and redox-active materials [65]. Because of the wide use of highly acidic electrolytes, stability is very important, especially with the high-oxidation redox-active species like Ce^{3+} in cerium flow battery. Currently, after these many years of research, there are few electrode materials to choose; graphite- or carbon-based felt, fabric, paper, powder and carbon black are most widely used.

To promote the redox reactions, some kinds of RFBs, such as iron-chromium battery (ICB), need catalyst loading. Plain graphite or carbon-based materials do not have reversible kinetic and electrochemical activity in VRBs; that's the reason for the treatment of electrode material. Chemical treatment [67], thermal treatment [68], electrochemical oxidations [69] and combining or depositing metallic catalysts [70] are the methods for improving the efficiency of carbon or graphite electrode materials. In Figure 12.5, comparative morphologies of the graphite felt (GF) after various treatment are presented. Figure 12.5a shows that pristine GF displays a 3D-architecture and consists of surface impurities on cross-linked fibers. Thermal treatment of graphite felt (TTGF) in the presence of air was carried out and shown in Figure 12.5b. It is apparent that the impurities disappeared and the smoother fibers obtained. The GF was placed in electrophoretic deposition (EPD) utilizing H_2SO_4 cell for anodic polarization, which results in rough GF fibers and attributed to apparent defects, work as active sites for active species to react as presented in Figure 12.5c. Further, the GF treated by EPD using graphene oxide water suspension exhibited sheets look alike graphene when placed over fiber surfaces in wrinkled (Figure 12.5d) or anchored (Figure 12.5e) configuration. Anchored configuration is abundantly available throughout the material. HRTEM image of the felt and the sheets deposited using EPD which are randomly arranged are shown in Figure12.5f [71].

Undoubtedly, heat treatment is the simplest, economic and most prominent ways of improving the operation of electrode materials based on carbon. This method improves the formation of

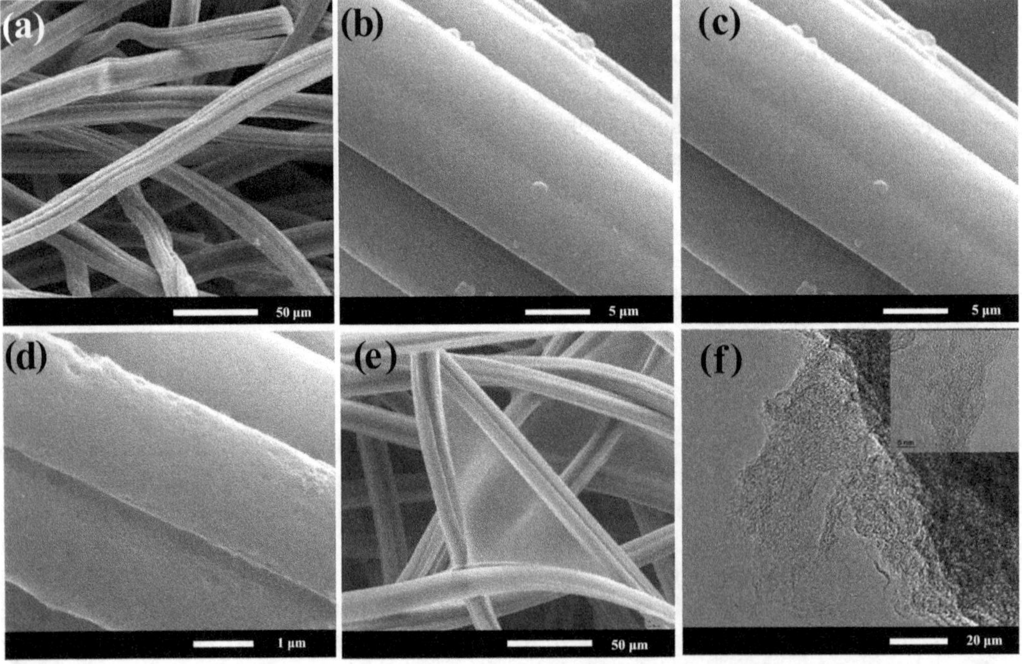

FIGURE 12.5 SEM images of (a) GF, (b) TTGF, (c) GF-H_2SO_4 and (d, e) GF-G. (f) HRTEM image of an interface region of GF-G. Adapted and reproduced from Ref. [71]. Copyright © 2017 Elsevier.

C-O-H and C=O functional oxygen-based groups onto the carbon surface, both of which help to enhance the hydrophilicity and accelerate the vanadium redox reactions [72]. Mostly, the mechanism of this catalytic process is studied. The chemistry of flow batteries is yet to be completely studied although the effect of these active species, electrocatalysts and carbon formations is evident.

After this studies, some other oxidative methods surfaced; Li et al. [72] conducted O_2 plasma on GF electrodes, for example, accompanied by H_2O_2. Their usage in a VRB at 150 mAcm^{-2} current density exhibited an improved EE by 8.2%. O-C=O groups were also found to improvise the performance of the RFB, while C-O and C=O groups reduced that [72]. Thermal treatment of carbon paper electrodes was performed in a specific operating environment having 42% oxygen and 58% nitrogen which resulted in a reduction in overpotential activation of nearly~140 mV and higher efficiency. Nitrogen, combined oxidative treating procedure on the electrode surface, mixing may again enhance the performance of both GF [73] and carbon cloth [74] electrode materials, specifically for VO^{2+}/VO_2^+ redox reaction [71].

The redox reaction reversibility of the VO^{2+}/VO_2^+ active species can be enhanced by operating with more surface area of electrode through flourishing carbon nanofibers via chemical vapor deposition [75] or by spraying graphene oxide on GF in the form of multiwalled nanosheets. Utilizing basal-exposed and edge-exposed graphite foil electrodes, it implies that the kinetics of V^{2+}/V^{3+} and V^{4+}/V^{5+} redox reactions can be impressively increased by applying more edge sites, preferably at low vanadium ion concentrations [76,77].

Catalysts research was going on for the modification in the vanadium flow battery system, along with oxidative and nitrogen-doped modification procedures. For the long duration use of an RFB cell, both the chemical and physical stability of the settled catalyst are significant. The reduced iridium from the pyrolysis of H_2IrCl_6 when added on carbon felt reduced overpotential, increased the activity of the redox couple VO^{2+}/VO_2^+ and reduced the inner resistivity of the single cell by 25% [78]. With comparison to the cell using electrolytes excluding Bi^{3+}, for the operating current density of 150 mA cm^{-2}, this expressively improved EE around ~11% more in all-vanadium flow batteries [79] (Figure 12.6).

A small amount of Nb_2O_5 catalyst on the GF can accelerate the transmission of charges and improve the performance of the electrode. However, the enhancing concentration of Nb in precursor solutions causes an intense agglomeration of Nb_2O_5 in large particles on GF surfaces, resulting in the catalyst being transmitted from GF surfaces to the circulating electrolyte. However, as a consequence of enhanced Nb levels, the agglomeration of catalysts is still occurring. Therefore, optimum Nb (0,05 M) concentration ensures the high-performance flow battery performance. The reason for this is that smaller nanoparticles on the surfaces of GF are uniformly distributed. EE values are improved when W-doped Nb_2O_5 nano-catalysts are equally distributed. In addition, there was no fading of EE on the surface of both negative and positive GF electrodes after over 50 cycles, and a significant number of nanoparticles were still present [80].

It was also found the fast kinetics of other RFB chemistries can be obtained by enhancing the surface area and the number of active functional groups on a carbon electrode. The Ce^{4+}/Ce^{3+} redox-active couple for a V/Ce RFB can be enhanced by nanoporous graphene oxide edges like electrocatalysts consisting hydroxyl and epoxy classes. Comparative SEM morphology of electrocatalyst is shown in Figure 12.7. The stability was markedly improved when the nanoporous graphene oxide edge (NP-GO_{el}) electrodes were changed with Nafion, with no alteration in electrode morphology [41].

Because of the higher oxidation voltage of Ce^{3+} in V/Ce RFBs, it was noted that the stabilization of carbon electrodes for extended operational time is debatable. However, polymer coatings on electrode materials can limit the breakdown of carbon particles into the electrolyte [40].

For zinc-cerium RFBs, a carbon-polymer anode [81] is required, while a pristine carbon paper [82], platinum-iridium coating [79] or platinum titanium mesh [83] can be used for the cathode. In ZN/Br, RFBs linked by a microporous polyolefin separator and carbon-polymer electrodes are used.

FIGURE 12.6 Surface morphology of graphite felt: (a) Raw felt as received and (b) Ir-modified graphite felt. Adapted and reproduced from Ref. [78]. Copyright © 2017 Elsevier.

Metal zinc dendrite is first electrically plated on a side while charging, the carbon-polymer combined electrocatalyst needs unique materials for this electrodeposition phase thus creating higher cost of the electrode [64]. The catalyst for the Cr (III)/Cr (II) is required in the ICB and should possess a higher overpotential for the H_2 evolution reaction, since this decreases the Coulombic efficiency (CE) and induces an imbalanced SOC over prolonged cycles in between the anolyte and catholyte, ultimately increasing the decay of capacity. As stated in the previous section, to control this, Bi^{3+} should be applied on the electrolyte and electrodeposited on the GF. Instead of using graphite-based electrodes with just a nominal amount of gold (Au), the inclusion of thallium-I-chloride will accelerate the Cr^{3+}/Cr^{2+} redox reaction and increase the overvoltage of hydrogen evolution more than Bi^{3+} as well [84].

In the aqueous Zn-polyiodide RFB, associated low reversible kinetics and electrochemical activity of I_3^-/I^- pair redox reaction on graphite lead to low EE. Two nanoporous metal-organic frameworks (MOFs) were introduced on GF shown in Figure 12.8. At first, felt electrode was integrated with UiO-66-CH_3 (Figure 12.8a), which increased the total surface area of electrode and intensified the I_3^-/I^- pair redox reaction. Then, MIL-125-NH_2 was integrated on GF (Figure 12.8b). The application of MOFs boosted the EE by approximately 2.7% and 6.4%, respectively (at 30 $mAcm^{-2}$

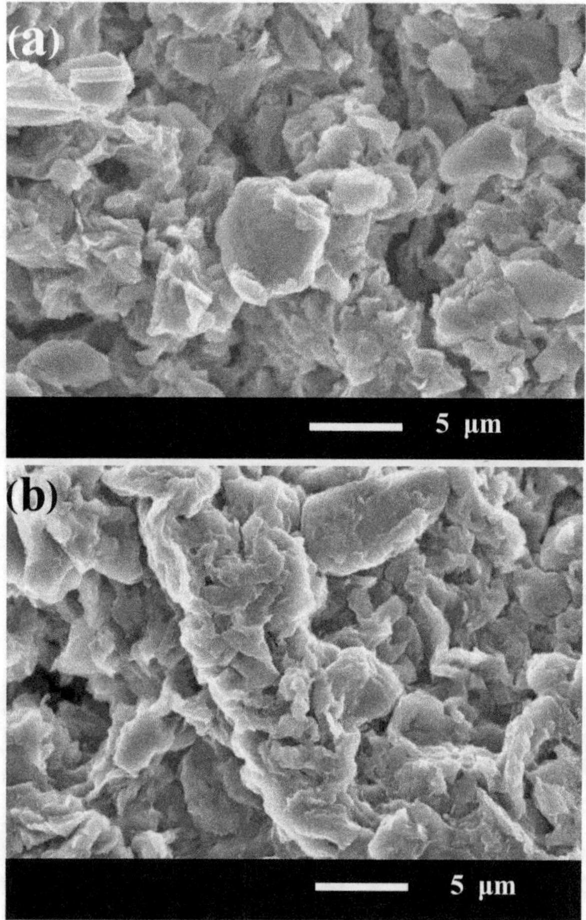

FIGURE 12.7 Morphological analysis of NP-GO_{el} and Nafion-coated NP-GO_{el}. Adapted and reproduced from Ref. [41]. Copyright © 2017 Elsevier.

current density). In a weak acidic electrolyte, UiO-66-CH_3 has high chemical sustainability [85]. In nonaqueous RFBs using redox couples $[Co\text{-}(bpy)_3]^{+/2+}$ and $[Fe(bpy)_3]^{2+/3+}$, the redox reactivity and extended durability for operations can be improved by using carbon with a coating of Ni-FeCrAl and Cu metal foam electrocatalysts. Without any noticeable loss of EE, the cell conducted more than 300 cycles [86]. Zn-polyiodide and other nonaqueous RFBs illustrate this by using alternate redox chemistries; alternative electrode materials may be used wherever the supporting electrolytes and redox-active materials are less acidic.

12.2.3 Membrane

In an RFB, the membrane isolates the cathode and anode as well as their compartments while enabling the completion of the circuit by transporting ions (e.g. H^+ and SO_4^{2-}). High ionic conductivity, chemical steadiness against the electrolyte for all state of charges, temperature window, physical stability and maximum ion selectivity are the soul property requirements of the membrane to reduce the crossover of redox species. This crossover of redox species leads to the loss in capacity and energy stored [65].

Ionic conductivity and ion selectivity incline to contrast with each other; the other can normally be decreased by increasing one. So, these properties need to be balanced, depending on the RFB

FIGURE 12.8 Field emission scanning electron microscope (FESEM) of morphologies of MOF-modified GF: (a) MIL-125-NH_2 and (b) UiO-66-CH_3. Adapted and reproduced from Ref. [85]. Copyright © 2016 American Chemical Society.

system. In addition, the membrane should be less expensive, and the future commercialization of the RFB technology should be increased [87]. Traditionally, the perfluorinated polymer membrane Nafion is used by most RFBs and VRBs. Because it can sustain against heavy acidic electrolytes and very oxidative V^{5+} ions. Nevertheless, Nafion membranes will be responsible for about 40% of the total cost of the cell stack; thus, it will reduce the cost of developing less corrosive flow battery like iron-chromium, which could use hydrocarbon membranes [88]. Current RFB membrane study, however, continues to concentrate on perfluorinated polymer membranes. The diffusion coefficients associated with the vanadium ions in VRB via Nafion membranes were calculated by Sun et al. [89] and the order is $V^{2+}>VO^{2+}>VO_2^+>V^{3+}$. Because of the clustered network structure of the hydrated pores that are of a sufficient size to enable the transport of all required ions and their hydration shells, this diffusion in Nafion is inevitable. A sequence of commercially available Nafion membranes (N117, N-115, N-1135 and N-112) of various thicknesses has recently been studied in detail in RFBs. It was determined that thick membrane gives higher CE and low permeability of the vanadium ions, whereas an opposite pattern was shown by the discharge power fading rate and the shift in electrolyte thickness. The thinner membrane results in the lower resistance to the area, permitting higher current densities to operate and reducing the time for one cycle.

The N-115 membrane exhibited the best EE and the good electrolyte consumption at the operating current density of 120–240 mA cm^{-2} because of a balance of ionic conductivity and selectivity. It was established that for VRB applications, the best overall performing membrane was N-115 [90]. In addition to the thickness, it was found that the equivalent weight (EW) of the Nafion membrane substantially changes the membrane morphology and thus influences the permeability of the vanadium ion. A recent study showed that the cycling performance of a vanadium RFB can be significantly upgraded for long-term stability and effectiveness by changing the thickness [91].

The thermal stability of the commercially available cation exchange membranes (CEMs) such as Nafion 115, Nafion 212, Selemion CMV and Sulfonated Poly(ether ether ketone) (SPEEK) alongside the same of Selemion apical membrane vesicles (AMV) and asahi polysulfone (APS) anion exchange membranes (AEMs) was studied in the temperature window –20°C and 50°C [92]. The SPEEK membranes exhibited low chemical durability, although the APS membrane gives lower retention potential at higher temperatures. The lowest capacity declining rate for the temperature range –15°C–50°C was demonstrated by N115.NR-212 and N-115 which are ideal for lower temperature application as shown in Figure 12.9 [93]. Via chemical modification, several researchers had worked to enhance the thermal stability and permeability of vanadium ions associated with the Nafion membranes. During membrane processing, the inclusion of a fluorocarbon surfactant (potassium nonafluoro-1-butanesulfonate) reduced significantly the permeability of the vanadium ions in the membrane [94].

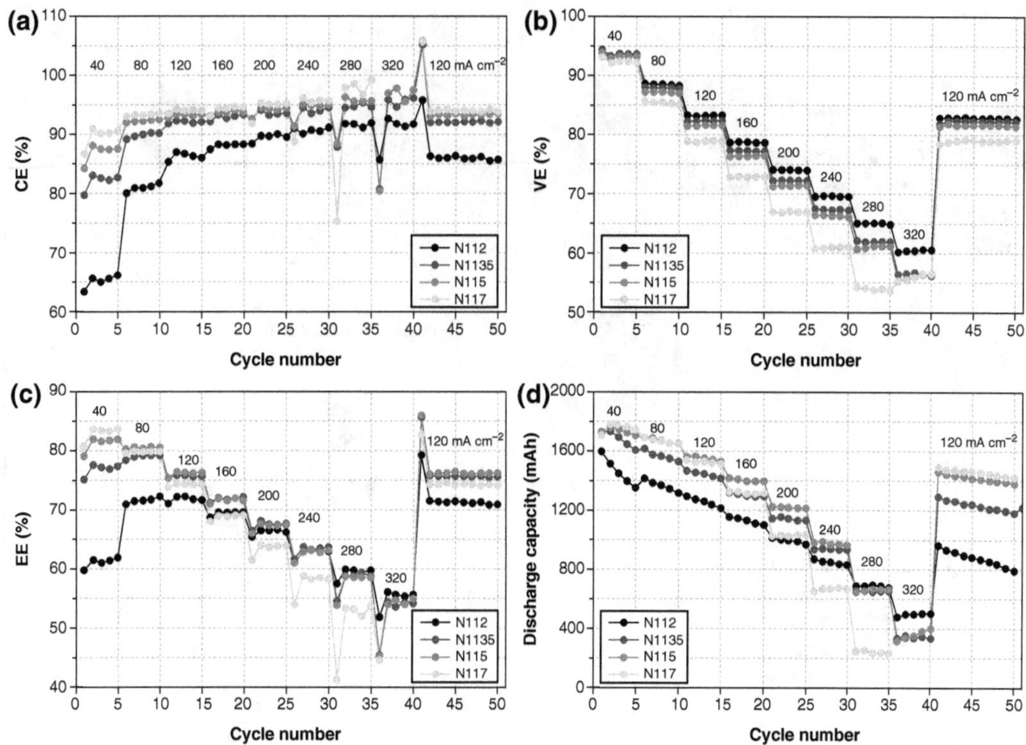

FIGURE 12.9 Cell performance of VRFBs with different Nafion membranes at different current densities: (a) Coulombic efficiency (CE) vs cycle number, (b) voltage efficiency (VE) vs cycle number, (c) energy efficiency (EE) vs cycle number and (d) discharge capacity vs cycle number. Adapted and reproduced from Ref. [93]. Copyright © 2016 Elsevier.

The phenomenon of vanadium transport has drawn a good interest in research studies. Mostly, the diffusion coefficient of the V^{4+}/V^{5+} ions is higher than the V^{2+}/V^{3+} ions for various forms of membranes studied. Apparently high diffusion coefficient of the redox-active material V^{4+} and V^{5+} may be due to the combination with SO_2^{4-} ions forming complex, although there is no such complex formation for the V^{2+} and V^{3+} ions reported yet. Various vanadium ions exhibit varying rates of diffusion over all membrane forms, such that the net transfer of vanadium ions results in the accumulation of vanadium ions in a half-cell and the drop in another half-cell. The adverse effect of this is a decrease in capacity, which could be recovered by regularly mixing the both half-cell electrolytes [93]. Ao et al. [95] revealed that crossover-related self-discharge reactions could lead to the excessive electrolyte heating inside a stack at temperatures above 55°C. The transport of vanadium active material, water transfer and electro-osmosis lead to a major effect on the crossover of reactive ions [96]. In one analysis, throughout the charge–discharge process of a VRFB using different membranes, researchers measured the transport number for water [97]. A variation in solution diffusion fluxes was found, leading to a net accumulation of AEMs at the negative half-cell and CEMs at the positive half-cell, respectively [98–100].

Aruna et al. [101] synthesized a proton exchange membrane for direct methanol fuel cell. As the fuel cell and flow batteries worked more or less on a same principle, the same membrane might give good result in the flow batteries. The membrane has chemically treated polytetrafluoroethylene (PTFE) support and pore infiltration of ZrP-PVA (poly vinyl alcohol), as shown in Figure 12.10. This membrane gives comparably higher ion exchange capacity (IEC, 1.28 meq g^{-1}) than Nafion series membranes. Another work from the same research group in which the inorganic silica immobilized phosphotungstic acid (PWA)-based (Si-PWA)-PVA/PTFE composite membrane was developed by

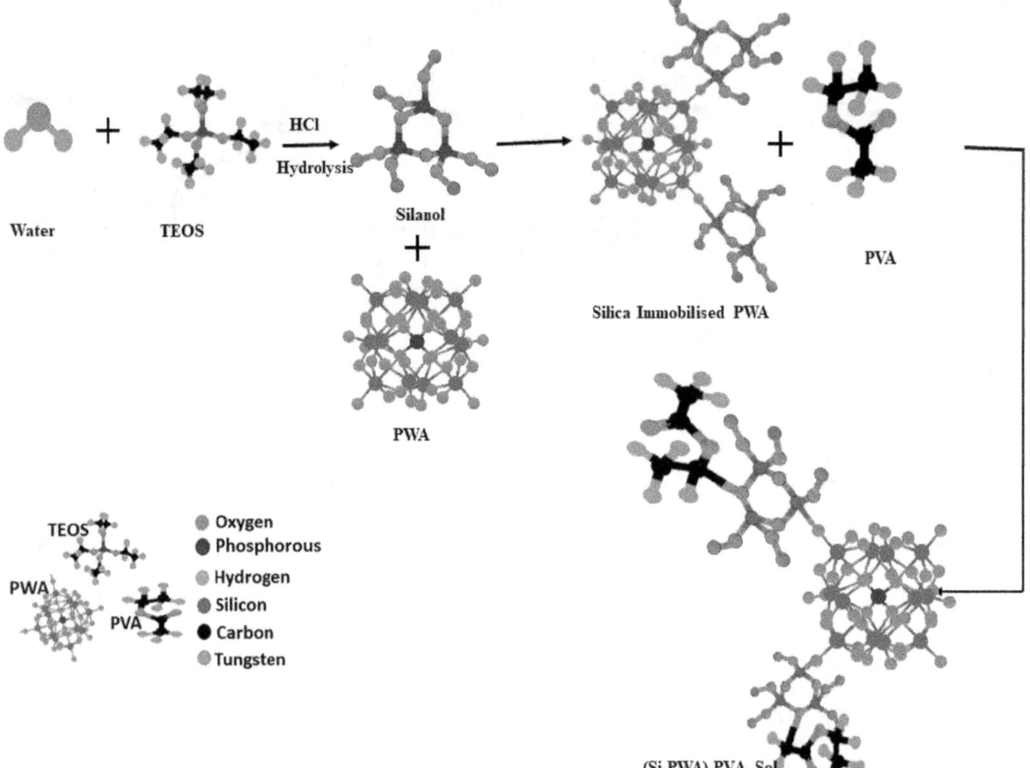

FIGURE 12.10 Structural networking of silica immobilized PWA-PVA sol formation. Adapted and reproduced from Ref. [101]. Copyright © 2020 Elsevier.

an amalgamation of pore filling and layer-by-layer (LBL) casting. The membrane has exhibited a maximum proton conductivity of 41.2 mS cm^{-1} at 100°C. As per the results, these membranes can be a potential component of a RFB.

In addition to the electrolyte fluxes throughout Nafion, when observing large-scale industrial application in RFBs, the cost was considered significant [102,103]. In a variety of tests, aromatic CEMs were found to have good impact to potentially lower the cost. Chen et al. [104] produced a stable fluorinated sulfonated poly(arylene ether) (SFPAE) CEM which possess aromatic characteristic with exceptional chemical durability for operations, increased CE and EE, and a unchanging discharge current capacity above 33% better than NR212 membrane above 75 cycles of charging and discharging an RFB.

SFPAE had equal proton conductivity to Nafion and it exhibits IEC of 1.8 $meqg^{-1}$, but much lower VO^{2+} permeability. As possible substitutes for expensive Nafion (perfluorinated-type) membranes, aromatic-based CEMs have been studied. NR-212 was compared to synthesized SPEEK of

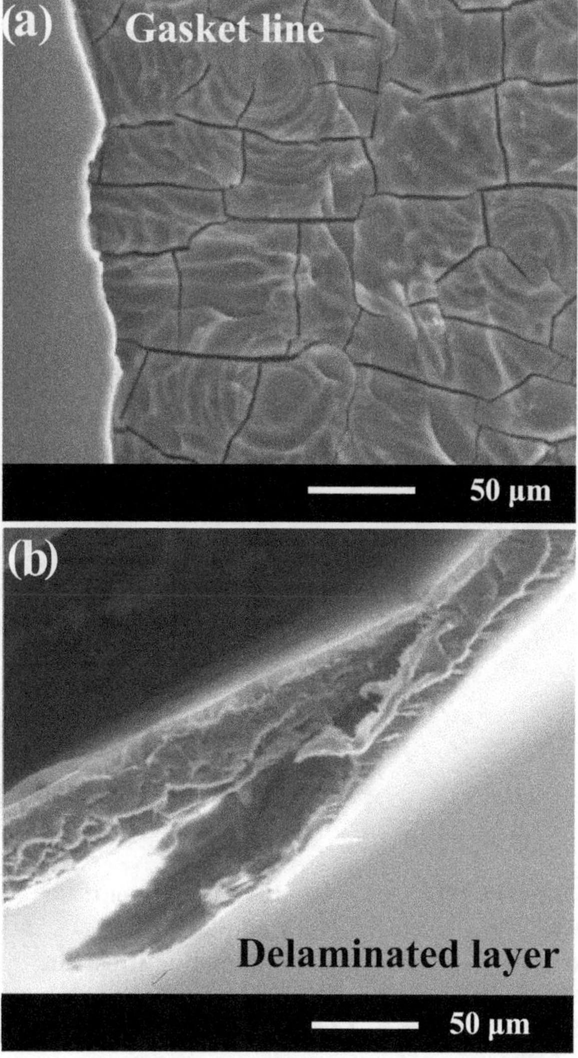

FIGURE 12.11 (a) Interface of the gasketed area and active area for positive facing surface and (b) delaminated layer on the surface facing the positive electrode. Adapted and reproduced from Ref. [101]. Copyright © 2011 Springer Nature.

various thicknesses. The SPEEK with 50 μm thickness displayed reduced overpotential, higher EE, high power density and a smaller reduction in capacity [100]. A sulfonated poly(sulfone) (S-Radel) membrane which possesses a higher proton conductivity and lower vanadium ion permeability demonstrated appreciable performance in a VRB, but exhibited mechanical and chemical deterioration in between cycles. However, the long-time durability of aromatic CEMs, mainly in V^{5+} consisting electrolytes, is a prospective challenge [101]. In the aqueous V^{5+} solution, the membrane was also submerged, and the sample fractured to tiny parts. The degrading represented the loss of the SO_2 group [105].

12.2.4 Battery Stack and System

Battery stack is a series of many cells linked to give energy and power to the source. A typical stack's components are presented in Figure 12.12 with stack's specifications differing from that of a single cell at the research level to the large scale. Although the previously comprehensive element constancies and costs are still applicable, engineering solutions for taking care of the flow circulation [104], shunt currents and flow field design are required. The surface of both electrodes should retain a continuous average linear flow of solution of electrolyte at a preferred velocity for ideal flow dispersal. In reality, in some areas of the electrode surface, unequal flow distribution occurs and stagnant zones are created [64].

The bypass or leakage currents [107] are self-discharging currents and are better minimized by raising the ionic resistance associated with the flow ports and manifolds. This is accomplished by

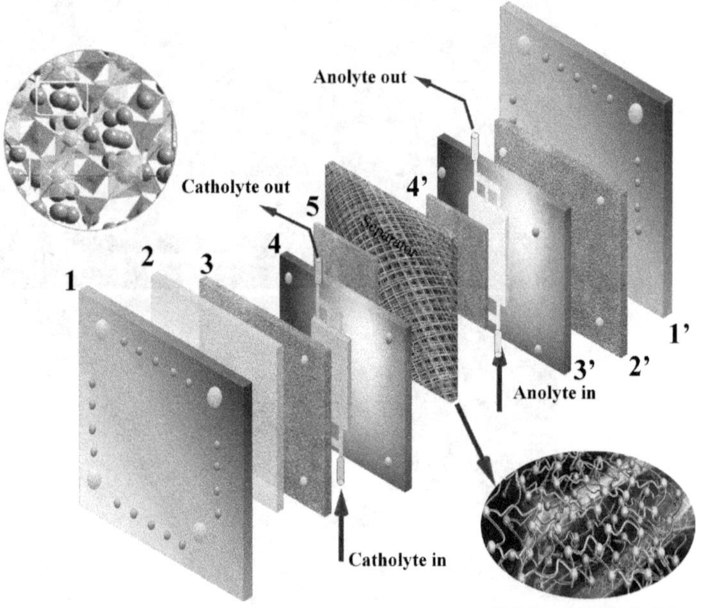

Cathode compartment:
(1) End plate
(2) Isolation plate
(3) Supporting substrate
(4) Flow frame
(5) Current collector

Anode compartment:
(1') End plate
(2') Supporting substrate
(3') Flow frame
(4') Current collector

FIGURE 12.12 Components of flow battery stack. Adapted and reproduced from Ref. [106]. Copyright © 2017 Elsevier.

two ways: increasing their length and reducing the port's cross-sectional area. However, increasing the manifold length increases the flow resistance of electrolyte, which means high pumping capacity and high cost of the device. Because of its straightforward redox electrochemistry, the VRB is frequently tested on the stack-level kW scale. VRB stack consisting of 31 cells having 2714 cm^2 electrode surface area with a commercial anion exchange membrane was measured at current densities of 60 and 90 $mAcm^{-2}$ with an electrolyte of 1.2 M $VOSO_4$ in 2 M sulfuric acid. For 60 mA cm^{-2}, the EE was 76%, and for 90 $mAcm^{-2}$, it was 70%. Input power was 11.4 kW and output power was 8.9 kW in the 90 $mAcm^{-2}$ stacks [108]. In the operating range 15%–85% SOC at 80 $mAcm^{-2}$, having an EE of 82% and an energy content of 1.4 kWh, the mixed-acid (H_2SO_4 and HCl) VRB was studied at 1.1 kW. Because of excellent kinetics and low electrolyte resistance, the device operated steadily without any sort of precipitation at solution temperatures above 45°C for high EE stack, and the viscosity associated with the mixed-acid electrolyte was also reduced [109,110].

A stack consisting of three-cell 1 kW-class mixed-acid VRFB was used in three separate flow designs. The standard flow-through configuration had the best stack [111,112] EE compared to two interdigitated models at the minimum flow rate of 400 cc/min/cell, limiting the pressure drop to <10 psi in the flow circuits. The stack EE of the interdigitated designs was better than the standard flow-through design, since the flow rate was 1200 cc/min/cell (at a maintained pressure drop <10 psi) [113]. NR-212 and NR-211 membranes are thinner than N-115 membrane. The application of thinner Nafion membranes in a mixed-acid VRB was observed to have an improved performance and more economic when compared with thicker membranes. For all Nafion membranes, the electrolyte temperature and the pressure drop were comparable. The electrical efficiency of NR-212 and NR-211 for various current densities was evaluated and it was comparable, but NR-212 was handy and easy to assemble into the stack. With improved methods of holding and assembling stacks, the NR-211 membrane can turn out to be beneficial at high current densities [114]. The subsequent cost study, however, also revealed that the distribution of system cost dramatically changed as the system's power capacity enhanced to about fourfold to its installed value; at that moment, the membrane's cost share in the system was significantly reduced [113].

12.3 TYPES OF REDOX FLOW BATTERIES

Various redox flow chemistries were developed with advances in RFB science [115,116]. These chemistries can typically be divided into four categories on the basis of the composition of the electrolyte, particularly solvents and redox-active materials:

- Aqueous solvent and metal-based redox-active materials
- Nonaqueous solvent and metal-based redox-active materials
- Aqueous solvent and organic redox-active materials
- Nonaqueous solvent organic redox-active materials

12.3.1 Aqueous Solvent and Metal-Based Redox-Active Materials

The most well-established systems so far are RFBs with aqueous metal-based electrolytes.

12.3.1.1 Iron-Chromium Flow Battery

The ICRFB developed by NASA was the first active RFB prototype, and the battery used the following redox reactions to transform energy (equation 12.2):

$$Fe^{3+} + Cr^{2+} \underset{\text{Charge}}{\overset{\text{Discharge}}{\rightleftharpoons}} Cr^{3+} + Fe^{2+} \tag{12.2}$$

A standard theoretical potential of 1.18 V for ICBs is provided by the potential difference between the two redox couples. Despite being the first prototype, intrinsic problems eventually hindered the work on ICBs [65]. In addition to the cross-contamination problem described earlier, ICBs suffer from two significant challenges arising from the Cr(III)/Cr(II) couple. The H_2 evolution is a significant side reaction occurring while charging the cell because of the low redox potential of Cr(III)/Cr(II) pair, which decreases the CE and restricts the charge depth. Another issue that stems in the Cr^{3+}/Cr^{2+} couple is sluggish kinetics, and consequently, catalyst loadings are needed to improve cell output along with elevated operating temperatures.

12.3.1.2 All-Vanadium Redox Flow Battery

VRFB is the most successful and prominent flow battery system and it uses four different vanadium oxidation states; the following reaction takes place during the process (equation 12.3):

$$VO_2^{+} + V^{2+} \underset{\text{Charging}}{\overset{\text{Discharging}}{\rightleftarrows}} V^{3+} + VO^{2+} \tag{12.3}$$

The VRB exhibited 1.26 V of the cell voltage (standard) based on the previous reactions. The mixing of active species is again a problem in VRBs, despite the use of the same element, and leads to frequent decreases in ability because of different diffusion rates and transfer rates of all four different vanadium ions throughout the membranes; the electrolyte volume also changes for both of the tanks [117,118].

However, such capacity losses can be recovered by techniques as timely remixing [119] and management of hydraulic pressure [117]. Gas evolution is also an obstacle for VRBs, since hydrogen or oxygen could be produced at the anode or cathode with high overpotentials, respectively. The higher corrosion rate of V^{5+}species and highly concentrated acids, as well as the comparatively high cost of Nafion membranes and vanadium solution, are other obstacles to VRB growth.

12.3.1.3 Other Flow Batteries with the $V^{(3+)}/V^{(2+)}$ Couple

By employing the $V^{(5+)}/V^{(4+)}$ pair rather than others on cathode side, derivatives of VRBs were produced. The iron-vanadium battery (IVB) [13,33], vanadium-polyhalide [34] and vanadium-cerium flow batteries [116] are examples of such RFBs. $V^{(3+)}/V^{(2+)}$ pair has stronger kinetics as compared to the pair of $Cr^{(3+)}/Cr^{(2+)}$. $Fe^{(3+)}/Fe^{(2+)}$ is having a lesser corrosion rate than the pair of $V^{(5+)}/V^{(4+)}$, although its low redox potential removes the probability of evolution of oxygen throughout the cell charging. For operation, IVBs also have a relatively large temperature range: 0°C–50°C. In IVBs, the cell reaction is the following (equation 12.4):

$$Fe^{3+} + V^{2+} \underset{\text{Charge}}{\overset{\text{Discharge}}{\rightleftarrows}} V^{3+} + Fe^{2+} \tag{12.4}$$

A theoretical cell voltage of 1.02 V is achieved by the reaction, which is quite lower than that of ICBs and VRBs, and it is the main disadvantage of IVBs. Lower voltage results in a lower density of energy, which can be partly balanced by enhancing the concentration of redox-active species and higher SOC activity [115].

12.3.1.4 Hybrid Flow Batteries Comprising Metallic Anodes

The previously described traditional RFBs on both sides of the electrolytes use soluble active materials and store those in external electrolyte tanks. Nevertheless, with several modern RFB systems emerging from the other energy storage systems, i.e., solid-state batteries and fuel cells, such a concept has been blurred [120]. Some of these systems use metals as anodes for example: zinc-halogen flow batteries, all-copper flow batteries and all-iron flow batteries (Fe-FB).

A zinc-bromine flow battery (ZBRFB) system is the well-known Zn-halogen RFB system that is to be shown at large scales (50–500 kWh) other than VRBs [117]. In ZBRFBs, the reaction between

Zn and Br_2 occurs and a high standard potential of 1.85 V is produced. However, the production of ZBBs is not that much because the bromine is extremely corrosive and toxic in nature. The emergence of zinc dendrite and the evolution of hydrogen gas are also the major concern [118,121,122]. Recently, zinc-polyiodide system has been surfaced to replace the corrosive bromine, with the following reaction [53] (equation 12.5):

$$Zn + I_3^- \underset{\text{Charge}}{\overset{\text{Discharge}}{\rightleftarrows}} Zn^{2+} + 3I^- \tag{12.5}$$

The standard voltage of the battery is about 1.30 V, and it is possible to achieve a significantly high energy density around ~322 Wh/L in the device based on high solubility of zinc iodide (around 7.0 M). Indeed, an energy density at discharge was 166.7 Wh/L as per demonstration, similar to the low-end $LiFePO_4$-based batteries containing lithium ions [53].

The iron flow battery FeFB is yet another noteworthy metal-based RFB device. FeFBs use the three different oxidation states of Fe for energy conversion, comparable to VRBs, and the battery shows a theoretical standard voltage of 1.2 V [123,124].

12.3.2 Nonaqueous Solvent and Organometallic Redox-Active Materials

Nonaqueous RFBs were designed to achieve higher battery voltages to enlarge the relatively small operational window of water. In nonaqueous RFBs, organometallic active complexes are a strong range of redox species. For use in RFBs, a vast library of organometallic complexes have been studied and can be categorized by their metallic centers (i.e., V, Mn and Fe) or different complexes (i.e., bipyridine and acetylacetonate) [63,115]. In RFBs [125] and all-copper RFBs, ionic liquids containing metal ions were also synthesized and proposed to use as the redox-active materials. A system of ionic liquids comprising copper has been introduced. Nonaqueous hybrid equivalents have formed along with lithium at the anodes and organometallic complexes at the cathode sides, including the aqueous hybrid RFBs with metal anodes. A nonaqueous membrane-less ferrocene-based hybrid RFB was constructed with lithium metal anodes, for instance [123].

12.3.3 Aqueous Solvent and Organic Redox-Active Materials

Recently, RFBs with aqueous solvents and organic redox-active materials have drawn a considerable wide range of research interests. Here, based on the backbones of its active components, a brief overview of their innovations is provided.

12.3.3.1 Quinone-Based Flow Chemistry

Quinones are categorized as redox-active carbonyl based (C=O) aromatic compounds which exhibit ring conjugate configurations. Quinones are biologically active molecules that are used for photosynthesis and aerobic respiration/ATP reaction, which forms redox-active centers and promotes electron movement [126–128]. Quinones have high electrochemical reversibility and rapid reaction rates. For quinones, $2e^-$ transfer typically occurs in a single step triggered by the speedy proton-coupled electron transfer (PCET). Quinones also have the limited activation energy to reshape the aromatic ring via an electron transfer reaction in the outer sphere [127]. A single PCET step produces dihydroquinone in aqueous acidic solutions, when the proton concentration is higher than the quinone concentration. Moreover, if the quinone concentration is greater than that of the protons (e.g., neutral pH), two $1e^-$ transfer steps occur stepwise. Oxyanions are selectively appeared through a single step of double e^- transfer in alkaline solutions. In this case, where PCET does not happen, the oxyanions are highly stabilized by hydrogen bonding by the surrounded H_2O molecules that would result in quick $2e^-$ transmission at the same time. Sulfonic (RSO_3H) and hydroxyl (OH^-) functional groups are usually used onto quinone structures through molecular modifications to advance the solubility of the quinone derivatives, and this increases the overall energy density associated with the RFB systems as well.

12.3.3.2 TEMPO-Based Flow Chemistry

A most researched redox mediator and radical marker is the organic radical TEMPO. Due to the delocalization of electrons between N and O and the steric protection of radicals from four methyl groups, TEMPO is incredibly stable, unlike a standard radical species. In the potential range of 0.8–1.1 V vs SHE, TEMPO exhibits a reversible and fast single-electron redox reaction and forms an oxo-ammonium salt after the anodic reaction [128].

(2,2,6,6-Tetramethylpiperidin-1-yl)oxyl (TEMPO) is the most stable radical, and usually, it undergoes a reversible single-electron transfer process. TEMPO derivatives, for example 4hydroxy-TEMPO (TEMPOL), are also used in RFBs because of the poor solubility of pure TEMPO molecules in aqueous solutions. TEMPOL, when coupled with MV dichloride in an all-organic RFB device, results in a battery voltage of 1.25 V [60]. At concentration levels of 0.1 and 0.5 M, the RFB system shows stable cycling performance, although capacity loss is apparent in the latter case. Some of the TEMPO-based derivatives are shown in Figure 12.13. This system's key benefit is the significant low cost, with an approximate amount of slightly less than \$180/kWh [60].

12.3.4 Nonaqueous Solvent Organic Redox-Active Materials

The active species used in an RFB are mostly metallic redox pairs dissolved in aqueous electrolytes. Recently, there have been a substantially growing number of organic redox materials reported [130,131]. Most of the organic active species are water resistant, and hence, the use of an organic solvent is important for flow battery operations. The aprotic organic solvent is electrochemically more stable than protic solvents, including water, and also has a larger operational window. This can result in high energy density batteries, as redox pairs having higher voltages can be employed. In organic solvents, however, the ion conductivity is much poorer, thereby limiting the existing densities. A few other organic materials are water soluble, especially when the molecule involves polar substituents. This phenomenon is reduced somewhat by higher voltage. High densities of current also apply, yet the voltage is limited for the two redox couples [3].

12.3.4.1 Nitroxide Radical Compound-Based Flow Chemistry

As active materials in nonaqueous RFBs, nitroxide radical compounds like TEMPO and 2-phenyl-4,4,5,5-tetramethylimidazoline-1oxyl 3-oxide (PTIO) were selected. With greater

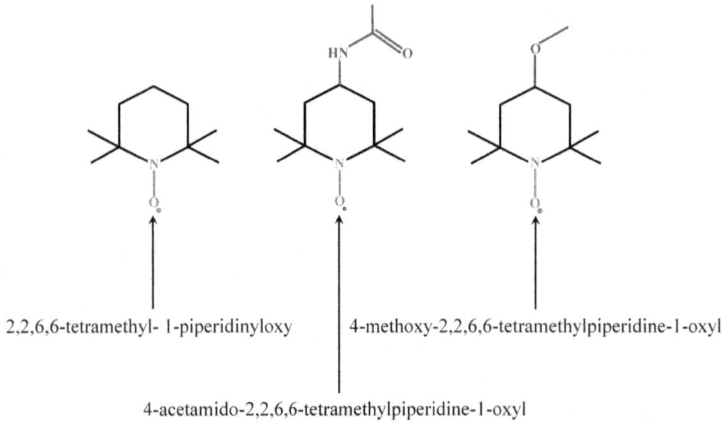

FIGURE 12.13 Various TEMPO derivatives for RFB applications. Adapted and reproduced from Ref. [129]. Copyright © 2020 Frontier Chemistry.

solubility in organic solvents as compared to aqueous solvents, mostly TEMPO is used as the catholyte substance in its pure form in nonaqueous flow chemistry. In one pioneer work on nonaqueous Organic RFBs, Liu et al. [131] employed TEMPO with N-methylphthalimide and obtained a cell voltage of 1.60V. For PTIO, electrochemically disproportionate reversible reactions can be conducted and therefore used on both the sides as active material to build a symmetrical RFB device having a cell voltage of 1.73V [132]. TEMPO was also paired with lithium metal in order to build a hybrid nonaqueous RFB system. The high theoretical volumetric energy density of 188 Wh/L was projected, and an energy density of 126 Wh/L was shown, with the higher solubility of TEMPO in carbonate solvents and a high cell voltage of 3.5V [133].

12.3.4.1.1 Quinone-Based Flow Chemistry

One of the examples of nonaqueous metal-organic hybrid RFBs is of quinone derivatives [134]. 1,5-bis(2-(2-(2-methoxyethoxy) ethoxy) ethoxy) anthracene-9,10-dione (15D3GAQ), a highly soluble derivative, was obtained by grating ether chains on the anthraquinone backbone. Two charge and two discharge plateaus were observed when coupled with lithium metals for the two-electron transfer process. A discharge power density of around 25Wh/L was supplied by the battery. Some of the well-known quinone derivatives are shown in Figure 12.14.

12.3.4.2 Dialkoxybenzene-Based Flow Chemistry

Dialkoxybenzene derivatives are widely used in nonaqueous RFBs as catholyte materials, and Brushett et al. [135,136] reported the first example by coupling 2,5-di-tert-butyl-1,4bis(2-methoxyethoxy)benzene (DBBB) with a 2,3,6-trimethylquinoxaline derivative of quinoxaline, 2,3,6-trimethylquinoxaline (TMQ). Since then, the tailoring structures of DBBB have made considerable efforts to increase the solubility and lower the molecular weight of the derivatives obtained.

12.3.4.3 Phenothiazine-Based Flow Chemistry

Phenothiazine derivatives are based on another noteworthy catholyte branch. The nonaqueous RFB can be formed when 3,7-bis(trifluoromethyl)-N-ethylphenothiazine (BCF3EPT) was coupled with TMQ molecules as anolytes. The solubility of these phenothiazine derivatives was also improved by phenothiazine alkylation with ether chains [137]. Despite the previously mentioned developments in the flow battery chemistry, nonaqueous RFBs, with many technological challenges, are still in their infancy. Because of the intrinsically low movement of ions in nonaqueous solvent, the utter most challenge for the nonaqueous RFB is its restricted power capacity. For now, the lack of suitable membranes is another major obstacle, and cross-contamination is a serious problem, leading to irreversible declines in ability. The use of active materials in polymer [138–140] or colloid forms [141] instead of small molecules is a promising way to conquer this challenge and their performance in a flow cell is shown in Figure 12.15.

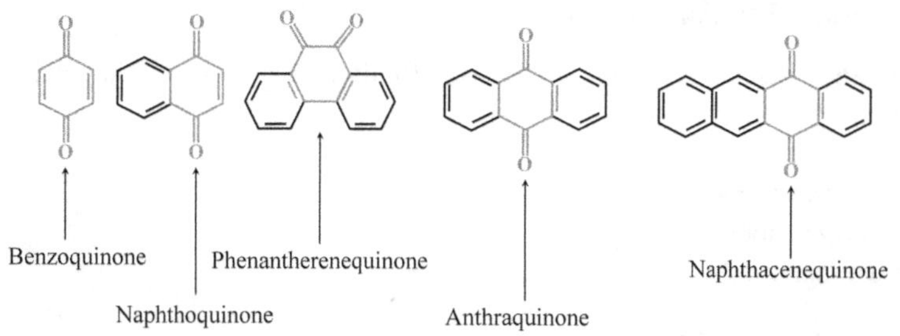

FIGURE 12.14 Various quinone derivatives for nonaqueous RFBs. Adapted and reproduced from Ref. [129]. Copyright © 2020 Frontier Chemistry.

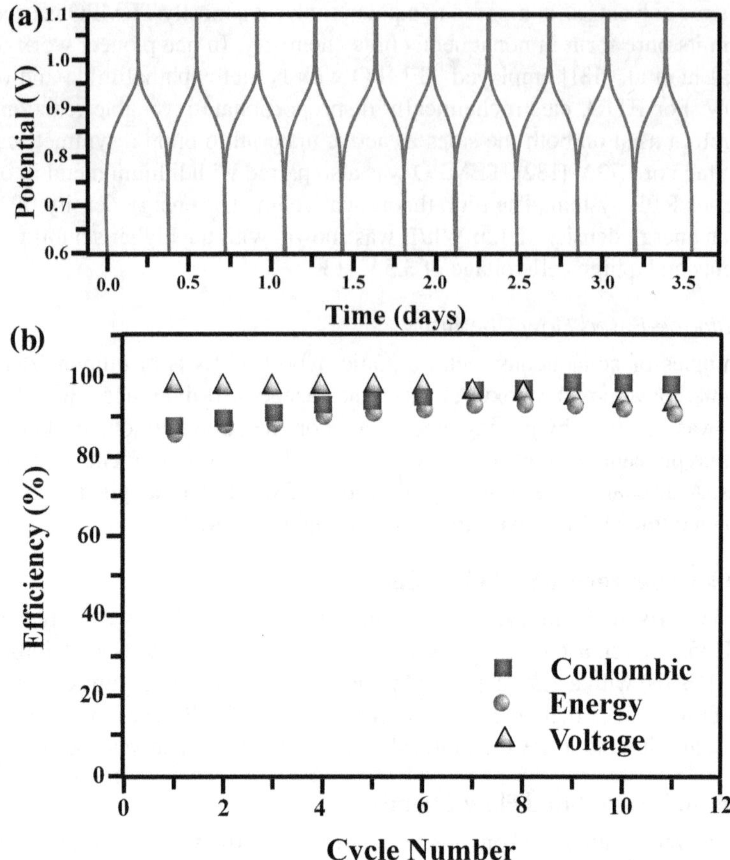

FIGURE 12.15 Performance of redox-active colloids in a flow cell. Adapted and reproduced from Ref. [141]. Copyright © 2016 American Chemical Society.

12.4 TESTING OF RFBs

Three key aspects include the assessment of RFBs: raw materials, stacks assembly and overall battery systems. Previously, the evaluation associated with raw materials, including electrode materials, membrane or separator materials, and electrolyte materials, is explained. The evaluation method for RFB stacks and systems is defined in this section.

12.4.1 Charging–Discharging Process

The charging and discharging technique is fundamentally used for RFB performance evaluation. The simplest ways to charge and discharge are

- Constant current mode
- Constant power mode
- Constant voltage mode

12.4.1.1 Constant Current Mode

During the charge and discharge procedures, the current is constant in the constant current mode. The power and voltage increase in the charging process, while voltage and power decrease during discharge process. The current and power move in the opposite directions during processes of charge and discharge as shown in Figure 12.16.

12.4.1.2 Constant Power Mode

The constant power mode refers to the constant power during the charge and discharge processes, and as the charge or discharge continues, the current and voltage change. The power is constant and represented as the voltage times the current. In the charging process, the voltage increases, but the current decreases. The voltage decreases in the discharge process as the current increases as shown in Figure 12.17.

12.4.1.3 Constant Voltage Mode

The mode of constant voltage is typically paired with the mode of constant current or constant power. The voltage is constant in the constant voltage mode, but during the charging or discharge processes, the current and power change. The stack or battery system is closer to being completely charged in the charging process, as the voltage is constant, and the difference between the charging voltage and the OCV is getting smaller and smaller; hence, as the process continues, the current and power decrease. The phenomenon during the discharge process is quite similar to the charge

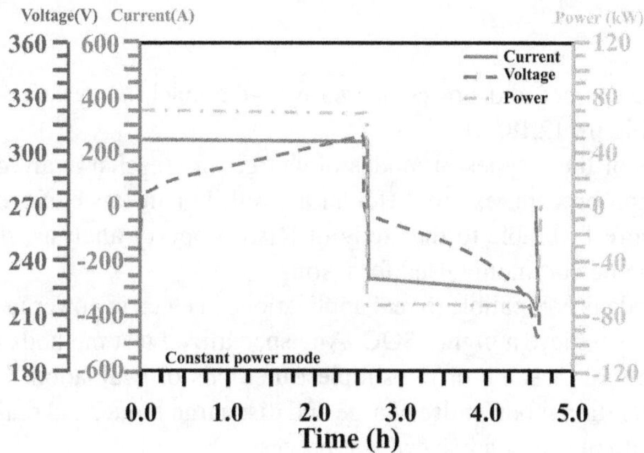

FIGURE 12.16 Constant current mode. Adapted and reproduced from Ref. [142]. Copyright © 2018 by Taylor & Francis Group.

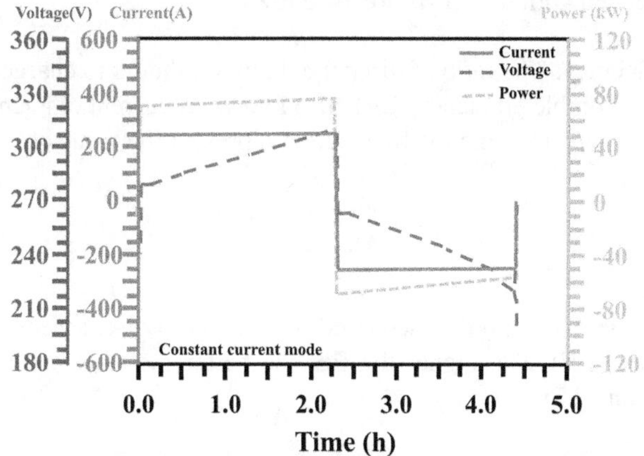

FIGURE 12.17 Constant power mode. Adapted and reproduced from Ref. [142]. Copyright © 2018 by Taylor & Francis Group.

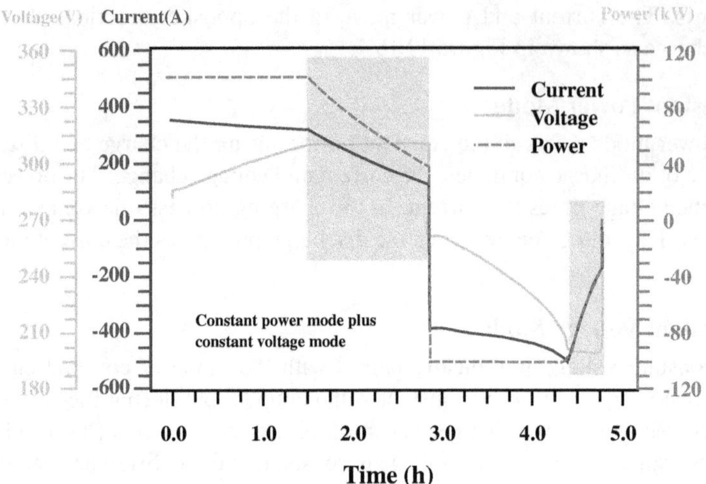

FIGURE 12.18 Constant voltage mode (marked in rectangle). Adapted and reproduced from Ref. [142]. Copyright © 2018 by Taylor & Francis Group.

process, and both the current and power decrease as the stack or system being completely discharged as shown in Figure 12.18.

The characteristics of these types of modes of charge and discharge are distinct, which means that the RFB evaluation procedures are different as well. For instance, since the constant current mode proves to be more favorable to the study of RFB property analysis, the properties of RFB stacks and systems can be commonly used for testing.

Constant power mode is compatible to real applications in electric power industry, while constant voltage mode is used to achieve a higher SOC. Any specialized test methods may be used for RFB evaluation by coupling all or some of the simplest methods of evaluation. Consecutive operation mode, intermittent operation mode, pulse charge and discharge mode, and real operation simulation mode are included in this form of advanced test process.

12.4.2 Efficiency

The efficiencies associated with the RFBs are as follows:

1. **Coulombic efficiency**: The ratio of discharged ampere-hours to charged ampere-hours is referred to as Coulombic efficiency (CE), also known as current efficiency. The crossover, the electrolyte osmotic rate and the by-reaction represent the value of CE (equation 12.6):

$$\eta_c = \frac{Ah_{discharge}}{Ah_{charge}} \times 100\% \tag{12.6}$$

2. **Energy efficiency:** The ratio of discharged watt-hours to discharged watt-hours refers to energy efficiency (EE). The higher the EE, the better the stack's electricity conversion capacity (equation 12.7):

$$\eta_c = \frac{Wh_{discharge}}{Wh_{charge}} \times 100\% \tag{12.7}$$

3. **Voltage efficiency:** The ratio of the average discharged voltage to the average charged voltage is voltage efficiency (VE). The VE will be significantly affected by the choice of electrode materials, the development process and the delivery of the solution. VE correlates, in other words, with the battery's polarization properties [65] (equation 12.8):

$$\eta_c = \frac{Vnh_{discharge}}{Vnh_{charge}} \times 100\% = \int \frac{V_{discharge}dt \big/ t_{discharge}}{V_{charge}dt \big/ t_{scharge}} \times 100\% \tag{12.8}$$

The following is the relationship between CE, EE and VE in a constant current mode (equation 12.9):

$$\eta_v = \int \frac{V_{discharge}dt \big/ t_{discharge}}{V_{charge}dt \big/ t_{charge}} \times 100\%$$

$$= \frac{I_{charge} t_{charge} \int V_{discharge}dt \big/ t_{discharge} \, dt}{I_{discharge} t_{discharge} \int V_{charge}dt \big/ t_{charge} \, dt} \times 100\% \tag{12.9}$$

$$= \frac{Ah_{charge}}{Ah_{discharge}} \times \frac{Wh_{discharge}}{Wh_{charge}} 100\% = \frac{\eta_E}{\eta_C} \times 100\%$$

As for RFB systems, EE is the primary indicator of the efficiency of the system. Its value represents the potential for energy conversion. The EE system can take into account more parameters, including pump consumption, cooling system consumption, ventilation system consumption and other auxiliary parameters, compared to the EE of the stacks.

For RFB systems, the calculation formula is as follows:

$$\eta = \frac{E_d - W_d}{E_d + W_d} \times 100\% \tag{12.10}$$

where η is the EE of the system (%), E_d is the number of watt-hours the system discharges on the DC side (kWh), W_d is the auxiliary consumption during discharge (kWh), E_c is the number of watt-hours the system charges on DC side (kWh) and W_c is the auxiliary consumption during charge (kWh). The EE can be called the rated EE for the system if the system charges and discharges with rated power.

In actual applications, users in practical usage modes are more concerned with EE on the AC side. Watt-hours should be registered on the AC side and the auxiliary usage concept should be greater than that on the DC side. Additionally, the operating mode is very different in actual applications from the rated conditions or test conditions.

For a typical Zn-Fe RFB, cell performance is shown in Figure 12.19. In Figure 12.19a, polarization curve is shown where the positive current density shows discharge and the negative current density displayed charge. Figure 12.19b exhibited the cell voltage vs SOC at different current densities 50, 100 and 150 mA cm^{-2}. Figure 12.19c presents cell voltage vs time for 12 charge–discharge cycles. In Figure12.19d, CE, VE and EE are shown with respect to the number of cycles.

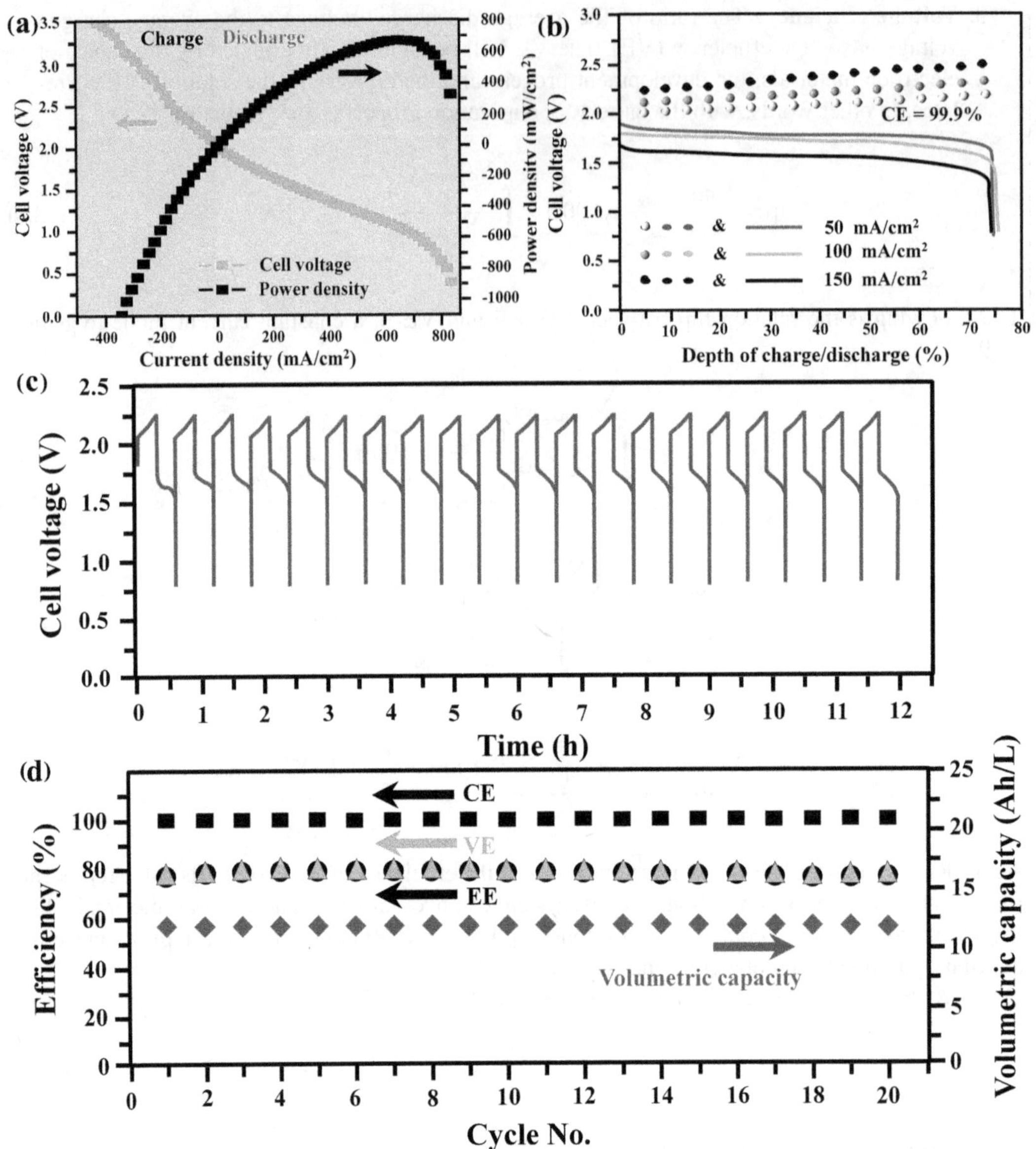

FIGURE 12.19 Cell performance of a Zn-Fe RFB: (a) charge and discharge polarization curve, (b) charge–discharge test at 50, 100 and 150 mA cm^{-2} current density, (c) voltage curve of a 20-cycle test at 80 mA cm^{-2} and (d) Coulombic efficiency (CE), voltage efficiency (VE), energy efficiency (EE) and volumetric capacity of the experimental Zn-Fe RFB. Adapted and reproduced from Ref. [143]. Copyright © 2017 Elsevier.

12.5 OTHER IMPORTANT FACTORS AFFECTING PERFORMANCE OF REDOX FLOW BATTERIES

12.5.1 Capacity and Electrolyte Utilization

A battery's capacity refers to its energy storage capability. Under such conditions, it communicates the sum of energy production. An RFB's capacity is characterized as the best capacity it can provide. It is determined by the electrolyte's volume and concentration. As RFBs are mainly used in the electricity industry, the capacity is typically expressed in kWh and represents the maximum energy that can be provided by the RFB. The capability and power are independent of each other entirely.

It is futile to alter the power of the device to increase the ability of the RFB system, which means increasing the area of the electrode or the number of the stack; rather, it is important to increase the volume or concentration of the electrolyte. There is currently no consensus on the best testing method for capacity. Another method in which the battery system is pre-charged to 100% SOC, then discharged to 30% SOC with rated power is given by the Chinese standard 'NB/T42040-2014'. Ultimately, once it exceeds the cutoff voltage, the device is discharged with 30% rated electricity. The capacity, approximately the full discharged capacity of 100% to 0% SOC, can therefore be developed.

Some other nations have various methods of testing by using a sufficiently low power to get the full discharge capacity on the DC side, while some use the constant power mode. Some use the constant power mode with the constant voltage mode to discharge until the capacity is established by the cutoff voltage. Others at various SOCs acquire the capacity with the constant power mode. Users are more interested with the capacity on the AC side in practical applications. The capacity on the AC side takes care of the efficiency of the AC/DC conversion and auxiliary usage, relative to the capacity on the DC side.

There is one more significant parameter: the electrolyte utilization that is defined as the ratio between the actual capacity and the theoretical capacity. The usage of the active substance in the electrolyte is reflected. The utilization of electrolytes can be measured using the following equations (12.11, 12.12):

$$\eta_{utilizatio} = \frac{Wh_{actual}}{Wh_{theoretical}} \times 100\% \tag{12.11}$$

where $\eta_{utilization}$ is the electrolyte utilization, Wh_{actual} is the actual capacity discharged from the battery and $Wh_{theoretical}$ is the theoretical capacity discharged from the battery:

$$Wh_{theroretical} = \frac{cVF}{3600} \times V \tag{12.12}$$

with c being the electrolyte concentration, V the electrolyte volume, F Faraday's constant and $\overline{v}$ the average voltage during discharge.

12.5.2 Uniformity of Stack and System

An RFB stack consists of a number of single cells. The same electrolyte pipeline is shared with all the single cells. The concentration polarization in the cell with a lower flow rate would be far more intense when the distribution of electrolytes to each cell is non-uniform. Compared to other single cells, in a single cell, it can trigger various working conditions and can be characterized by a higher or lower voltage. One of the most significant criteria for assessing the efficiency of both the stack and the device is uniformity. Uniformity can impact stack efficiency and life on the stack scale.

For instance, some cell voltages might be significantly high than others when a stack has lower uniformity charges, within the single cell, this phenomenon can damage electrode and BP material, and the service life will be shortened. The voltage range and the standard voltage deviation will also usually measure the homogeneity of the stack, making sure that no single cell with a differently high voltage is present. With system size, due to difference in installation locations, the stack voltage and the current from the stack will not be same. This will affect the distribution of voltage and current between the stacks. The efficiency of the system will also be affected by uniformity during charging or discharging.

12.5.3 Power Characteristics

Power characteristics are another significant feature of RFBs. These reflect the capabilities of the battery's input and output. Rated power is a key feature of power properties, since energy storage systems are defined with energy rating and a nominal power. Rated power is described as 'the maximal power which could achieve EE demands'. The Chinese standard emphasizes that when defining the rated power, the EE should achieve a fair level. Although there are some variations, on certain issues, various guidelines have already come to an agreement. First, for each device, the rated power has to be specified by the manufacturer. Some other power properties, in addition to rated power, also concern clients.

12.5.4 Self-Discharge

For a long period of time if a battery is not in operation, it will lose the power that it has stored. Assuming that the battery has been fully charged, the battery will not discharge all the power after 1 month, for example, it is called self-discharge. The self-discharge rate can be called the power loss during self-discharge process over a certain span of time. There are several explanations for self-discharge for traditional batteries, the main one is irreversible reactions, like those in the anode, cathode, electrolyte and separator, as well as manufacturing impurities within the battery's and miniature short circuits.

Self-discharge is a little more complex for RFBs. In RFBs, there are two important self-discharge working conditions: idle running mode and standby mode. Although there is no power input or output, when an RFB runs in idle mode, the pumps are still working and the auxiliary devices are still operating. Self-discharge can be interpreted in this mode as a review of the self-discharge of the stack and auxiliary usage, including refrigeration and ventilation. There is still no power input or output while operating in standby mode; auxiliary energy consumption is almost zero. Only inside the stacks can the self-discharge due to electrolyte crossover via the membrane happens. It will not affect the electrolytes inside the tanks. The self-discharge rate of the entire system would be nearly zero after the stack completely self-discharges. That is to say, after the stack completely self-discharges in standby mode, the capacity will not alter.

12.5.5 Capacity Reduction and Life Evaluation

For traditional batteries, under certain charging and discharge conditions, the life cycle can be described as the number of cycles until the battery capacity reduces to 60%–80%. In particular charge and discharge operations, the calendar life can be described as the time until the battery capacity reduces to 60%–80%. The RFB is a battery system which is very unusual. Cycle life and calendar life are hard to articulate. Some suppliers have also indicated that the life of the RFB is indefinite. The explanation why it is so hard to describe the life of the RFB is that the power system and the energy system are autonomous. The reduction in the capacity of conventional batteries is because of the changes in the physical structure of the material and reliability, so the lifetime of the capacity drop can be measured. The power and energy are already determined once a battery is manufactured. But for an RFB, the strength of the system is determined by the stack since the capacity is determined by the electrolyte's concentration and volume. This implies that it is possible to retrospectively adjust the power and capacity of the RFB. Even though the ability of the RFB will reduce during operation, the structure of the materials will be the same because they are not involved in the electrochemical reaction. The power can be taken back to its previous level without incorporating any electrolyte through the use of simple chemical methods.

To test RFB service life, it is not necessary to use capacity decay. Additionally, equivalent life testing methods can be employed to determine cycle life and calendar life for lithium-ion batteries and lead-acid batteries. But for RFBs, since the operating conditions of RFBs are yet to be determined as there aren't sufficient data to validate that the equivalent life evaluation procedure is acceptable to RFBs, it is difficult to find an effective method for these types of tests. Only through damage to key materials can an RFB stack meets its end. But the reliability of the materials is relatively high, based on recent test results. The life cycle and calendar life of RFBs may be very long, but at this point, it is difficult to reliably predict them.

12.5.6 Safety, Usability and Environmental Issues

Here are several other significant problems about which customers are often concerned. In the electrical power industry, safety is a primary concern. RFBs in actual implementations would not cause safety issues. RFBs have no combustion and explosion threats relative to lithium-ion batteries. The primary materials of RFBs are less toxic to the nature and to humans compared with lead-acid batteries. They can quickly recycle most of the materials. It is also necessary to note that in soil or water, RFB electrolytes should not be carelessly abandoned as this can cause pollution. Most consumers want the battery to be charged and discharged at the same time. Another parameter that clients consider before purchase is the charge rate to discharge rate ratio. The capabilities of lead-acid battery systems are always unnecessary to achieve the same charge and discharge rate. The charge and discharge rates are the same for RFB systems. To fulfill the power requirement, there is no need to use more capacity. For the ease of service and repair, caution should be taken as well. The operation and maintenance of RFBs is a bit complicated compared with other traditional batteries, as there are so many sensors and detectors in the device. Operation and maintenance would become more and simpler with the advancement of automatic techniques.

12.6 CONCLUSIONS

While significant progress has been made in recent years, in the production of new RFB chemicals, available technologies do not meet all of the large-scale energy storage requirements. With respect to energy density, power density, protection and cost, further improvements are required. At present, the aqueous system based on metal redox couples, especially the VRB, is still the most well-developed and advanced system for large-scale industrial assemblies. However, the ability to further lower the overall device cost has been severely limited by the cost of vanadium. In the hope of finding economic redox materials, which have shown very prominent progress, organic aqueous RFB chemistry was largely established. Organic systems, however, face their own challenge, namely the structural stability which is directly linked to the RFB system's long-term service. Nonaqueous flow battery technology, on the other hand, is at an early stage.

REFERENCES

1. J, assumptions Vipin Arora M, Singer LE (2016) *International Energy Outlook 2016 Liquid Fuels.* DOE/EIA-0484, Washington, DC 20585.
2. Thaller LH (1974) Electrically Rechargeable Redox Flow Cells. *Intersociety Energy Conversion Engineering Conference*, 9th, San Francisco, CA.
3. Winsberg J, Hagemann T, Janoschka T, et al (2017) Redox-flow batteries: From metals to organic redox-active materials. *Angew Chemie – Int Ed* 56:686–711. https://doi.org/10.1002/anie.201604925.
4. Wang W, Sprenkle V (2016) Energy storage: Redox flow batteries go organic. *Nat. Chem.* 8:204–206.
5. Zhao Y, Hong M, Bonnet Mercier N, et al (2014) A 3.5 V lithium-iodine hybrid redox battery with vertically aligned carbon nanotube current collector. *Nano Lett.* https://doi.org/10.1021/nl404784d.
6. Zhao Y, Ding Y, Li Y, et al (2015) A chemistry and material perspective on lithium redox flow batteries towards high-density electrical energy storage. *Chem Soc Rev.* 44:7968–7996.
7. Li W, Liang Z, Lu Z, et al (2015) Magnetic field-controlled lithium polysulfide semiliquid battery with ferrofluidic properties. *Nano Lett.* https://doi.org/10.1021/acs.nanolett.5b02818.

8. Chen X, Hopkins BJ, Helal A, et al (2016) A low-dissipation, pumpless, gravity-induced flow battery. *Energy Environ Sci*. https://doi.org/10.1039/c6ee00874g.
9. Braff WA, Bazant MZ, Buie CR (2013) Membrane-less hydrogen bromine flow battery. *Nat Commun*. https://doi.org/10.1038/ncomms3346.
10. Yu M, McCulloch WD, Huang Z, et al (2016) Solar-powered electrochemical energy storage: An alternative to solar fuels. *J Mater Chem A*. https://doi.org/10.1039/c5ta06950e.
11. Zhu YG, Jia C, Yang J, et al (2015) Dual redox catalysts for oxygen reduction and evolution reactions: Towards a redox flow Li-O_2 battery. *Chem Commun*. https://doi.org/10.1039/c5cc01616a.
12. Bockelmann M, Kunz U, Turek T (2016) Electrically rechargeable zinc-oxygen flow battery with high power density. *Electrochem Commun*. https://doi.org/10.1016/j.elecom.2016.05.013.
13. Wang W, Luo Q, Li B, et al (2013) Recent progress in redox flow battery research and development. *Adv Funct Mater*. https://doi.org/10.1002/adfm.201200694.
14. Vijayakumar M, Li L, Graff G, et al (2011) Towards understanding the poor thermal stability of V^{5+} electrolyte solution in vanadium redox flow batteries. *J Power Sources*. https://doi.org/10.1016/j.jpowsour.2010.11.126.
15. Weber AZ, Mench MM, Meyers JP, et al (2011) Redox flow batteries: A review. *J Appl Electrochem* 41, 1137–1164.
16. Skyllas-Kazacos M, Chakrabarti MH, Hajimolana SA, et al (2011) Progress in flow battery research and development. *J Electrochem Soc*. https://doi.org/10.1149/1.3599565.
17. Viswanathan V, Crawford A, Stephenson D, et al (2014) Cost and performance model for redox flow batteries. *J Power Sources*. https://doi.org/10.1016/j.jpowsour.2012.12.023.
18. Bon M, Laino T, Curioni A, Parrinello M (2016) Characterization of vanadium species in mixed chloride-sulfate solutions: An Ab initio metadynamics study. *J Phys Chem C*. https://doi.org/10.1021/acs.jpcc.6b02642.
19. Kim S, Vijayakumar M, Wang W, et al (2011) Chloride supporting electrolytes for all-vanadium redox flow batteries. *Phys Chem Chem Phys*. https://doi.org/10.1039/c1cp22638j.
20. Li L, Kim S, Wang W, et al (2011) A stable vanadium redox-flow battery with high energy density for large-scale energy storage. *Adv Energy Mater*. https://doi.org/10.1002/aenm.201100008.
21. Roe S, Menictas C, Skyllas-Kazacos M (2016) A high energy density vanadium redox flow battery with 3 M vanadium electrolyte. *J Electrochem Soc*. https://doi.org/10.1149/2.0041601jes.
22. Vijayakumar M, Wang W, Nie Z, et al (2013) Elucidating the higher stability of vanadium(V) cations in mixed acid based redox flow battery electrolytes. *J Power Sources*. https://doi.org/10.1016/j.jpowsour.2013.04.072.
23. Li S, Huang K, Liu S, et al (2011) Effect of organic additives on positive electrolyte for vanadium redox battery. *Electrochim Acta*. https://doi.org/10.1016/j.electacta.2011.03.048.
24. Zhang J, Li L, Nie Z, et al (2011) Effects of additives on the stability of electrolytes for all-vanadium redox flow batteries. *J Appl Electrochem*. https://doi.org/10.1007/s10800-011-0312-1.
25. Wu X, Liu S, Wang N, et al (2012) Influence of organic additives on electrochemical properties of the positive electrolyte for all-vanadium redox flow battery. *Electrochim Acta*. https://doi.org/10.1016/j.electacta.2012.06.065.
26. Wang G, Chen J, Xu Y, et al (2014) Several ionic organic compounds as positive electrolyte additives for a vanadium redox flow battery. *RSC Adv*. https://doi.org/10.1039/c4ra09108f.
27. Lee JG, Park SJ, Cho Y Il, Shul YG (2013) A novel cathodic electrolyte based on H_2C_2O 4 for a stable vanadium redox flow battery with high charge-discharge capacities. *RSC Adv*. https://doi.org/10.1039/c3ra44234a.
28. Wang G, Zhang J, Zhang J, et al (2016) Effect of different additives with -NH_2 or -NH_4^+ functional groups on V(V) electrolytes for a vanadium redox flow battery. *J Electroanal Chem*. https://doi.org/10.1016/j.jelechem.2016.02.029.
29. Wang G, Chen J, Wang X, et al (2014) Study on stabilities and electrochemical behavior of V(V) electrolyte with acid additives for vanadium redox flow battery. *J Energy Chem*. https://doi.org/10.1016/S2095-4956(14)60120-0.
30. Zeng YK, Zhao TS, An L, et al (2015) A comparative study of all-vanadium and iron-chromium redox flow batteries for large-scale energy storage. *J Power Sources*. https://doi.org/10.1016/j.jpowsour.2015.09.100.
31. Wu CD, Scherson DA, Calvo EJ, et al (1986) A bismuth-based electrocatalyst for the chromous-chromic couple in acid electrolytes. *J Electrochem Soc*. https://doi.org/10.1149/1.2108351.
32. Zeng YK, Zhou XL, An L, et al (2016) A high-performance flow-field structured iron-chromium redox flow battery. *J Power Sources*. https://doi.org/10.1016/j.jpowsour.2016.05.138.

33. Wang W, Kim S, Chen B, et al (2011) A new redox flow battery using Fe/V redox couples in chloride supporting electrolyte. *Energy Environ Sci* 4:4068–4073. https://doi.org/10.1039/c0ee00765j.
34. Li B, Li L, Wang W, et al (2013) Fe/V redox flow battery electrolyte investigation and optimization. *J Power Sources* 229:1–5. https://doi.org/10.1016/j.jpowsour.2012.11.119.
35. Skyllas-Kazacos M (2003) Novel vanadium chloride/polyhalide redox flow battery. *J Power Sources* 124:299–302. https://doi.org/10.1016/S0378-7753(03)00621-9.
36. Vafiadis H, Skyllas-Kazacos M (2006) Evaluation of membranes for the novel vanadium bromine redox flow cell. *J Memb Sci.* https://doi.org/10.1016/j.memsci.2005.12.028.
37. Xia X, Liu H-T, Liu Y (2002) Studies of the feasibility of a Ce^{4+}/Ce^{3+}-V^{2+}/V^{3+} *redox cell. J Electrochem Soc.* https://doi.org/10.1149/1.1456534.
38. Liu Y, Xia X, Liu H (2004) Studies on cerium (Ce^{4+}/Ce^{3+})-vanadium(V^{2+}/V^{3+}) redox flow cell - Cyclic voltammogram response of Ce^{4+}/Ce^{3+} redox couple in H_2SO_4 solution. *J Power Sources* 130:299–305. https://doi.org/10.1016/j.jpowsour.2003.12.017.
39. Paulenova A, Creager SE, Navratil JD, Wei Y (2002) Redox potentials and kinetics of the Ce^{3+}/Ce^{4+} redox reaction and solubility of cerium sulfates in sulfuric acid solutions. *J Power Sources* 109:431–438. https://doi.org/10.1016/S0378-7753(02)00109-X.
40. Govindan M, He K, Moon IS (2013) Evaluation of dual electrochemical cell design for cerium-vanadium redox flow battery to use different combination of electrodes. *Int J Electrochem Sci.* 8:10265–10279.
41. Govindan M, Moon IS (2013) Improved electrochemical preparation of nano porous-graphene oxide edge like electrode for cerium/vanadium redox flow batteries. *Int J Electrochem Sci.* 8:12172–12183.
42. Leung PK, Mohamed MR, Shah AA, et al (2015) A mixed acid based vanadium-cerium redox flow battery with a zero-gap serpentine architecture. *J Power Sources.* https://doi.org/10.1016/j.jpowsour.2014.10.034.
43. Leung PK, Ponce De Leon C, Walsh FC (2012) The influence of operational parameters on the performance of an undivided zinc-cerium flow battery. *Electrochim Acta.* https://doi.org/10.1016/j.electacta.2012.06.074.
44. Xie Z, Xiong F, Zhou D (2011) Study of the Ce^{3+}/Ce^{4+} redox couple in mixed-acid media (CH_3SO_3H and H_2SO_4) for redox flow battery application. *Energy Fuels.* 25:2399–2404. https://doi.org/10.1021/ef200354b.
45. Peng S, Wang NF, Wu XJ, et al (2012) Vanadium species in CH_3SO_3H and H_2SO_4 mixed acid as the supporting electrolyte for vanadium redox flow battery. *Int J Electrochem Sci.* 7:643–649.
46. Walsh FC, Poncedeléon C, Berlouis L, et al (2015) The development of Zn-Ce hybrid redox flow batteries for energy storage and their continuing challenges. *Chempluschem.* 80:288–311. https://doi.org/10.1002/cplu.201402103.
47. Ludek J, Wei Y, Kumagai M (2006) Electro-oxidation of concentrated Ce(III) at carbon felt anode in nitric acid media. *J Rare Earths.* https://doi.org/10.1016/S1002-0721(06)60105-1.
48. Gernon MD, Wu M, Buszta T, Janney P (1999) Environmental benefits of methanesulfonic acid: Comparative properties and advantages. *Green Chem.* https://doi.org/10.1039/a900157c.
49. Nikiforidis G, Berlouis L, Hall D, Hodgson D (2013) Impact of electrolyte composition on the performance of the zinc-cerium redox flow battery system. *J Power Sources.* https://doi.org/10.1016/j.jpowsour.2013.06.045.
50. Nikiforidis G, Daoud WA (2014) Effect of mixed acid media on the positive side of the hybrid zinc-cerium redox flow battery. *Electrochim Acta.* https://doi.org/10.1016/j.electacta.2014.06.142.
51. Xie Z, Liu Q, Chang Z, Zhang X (2013) The developments and challenges of cerium half-cell in zinc-cerium redox flow battery for energy storage. *Electrochim Acta.* 90:695–704.
52. Leung PK, Ponce-De-León C, Low CTJ, et al (2011) Characterization of a zinc-cerium flow battery. *J Power Sources.* https://doi.org/10.1016/j.jpowsour.2011.01.095.
53. Rajarathnam GP, Schneider M, Sun X, Vassallo AM (2016) The influence of supporting electrolytes on zinc half-cell performance in zinc/bromine flow batteries. *J Electrochem Soc* 163:A5112–A5117. https://doi.org/10.1149/2.0151601jes.
54. Li B, Nie Z, Vijayakumar M, et al (2015) Ambipolar zinc-polyiodide electrolyte for a high-energy density aqueous redox flow battery. *Nat Commun.* https://doi.org/10.1038/ncomms7303.
55. Jeon JD, Yang HS, Shim J, et al (2014) Dual function of quaternary ammonium in Zn/Br redox flow battery: Capturing the bromine and lowering the charge transfer resistance. *Electrochim Acta* 127:397–402. https://doi.org/10.1016/j.electacta.2014.02.073.
56. Farkas L, Perlmutter B, Schächter O (1949) The reaction between ethyl alcohol and bromine. *J Am Chem Soc.* https://doi.org/10.1021/ja01176a069.

57. Shin SH, Yun SH, Moon SH (2013) A review of current developments in non-aqueous redox flow batteries: Characterization of their membranes for design perspective. *RSC Adv.* 3:9095–9116.
58. Lin K, Chen Q, Gerhardt MR, et al (2015) Alkaline quinone flow battery. *Science* 80. https://doi.org/10.1126/science.aab3033.
59. Chen Q, Eisenach L, Aziz MJ (2016) Cycling Analysis of a quinone-bromide redox flow battery. *J Electrochem Soc.* https://doi.org/10.1149/2.0081601jes.
60. Liu T, Wei X, Nie Z, et al (2016) A total organic aqueous redox flow battery employing a low cost and sustainable methyl viologen anolyte and 4-HO-TEMPO catholyte. *Adv Energy Mater.* https://doi.org/10.1002/aenm.201501449.
61. Singh V, Kim S, Kang J, Byon HR (2019) Aqueous organic redox flow batteries. *Nano Res* 12:1988–2001. https://doi.org/10.1007/s12274-019-2355-2.
62. Huskinson B, Nawar S, Gerhardt MR, Aziz MJ (2013) Novel quinone-based couples for flow batteries. *ECS Trans* 53:101–105. https://doi.org/10.1149/05307.0101ecst.
63. Janoschka T, Morgenstern S, Hiller H, et al (2015) Synthesis and characterization of TEMPO- and viologen-polymers for water-based redox-flow batteries. *Polym Chem.* https://doi.org/10.1039/c5py01602a.
64. Gong K, Fang Q, Gu S, et al (2015) Nonaqueous redox-flow batteries: Organic solvents, supporting electrolytes, and redox pairs. *Energy Environ. Sci.* 8:3515–3530.
65. Tomazic G, Skyllas-Kazacos M (2015) *Redox Flow Batteries.* Elsevier, Amsterdam.
66. Singh P (1984) Application of non-aqueous solvents to batteries. *J Power Sources.* https://doi.org/10.1016/0378-7753(84)80079-8.
67. Ponce de León C, Frías-Ferrer A, González-García J, et al (2006) Redox flow cells for energy conversion. *J Power Sources.* 160: 716–732.
68. Wang W, Luo Q, Li B, et al (2013) Recent progress in redox flow battery research and development. *Adv Funct Mater* 23:970–986. https://doi.org/10.1002/adfm.201200694.
69. Tokuda N, Kanno T, Hara T, et al (2000) Development of a redox flow battery system. *SEI Tech Rev.* 50(50):88–94.
70. Sun B, Skyllas-Kazacos M (1992) Chemical modification of graphite electrode materials for vanadium redox flow battery application-Part II. Acid treatments. *Electrochim Acta.* https://doi.org/10.1016/0013-4686(92)87084-D.
71. González Z, Flox C, Blanco C, et al (2017) Outstanding electrochemical performance of a graphene-modified graphite felt for vanadium redox flow battery application. *J Power Sources* 338:155–162. https://doi.org/10.1016/j.jpowsour.2016.10.069.
72. Sun B, Skyllas-Kazacos M (1992) Modification of graphite electrode materials for vanadium redox flow battery application-I. Thermal treatment. *Electrochim Acta.* https://doi.org/10.1016/0013-4686(92)85064-R.
73. Pittman CU, Jiang W, Yue ZR, et al (1999) Surface properties of electrochemically oxidized carbon fibers. *Carbon N Y.* https://doi.org/10.1016/S0008-6223(99)00048-2.
74. Sun B, Skyllas-Kazakos M (1991) Chemical modification and electrochemical behaviour of graphite fibre in acidic vanadium solution. *Electrochim Acta.* https://doi.org/10.1016/0013-4686(91)85135-T.
75. Estevez L, Reed D, Nie Z, et al (2016) Tunable oxygen functional groups as electrocatalysts on graphite felt surfaces for all-vanadium flow batteries. *ChemSusChem.* https://doi.org/10.1002/cssc.201600198.
76. Park S, Kim H (2015) Fabrication of nitrogen-doped graphite felts as positive electrodes using polypyrrole as a coating agent in vanadium redox flow batteries. *J Mater Chem A.* https://doi.org/10.1039/c5ta02674a.
77. Huang Y, Huo J, Dou S, et al (2016) Graphitic C_3N_4 as a powerful catalyst for all-vanadium redox flow batteries. *RSC Adv.* https://doi.org/10.1039/c6ra11381h.
78. He Z, Chen Z, Meng W, et al (2016) Modified carbon cloth as positive electrode with high electrochemical performance for vanadium redox flow batteries. *J Energy Chem.* https://doi.org/10.1016/j.jechem.2016.04.002.
79. Li B, Gu M, Nie Z, et al (2013) Bismuth nanoparticle decorating graphite felt as a high-performance electrode for an all-vanadium redox flow battery. *Nano Lett.* https://doi.org/10.1021/nl400223v.
80. He Z, Liu L, Gao C, et al (2013) Carbon nanofibers grown on the surface of graphite felt by chemical vapour deposition for vanadium redox flow batteries. *RSC Adv.* https://doi.org/10.1039/c3ra22631j.
81. Pour N, Kwabi DG, Carney T, et al (2015) Influence of edge-and basal-plane sites on the vanadium redox kinetics for flow batteries. *J Phys Chem C.* https://doi.org/10.1021/jp5116806.
82. Wang WH, Wang XD (2007) Investigation of Ir-modified carbon felt as the positive electrode of an all-vanadium redox flow battery. *Electrochim Acta.* https://doi.org/10.1016/j.electacta.2007.04.121.

83. Li B, Gu M, Nie Z, et al (2014) Nanorod niobium oxide as powerful catalysts for an all vanadium redox flow battery. *Nano Lett.* https://doi.org/10.1021/nl403674a.
84. Xie Z, Yang B, Yang L, et al (2015) Addition of graphene oxide into graphite toward effective positive electrode for advanced zinc-cerium redox flow battery. *J Solid State Electrochem.* https://doi.org/10.1007/s10008-015-2958-9.
85. Nikiforidis G, Xiang Y, Daoud WA (2015) Electrochemical behavior of carbon paper on cerium methanesulfonate electrolytes for zinc-cerium flow battery. *Electrochim Acta.* https://doi.org/10.1016/j.electacta.2014.11.134.
86. Nikiforidis G, Berlouis L, Hall D, Hodgson D (2014) An electrochemical study on the positive electrode side of the zinc-cerium hybrid redox flow battery. *Electrochim Acta.* https://doi.org/10.1016/j.electacta.2013.09.081.
87. Us CA, Harrison S (2003) Battery with bifunctional electrolyte. US6986966B2 2.
88. Cheng DS, Hollax E (1985) The influence of thallium on the redox reaction Cr^{3+}/Cr^{2+}. *J Electrochem Soc.* 132:269–273. https://doi.org/10.1149/1.2113807.
89. Li B, Liu J, Nie Z, et al (2016) Metal-organic frameworks as highly active electrocatalysts for high-energy density, aqueous zinc-polyiodide redox flow batteries. *Nano Lett.* https://doi.org/10.1021/acs.nanolett.6b01426.
90. Park MS, Lee NJ, Lee SW, et al (2014) High-energy redox-flow batteries with hybrid metal foam electrodes. *ACS Appl Mater Interfaces.* https://doi.org/10.1021/am5025935.
91. Schwenzer B, Zhang J, Kim S, et al (2011) Membrane development for vanadium redox flow batteries. *ChemSusChem.* 4:1388–1406.
92. Kamath H, Rajagopalan S, Zwillenberg M (2007) *Vanadium Redox Flow Batteries: An In-Depth Analysis*, 1–102.
93. Jiang B, Wu L, Yu L, et al (2016) A comparative study of nafion series membranes for vanadium redox flow batteries. *J Memb Sci.* https://doi.org/10.1016/j.memsci.2016.03.007.
94. Teng X, Dai J, Su J, Yin G (2015) Modification of Nafion membrane using fluorocarbon surfactant for all vanadium redox flow battery. *J Memb Sci.* https://doi.org/10.1016/j.memsci.2014.11.014.
95. Vijayakumar M, Luo Q, Lloyd R, et al (2016) Tuning the perfluorosulfonic acid membrane morphology for vanadium redox-flow batteries. *ACS Appl Mater Interfaces.* https://doi.org/10.1021/acsami.6b10744.
96. Xi J, Jiang B, Yu L, Liu L (2017) Membrane evaluation for vanadium flow batteries in a temperature range of –20-50°C. *J Memb Sci.* https://doi.org/10.1016/j.memsci.2016.09.012.
97. Kazacos MK and Kazacos MS (2006) High energy density vanadium electrolyte solutions, methods of preparation thereof and all-vanadium redox cells and batteries containing high energy vanadium electrolyte solutions. US Pat 7,078,123 2006.
98. Sun C, Chen J, Zhang H, et al (2010) Investigations on transfer of water and vanadium ions across Nafion membrane in an operating vanadium redox flow battery. *J Power Sources.* https://doi.org/10.1016/j.jpowsour.2009.08.041.
99. Chieng SC (1993) Membrane processes and membrane modification for redox flow battery applications. Univ New South Wales Thesis.
100. Mohammadi T, Chieng SC, Kazacos MS (1997) Water transport study across commercial ion exchange membranes in the vanadium redox flow battery. *J Memb Sci.* https://doi.org/10.1016/S0376-7388(97)00092-6.
101. Pagidi A, Seepana MM (2020) Synthesis of (Si-PWA)-PVA/PTFE high-temperature proton-conducting composite membrane for DMFC. *Int J Hydrogen Energy.* 45:25851–25861. https://doi.org/10.1016/j.ijhydene.2020.02.113.
102. Tang A, Bao J, Skyllas-Kazacos M (2012) Thermal modelling of battery configuration and self-discharge reactions in vanadium redox flow battery. *J Power Sources.* https://doi.org/10.1016/j.jpowsour.2012.06.052.
103. Darling RM, Weber AZ, Tucker MC, Perry ML (2016) The influence of electric field on crossover in redox-flow batteries. *J Electrochem Soc.* https://doi.org/10.1149/2.0031601jes.
104. Chen D, Kim S, Li L, et al (2012) Stable fluorinated sulfonated poly(arylene ether) membranes for vanadium redox flow batteries. *RSC Adv.* https://doi.org/10.1039/c2ra20834b.
105. Varcoe JR, Atanassov P, Dekel DR, et al (2014) Anion-exchange membranes in electrochemical energy systems. *Energy Environ Sci.* 7:3135–3191.
106. Arenas LF, Ponce de León C, Walsh FC (2017) Engineering aspects of the design, construction and performance of modular redox flow batteries for energy storage. *J Energy Storage.* 11:119–153. https://doi.org/10.1016/j.est.2017.02.007.

107. Jung H-Y, Jeong S, Kwon Y (2016) The effects of different thick sulfonated poly (ether ether ketone) membranes on performance of vanadium redox flow battery. *J Electrochem Soc.* https://doi.org/10.1149/2.0121601jes..
108. Sukkar T, Skyllas-Kazacos M (2004) Membrane stability studies for vanadium redox cell applications. *J Appl Electrochem.* https://doi.org/10.1023/B:JACH.0000009931.83368.dc.
109. Kim S, Tighe TB, Schwenzer B, et al (2011) Chemical and mechanical degradation of sulfonated poly(sulfone) membranes in vanadium redox flow batteries. *J Appl Electrochem.* https://doi.org/10.1007/s10800-011-0313-0.
110. Jansson REW, Marshall RJ (1982) Axial dispersion in parallel channel electrochemical cells. *Electrochim Acta.* https://doi.org/10.1016/0013-4686(82)80203-X.
111. White RE, Walton CW, Burney HS, Beaver RN (1986) Predicting shunt currents in stacks of bipolar plate cells. *J Electrochem Soc.* https://doi.org/10.1149/1.2108606.
112. Park DJ, Jeon KS, Ryu CH, Hwang GJ (2017) Performance of the all-vanadium redox flow battery stack. *J Ind Eng Chem.* https://doi.org/10.1016/j.jiec.2016.10.007.
113. Reed D, Thomsen E, Li B, et al (2016) Performance of a low cost interdigitated flow design on a 1 kW class all vanadium mixed acid redox flow battery. *J Power Sources.* https://doi.org/10.1016/j.jpowsour.2015.11.089.
114. Reed D, Thomsen E, Wang W, et al (2015) Performance of Nafion(r) N115, Nafion(r) NR-212, and Nafion(r) NR-211 in a 1 kW class all vanadium mixed acid redox flow battery. *J Power Sources.* https://doi.org/10.1016/j.jpowsour.2015.03.099.
115. Noack J, Roznyatovskaya N, Herr T, Fischer P (2015) The chemistry of redox-flow batteries. *Angew Chemie – Int Ed.* https://doi.org/10.1002/anie.201410823.
116. Pan F, Wang Q (2015) Redox species of redox flow batteries: A review. *Molecules.* 20(11):20499–20517.
117. Li B, Luo Q, Wei X, et al (2014) Capacity decay mechanism of microporous separator-based all-vanadium redox flow batteries and its recovery. *ChemSusChem.* https://doi.org/10.1002/cssc.201300706.
118. Kim S, Yan J, Schwenzer B, et al (2010) Cycling performance and efficiency of sulfonated poly(sulfone) membranes in vanadium redox flow batteries. *Electrochem Commun.* https://doi.org/10.1016/j.elecom.2010.09.018.
119. Kim S, Thomsen E, Xia G, et al (2013) 1 kW/1 kWh advanced vanadium redox flow battery utilizing mixed acid electrolytes. *J Power Sources* 237:300–309. https://doi.org/10.1016/j.jpowsour.2013.02.045.
120. Luo Q, Li L, Wang W, et al (2013) Capacity decay and remediation of nafion-based all-vanadium redox flow batteries. *ChemSusChem.* https://doi.org/10.1002/cssc.201200730.
121. Fang B, Iwasa S, Wei Y, et al (2002) A study of the Ce(III)/Ce(IV) redox couple for redox flow battery application. *Electrochim Acta* 47:3971–3976. https://doi.org/10.1016/S0013-4686(02)00370-5.
122. Park M, Ryu J, Wang W, Cho J (2016) Material design and engineering of next-generation flow-battery technologies. *Nat. Rev. Mater.* 2:16080.
123. Tucker MC, Phillips A, Weber AZ (2015) All-iron redox flow battery tailored for off-grid portable applications. *ChemSusChem.* https://doi.org/10.1002/cssc.201500845.
124. Manohar AK, Kim KM, Plichta E, et al (2016) A high efficiency iron-chloride redox flow battery for large-scale energy storage. *J Electrochem Soc.* https://doi.org/10.1149/2.0161601jes.
125. Ravikumar MK, Rathod S, Jaiswal N, et al (2017) The renaissance in redox flow batteries. *J. Solid State Electrochem.* 21:2467–2488.
126. Pratt HD, Rose AJ, Staiger CL, et al (2011) Synthesis and characterization of ionic liquids containing copper, manganese, or zinc coordination cations. *Dalt Trans.* https://doi.org/10.1039/c1dt10973a.
127. Ding Y, Zhao Y, Yu G (2015) A membrane-free ferrocene-based high-rate semiliquid battery. *Nano Lett.* https://doi.org/10.1021/acs.nanolett.5b01224.
128. Kurreck H, Huber M (1995) Model reactions for photosynthesis-photoinduced charge and energy transfer between covalently linked porphyrin and quinone units. *Angew. Chemie Int. Ed. English* 34:849–866.
129. Zhong F, Yang M, Ding M, Jia C (2020) Organic electroactive molecule-based electrolytes for redox flow batteries: Status and challenges of molecular design. *Front Chem* 8:1–14. https://doi.org/10.3389/fchem.2020.00451.
130. Quan M, Sanchez D, Wasylkiw MF, Smith DK (2007) Voltammetry of quinones in unbuffered aqueous solution: Reassessing the roles of proton transfer and hydrogen bonding in the aqueous electrochemistry of quinones. *J Am Chem Soc.* https://doi.org/10.1021/ja0743083.
131. Janoschka T, Martin N, Hager MD, Schubert US (2016) Aqueous redox-flow battery with high capacity and power : The TEMPTMA/MV system. *Angew Chem* 55:14427–14430. https://doi.org/10.1002/anie.201606472.

132. Li Z, Li S, Liu S, et al (2011) Electrochemical properties of an all-organic redox flow battery using 2,2,6,6-tetramethyl-1-piperidinyloxy and N-Methylphthalimide. *Electrochem Solid-State Lett.* 14:1–4. https://doi.org/10.1149/2.012112esl.
133. Wei X, Xu W, Vijayakumar M, et al (2014) TEMPO-based catholyte for high-energy density nonaqueous redox flow batteries. *Adv Mater.* https://doi.org/10.1002/adma.201403746.
134. Wang W, Xu W, Cosimbescu L, et al (2012) Anthraquinone with tailored structure for a nonaqueous metal-organic redox flow batter. *Chem Commun.* https://doi.org/10.1039/c2cc32466k.
135. Huang J, Cheng L, Assary RS, et al (2015) Liquid catholyte molecules for nonaqueous redox flow batteries. *Adv Energy Mater.* https://doi.org/10.1002/aenm.201401782
136. Huang J, Su L, Kowalski JA, et al (2015) A subtractive approach to molecular engineering of dimethoxybenzene-based redox materials for non-aqueous flow batteries. *J Mater Chem A.* https://doi.org/10.1039/c5ta02380g.
137. Duan W, Vemuri RS, Milshtein JD, et al (2016) A symmetric organic-based nonaqueous redox flow battery and its state of charge diagnostics by FTIR. *J Mater Chem A.* https://doi.org/10.1039/c6ta01177b.
138. Nagarjuna G, Hui J, Cheng KJ, et al (2014) Impact of redox-active polymer molecular weight on the electrochemical properties and transport across porous separators in nonaqueous solvents. *J Am Chem Soc.* https://doi.org/10.1021/ja508482e.
139. Burgess M, Chénard E, Hernández-Burgos K, et al (2016) Impact of backbone tether length and structure on the electrochemical performance of viologen redox active polymers. *Chem Mater.* https://doi.org/10.1021/acs.chemmater.6b02825.
140. Burgess M, Moore JS, Rodríguez-López J (2016) Redox Active polymers as soluble nanomaterials for energy storage. *Acc Chem Res.* https://doi.org/10.1021/acs.accounts.6b00341.
141. Montoto EC, Nagarjuna G, Hui J, et al (2016) Redox active colloids as discrete energy storage carriers. *J Am Chem Soc.* https://doi.org/10.1021/jacs.6b06365.
142. Kamat PV, Schanze KS, Buriak JM (2017) *Redox Flow Batteries.* ACS Pub. Washington.
143. Gong K, Ma X, Conforti KM, et al (2015) A zinc-iron redox-flow battery under $100 per kW h of system capital cost. *Energy Environ Sci.* 8:2941–2945. https://doi.org/10.1039/c5ee02315g.

13 Vanadium Redox Flow Batteries
Electrode Materials

Nawin Ra, Hiranmay Saha, and Ankur Bhattacharjee

13.1 INTRODUCTION

VRFB utilizes highly acidic electrolyte. So, the corrosion of metals in VRFB cells is bound to occur owing to the flow of the acidic electrolyte. In the existing literature, various materials including lead, platinum and gold were reported as the possible electrode material. Among the several available materials for VRFB electrode, materials made of carbon showed better performance [1]. Researchers had the quest to find the electrode material that was suitable for utilization as both positive and negative electrodes. Graphite felts were one among the various materials chosen for both the positive and negative electrodes. The materials such as graphite felts and carbon felts were able to achieve mechanical and chemical stability [2]. Figure 13.1 represents the schematic diagram of VRFB. VRFB consists of components such as electrodes, membrane, bipolar plate, flow frame, cell frame, pump and electrolyte tanks.

Among these components, electrodes play a significant role in maximizing the system efficiency. Figure 13.2 illustrates the three different forms of commercialized carbon electrodes, namely carbon cloth electrode, carbon paper electrode and carbon felt electrode.

In this chapter, electrode materials utilized for negative electrode component, positive electrode component and both positive–negative electrode components were discussed in Sections 13.2–13.4.

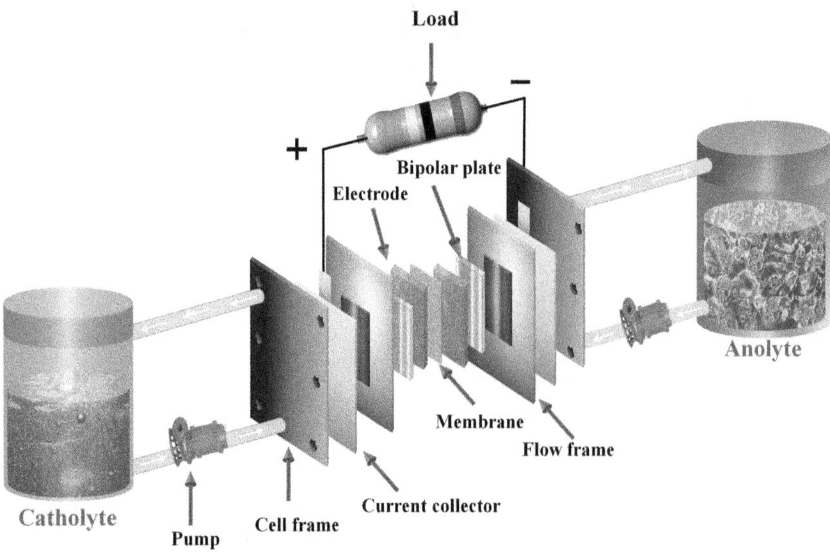

FIGURE 13.1 Schematic representation of vanadium redox flow battery. Adapted and reproduced with permission from Ref. [3]. Copyright © 2013 The Royal Society of Chemistry.

DOI: 10.1201/9781003310167-13

FIGURE 13.2 SEM images of (a) carbon cloth electrode, (b) carbon paper electrode and (c) carbon felt electrode. Reproduced with permission from [3]. Copyright 2013, The Royal Society of Chemistry.

Potential development in VRFB electrodes is discussed in Section 13.5. The conclusion of the chapter is given in Section 13.6.

13.2 NEGATIVE ELECTRODE MATERIALS OF VRFB

The widely used negative electrode materials of VRFB are carbon-based materials. Initially, the carbon-based materials used as negative electrodes were lacking chemical stability. So, the researchers tried to improve the chemical stability of the electrodes and ultimately increase the performance of VRFB. Researchers modified the carbon-based material by combining it with metal oxides. They also combined carbon-based material with heteroatoms such as oxygen and nitrogen for the utility in VRFB negative electrode. The modified electrodes operated with decreased ohmic losses. Electrospinning technique was used for modification of carbon fiber electrode by rutile phase TiO_2 [4]. He et al. [4] reported that the TiO_2-modified electrode reduced the potential difference between anodic and cathodic peaks. Wei et al. [5] reported the modification of graphite felt electrode by titanium nitrite nanowire. The preparation of titanium nitrite-coated graphite felt involved two-step process. The first process included the deposition of TiO_2 on graphite felt electrode and the second process involved the thermal treatment of TiO_2 to obtain TiN. With these modifications, researchers improved the capacity retention of VRFB. Similar approach with two-step process was carried by Ghimire et al. [6]: the first step involved the decoration of graphite felt by TiO_2 and the second step involved the carbothermal reaction of TiO_2 to TiC by carbothermal reaction. With these modifications, Ghimire et al. [6] achieved the improvement in the energy efficiency of VRFB. Hou et al. [7] made use of hydrothermal treatment for modification of carbon electrode by TiO_2.

Figures 13.3 and 13.4 display the morphological characterization of the modified electrodes based on the scanning electron microscope (SEM) analysis. Figures 13.3 and 13.4 show the increase in the deposition of TiO_2 on a bare carbon paper with an increase in the time intervals (3 hours, Figure 13.3c,d; 6 hours, Figure 13.4a,b and 12 hours, Figure 13.4c,d). As per the findings of Hou et al. [7], carbon paper electrode modified by TiO_2 exhibited better electrochemical performance while operated toward redox couples (V^{3+}/V^{2+}). Among three different time treatments of sample, the sample treated for 6 hours (Figure 13.4a,b) exhibited the best electrochemical properties. The accelerated transfer of vanadium ions and electrons was the reason for the enhancement of electrochemical performance of the battery.

13.3 POSITIVE ELECTRODE MATERIALS OF VRFB

The carbon-based materials are widely used in the electrode components of VRFB. But for the utility in positive electrode components, the carbon-based materials pose the problem of oxidation during charging. As discussed in the previous section for negative electrode components, the

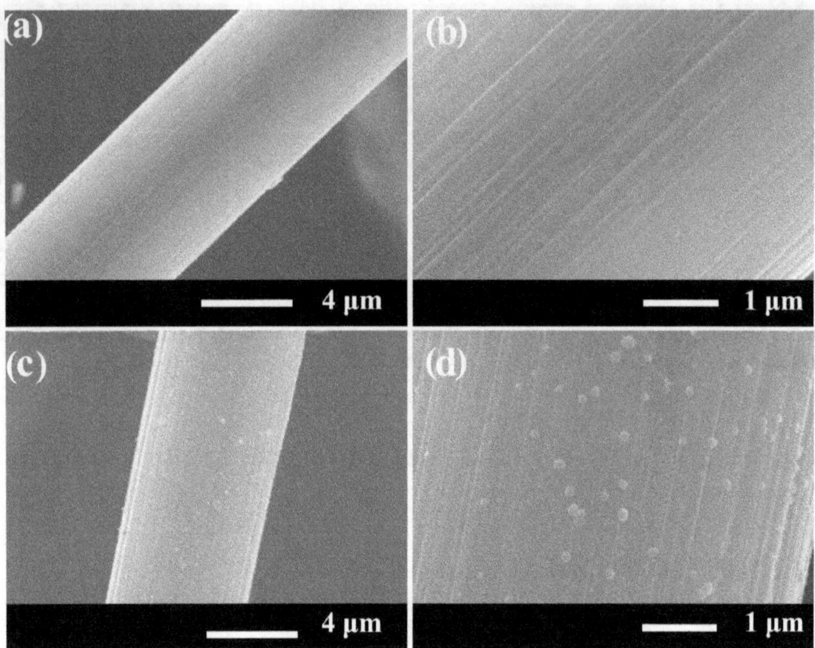

FIGURE 13.3 (a, b) SEM images of bare carbon paper. (c, d) SEM images of TiO_2-decorated carbon paper electrodes through hydrothermal process after 3 hours. Adapted and reproduced with permission from Ref. [7]. Copyright © 2018 Elsevier.

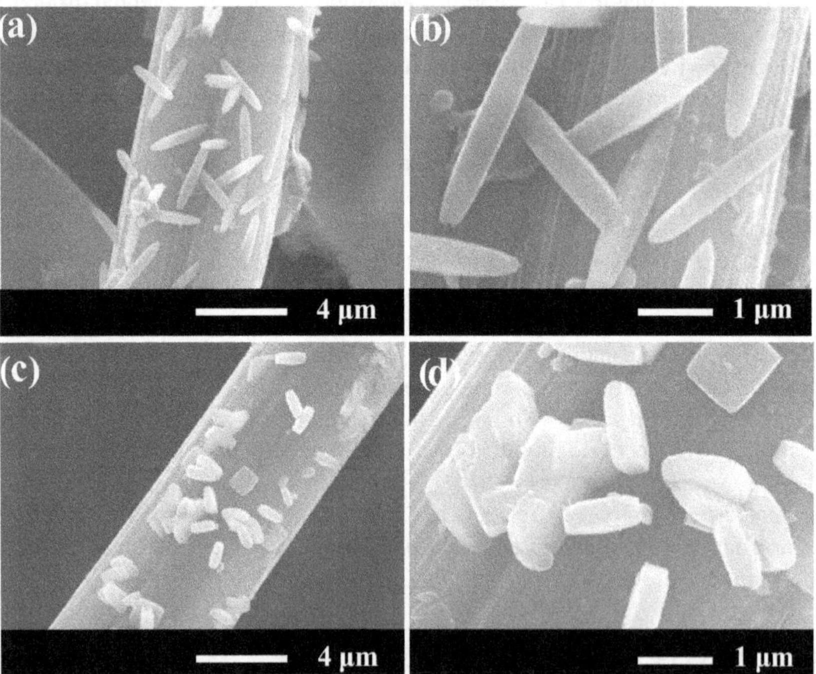

FIGURE 13.4 (a, b) The status after time interval of 6 hours. (c, d) The status after time interval of 12 hours. Adapted and reproduced with permission from Ref. [7]. Copyright © 2018 Elsevier.

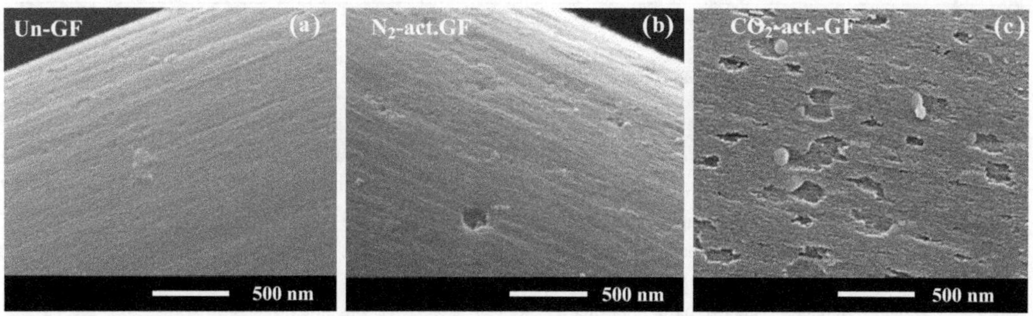

FIGURE 13.5 SEM images of (a) bare graphite felt, (b) N_2-activated graphite felt and (c) CO_2-activated graphite felt. Adapted and reproduced with permission from Ref. [10]. Copyright © 2017, Elsevier.

modification of carbon-based materials for the utility in positive electrode component plays a crucial role in determining the performance of the battery. For the utility as positive electrode component, carbon felt decorated with 3D graphene nanowell was used [8]. The modification improved the energy efficiency by 11% in comparison with the bare carbon felt electrode.

Solution coating method was utilized to deposit graphene on a carbon felt electrode [9]. This modification helped in improving both the voltage efficiency and energy efficiency of the battery. On the positive electrode component, there was a decrease in the polarization and an increase in the electrocatalytic activity due to these modifications. For increasing the porosity of graphite felt as positive electrode component of VRFB, Chang et al. [10] used CO_2 to activate graphite felt through CO_2 activation process. Figure 13.5 displays the SEM images indicating the increase in the porosity of the CO_2-activated graphite felt in comparison with N_2 activated graphite felt and unmodified graphite felt. The CO_2 activation process creates oxygen-containing functional groups, thus increasing the electrochemical activity of the resultant VRFB. This also increases the energy efficiency of the battery.

13.4 POSITIVE AND NEGATIVE ELECTRODE MATERIALS OF VRFB

In order to develop a cost-effective VRFB system and make the commercialization of VRFB smoother, the utility of electrode materials in both the electrode components must be the same. Carbon-based materials are the widely preferred materials for the VRFB electrodes. The researchers tried to fabricate a material that will be suitable for both the electrode components of VRFB. Carbon-based materials modified by metal oxides such as CeO_2, Nd_2O_3, NiO and Mn_3O_4 were used for the positive and negative electrode components of VRFB. Table 13.1 describes the recent trends in electrode material composition utilized as both positive and negative electrodes of VRFB [11]. Park et al. [12] developed graphene-based graphite felt electrode with an energy efficiency of 71%. The charge/discharge potential range amounted to 0.8 and 1.6 V. Carbon felt electrode decorated with oxygen-rich phosphate functional group performed with an energy efficiency of 88% [13].

Carbon felt electrode modified with bi-functionalized graphene also exhibited a similar range of energy efficiency, ranging around 88% [14]. Among several available electrode materials in the existing literature, the carbon felt electrodes developed by Hu et al. [14] and Kim et al. [13] were found to be superior with respect to energy efficiency. Considering the electrochemical activity of electrodes, graphite electrode fabricated by Shah et al. [15] was found to be working well.

TABLE 13.1
Recent Trends in Electrode Material Composition Utilized as Both Positive and Negative Electrodes of VRFB

Reference	Electrode	Charge and Discharge Potential Range (V)	Energy Efficiency (%)
[12]	Graphene-based graphite felt electrode	0.8 and 1.6	71
[13]	Carbon felt electrode decorated with oxygen-rich phosphate functional groups	0.8 and 1.7	88
[16]	Reduced graphene oxide modified graphene felt electrode	0.8 and 1.6	72
[17]	Carbon felt electrode co-functionalized with nitrogen and oxygen atoms	1 and 1.6	74
[14]	Carbon felt electrode modified with bi-functionalized graphene	0.75 and 1.65	88
[18]	Boron-doped graphite felt electrode	0.8 and 1.65	82
[15]	Sulfur and nitrogen functional groups to graphite felt electrodes	0.8 and 1.7	83

13.5 POTENTIAL DEVELOPMENT IN VRFB

In an electrochemical cell, the total cell reaction is composed of two half reaction independent of each other that describes the chemical change in both electrodes. Electrode reactions are heterogeneous and take place at the electrode–electrolyte interface. The interfacial potential is the driving force for charge transfer and it influences the electrode reaction [19]. Figure 13.6 represents the internal electrical parameters of VRFB.

The charge transfer resistance of VRFB is given by the following equation:

$$R_{ct} = {}^{RT}\!/_{n_e}\, FI_o \tag{13.1}$$

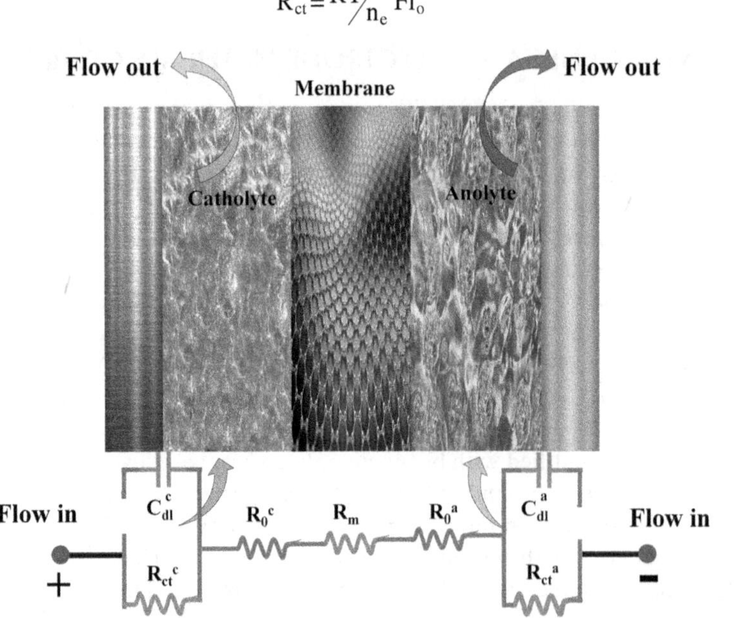

FIGURE 13.6 Internal electrical equivalent circuit elements of a VRFB cell. Adapted and reproduced with permission from Ref. [20]. Copyright © 2018, Elsevier.

where 'R' represents the universal gas constant whose magnitude is given by 8.314 $JK^{-1}mol^{-1}$. 'T' represents ambient temperature (K). 'n_e' represents the number of electrons transferred per mole of electrolyte solution. 'F' represents the faraday's constant whose magnitude is given by 96485 C. mol^{-1}. 'I_0' signifies exchange current (A) and it is given in the following equation:

$$I = I_0 \left\{ \exp\left[\frac{\alpha_a n_e F}{RT} (E - E_{eq}) \right] - \exp\left[-\frac{\alpha_c n_e F}{RT} (E - E_{eq}) \right] \right\} \tag{13.2}$$

where 'I' represents terminal current (A). α_a represents charge transfer coefficient in the anolyte side and α_c represents charge transfer coefficient in the catholyte side. 'E' represents electrode potential (Volts) and 'E_{eq}' represents the equilibrium potential (Volts) [21]. The difference between the equilibrium cell potentials of the positive and negative electrodes is expressed in the following equations:

$$E^0 = E^{0+} - E^{0-} \tag{13.3}$$

$$E^+ = E^{0+} - \frac{RT}{F} \ln \left[\frac{[VO^{2+}]}{[VO_2^+][H^+]^2} \right] \tag{13.4}$$

$$E^- = E^{0-} - \frac{RT}{F} \ln \left[\frac{[V^{2+}]}{[V^{3+}]} \right] \tag{13.5}$$

where E^+ represents the positive electrode potential (Volts) and E^- represents the negative electrode potential (Volts). $[VO^{2+}]$, $[VO_2^+]$, $[V^{2+}]$ and $[V^{3+}]$ represent the concentration of VO^{2+}, VO_2^+, V^{2+} and V^{3+}, respectively (in mol L^{-1}).

13.6 CONCLUSION

The widely used electrode materials of VRFB are carbon-based materials. Initially, the carbon-based materials used for VRFB electrodes were lacking chemical stability. So, the researchers tried to improve the chemical stability of the electrodes and ultimately increase the performance of VRFB. Researchers modified the carbon-based material by combining it with metal oxides. They also combined carbon-based material with heteroatoms such as oxygen and nitrogen for the utility in VRFB electrodes. Two separate electrode materials for negative and positive electrode components were used in the initial days of VRFB system development. In order to make the fabrication of electrode materials cost-effective, there was a quest for the suitable electrode material which can be used in both positive and negative electrodes. Due to various improvements in VRFB electrode fabrication over the years, the development of VRFB system has become cost-effective. The energy efficiency of VRFB system has also increased significantly.

ACKNOWLEDGMENTS

The work has been executed at the Birla Institute of Technology and Science – Pilani, Hyderabad Campus. The first author is thankful to the University for granting his research fellowship.

REFERENCES

1. Zhang Z, Bai B, Zeng L, et al (2019) Aligned electrospun carbon nanofibers as electrodes for vanadium redox flow batteries. *Energy Technol* 7:1–8. https://doi.org/10.1002/ente.201900488.
2. Zhong S, Padeste C, Kazacos M, Skyllas-Kazacos M (1993) Comparison of the physical, chemical and electrochemical properties of rayon- and polyacrylonitrile-based graphite felt electrodes. *J Power Sources* 45:29–41. https://doi.org/10.1016/0378-7753(93)80006-B.
3. Kim KJ, Park MS, Kim YJ, et al (2013) A technology review of electrodes and reaction mechanisms in vanadium redox flow batteries. *J Mater Chem A* 3:16913–16933. https://doi.org/10.1039/c5ta02613j.
4. He Z, Li M, Li Y, et al (2019) ZrO_2 nanoparticle embedded carbon nanofibers by electrospinning technique as advanced negative electrode materials for vanadium redox flow battery. *Electrochim Acta* 309:166–176. https://doi.org/10.1016/j.electacta.2019.04.100.
5. Wei L, Zhao TS, Zeng L, et al (2017) Highly catalytic and stabilized titanium nitride nanowire array-decorated graphite felt electrodes for all vanadium redox flow batteries. *J Power Sources* 341:318–326. https://doi.org/10.1016/j.jpowsour.2016.12.016.
6. Ghimire PC, Schweiss R, Scherer GG, et al (2018) Titanium carbide-decorated graphite felt as high performance negative electrode in vanadium redox flow batteries †. https://doi.org/10.1039/c8ta00464a.
7. Hou B, Cui X, Chen Y (2018) In situ TiO_2 decorated carbon paper as negative electrode for vanadium redox battery. https://doi.org/10.1016/j.ssi.2018.08.006.
8. Li W, Zhang Z, Tang Y, et al (2016) Graphene-nanowall-decorated carbon felt with excellent electrochemical activity toward VO^{2+}/VO^{2+} couple for all vanadium redox flow battery. *Adv Sci* 3:1–7. https://doi.org/10.1002/advs.201500276.
9. Xia L, Zhang Q, Wu C, et al (2019) Graphene coated carbon felt as a high-performance electrode for all vanadium redox flow batteries. *Surf Coatings Technol* 358:153–158. https://doi.org/10.1016/j.surfcoat.2018.11.024.
10. Chang Y-C, Chen J-Y, Kabtamu DM, et al (2017) High efficiency of CO_2-activated graphite felt as electrode for vanadium redox flow battery application. *J Power Sources*. https://doi.org/10.1016/j.jpowsour.2017.07.103.
11. Gencten M, Sahin Y (2020) A critical review on progress of the electrode materials of vanadium redox flow battery. *Int J Energy Res* 44:7903–7923. https://doi.org/10.1002/er.5487.
12. Park JJ, Park JH, Park OO, Yang JH (2016) Highly porous graphenated graphite felt electrodes with catalytic defects for high-performance vanadium redox flow batteries produced via NiO/Ni redox reactions. *Carbon N Y* 110:17–26. https://doi.org/10.1016/j.carbon.2016.08.094.
13. Kim KJ, Lee HS, Kim J, et al (2016) Superior electrocatalytic activity of a robust carbon-felt electrode with oxygen-rich phosphate groups for all-vanadium redox flow batteries. *ChemSusChem* 9:1329–1338. https://doi.org/10.1002/cssc.201600106.
14. Hu G, Jing M, Wang DW, et al (2018) A gradient bi-functional graphene-based modified electrode for vanadium redox flow batteries. *Energy Storage Mater* 13:66–71. https://doi.org/10.1016/j.ensm.2017.12.026.
15. Shah AB, Wu Y, Joo YL (2019) Direct addition of sulfur and nitrogen functional groups to graphite felt electrodes for improving all-vanadium redox flow battery performance. *Electrochim Acta* 297:905–915. https://doi.org/10.1016/j.electacta.2018.12.052.
16. Deng Q, Huang P, Zhou WX, et al (2017) A high-performance composite electrode for vanadium redox flow batteries. *Adv Energy Mater* 7:1–7. https://doi.org/10.1002/aenm.201700461.
17. Kim J, Lim H, Jyoung JY, et al (2017) High electrocatalytic performance of N and O atomic co-functionalized carbon electrodes for vanadium redox flow battery. *Carbon N Y* 111:592–601. https://doi.org/10.1016/j.carbon.2016.10.043.
18. Jiang HR, Shyy W, Zeng L, et al (2018) Highly efficient and ultra-stable boron-doped graphite felt electrodes for vanadium redox flow batteries. *J Mater Chem A* 6:13244–13253. https://doi.org/10.1039/c8ta03388a.
19. Bard AJ, Faulkner LR, White HS (2001) *Electrochemical Methods Fundamentals and Applications*, Third edition. Wiley, UK.
20. Bhattacharjee A, Roy A, Banerjee N, et al (2018) Precision dynamic equivalent circuit model of a vanadium redox flow battery and determination of circuit parameters for its optimal performance in renewable energy applications. *J Power Sources* 396:506–518. https://doi.org/10.1016/j.jpowsour.2018.06.017.
21. Bhattacharjee A, Saha H (2017) Design and experimental validation of a generalised electrical equivalent model of vanadium redox flow battery for interfacing with renewable energy sources. *J Energy Storage* 13:220–232. https://doi.org/10.1016/j.est.2017.07.016.

14 Electrolytes for Vanadium Redox Flow Batteries

Roshny Joy, Thomas George, Akhila Das, Jabeen Fatima M. J., Abhilash Pullanchiyodan, and Prasanth Raghavan

14.1 INTRODUCTION

In the present scenario, climate changes are the most critical issue. Energy is playing an important role in the day-to-day life. Renewable energy sources such as wind, tide, and solar are integral part of human life. Without energy, the survival of the entire sustainable system will be imbalanced. In the case of renewable energy, there is no proper way or they possess technical problems for emergency needs. A safe and reliable system is very necessary to store the energy generated during the peak time. The interest in the advancement of energy storage methods has paved new trends in renewable energy source. The discovery of battery creates an explosion in the energy storage methods. Redox flow batteries are one of the most favorable ways out for energy storage. In energy storage trends, vanadium redox flow batteries (VRFBs) are one of the emerging energy storage techniques and are getting a special attention owing to its long life. The main striking features are stable capacity, flexible structural designs, independently tunable, and safety [1].

In the 1980s, Skyllas-Kazacos has introduced vanadium redox flow batteries and it is considered as good candidate for grid-scale energy storage in the present time [2]. The important component in VRFB is electrolyte. The electrode, bipolar plate, gasket, current collector, pumps, and storage tanks are the other parts. Apart from the storage of electrolytes in redox flow batteries, electrochemical batteries are almost same (represented in Figure 14.1) [3]. Redox flow batteries store electrolytes in external tanks which are away from the battery center. In VRFB, vanadium poses four different oxidation states such as V^{2+}, V^{3+}, VO^{2+}[or V^{4+}], and VO_2^+[or V^{5+}]. The half-cell reactions for the positive and negative electrodes are given in equations 14.1 and 14.2 separately. The overall cell reaction is shown in equation 14.3:

$$\text{Positive cell reaction: } VO_2^+ + 2H^+ + e^- \leftrightarrow VO^{2+} + H_2O \tag{14.1}$$

$$\text{Negative cell reaction: } V^{2+} \leftrightarrow V^{3+} + e^- \tag{14.2}$$

$$\text{Overall cell reaction: } VO_2^+ + V^{2+} + 2H^+ \leftrightarrow VO^{2+} + V^{3+} + H_2O \tag{14.3}$$

In the discharge state, positive and negative electrolytes have V^{3+} and V^{4+} (VO^{2+}), respectively; during charging, the negative electrolyte is reduced to V^{2+}, while the positive electrolyte is oxidized to V^{5+} (VO_2^+). In discharging, battery moves electron through the bipolar plate from the negative side to the positive side and causes hydrogen (H^+) ions to diffuse through the membrane to the positive side. The same reaction occurs in reverse during recharge. Early proposals of VRFB had energy efficiencies around 70% and earlier advances led to efficiencies of over 80% [3,4].

Electrolytes in the VRFB function as an energy storage medium, and they are composed of vanadium ions of different valences in the supporting electrolytes. The volume and concentration of vanadium in electrolytes are in authority for the capacity and energy density of a VRFB. Their electrochemical activity and stability have a significant influence on the performance of the VRB.

DOI: 10.1201/9781003310167-14

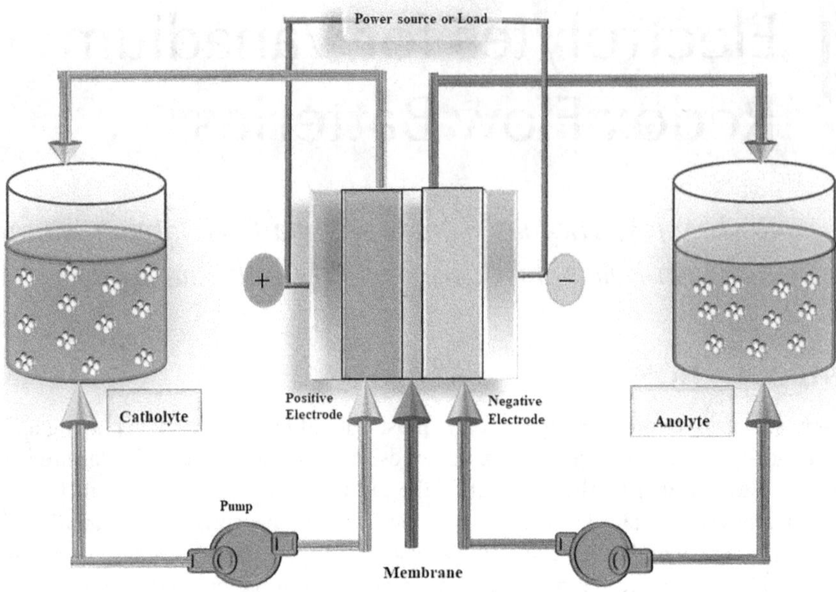

FIGURE 14.1 Redox flow battery diagram. Adapted and reproduced with permission from Ref. [3]. Copyright © 2019, Elsevier.

Hence, the supporting electrolyte used in the VRFB should have quite a few necessities together with high solubility, good electrochemical kinetics, and high stability, low cost and wide operation potential. Here, the major work focuses on the electrolytes that use sulfuric acid as a supporting electrolyte, along with other new electrolytes [2].

14.2 SUPPORTING ELECTROLYTE

14.2.1 Sulfuric Acid

Energy density of VRFB depends upon the concentration of vanadium ions in the electrolyte solution. Even though aqueous hydrochloric acid (HCl) has high solubility of vanadium ions, it cannot be used as a supporting electrolyte due to the emission of chlorine gas formed by the reduction of chloride ions by V^{5+} [5]. When sulfuric acid (H_2SO_4) is used as a solvent, all four vanadium species are soluble in it. It also increases the ionic conductivity and provides protons for the reduction of VO^{2+}. Vanadium ion solubility and temperature stability are the two important factors that affect the vanadium ion concentration in the VRFB electrolyte which in turn affects the energy density.

In the positive electrolyte, 2 M V^{5+} in sulfuric acid having concentration 3–4 M undergoes thermal precipitation of V^{5+} at a temperature above 40°C if the VRFB remains fully charged. In 1990, Kazacos et al. [6] reported that 1.5 M V ions in 3–4 M H_2SO_4 solution is suitable for the batteries that remain fully charged at high temperatures for more days. The reason for the thermal precipitation was reported by Vijayakumar et al. [7]. According to their conclusions, V^{5+} remains as a penta-coordinated ion $[VO_2(H_2O)_3]^+$ in the electrolyte (represented in Figure 14.2).

The conversion of this penta-coordinated vanadium ion to H_3VO_4 was an endothermic reaction. As it is an endothermic reaction at high temperatures, reaction takes place in the forward direction. The H_3VO_4 formed is converted to vanadium pentoxide (V_2O_5) by a condensation reaction. Equations 14.4 and 14.5 represent this reaction.

$$[VO_2(H_2O)_3]^+ \xrightarrow{\Delta} H_3VO_4 + H_3O^+ \tag{14.4}$$

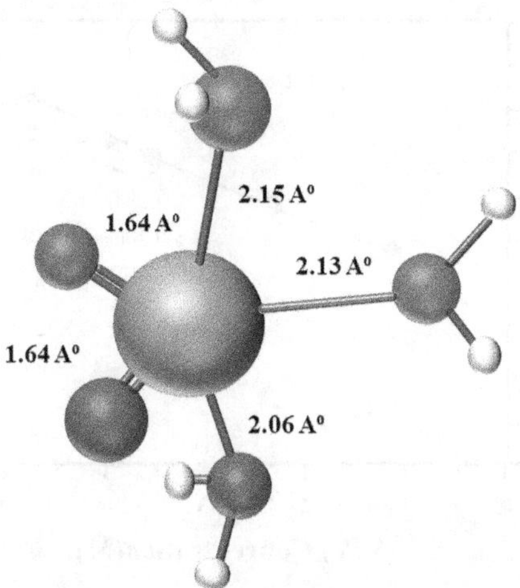

FIGURE 14.2 Structure of V^{5+} in sulfate solutions. Adapted and reproduced with permission from Ref. [8]. Copyright © 2011, Wiley.

$$2H_3VO_4 \xrightarrow{\Delta} V_2O_5 \cdot 3H_2O \tag{14.5}$$

$$1/2\ V_2O_5 + H^+ \xleftrightarrow{\Delta} VO_2^+ + 1/2\ H_2O \tag{14.6}$$

The increase in the concentration of sulfuric acid increases the solubility of V_2O_5 which is formed by thermal precipitation of V^{5+}. Equation 14.3 represents the solubility of V_2O_5. When H_2SO_4 concentration increases, the concentration of H^+ ions increases and the reaction takes place in the forward direction. This will increase the stability of V^{5+} ions [6]. V^{4+} ions are precipitated as Vanadyl sulfate. The increase in H_2SO_4 concentration increases the sulfate concentration. Due to the common ion effect, it will lead to the precipitation of V^{4+} ions [9]. The following equation represents the reaction:

$$VOSO_4 \cdot xH_2O \longleftrightarrow VO_2^+\ SO_4^{2+} + xH_2O \tag{14.7}$$

Rahman et al. [10] in their viscosity studies of V^{5+} solutions on 5–7 M total sulfate solution show that viscosity of the electrolyte increases rapidly on increasing vanadium concentration to 5 M. The effect of vanadium concentration on density of the supersaturated V(V) solution at 20°C along with variation of viscosity of V^{5+} total sulfate/bisulfate concentration at 20°C is shown in Figure 14.3. The viscosity increases due to the formation of vanadate polymers [11]. This viscosity increase will affect the flow of electrolytes and loss of energy. It also leads to electrode polarization.

In negative electrolytes as in V^{4+}, the V^{2+} and V^{3+} sulfate solubility decrease with an increase in the concentration of sulfuric acid due to the common ion effect. An increase in temperature is a favorable condition for solubility. V(II)sulfate is more soluble than V(III)sulfate at the same temperature and same concentration of sulfuric acid [12]. Short range of operational temperature causes an additional loss in energy and increases the cost.

14.2.2 Hydrochloric Acid

Kim et al. [13] reported that the vanadium chloride flow battery has superior performance and enhanced capacity than VRFB using sulfuric acid as a supporting electrolyte. The improved performance of VRFB using HCl as a supporting electrolyte is due to the chloride solution that can dissolve greater than 2.3 M vanadium and has an operating temperature 0°C–50°C without any precipitations. Chloride electrolyte has 30% more energy density than the sulfate system. Low

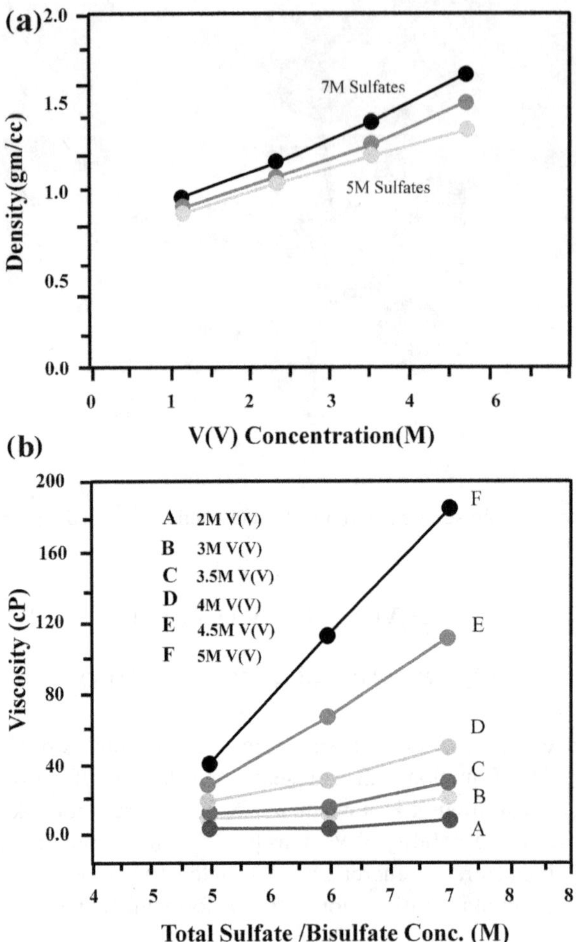

FIGURE 14.3 (a) Effect of vanadium concentration on density of the supersaturated V(V) solution at 20°C. (b) Variation of viscosity of V^{5+} total sulfate/bisulfate concentration at 20°C. Adapted and reproduced with permission from Ref. [10]. Copyright © 2009 Elsevier.

viscosity and high operation temperature make it cost-effective. Chlorine gas evolution is a problem. The better stability of V^{5+} ion is due to the formation of vanadium binuclear-chloro complex and binuclear complex.

14.2.3 Chloride Sulfate Mixed Acid Electrolytes

The deprotonation of $[VO_2(H_2O)_3]^+$ leads to the precipitation of vanadium pentoxide in the case of sulfate electrolyte. Li et al. [8] reported that the addition of chloride to the electrolyte solution containing sulfate increases both the energy capacity and the working temperature. The new mixed electrolyte is able to dissolve 2.5 M vanadium that is about 70% increase. The operation temperature also increases from 10°C–40°C to −5°C–50°C. the stability of V^{5+} is due to the conversion of penta-coordinated vanadium ion to neutral species $VO_2Cl(H_2O)_2$ (represented in Figure 14.4). The reaction is represented in the following equation:

$$[VO_2(H_2O)_3]^+ + Cl^- + H_2O + V^{3+} \longleftrightarrow VO_2Cl(H_2O)_2 + H_3O^+ \tag{14.8}$$

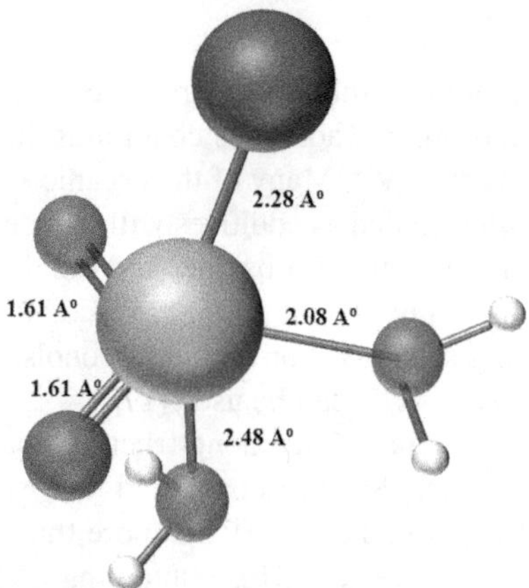

FIGURE 14.4 Structure of V^{5+} in chloride and sulfate mixed solutions. Adapted and reproduced with permission from Ref. [8]. Copyright © 2011, Wiley.

The stability of V^{2+} and V^{3+} is due to the decrease in the concentration of sulfate ions. The overall cell reaction in the mixed electrolyte system is represented in the following equation:.

$$VO_2^+ + Cl^- + H_2O + V^{3+} \underset{\text{discharge}}{\overset{\text{charge}}{\rightleftharpoons}} VO_2Cl + 2H^+ + V^{2+} \tag{14.9}$$

The demonstration of the mixed electrolyte (chloride-sulfate) VRFB in kW scale was reported by Kim et al. [14]. With 82% energy efficiency and 15%–85% state of charge (SOC), VRFB delivers greater than 1 kW. It also operated steadily at high temperatures.

14.2.4 Methanesulfonic Acid–Sulfuric Acid Mixed Acid Electrolytes

Methanesulfonic acid (CH_3SO_3H), which is stronger than sulfuric acid, is used as a supporting electrolyte for other RFBs, e.g., zinc-vanadium redox battery [15]. Peng et al. [16] reported the use of methanesulfonic acid (MSA) sulfuric acid mixed acid solution as electrolyte for VRFB. The battery using this mixed acid solution has stable cycling performance for about 30 cycles. The cyclic voltammetry studies of the mixed acid solution containing 2 M V^{4+} in 1.5 M CH_3SO_3H and 1.5 M H_2SO_4 has an upper hand in redox reaction kinetics of V^{4+} and V^{5+} than electrolyte solution containing only H_2SO_4. The improved thermal stability of the V^{5+} ion is due to the formation of $VO(CH_3SO_3)_2$, which prevents further precipitation [17].

14.3 ADDITIVES AND IMPURITIES IN THE ELECTROLYTE

For practical usage of VRFB, various methods are proposed to increase the energy density and the range of working temperature. One of the methods is the addition of foreign chemical substances that can improve the stability of electrolyte solutions which in turn increase the overall performances and reduces the cost of production of VRFB. The term stabilizing agents was first put forward by Skyllas-Kazacos and co-workers in the mid-1990s. These are the compounds introduced to the electrolyte solution to prevent precipitation at low and high temperatures. The stabilizing agents do not increase solubility, but they work by increasing the time interval between the formation of an appreciable amount of the solid phase and the initial moment of supersaturation of the solution. For an additive to be suitable for use in a VRFB electrolyte, it must be functional and do not have any ruinous effects on the vanadium oxidation states [18].

14.3.1 Organic Additives

Apart from acids, organic and inorganic substances are used as additives to increase the stability of VRFBs. Many of the suggested organic additives containing functional groups -NH_2, -COOH, -COO^-, -COH, -CO^-, -OH, -SH, reduce V^{5+}. Many of the organic additives are not a good option to stick on. Most carboxylic acids when added as additives will reduce the V^{5+} and produce CO_2. Slow oxidation takes place while adding ammonium oxalate and rapid oxidation while using EDTA or malic acid. At low concentrations of glycerin, it doesn't reduce V^{5+}. The one which can be used as organic additives is the one having secondary or tertiary alcohols. Likewise, compounds having 2 or more secondary or tertiary –SH or –NH can be used [19].

The studies were conducted by Li et al. [20] among fructose, mannitol, glucose, and D-sorbitol; the EIS (Electrochemical Impedance Spectroscopy) and CV studies suggest that the use of D-sorbitol increases the overall performance of VRFB more than others. The sorbitol containing electrolyte has higher solubility of V^{5+}. The VRFB containing additive such as D-sorbitol has an energy efficiency of 81.8%. The D-sorbitol that gets absorbed on the electrode surface by hydrogen bond provides more -OH groups for ion exchange. The charge–discharge curve was represented in Figure 14.5. The increase in output current density is due to the increase in the number of ion exchange methods between the vanadium ion and –OH group.

Trishydroxymethyl aminomethane (Tris) additive on VRFB electrolyte has less discharge capacity fade rate, high thermal stability and good cycle stability. Tris additive also does not have an oxidation effect on carbon felt electrode [21]. The use of hexadecyl trimethyl ammonium bromide (CTAB) as an additive increases the electrochemical performance and stability of VRFB. CTAB is a cationic surfactant whose quaternary ammonium head groups interact with V^{5+}, which prevents the vanadium ion from crystallization. Micellar catalysis catalyzes the redox reaction of V^{5+}/V^{4+} and the catalyst is the stable hemispherical particle formed [17].

Among inositol and phytic acid used as an organic additive by Wu et al. [22], inositol is a good additive for the positive electrolyte. Both inositol and phytic acid increase thermal stability, better reaction kinetics, exchange current density, and electrochemical reaction constant. But inositol as an additive shows an enhanced charge–discharge cycle than using phytic acid. Due to the formation of sediments in the positive electrode, phytic acid is not a favorable additive. The addition of L-Glutamic acid to the positive electrolyte reduces the precipitation of V_2O_5. The advancement in

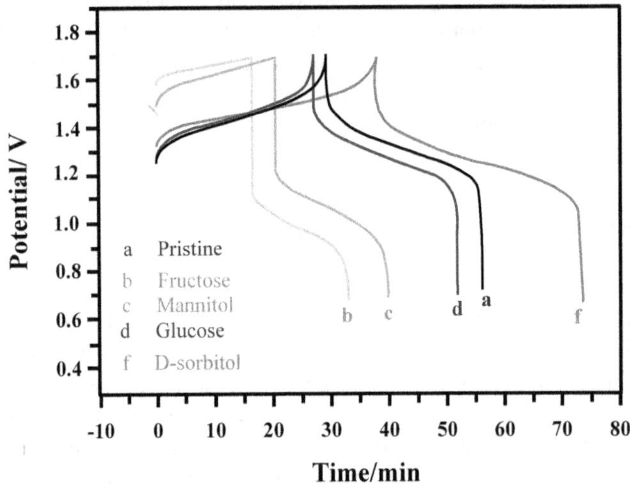

FIGURE 14.5 Charge–discharge curve for vanadium redox flow cell employed electrolyte with different additives at a current density of 20 mA cm^2. Adapted and reproduced with permission from Ref. [20]. Copyright © 2011, Elsevier.

electrochemical performance is due to the combined effect of oxygen- and nitrogen-containing groups introduced on the electrode surface [23,24]. The comparative study of MSA and amino methyl sulfonic acid (AMSA) as additives reported by Zhangxing et al. [25] shows that AMSA has higher-level electrochemical activity and kinetics than MSA. This superior performance is attributed to more active sites yield by $-NH_2$ group and the increased hydrophilicity of the electrode provided by NH_2 and SO_3H groups. Anthraquinone-2,6-disulfonic acid (AQDS) is used as an additive instead of expensive heavy metals and expensive catalysts by Mulcahy et al. [26]. It increases the capacity efficiency and reduces the overall cycle duration.

14.3.2 Inorganic Additives

Inorganic additives can also be used to increase the overall efficiency of VRFBs. Roe et al. [27] reported 3 M vanadium electrolyte battery formed by adding additives phosphoric acid (H_3PO_4) and ammonium sulfate which is stable over 90 cycles. The addition of H_3PO_4 and ammonium sulfate increases the concentration of SO_4^{2-} and H^+, which shift the reaction (3) toward the right and increase the solubility of precipitate V_2O_5. The presence of ionic species like NH^{4+}, PO_4^{3-}, HPO_4^{2-}, and $H_2PO_4^-$ also increases the temperature stability by ion pairing with VO_2^+ or by adsorbing them on the nucleation site [28]. Mousa et al. [29] conducted a study among different additives that prevent the precipitation to decrease the lower limit of VRFB. Ammonium phosphate $(NH_4)_3PO_4$ and ammonium sulfate $(NH_4)_2SO_4$ show a good stabilizing effect among others. Experiments carried out in 4 M Vanadyl sulfate in 3 M H_2SO_4 at 4°C with additives 5wt. % potassium sulfate (K_2SO_4) or 3wt.% sodium hexametaphosphate (SHMP) do not form any precipitate after 80 days [30]. Rahman et al. [31] used the inorganic additive formulation of 1wt.% of Potassium phosphate (K_3PO_4),1 wt.% of SHMP in 3.5 M V^{5+} in total sulfate/bisulfate concentration of 5.7 M which is stable up to 40°C. The role of phosphates as an additive was studied using nuclear magnetic resonance (NMR) and dynamic light scattering (DLS) by Roznyatovskaya et al. [32]. They reported that phosphates can delay the induction time of V_2O_5 formation. This stabilization is due to the reaction of the formed monomer and dimer of V^{5+} with phosphates. If the phosphate concentration in the solution increases above 0.15 M, precipitation of $VOPO_4$ occurs. In recent studies, sodium chloride (NaCl) is used as a low-cost additive in VRFBs. The chloride ions in the solution act as a catalyst for the redox reactions of VO_2^+/VO^{2+}. The viscosity of negative electrolyte also decreases. The solubility of $VOSO_4$ also increases as the $[VO(Cl)\cdot(H_2O)_4]^+$ formed exhibits longer life time than $[VO(H_2O)_5]^{2+}$. The electrolyte having 0.04 M concentrations of chloride ions exhibits 83.1% capacity retention rate after 100 cycles. The cyclic voltammetry studies of the electrolyte having chloride ion concentration less than 0.08 M have no peak corresponding to Cl^-/Cl_2. The CV curve of electrolyte containing 0.04 M chloride ions does not show any Cl^-/Cl_2 peak (represented in Figure 14.6). It shows that there is no hazardous elimination of Cl_2 gas [33].

14.3.3 Immobilizing Agents

To make sure about the high concentration of vanadium ions in the RFBs and prevent the precipitation of these vanadium ions, another way suggested was the use of agents like gels, starch, alkaline pectates, and gums. It can eliminate the problem caused by electrolyte leakage. In spite of the fact that the practical application of such static VRFBs would be more confined, the use of such a device in frequency regulation and voltage control could be particularly important [18].

14.3.4 Electrolyte Impurities

Vanadyl sulfate ($VOSO_4$) or vanadium pentoxide (V_2O_5) are used for the preparation of vanadium electrolytes. V_2O_5 has ten times less solubility than $VOSO_4$ [34,35]. In order to reduce the cost of production of vanadium electrolytes, V_2O_5 is preferred. Vanadium is the 22nd abundant element in

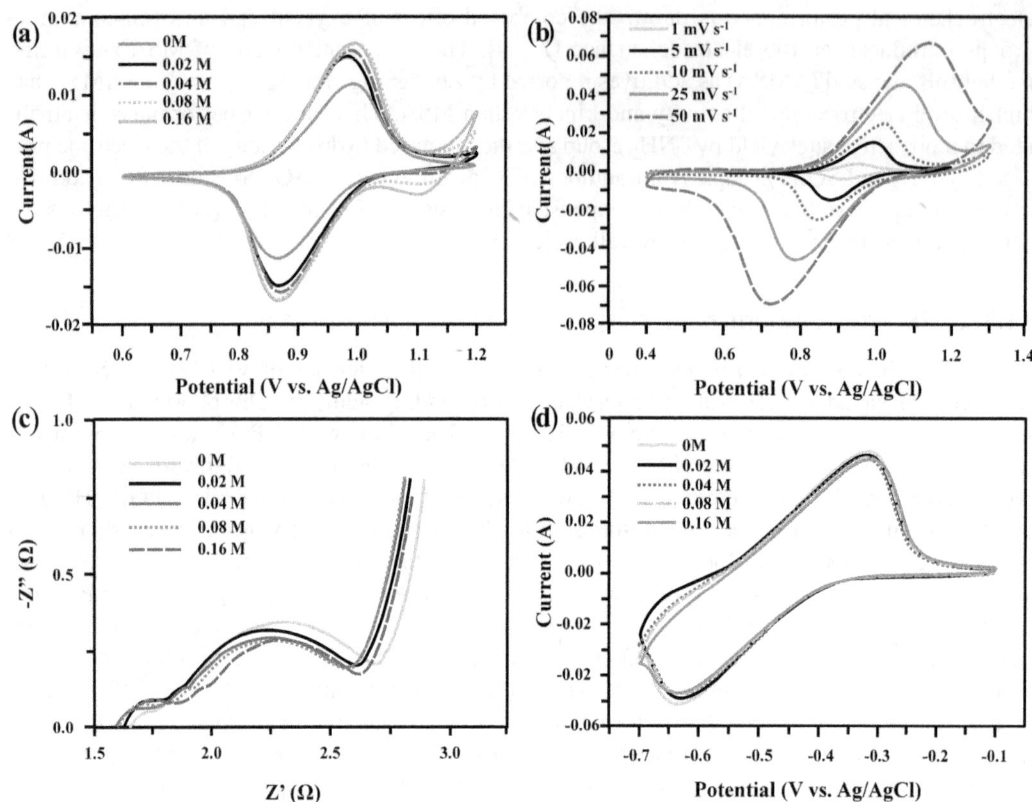

FIGURE 14.6 CV results for VO^{2+}/VO_2^+ reaction (a) with various contents of Cl^- at a scan rate of 5 mV s^{-1}, (b) with 04 M Cl^- under scan rates from 1 to 50 mV s^{-1}, (c) Nyquist plots comparison with various concentrations of Cl^{-1} at 0.9 V, and (d) CV results for V^{2+}/V^{3+} reaction at 25 mVs^{-1+}. Adapted and reproduced with permission from Ref. [33]. Copyright © 2021, Elsevier.

the earth's crust, and the sources of V_2O_5 are ores, mining slag, oil field sludge, metallurgical slags, and petroleum residues. These V_2O_5 contain Fe, Si, K, Na, Ca, Mg, Cr, Cu, Mo, Zn, Mg, and Ni as impurities [36], and these impurities remain in the V_2O_5 and enter into the electrolyte. Many of the impurities do not have any harmful effects on the battery, while others have a detrimental effect on the performance. For example, the increased hydrogen evolution due to the deposition of Cu and Ni in the negative electrode and the carbon electrode is blocked by the contaminants like Si and NH_4^+ during repeated charge–discharge cycle which retard the circulation of electrolytes [37]. Si can be removed by filtration [38] and ammonium ions can be removed by heating the compound at 400°C–690°C for 1–4 hours. High purity V_2O_5 is preferred and it will increase the cost of production [18]. Cr^{3+} as an impurity does not cause any side reactions. But when the concentration of Cr^{3+} increases from 0 to 30 g L^{-1}, the reversibility of the V^{5+}/V^{4+} reaction and diffusion coefficient was raised. But undesirable consequences were observed in diffusion resistance and the diffusion coefficient when the concentration of Cr^{3+} goes above 30 g L^{-1} [39].

To evaluate the effects of impurities Na^+, K^+, and NH_4^+, Wen et al. [40] deliberately added K_2SO_4, Na_2SO_4, and $(NH_4)_2SO_4$ into the electrolyte solutions. The addition of three impurities to the 2 M V^{4+} in 5 M H_2SO_4 badly affects the stability of V^{4+}. When the concentration of supporting electrolyte is reduced to 4 M, K^+ increases the precipitation time, while NH_4^+ and Na^+ decrease the precipitation time of V^{4+}. After charge–discharge cycles, the impurities like Al, Ca, K, Fe, Mg, Na, and B may be present in the electrolyte solution (<100 ppm). These impurities are introduced from the

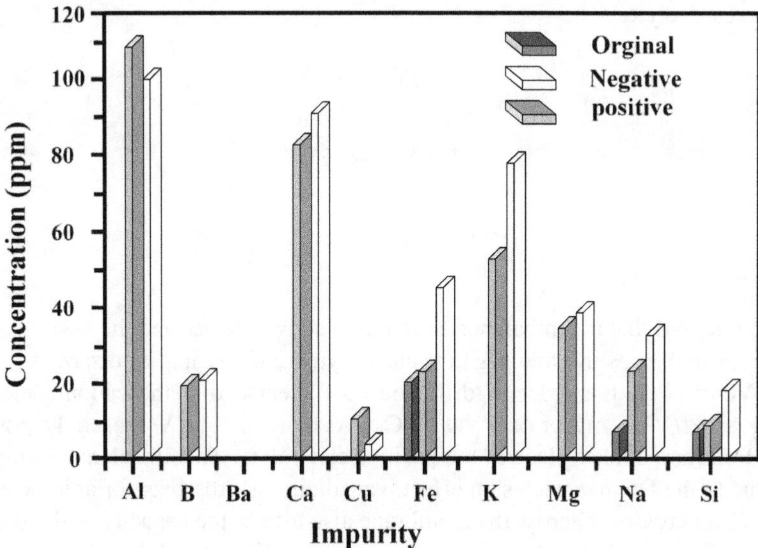

FIGURE 14.7 The variation of impurities in positive, negative, and original electrolytes after charge–discharge cycles. Adapted and reproduced with permission from Ref. [40]. Copyright © 2013, Elsevier.

framework materials. The variation of impurities in positive, negative, and original electrolytes with V_2O_5 of high purity after charge–discharge cycles is represented in Figure 14.7.

14.4 ELECTROLYTE IMBALANCE

The capacity of VRFBs decreases due to the imbalance of active species in the negative and positive electrolytes during long-term operations. This loss of capacity is ascribed to (i) hydrogen and oxygen evolution reaction (HER and OER), (ii) vanadium ion and water crossover, and (iii) air oxidation in the negative electrolyte solution.

14.5 VANADIUM ION AND WATER CROSSOVER

Among the transport phenomenon across the membrane in VRFBs, the important issue which affects the performance is the water and vanadium ion crossover. In order to prevent cross-contamination, vanadium ion in different oxidation is used in VRFBs. The role of the membrane is to prevent the cross-contamination of the catholyte and anolyte while allowing the protons to maintain electro-neutrality. This is due to the fact that the unwanted crossover of vanadium ion through the ion exchange membrane results in the undesired side reactions leading to the capacity loss of VRFBs. This crossover is owing to the concentration difference that exists in the two sides of the membrane [41]. The transferred vanadium ions across the membrane result in the following reactions [42].

In the positive electrolyte,

$$V^{2+} + 2V^{5+} \rightarrow 3V^{4+} \tag{14.10}$$

$$V^{3+} + V^{5+} \rightarrow 2V^{4+} \tag{14.11}$$

$$V^{2+} + V^{4+} \rightarrow 2V^{3+} \tag{14.12}$$

In the negative electrolyte,

$$V^{5+} + 2V^{2+} \rightarrow 3V^{3+} \quad (14.13)$$

$$V^{4+} + V^{2+} \rightarrow 2V^{3+} \quad (14.14)$$

$$V^{5+} + V^{3+} \rightarrow 2V^{4+} \quad (14.15)$$

Luo et al. [43] reported that the imbalanced vanadium active species and the asymmetrical valence of vanadium ions in the positive and negative electrolytes lead to capacity decay.

V^{2+}>VO^{2+}>VO_2^+>V^{3+} is the order of diffusion coefficient across the cationic membrane [44]. Transport rate of V^{2+}/V^{3+} is higher than VO^{2+}/VO_2^+. As more V^{2+} and V^{3+} reach the positive electrolyte, more self-discharge takes place in the positive electrolyte. This imbalance results in capacity decay [43]. Due to the Donnan exclusion effect, vanadium ion crossover for anion exchange membrane is less. Water crossover across the membrane also affects the capacity of the battery [2]. The transfer of water from one-half cell to the other makes one diluted and the other concentrated. Sun et al. [44] conducted a study on water transfer across the membrane during the charge–discharge cycle and self-discharging. In the case of the cycle process, the water transfer is ascribed to the transfer of vanadium ions bound with water, to balance the charges and for the formation of internal circuit transfer of protons with dragged water, and to osmosis. All these factors that caused water transfer are represented in Figure 14.8. They also observe the water crossover in Nafion 115 membranes. During the charging process, the amount of water in the positive electrolyte decreased. This shows that the water transferred from positive to negative. And while discharging, it is from negative to positive.

The amount of water transferred to positive is more than negative. In long-term cycles, the transfer of water is in the direction of vanadium ions. Expect that the transfer of protons with the dragged water for the formation of external circuits in solution plays a role in water transfer in self-discharge (represented in Figure 14.9).

In the self-discharge process, 75% of the total water transfer is by osmosis. To reduce the water transfer, many modifications are made in the commercially available membrane. The neutralization of the anion exchange group with a cationic polyelectrolyte is an example. A notable reduction in transfer of water was observed while the anion exchange membrane Selemion AMV was sulfonated using sulfuric acid, which includes poly(sodium-4-styrenesulfonate) (PSSS) a cationic polyelectrolyte [45].

14.6 HYDROGEN EVOLUTION REACTION AND OXYGEN EVOLUTION REACTION

Gassing side reactions such as OER and HER take place only through charging both the negative and positive electrolytes when the polarization potential is higher than the formal potential of the reactions. The formation of gas bubbles affects the performance by reducing the liquid electrolyte volume, the active surface area for electrochemical reaction, and the diffusion coefficient. The bubbles also obstruct the flow distribution of electrolyte in electrode which results in the reduction of mass and charge transfer. The charging time of the battery increases as the side reaction uses the applied current partially [46,47].

Oxygen evolution at the positive electrolyte:

$$2H_2O \leftrightarrow O_2 + 4H^+ + 4e^- \quad (14.16)$$

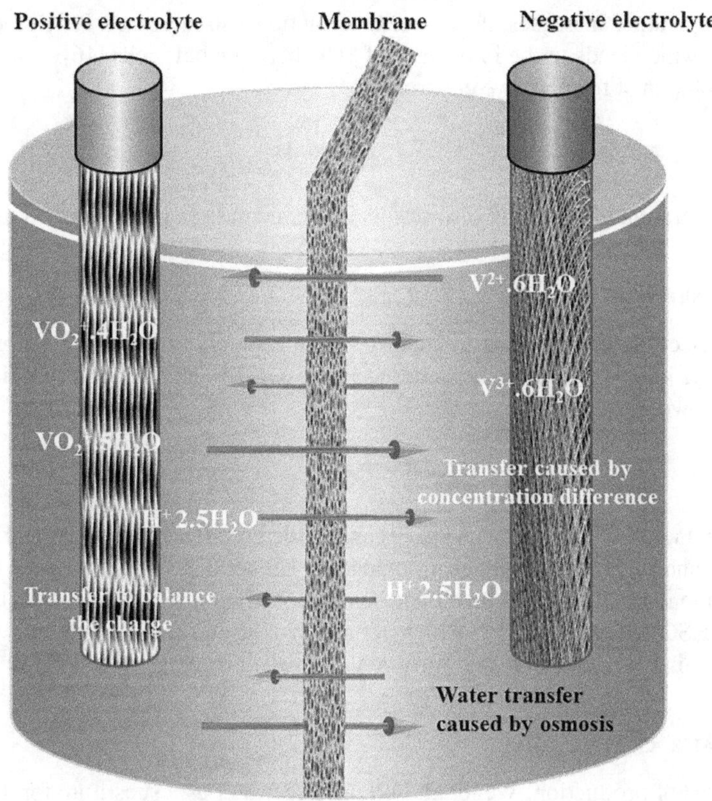

FIGURE 14.8 Transfer of water and vanadium ions in charge–discharge cycles. Adapted and reproduced with permission from Ref. [44]. Copyright © 2010, Elsevier.

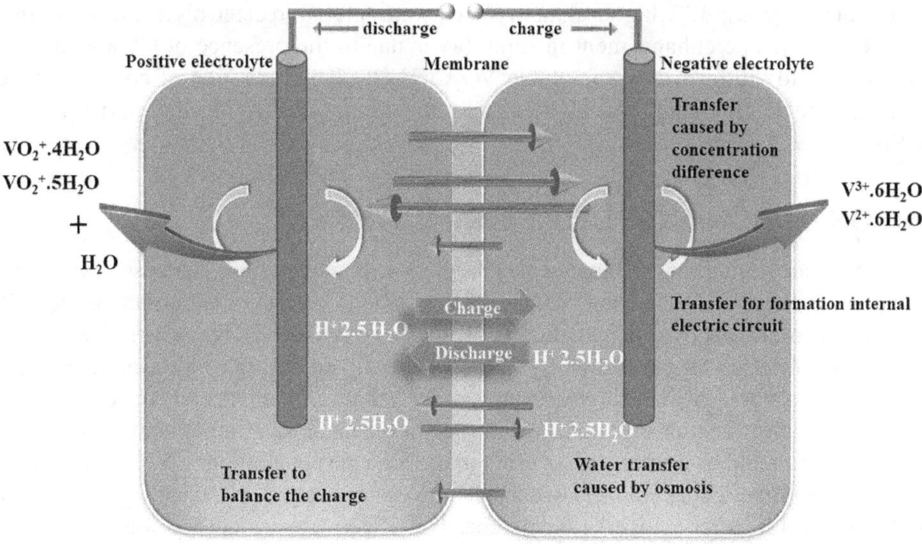

FIGURE 14.9 Transfer of water and vanadium ions in self-discharge cycles. Adapted and reproduced with permission from Ref. [44].

As operating temperature increases, the rate of O_2 evolution also increases. It will reduce the rate of oxidation of V^{4+}, which leads to the imbalance of SOC between half cells [46].

Hydrogen evolution at the negative electrolyte:

$$2H^+ + 2e^- \leftrightarrow H_2 \tag{14.17}$$

An increase in operation temperature also results in the increase in hydrogen evolution [47].

14.6.1 Air Oxidation

In the negative electrolyte, the V^{2+} is in contact with air undergo oxidation. Air oxidation of V^{2+} leads to electrolyte imbalance and limits V^{2+} ions in the negative electrolyte. Air oxidation is represented in the following equation:

$$O_2 + 4H^+ + 4V^{2+} \rightarrow 4V^{3+} + 2H_2O \tag{14.18}$$

To reduce the contact of V^{2+} ions, vacant space can be filled with inert gases. Higher concentrations of vanadium ion and sulfuric acid reduce air oxidation. Higher concentration of vanadium ions that is V^{3+} concentration is high, and high sulfuric acid concentration results in the high concentration of ions such as HSO_4^-, SO_4^{2-}, and H^+, which act as a barrier in the interface, and vanadium salts formed by combining V^{2+} ions with free sulfuric ions present increases the stability [48].

14.6.2 Seawater as Alternative

To reduce the cost of production, Wei et al. [49] used seawater as a substitute for deionized water (DI) as the use of DI water increases the overall cost of production. The impurities present in the electrolyte will decrease the diffusion coefficient of V^{3+} ions which in turn decreases the voltage efficiency of the battery but the battery achieves high Columbic efficiency and capacity retention. The comparison of Columbic efficiency, energy efficiency, and discharge capacities of VRFBs when using electrolytes made by seawater and DI water is shown in Figure 14.10. The solubility of $VOSO_4$ on seawater is 12% higher than in the DI water-prepared electrolyte and it will increase the energy density. The enhancement in solubility is due to the presence of Cl^- ions in seawater which will lead to the formation of soluble $VO_2Cl(H_2O)_2$. The reduction of cost of production, increased Columbic efficiency, and high capacity retention will increase the market penetration of VRFBs.

14.7 CONCLUSION AND FUTURE PERSPECTIVES

VRFBs are a good option for the large-scale energy storage applications. The use of the same element with a different oxidation state removes the possibility of cross-contamination. VRFBs are under evolution toward a most efficient, most dependable, and cost-effective system. The use of additives and mixed electrolytes can increase the energy density and overall performance of batteries. Output voltage increased while using nonaqueous electrolytes in the place of aqueous electrolytes. For effective working, the transport phenomenon across the membrane, air oxidation, hydrogen, and oxygen evolution should be minimized. In order to increase the penetration of VRFBs to the market, the two most important issues we have to consider are cost reduction and vanadium supply. Further research is needed in other areas like membranes that effectively prevent the water and vanadium crossover, the study about the effects of impurities in electrolytes, and the study of the use of lower purity vanadium raw materials.

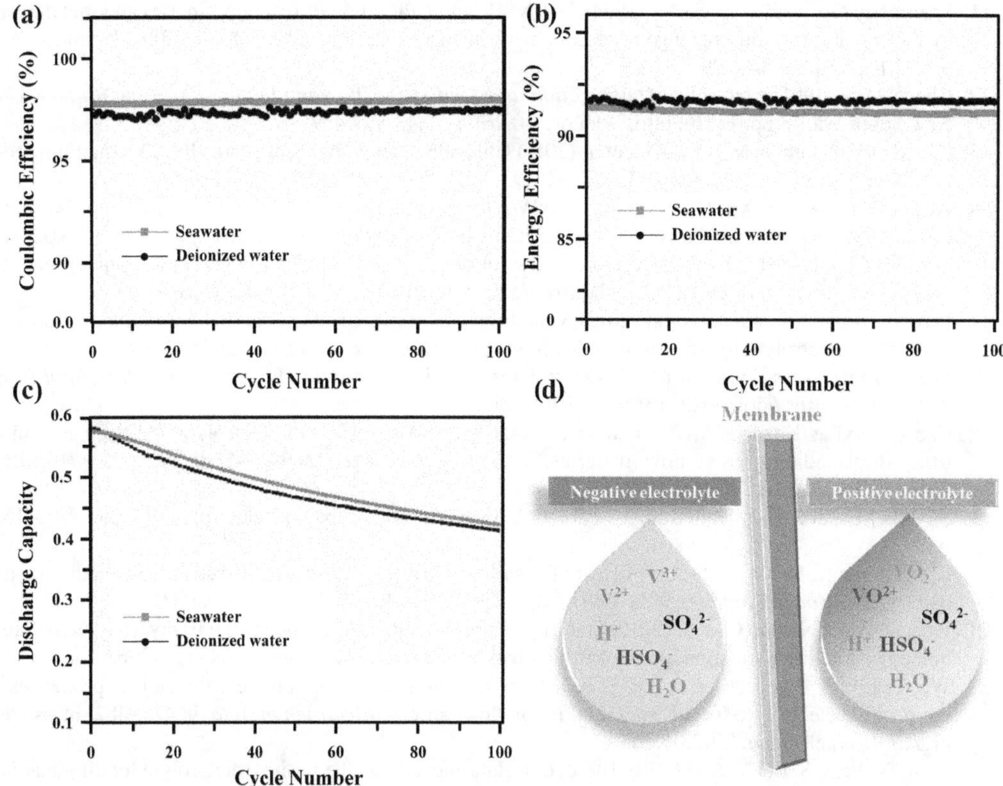

FIGURE 14.10 Cycling tests of VRFBs using different electrolytes: (a) Coulombic efficiency, (b) energy efficiency, (c) discharge capacities of VRFBs at a current density of 100 mA cm^{-2}, and (d) schematic of the electrolyte composition in vanadium redox flow batteries based on the H_2SO_4 electrolyte. Adapted and reproduced with permission from Ref. [49]. Copyright © 2019, Elsevier.

REFERENCES

1. Wu X, Liu J, Xiang X, et al (2014) Conference paper: Electrolytes for vanadium redox flow batteries. *Pure Appl Chem* 86:661–669. https://doi.org/10.1515/pac-2013-1213.
2. Choi C, Kim S, Kim R, et al (2017) A review of vanadium electrolytes for vanadium redox flow batteries. *Renew Sustain Energy Rev* 69:263–274. https://doi.org/10.1016/j.rser.2016.11.188.
3. Lourenssen K, Williams J, Ahmadpour F, et al (2019) Vanadium redox flow batteries: A comprehensive review. *J Energy Storage* 25. https://doi.org/10.1016/j.est.2019.100844.
4. Poli N, Schäffer M, Trovò A, et al. (2021) Novel electrolyte rebalancing method for vanadium redox flow batteries. *Chem Eng J* 405:126583. https://doi.org/10.1016/j.cej.2020.126583.
5. Qasem EA (2020) (12) United States Patent. 2.
6. Kazacos M, Cheng M, Skyllas-Kazacos M (1990) Vanadium redox cell electrolyte optimization studies. *J Appl Electrochem* 20:463–467. https://doi.org/10.1007/BF01076057.
7. Vijayakumar M, Li L, Graff G, et al (2011) Towards understanding the poor thermal stability of V^{5+} electrolyte solution in vanadium redox flow batteries. *J Power Sources* 196:3669–3672. https://doi.org/10.1016/j.jpowsour.2010.11.126.
8. Li L, Kim S, Wang W, et al (2011) A stable vanadium redox-flow battery with high energy density for large-scale energy storage. *Adv Energy Mater* 1:394–400. https://doi.org/10.1002/aenm.201100008.
9. Rahman F, Skyllas-Kazacos M (1998) Solubility of vanadyl sulfate in concentrated sulfuric acid solutions. *J Power Sources* 72:105–110. https://doi.org/10.1016/S0378-7753(97)02692-X.
10. Rahman F, Skyllas-Kazacos M (2009) Vanadium redox battery: Positive half-cell electrolyte studies. *J Power Sources* 189:1212–1219. https://doi.org/10.1016/j.jpowsour.2008.12.113.

11. Pozarnsky GA, McCormick AV (1994) 51V NMR and EPR study of reaction kinetics and mechanisms in V_2O_5 gelation by ion exchange of sodium metavanadate solutions. *Chem Mater* 6:380–385. https://doi.org/10.1021/cm00040a009.
12. Cheng MIN (1991) *Electrolyte Optimization and Studies for the Vanadium Redox Flow Battery.* Univ New South Wales, pp. 1–216. https://doi.org/10.26190/unsworks/4591.
13. Kim S, Vijayakumar M, Wang W, et al (2011) Chloride supporting electrolytes for all-vanadium redox flow batteries. *Phys Chem Phys* 13:18186–18193. https://doi.org/10.1039/c1cp22638j.
14. Kim S, Thomsen E, Xia G, et al (2013) 1 kW/1 kWh advanced vanadium redox flow battery utilizing mixed acid electrolytes. *J Power Sources* 237:300–309. https://doi.org/10.1016/j.jpowsour.2013.02.045.
15. Tang C, Zhou D (2012) Methanesulfonic acid solution as supporting electrolyte for zinc-vanadium redox battery. *Electrochim Acta* 65:179–184. https://doi.org/10.1016/j.electacta.2012.01.036.
16. Peng S, Wang NF, Wu XJ, et al (2012) Vanadium species in CH_3SO_3H and H_2SO_4 mixed acid as the supporting electrolyte for vanadium redox flow battery. *Int J Electrochem Sci* 7:643–649.
17. Wu X, Liu J, Xiang X, et al (2014) Electrolytes for vanadium redox flow batteries. *Pure Appl Chem* 86:661–669. https://doi.org/10.1515/pac-2013-1213.
18. Cao L, Skyllas-Kazacos M, Menictas C, Noack J (2018) A review of electrolyte additives and impurities in vanadium redox flow batteries. *J Energy Chem* 27:1269–1291. https://doi.org/10.1016/j.jechem.2018.04.007.
19. Skyllas M (2002) (12) United States Patent. 2. Vanadium redox battery electrolyte. US 2004/0241552 A1.
20. Li S, Huang K, Liu S, et al (2011) Effect of organic additives on positive electrolyte for vanadium redox battery. *Electrochim Acta* 56:5483–5487. https://doi.org/10.1016/j.electacta.2011.03.048.
21. Peng S, Wang N, Gao C, et al (2012) Stability of positive electrolyte containing trishydroxymethyl aminomethane additive for vanadium redox flow battery. *Int J Electrochem Sci* 7:4388–4396.
22. Wu X, Liu S, Wang N, et al (2012) Influence of organic additives on electrochemical properties of the positive electrolyte for all-vanadium redox flow battery. *Electrochim Acta* 78:475–482. https://doi.org/10.1016/j.electacta.2012.06.065.
23. Liang X, Peng S, Lei Y, et al (2013) Effect of l-glutamic acid on the positive electrolyte for all-vanadium redox flow battery. *Electrochim Acta* 95:80–86. https://doi.org/10.1016/j.electacta.2013.01.138.
24. Lei Y, Liu S, Gao C, et al (2013) Effect of amino acid additives on the positive electrolyte of vanadium redox flow batteries. *J Electrochem Soc* 160:A722–A727. https://doi.org/10.1149/2.006306jes.
25. He Z, Liu J, Han H, et al (2013) Effects of organic additives containing-NH_2 and-SO_3H on electrochemical properties of vanadium redox flow battery. *Electrochim Acta* 106:556–562. https://doi.org/10.1016/j.electacta.2013.05.086.
26. Mulcahy J, Summers K, Chidambaram D (2017) Effect of quinone additives on the performance of electrolytes for vanadium redox flow Batteries. *J Appl Electrochem* 47:1173–1178. https://doi.org/10.1007/s10800-017-1109-7.
27. Roe S, Menictas C, Skyllas-Kazacos M (2016) A high energy density vanadium redox flow battery with 3 M vanadium electrolyte. *J Electrochem Soc* 163:A5023–A5028. https://doi.org/10.1149/2.0041601jes.
28. Kausar N, Mousa A, Skyllas-Kazacos M (2016) The effect of additives on the high-temperature stability of the vanadium redox flow battery positive electrolytes. *ChemElectroChem* 3:276–282. https://doi.org/10.1002/celc.201500453.
29. Mousa A, Skyllas-Kazacos M (2015) Effect of additives on the low-temperature stability of vanadium redox flow battery negative half-cell electrolyte. *ChemElectroChem* 2:1742–1751. https://doi.org/10.1002/celc.201500233.
30. Skyllas-Kazacos M, Peng C, Cheng M (1999) Evaluation of precipitation inhibitors for supersaturated vanadyl electrolytes for the vanadium redox battery. *Electrochem Solid-State Lett* 2:121–122. https://doi.org/10.1149/1.1390754.
31. Rahman F, Skyllas-Kazacos M (2017) Evaluation of additive formulations to inhibit precipitation of positive electrolyte in vanadium battery. *J Power Sources* 340:139–149. https://doi.org/10.1016/j.jpowsour.2016.11.071.
32. Fühl M, Gerber T, Küttinger M, et al (2017) The role of phosphate additive in stabilization of sulphuric-acid-based vanadium (V) electrolyte for all-vanadium redox- flow batteries. 363:234–243. https://doi.org/10.1016/j.jpowsour.2017.07.100.
33. Zhang ZH, Wei L, Wu MC, et al (2021) Chloride ions as an electrolyte additive for high performance vanadium redox flow batteries. *Appl Energy* 289. https://doi.org/10.1016/j.apenergy.2021.116690.
34. Skyllas-Kazacos M, Grossmith F (1987) Efficient vanadium redox flow cell. *J Electrochem Soc* 134:2950–2953. https://doi.org/10.1149/1.2100321.

35. Oei DG (1985) Permeation of vanadium cations through anionic and cationic membranes. *J Appl Electrochem* 15:231–235. https://doi.org/10.1007/BF00620938.
36. Moskalyk RR, Alfantazi AM (2003) Processing of vanadium: A review. *Miner Eng* 16:793–805. https://doi.org/10.1016/S0892-6875(03)00213-9.
37. Kubata M, Nakaishi O, Nobuyuki T (2012) Electrolyte for redox flow battery (US patent). 2.
38. Burch AW (2015) Impurity Effects in All-Vanadium Redox Flow Batteries. Master Theses, TRACE: Tennessee Research and Creative Exchange.
39. Huang F, Zhao Q, Luo CH, et al (2012) Influence of Cr^{3+} concentration on the electrochemical behavior of the anolyte for vanadium redox flow batteries. *Chinese Sci Bull* 57:4237–4243. https://doi.org/10.1007/s11434-012-5302-0.
40. Wen Y, Xu Y, Cheng J, et al (2013) Investigation on the stability of electrolyte in vanadium flow batteries. *Electrochim Acta* 96:268–273. https://doi.org/10.1016/j.electacta.2013.02.091.
41. Agar E, Knehr KW, Chen D, et al (2013) Species transport mechanisms governing capacity loss in vanadium flow batteries: Comparing nafion(r) and sulfonated radel membranes. *Electrochim Acta* 98:66–74. https://doi.org/10.1016/j.electacta.2013.03.030.
42. Tang A, Bao J, Skyllas-kazacos M (2011) Dynamic modelling of the effects of ion diffusion and side reactions on the capacity loss for vanadium redox flow battery. *J Power Sources* 196:10737–10747. https://doi.org/10.1016/j.jpowsour.2011.09.003.
43. Luo Q, Li L, Wang W, et al (2013) Capacity decay and remediation of nafion-based all-vanadium redox flow batteries. *ChemSusChem* 6:268–274. https://doi.org/10.1002/cssc.201200730.
44. Sun C, Chen J, Zhang H, et al (2010) Investigations on transfer of water and vanadium ions across Nafion membrane in an operating vanadium redox flow battery. *J Power Sources* 195:890–897. https://doi.org/10.1016/j.jpowsour.2009.08.041.
45. Ulaganathan M, Aravindan V, Yan Q, et al (2016) Recent advancements in all-vanadium redox flow batteries. *Adv Mater Interfaces* 3:1–22. https://doi.org/10.1002/admi.201500309.
46. Shah AA, Walsh FC (2010) Modelling the effects of oxygen evolution in the all-vanadium redox flow battery. *Electrochim Acta* 55:3192–3205. https://doi.org/10.1016/j.electacta.2009.12.085.
47. Shah AA, Walsh FC (2010) Dynamic modelling of hydrogen evolution effects in the all-vanadium redox flow battery. *Electrochim Acta* 55:1125–1139. https://doi.org/10.1016/j.electacta.2009.10.022.
48. Jirabovornwisut T, Arpornwichanop A (2019) A review on the electrolyte imbalance in vanadium redox flow batteries. *Int J Hydrogen Energy* 44:24485–24509. https://doi.org/10.1016/j.ijhydene.2019.07.106.
49. Wei L, Zeng L, Wu MC, et al (2019) Seawater as an alternative to deionized water for electrolyte preparations in vanadium redox flow batteries. *Appl Energy* 251. https://doi.org/10.1016/j.apenergy.2019.113344.

15 Aqueous Acidic Redox Flow Batteries

Kothandaraman Ramanujam and Chinmaya Mirle

15.1 INTRODUCTION

The electricity demand is set to increase up to 36.5 trillion kWh by 2040 with the ever-increasing population on earth. Limited availability of fossil resources coupled with global warming drives the nations across the world to shift from carbon-rich fossil fuels to the clean, renewable energy sources. Therefore, electricity production from the solar and wind, two of the earth's most abundant energy sources, is growing exponentially. As the energy production from these sources is intermittent, often there is a mismatch in the time of supply and demand, requiring to deploy energy storage systems to iron out these fluctuations to improve the grid stability and service quality. The urge to use energy storage devices gains prominence, especially when renewable energy to the grid crosses 20%. For example, the electrical form of energy from the solar farm is associated with sunlight availability, which can only happen in the day time. Therefore, having an energy storage system that effectively stores and uses the energy derived from renewable energy sources is mandatory. The anticipated mega-scale of stored energy requires next-generation technologies that offer the utmost safety, affordability, environmental benignity and operational reliability. Towards this end, rechargeable batteries are considered superior among various energy storage technologies, as they can cover a wide range of storage time and capacity to satisfy different grid applications. In principle, any rechargeable battery technologies can be adopted for energy storage application. Given the 4–12 hours of continuous operation required per day, it necessitates the availability of systems that offer substantially higher durability and cycle life than lithium-ion batteries (LIBs). The prime concerns of LIBs are irreversible loss of capacity due to phase transformations even under ideal conditions and thermal runaway induced fire hazards associated with the organic electrolytes are almost absent in redox flow batteries (RFBs).

RFBs are shining star for large-scale energy storage applications. It can decouple energy and power, meaning the power-producing element (electrode) is physically away from the energy-containing electrolytes stored in the tanks. Energy conversion occurs when the redox electrolyte is flown through the porous electrodes in the stack (Figure 15.1). The electrodes merely serve as an active site to carry out the charge transfer reaction without taking part in electrochemical reactions. This feature of RFBs makes it easily scalable while maintaining intrinsic safety. As the charged electrolytes (positive electrolyte and negative electrolyte) are being confined in separate tanks outside the electrode, the chance of thermal runaway is meagre, and it is a prime advantage of RFBs and more so when water is used as a solvent in preparing the electrolytes. Besides, the capacity degradation associated with phase changes is averted in RFBs. The redox-active compound is dissolved in the electrolyte and remains in a solution state in both charged and discharged conditions. This option is absent in hybrid RFBs wherein metal deposition at the negative electrode is involved during charging. Further, RFBs are energy-efficient (EE >80%), requiring a simple control and monitoring system in contrast to LIBs. The vital part of RFB is the redox-active materials. They need to be profusely soluble, possess suitable redox potential, and show good stability and be able to procure at low cost. These metrics directly affect the energy density, cell voltage, cycle life and economics of RFB.

 DOI: 10.1201/9781003310167-15

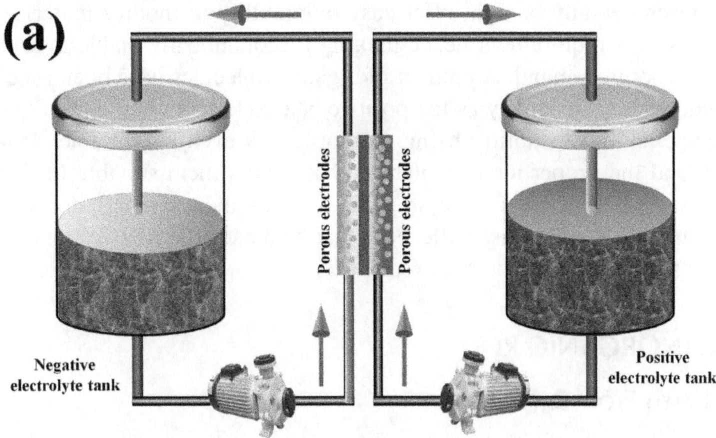

(a) Higher capacity with low power

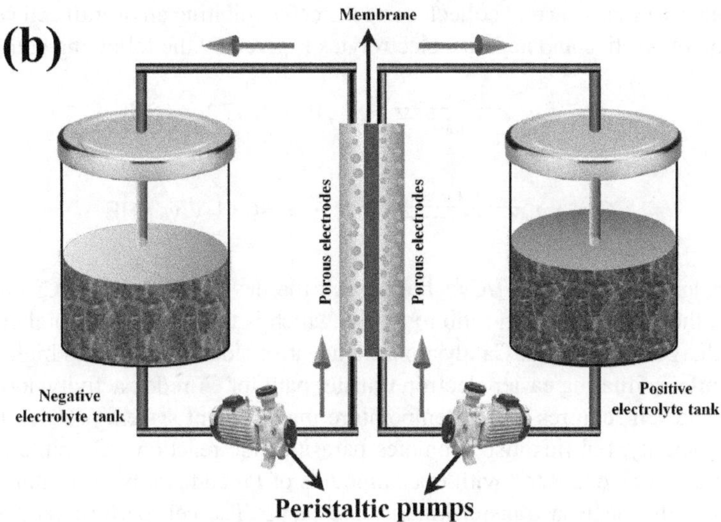

(b) Higher power with low capacity

FIGURE 15.1 The energy and power decoupling advantage of RFB by modulating solution tank (capacity) and electrode area of the cell, respectively.

Among all the RFBs, vanadium redox flow battery (VRFB) and zinc-Br_2 RFB have been extensively researched for the past two decades, and many large-scale installations of VRFBs are under service [1–3]. For the same reason, this chapter will delve into the working and development of VRFB in detail. The cost of VRFB is projected to be \$300–500 per kWh, with a projected lifetime of about 20 years. Although VRFB is mature, high cost of redox-active materials and system components such as Nafion membrane, lower energy density (<50 Wh L^{-1}) and dangers of handling toxic active material limit their widespread use. Though Zn-based RFBs provide energy density >50 Wh L^{-1}, other concurrent drawbacks such as Zn dendrite growth, the crossover of Br_2, gas evolutions and sluggish kinetics dim its popularity. Hence, contemporary research is on identifying new redox molecules which are low cost and abundant besides being durable and safe. In this pursuit, redox-active organic molecules have attracted a lot of attention in the recent past. Organic molecules are environmentally benign, offer easy tailorability to modify material properties and be synthesised as per the requirement hence making it economically viable to scale it up to meet the ever-increasing energy demand. A number of organic molecules have been used as the negative electrode material, whereas its utility as the positive electrode is limited due to sparse availability of suitable molecule and also stability being the bottleneck even if available. This work aims to study and understand the properties of materials that entails them suitable for long-term energy storage. The following sections deal in-depth about the prominent research done so far in aqueous acidic RFBs. We unearth the reasons/bottlenecks that hold back the RFB systems not reaching the large-scale energy storage application requirements.

15.2 ACIDIC INORGANIC RFBs

15.2.1 Iron-Based Flow Batteries

Iron and chromium (Fe-Cr) flow cell was the first ever aqueous RFB system reported as early as 1974 by Thaller in NASA [3]. But the development was cut short due to extensive crossover of the electrolytes across the membrane, poor redox kinetics and instability of Cr^{3+}. The electrolytes comprised of 1 M $CrCl_3$ and $FeCl_2$ in 3 M HCl in the negative and positive sides of the cell, respectively, with graphite felt acting as a current collector and the cell exhibiting an overall cell voltage of 1.18 V. The cell reaction of positive and negative electrolytes is given by the following equations:

$$Fe^{3+} + e^- \underset{\text{Charge}}{\overset{\text{Discharge}}{\rightleftarrows}} Fe^{2+}, E_0 = 0.77 \text{ V vs SHE} \tag{15.1}$$

$$Cr^{3+} + e^- \underset{\text{Charge}}{\overset{\text{Discharge}}{\rightleftarrows}} Cr^{2+}, E_0 = -0.41 \text{ V vs SHE} \tag{15.2}$$

The sluggish redox kinetics of Cr^{2+}/Cr^{3+} holds back the development of Fe-Cr cell. Parasitic H_2 evolution limits the cell's utilisation ratio to ~60%, which is the ratio of the total electrolyte available to the discharged electrolyte. Catalyst on the negative side, which offers high overvoltage for H_2 evolution while facilitating easier electron transfer path for Cr redox activity, looks like a promising strategy. The cell requires a high-temperature management system (~60°C) to facilitate the Cr^{3+}/Cr^{3+} redox activity, but this also promotes parasitic side reactions. Graphite felt current collector of the Cr side was deposited with trace amounts of Pb and Au, but no catalyst was used for the Fe reaction as the electron transfer kinetics are facile. The cell performance is hampered by the sluggish redox activity of Cr^{3+}, as it exists in two different speciation states, $Cr(H_2O)_6^{3+}$ and $Cr(H_2O)_5Cl^{2+}$. $Cr(H_2O)_5Cl^{2+}$ is the predominant species involved in the reduction reaction at the electrode surface. The equilibrium constant for the formation of $Cr(H_2O)_5Cl^{2+}$ is 0.27 as in equation 15.3, rendering sluggish kinetics at the negative electrode surface due to concentration polarisation. In an attempt to enhance the cycle life, all iron flow battery was tried using economically viable Fe^{2+} as starting material on both the sides, as shown in equations 15.3 and 15.4. A cell voltage of 1.21 V is expected of the cell.

$$\underset{\text{Electrochemically inactive}}{Cr(H_2O)_6^{3+}} + Cl^- \rightleftarrows \underset{\text{Electrochemically active}}{Cr(H_2O)_5Cl^{2+}} + H_2O \tag{15.3}$$

$$Fe^{2+} + 2e^- \underset{\text{Discharge}}{\overset{\text{Charge}}{\rightleftarrows}} Fe, E_0 = -0.44 \text{ V vs SHE} \tag{15.4}$$

But as seen from the Pourbaix diagram in Figure 15.2a, the potential of Fe-containing species is pH-dependent. Any change in pH during the charge/discharge would alter the potential of the redox couple due to varying H^+ ion concentrations apart from the voltage variation accompanying

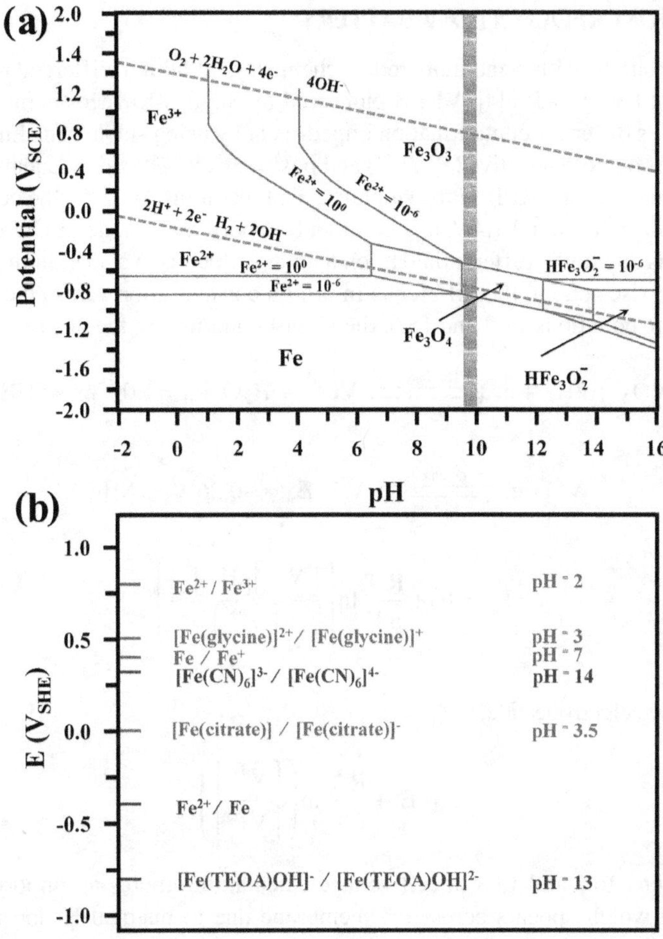

FIGURE 15.2 (a) Pourbaix diagram of Fe compounds showing the variation of potential in a different pH medium. (b) Variation of potential of Fe ions upon complexation with various ligands. Adapted and reproduced from ref. [4]. Copyright 2019 Elsevier.

polarisation phenomena. Fe/Fe^{2+} would show a stable redox potential around -0.6 V from pH of 0 to 10. At pH < 0, hydrogen evolution reaction (HER) is thermodynamically favoured in comparison to Fe/Fe^{2+} redox activity. Moreover, free Fe^{3+} would form electrochemically irreversible higher oxides of iron and rapidly drop the potential beyond pH of 2, making it inappropriate for cell operation. Hence, the strategy of masking the Fe centre by complexing with a suitable chelating ligand like glycine, citrate and cyanide would render stability from the external environment. It has been observed that chelation helps manipulate the redox potential, solubility and stability of the redox-active metal centre, as shown in Figure 15.2b. In addition, the size of the metal centre is increased upon chelation which can effectively minimise the transfer of redox-active ions across the membrane pores, thereby inhibiting the crossover of ions. For instance, Fe^{2+}/Fe^{3+} can be complexed to yield ferrocyanide and ferricyanide, which can be operated in a neutral and highly basic medium which was otherwise unstable in pH above 2. Using the similar strategy of having the same element on both the sides for eliminating crossover issue, vanadium can be explored as it exists in 4 different oxidation states in aqueous acidic medium. VRFB fabricated using vanadium ions on both the sides took a strong foothold in the energy sector with its development in the 1980s.

15.3 VANADIUM REDOX FLOW BATTERY

RFB literature abounds with vanadium redox chemistry, although different redox couples are unveiled in the last few decades [4]. VRFB pioneered by Skyllas-Kazacos is most popular among RFBs as it addresses the cross-contamination impediment by using stable vanadium species at various oxidation states on both negative (V^{3+}/V^{2+}) and positive (V^{4+}/V^{5+}) sides. Equations 15.5 and 15.6 depict the redox reaction in VRFB in the sulphuric acid medium. At 25°C, the cell's standard voltage would be 1.26 V, but around 1.4 V, it is commonly reported at a state of charge (SOC) of 50%. This discrepancy is due to the differential H^+ ion activity, which is excluded in the Nernst equation. Donnan potential arises due to the difference in proton concentration across the electrolyte membrane. According to equations 15.5 and 15.6, the Nernst equation for the positive electrode side is

$$VO_2^+ + 2H^+ + e^- \underset{\text{Charge}}{\overset{\text{Discharge}}{\rightleftarrows}} VO^{2+} + H_2O,\ E_0 = 1.00\text{ V vs SHE} \tag{15.5}$$

$$V^{3+} + e^- \underset{\text{Charge}}{\overset{\text{Discharge}}{\rightleftarrows}} V^{2+},\ E_0 = -0.26\text{ V vs SHE} \tag{15.6}$$

$$E_+ = E_+^0 + \frac{RT}{nF}\ln\left(\frac{[V^{5+}][H^+]_+^2}{[V^{4+}]}\right) \tag{15.7}$$

And for the negative electrode side,

$$E_- = E_-^0 + \frac{RT}{nF}\ln\left(\frac{[V^{3+}]}{[V^{2+}]}\right) \tag{15.8}$$

The use of equations 15.7 and 15.8 in cell voltage calculation ignores an important factor due to Donnan potential, which appears across the membrane due to unequal H^+ ion concentrations of positive and negative electrolytes. Figure 15.3 shows the origin of the Donnan potential. The difference in H^+ ion concentration is originating from the method adopted in preparing the electrolyte: (i) chemical dissolution method involving V_2O_5 and V_2O_3, (ii) from V_2O_5 suspended electrolysis in sulphuric acid and (ii) from electrolysis of $VOSO_4$.

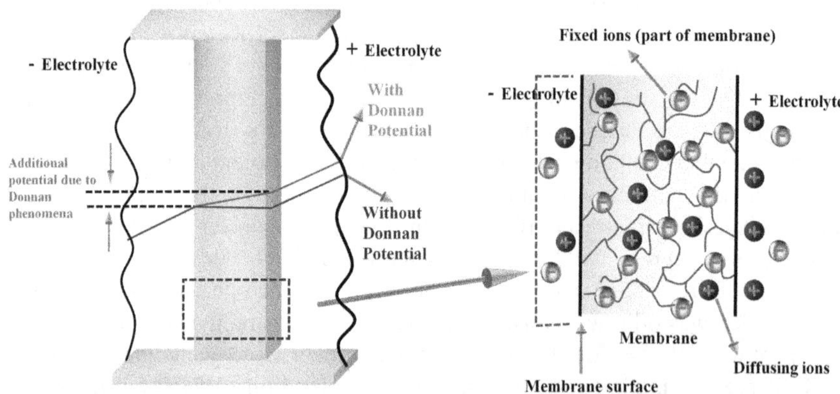

FIGURE 15.3 (a) Open circuit voltage (OCV) with and without membrane potential. (b) Qualitative representation of double layer on either side of the ion-exchange membrane.

i. In the chemical dissolution method, V_2O_3 is dissolved in H_2SO_4 to prepare a solution of V^{3+}. This is followed by the addition of a predetermined quantity of V_2O_5 as to chemically react with V_2O_5 to form an equimolar mixture of V^{3+}/VO^{2+} ($V^{3.5+}$). The same solution is taken on both sides of VRFB, where it undergoes initial charging to convert $V^{3.5+}$ to V^{4+} in the positive side and to V^{3+} in the negative side to acquire 0% SOC [5].
ii. The suspended powder electrolysis of V_2O_5 in H_2SO_4 is used to obtain $V^{3.5+}$ (an equimolar mixture of VO^{2+} and V^{3+} as in equation 15.9. $V^{3.5+}$ solution is kept in both positive and negative sides for initial charging to convert $V^{3.5+}$ to V^{4+} in the positive side and to V^{3+} in the negative side, as shown in equations 15.10 and 15.11.

$$V_2O_5 + 5H_2SO_4 + 3e^- + 2\,H^+ \rightarrow VOSO_4 + V^{3+} + 4H_2O \qquad (15.9)$$

At negative side:

$$\begin{aligned} &VOSO_4 + V^{3+} + 4HSO_4^- + xH_2O + e^- + H^+ \left(\text{from positive side}\right) \\ &\quad \rightarrow 2V^{3+} + 2\,SO_4^{2-} + 3HSO_4^- + (x+1)H_2O \end{aligned} \qquad (15.10)$$

At positive side:

$$VOSO_4 + V^{3+} + 4HSO_4^- + xH_2O \rightarrow 2\,SO_4^{2-} + 2\,H_2SO_4 + HSO_4^- + H^+ + (x-1)H_2O + e^- \qquad (15.11)$$

There is a difference in the proton concentration between negative and positive sides during charging, causing the development of Donnan potential.

iii. In the third method of preparation, a solution of 2M $VOSO_4$ in 2M H_2SO_4 is used in both the negative and positive sides. At the end of galvanostatic charging, vanadium in the positive side goes to V^{5+}, while the negative side produces V^{3+}.

$$2VOSO_4 + xH_2O + 2H_2SO_4 \rightarrow 2V^{3+} + 3SO_4^{2-} + (x+2)H_2O + H_2SO_4 \qquad (15.12)$$

Following this, V^{5+} is completely replaced with the original $VOSO_4$ solution ($2VOSO_4+xH_2O+2H_2SO_4$) in the positive side, retaining generated V^{3+} in the negative side to achieve VRFB with 0% SOC. In this method, there is an imbalance in the proton concentration between negative and positive sides to start with, which probably triggers water movement from the negative to the positive side during operation while using the cation exchange membrane.

Since the proton concentration at the negative and positive electrolytes differs, the potential generated at each membrane/electrolyte interface will be asymmetrical. This is the origin of Donnan potential as shown in equation 15.13. Equation 15.14 predicts the OCV including the proton concentration and the Donnan potential [6].

$$E_m = E_m^+ - E_m^- = \frac{R\,T}{F}\ln\left(\frac{[H^+]_+}{[H^+]_-}\right) \qquad (15.13)$$

$$E_{OC} = E_0 + \frac{R\,T}{n\,F}\ln\left(\frac{[V^{5+}][V^{2+}][H^+]_+[H^+]_+^2}{[V^{4+}][V^{3+}][H^+]_-}\right) \qquad (15.14)$$

where the subscripts "+" and "−" denote the positive and negative half-cells, respectively, and "m" stands for membrane potential. During the charging of VRFB, H_2 evolution at the negative electrode side and O_2 evolution at the positive electrode side are two known major side reactions which alter the solution composition and eventually cause electrolyte imbalance. Unlike lead-acid and Li-ion batteries, VRFB does not generate any toxic gases and possess a low risk of explosion. Moreover, as water is the major component of the electrolyte, it acts as a secondary cooling system, reducing the thermal management systems' burden. Most importantly, as same element at its different oxidation states is used in the negative and positive electrolyte sides, it requires minimum concern towards the crossover issue. In theory, vanadium electrolyte recovered from the end-of-life VRFB can be recycled and used for a new VRFB, making the lifetime cost of VRFB lesser than incumbent energy storage systems.

Skyllas-Kazacos et al. [5] explored VRFB with various acidic supporting electrolytes such as HCl, H_2SO_4 and $HClO_4$. Among them, redox couple reversibility and solubility were best in the H_2SO_4 medium. The capacity of VRFB is dependent on the concentration of vanadium ions, which in turn depends on the solubility of vanadium redox couples. The extent of solubility varies with temperature, sulphate ion concentration and the SOC of the VRFB. Positive electrolyte suffers due to V_2O_5 precipitation at temperatures >40°C. Other vanadium salts such as VSO_4, $V_2(SO_4)_3$ and $VOSO_4$ exhibit low solubility at low temperatures. Therefore, the concentration of vanadium, total sulphate ion concentration and operating temperature range were limited to 1.6–2 M, 4–5 M and 10°C–40°C, respectively. V^{5+} species exists in aqueous H_2SO_4 as hydrated penta-coordinated ion [7]. $VO_2(H_2O)_3^+$ undergoes deprotonation at temperatures >40°C forming neutral $VO(OH)_3$ species and precipitates out as V_2O_5 as shown in equations 15.15 and 15.16. This precipitation of V_2O_5 can be avoided by using a high concentration of H_2SO_4.

$$[VO_2(H_2O)_3]^+ \xrightarrow{\Delta} VO(OH)_3 + H_3O^+ \tag{15.15}$$

$$2VO(OH)_3 \rightarrow V_2O_5 \cdot 3H_2O \downarrow \tag{15.16}$$

15.3.1 Strategies for Minimising Electrolyte Degradation

One of the most common ways of degradation of electrolyte is vanadium precipitation as V_2O_5 in the positive electrolyte and as vanadium sulphate salts in the negative electrolyte. Therefore, there were numerous attempts to mitigate the precipitation. A total sulphate concentration of 5 M is suggested to stabilise V^{5+} for warm climates. A lower sulphate concentration of 4 M is suggested to minimise the precipitation of V^{2+} and V^{3+} ions in cold climates. Another fruitful method was the use of additives or complexing agents to improve the saturated electrolyte's solubility and thermal stability. Compared to the organic additives, mostly inorganic additives were preferred due to their stability in highly oxidising V^{5+} solution. Commonly reported additives are ammonium phosphate, potassium phosphate, disodium pyrophosphate, sodium hexametaphosphate, potassium sulphate, ammonium sulphate and Flocon-100 (a polyacrylate material) [8–11].

15.3.2 Role of Membranes and their Properties

The membrane in VRFB is an important component that is generally made from ionomers or polymers. Its role is to act as a separator between the positive and negative electrodes/compartments to physically separate the active species. Above all, it is an electronic insulator and ionic conductor allowing only the transport of protons or sulphate ions, maintaining charge balance within the cell. For the longevity of the cell, there are some stringent requirements followed in selecting the membrane. It should exhibit high ionic conductivity, have low vanadium ion permeability, be cost-effective, be chemically stable against oxidising VO_2^+ (V^{5+}) ions and show optimal water transport properties. High Coulombic efficiency (CE) is associated with high proton conductivity (in the case of cation exchange membrane) and low vanadium crossover (magnitude of vanadium diffusion across the membrane determines the extent of self-discharge of VRFB). As vanadium ion crossover often occurs with other ions to maintain electroneutrality, capacity decay becomes inevitable in VRFB. Sun et al. [12] explored vanadium crossover behaviour with two different membranes, VX-20 anion exchange membrane and Nafion 117 cation exchange membrane. In the case of VX-20, V^{4+} ions are accumulated at the positive side and requires the movement of SO_4^{2-} from the negative side to the positive side during charging to balance charges. Similarly, with Nafion-117, V^{3+} gets accumulated in the negative side, necessitating H^+ movement from the negative side to positive side. Given these environments where diffusion of V^{4+} is expected across the membrane, the membrane is sandwiched between two half-cells to measure the diffusion rate of V^{4+} ions. In one side of the cell, $VOSO_4$ solution was used, and in the other half-cell, $MgSO_4$ solution (same concentration as $VOSO_4$ to avoid osmotic pressure effect) is filled for measuring the permeability of the vanadium ion. Absorbance of $MgSO_4$ solution was periodically measured using UV/Vis spectrophotometer to understand the change in absorption of the solution due to the continuous arrival of V^{4+}. Using the following equation, the diffusion rate of V^{4+} is determined [13]:

$$\ln(B_0 - 2\,C) = \ln B_0 - \frac{2\,k_s\,A\,t}{v} \tag{15.17}$$

where B_0 is the initial absorbance of 1 M $VOSO_4$ and C is the absorbance of the solution containing 1 M $MgSO_4$ at any time. A, t and v are the membrane area exposed to the solution, time and volume of solution, respectively. The slope of the plot of $\ln(B_0 - 2\,C)$ vs t provides the mass transfer coefficient k_s of the V^{4+} ions through the membrane. The diffusion coefficient of the V^{4+} ions is calculated by multiplying the membrane thickness with k_s. In the literature, the Nafion-117 cation exchange membrane was characterised with a V^{4+} diffusion coefficient in the order of $\times 10^{-5}$ to $\times 10^{-4}\,cm^2 min^{-1}$. In general, fluorinated cation exchange membrane was identified with low vanadium crossover and exhibits better chemical stability. For example, polyvinylidene fluoride (PVDF) ultrafiltration membranes exhibited comparable performance to Nafion-117 and exceptionally stable over 1000s of VRFB cycles [14]. Another example is PVDF-graft-poly(styrene sulfonic acid) cation exchange membrane, which revealed higher performance than Nafion-117 due to its high conductivity and extremely low V^{4+} ion permeability. Qiu et al. [15] have sulfonated a styrene-grafted maleic anhydride onto PVDF, which exhibited lower V^{4+} permeability than Nafion-117 and maintained its OCV above 1.3 V in dynamic self-discharge mode for more than twice as long as Nafion-117.

Optimal water transport would avoid precipitation of the vanadium salt in the cell and flooding of the positive electrolyte (in the case of cation exchange membrane) and the negative electrolyte side (in the case of anion exchange membrane). According to Sukkar et al. [16], the direction of water transport is also a function of SOC. They reported water migration to the positive half-cell between 100% and 50% SOC and migration towards the negative side between 50% and 0% SOC. Water transport happens through hydrated vanadium ion crossover, electro-osmotic drag and diffusion via osmotic pressure difference between the half-cell solutions. The extent of water transfer by each of those process is a function of the membrane used. Although electro-osmotic drag due to proton movement across the cation exchange membrane causes a large change in volume, the net transfer per cycle is zero. Therefore, unidirectional water transport by diffusion is considered the primary cause for the electrolyte imbalance. Generally, cation exchange membrane contains groups such as $-SO_3^-$, $-C_6H_4O^-$, $-COO^-$, $-PO_3H^-$ and $-PO_3^{2-}$, and anion exchange membrane contains positively charged groups such as $-NH_3^+$, alkyl-substituted quaternary amines and $-SR_2^+$. Bipolar membranes having a sandwich of cation exchange membrane and anion exchange membrane are called amphoteric ion-exchange membranes. Sukkar and Skyllas-kazacos noted a drop in the resistivity of the membrane during the extended exposure of the membrane to the solution of 0.1 M V^{5+} in 0.25 M H_2SO_4, which they attributed to the erosion of polymeric material of the membrane and embedding of SO_4^{2-} and HSO_4^- ions into the pores of the membrane [17]. Although the decrease in resistivity is a preferable change, the concomitant increase in vanadium crossover causes loss in capacity due to reduced membrane thickness. Huang et al. [18] noted visible fracture in sulfonated polyimide-based membrane after 800 cycles. Through the optical and SEM examinations, they observed no change in the membrane surface facing the negative electrolyte but the presence of micro defects on the membrane surface facing the positive side. Based on the recent reports, it is clear that the chemical stability test in V^{5+} solution is the standard way of testing the new membrane for VRFB application. Membrane oxidation by V^{5+} produces blue-coloured V^{4+}, which is quantified using UV/Vis spectrophotometry by monitoring the change in absorbance with time to understand the rate of degradation of the membrane [17].

15.3.3 Electrode Materials and their Degradation Pathways

VRFB was first commercialised in 1997 in Japan, but continues to suffer from issues which are intertwined such as contamination, electrolyte imbalance and cell component corrosion due to corrosive acid and OER at the positive electrode side leading to a loss in capacity and power density [20]. Electrodes are a vital component of VRFB on which redox couple undergoes charge-transfer reaction. Therefore, it has to be stable in the operating potential window and has to sustain highly oxidising acidic environment exhibiting high overpotential for OER and HER at the positive and negative electrodes, respectively. Acidic electrolyte mandates the requirement of corrosion-proof carbon materials to serve as the current collectors and electrodes. But in reality, the graphite felt electrodes undergo oxidation during the charging event of the VRFB, which necessitates modification to the graphite electrodes to improvise the lifetime and the electrical performance of the battery [19]. The maximum charge voltage reported in the literature is around 1.6 to 1.8 V and the lowest discharging voltage reported is about 1 to 0.8 V. Charging is done either in constant current mode or with the combination of constant current (up to a certain voltage) and constant voltage mode to ensure complete charging of the VRFB. The cell voltage of charging differs from the discharging due to overpotential associated with sluggish kinetics of the redox reactions at the felt electrode and electrolyte interface, ohmic resistance associated with electrolyte resistance and poor contact between membrane, felt electrode, current collector, etc. CE within the cycle is calculated from the ratio of charge recovered during discharging and charge spent during charging. Voltage efficiency (VE) is the ratio between the mean discharging voltage and the mean charging voltage at constant current. The product of CE and VE indicates the energy efficiency (EE).

To improve the EE, the voltage losses associated with various types of polarisation, namely activation, ohmic and concentration overpotential, need to be minimised as much as possible. Among these overpotentials, activation overpotential wholly depends on the electrode material used. Electrode performance is generally studied using Tafel analysis calculating the transfer coefficient and exchange current density. Graphite disc electrode exhibited a transfer coefficient of 0.7 and an exchange current density on the order of 10^{-4} A cm^{-2} for V^{5+}/V^{4+} redox reaction [21]. Further, the exchange current density values of 0.9×10^{-2} and 0.8×10^{-2} A cm^{-2} were reported for graphite felt and reticulated vitreous carbon electrodes. This enhancement is credited to the higher surface area of these electrodes. In 1993, Skyllas-Kazacos et al. [22] examined rayon- and polyacrylonitrile-based carbon felts. Although both rayon and polyacrylonitrile were subjected to a similar process to obtain graphite felt, the resultant felts were different in their surface area and surface microstructure. Their X-ray photoelectron spectroscopy analysis indicated rayon-based felt was vulnerable to oxidation due to the presence of more edge planes and defects leading to higher surface-oxygen concentration (carbonyl groups) when subjected to thermal treatment. Although carbon degrades at high anodic potentials due to OER, Skyllas-Kazacos established that with strict voltage control during charging, this problem could be jettisoned, permitting a very long cycle life to be realised. Following this, Sumitomo in Japan demonstrated a 6 MWh VRFB system with an integrated wind farm in Hokkaido in 2005 for more than 200,000 cycles using carbon-based electrodes [23]. Other carbon materials such as carbon paper, carbon cloth and carbon nanotubes were also reported for VRFB application [24]. Among these, carbon cloth exhibited a high rate of mechanical degradation

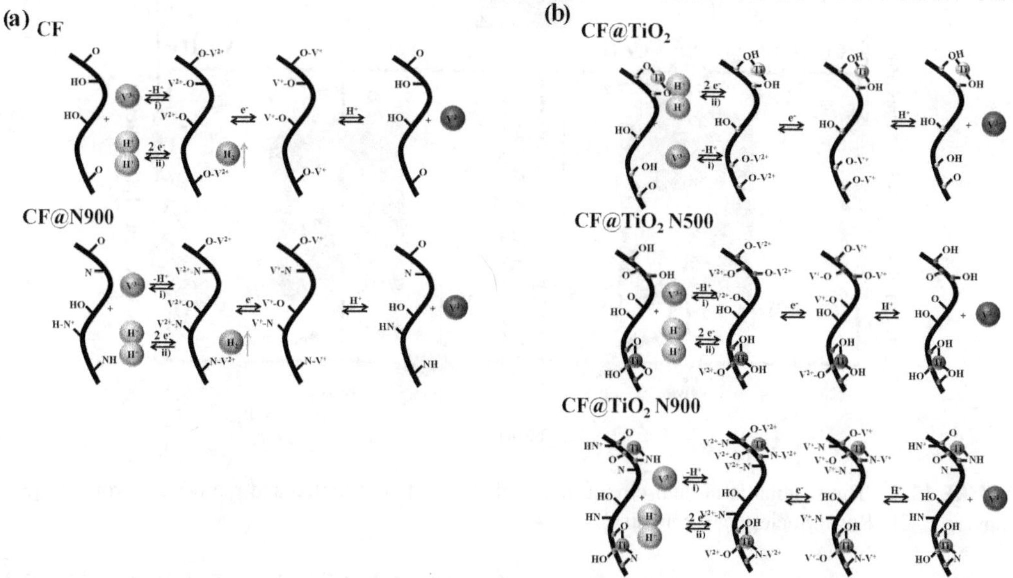

FIGURE 15.4 A plausible way for the redox reactions of V^{2+}/V^{3+} proposed by Vazquez-Galvan et al. [25]: (a) graphite felt and nitrogen modified (at 900°C) graphite felt and (b) TiO_2 and TiO_2-nitrogen modified (at 500°C and 900°C) graphite felt. Adapted and reproduced from ref. [25]. Copyright 2019 Elsevier.

especially during charging, which was attributed to the low surface area of the cloth materials and limited vanadium diffusion into the cloth material due to the flow-by configuration of the electrolyte leading to increased polarisation and OER.

In general, modifications to negative electrodes involved electrochemical activation of graphite, thermal activation of graphite felt in air, modification of carbon material with metal oxides and doping of heteroatoms such as nitrogen, oxygen, and phosphorous [27,28]. Graphite felt doped with nitrogen, oxygen and decorated with rutile TiO_2 was developed by Vazquez-Galvan et al. [25] which promoted negative electrode reaction with concomitant suppression of HER during charging. Figure 15.4 shows the proposed mechanism by Vazquez-Galvan et al. [26]. Similarly, flexible negative electrode synthesized by carbon nanofiber that is embedded in TiO_2 via electrospinning technique was also reported [29]. Nitrogen-doped carbon material contains various nitrogen functional groups attached to carbon, such as pyridinic-N, pyrrolic N-, quaternary N- and N-oxides, and the presence of these functional groups improves the discharge capacity and EE of VRFB in comparison to the bare carbon nanofibre electrodes.

15.3.4 Dual-Circuit Flow Battery

Dual-circuit flow battery was first introduced by Amstutz et al. [30] to realise concomitant energy storage and conversion. In a typical RFBs, electrical energy is stored as chemical energy in the positive and negative electrolytes until connected to a load. In the dual-circuit model, the charged electrolytes were chemically discharged by passing through catalytic beds in separate external tanks. This mode of operation alleviates the low energy density associated with the RFBs as it could be used to continuously produce H_2, a high energy density material. It is an alternative to water electrolysis, where expensive catalytic materials such as Pt are used and the electrolyzers suffer due to H_2–O_2 recombination and electrode degradation during O_2 evolution. Additionally, in dual-circuit mode, non-precious catalysts were used to obtain H_2, which is expected to bring down the H_2 production infrastructure's capital cost. The redox couples of RFB were selected according to their potential to oxidise H_2O and reduce H^+ in acidic condition on a catalyst bed.

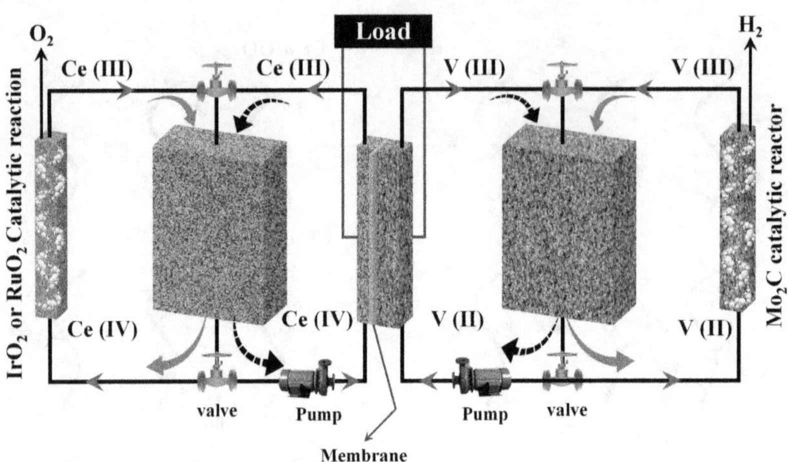

FIGURE 15.5 The schematic of dual-circuit model of V-Ce RFB. Adapted and reproduced from ref. [30]. Copyright 2014 Royal Society of Chemistry.

RFB designed by Amstutz et al. [30] consists of Ce^{4+}/Ce^{3+} in the positive electrolyte and V^{2+}/V^{3+} in the negative electrolyte. In this system, indirect water splitting occurs on a separate catalytic bed in parallel with the RFB unit using the charged redox species of V-Ce RFB. Figure 15.5 shows the dual-circuit V-Ce RFB, which can deposit the surplus energy beyond the capacity of the RFB in the form of H_2 gas. It is thus an add-on to the conventional RFBs to tap the electricity that would otherwise be lost when the RFB is at full capacity. They have demonstrated the production of H_2 by chemically discharging acidic V^{2+} solution over molybdenum carbide catalyst. Equation 15.18 shows the chemical reaction pertaining to H_2 production. The HER was monitored by following the intensity of absorption peak at 850 nm, which relates to V^{2+}, using a UV–Visible spectrometer. In a separate experiment, the amount of H_2 evolved was calculated by mixing 40 mM V^{2+} solution with Mo_2C catalyst. From the data of the first 3 minutes of the reaction, the reaction order with respect to V^{2+} was found to be one. In a similar way, Ce^{4+} was used to decompose water into oxygen and protons in the presence of IrO_2 or RuO_2 (equation 15.19). The choice of acid electrolyte changes the redox potential of Ce from 1.28 V in 1 M HCl to 1.7 V vs SHE in 1 M $HClO_4$. The redox potential of V^{2+}/V^{3+} is −0.26 V vs SHE and vanadium is soluble up to 5 M in sulphuric acid [31]. From these redox potentials, it is evident that thermodynamically both V^{2+} and Ce^{4+} are capable of splitting water (Figure 15.6), but the lack of H_2 and O_2 evolution indicates that the process is kinetically an uphill one that requires catalysts such as Mo_2C and IrO_2.

$$2H^+ + 2V^{2+} \xrightarrow{Mo_2C} H_2 + 2V^{3+} \tag{15.18}$$

$$4\,Ce^{4+} + 2H_2O \xrightarrow{Iro_2/Sio_2} O_2 + 4\,Ce^{3+} + 4\,H^+ \tag{15.19}$$

15.4 Fe/V CHEMISTRY

Operating VRFB at 50°C, V^{5+} would precipitate as V_2O_5, while V^{2+} would precipitate as VSO_4 when operated at lower temperature limiting the cell operation to 10°C–50°C. Fe/V system aims to use best of both the worlds of Fe/Cr and VRFB chemistry in pursuit of making better battery for grid-level energy storage application. Cr^{2+}/Cr^{3+} has a redox potential of −0.41 V, which is lower than that of V^{2+}/V^{3+} having −0.25 V thereby enhancing the possibility of H_2 evolution. Intrinsically, V^{2+}/V^{3+} has better redox activity than Cr^{2+}/Cr^{3+}, which eliminates the use of catalyst and the necessity of high-temperature management system. Using electrochemically reversible Fe^{2+}/Fe^{3+}, highly

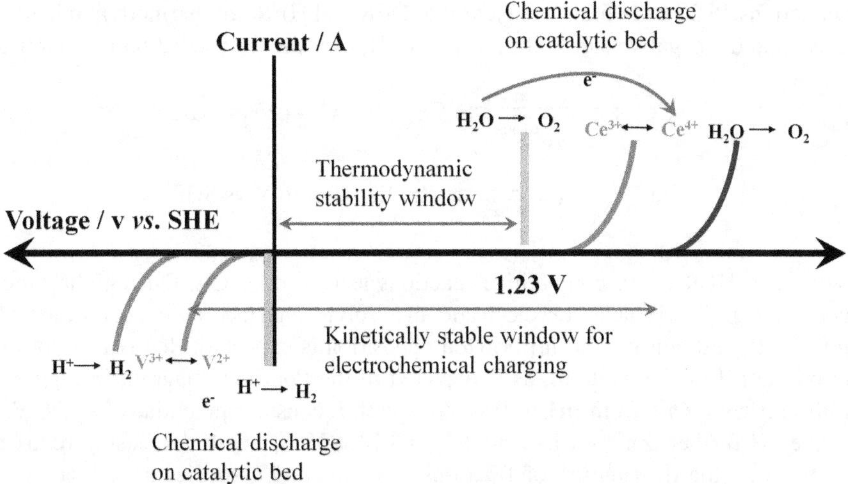

FIGURE 15.6 Thermodynamic stability window for water (dotted line). One moving outwards from the thermodynamic stability window is the electrochemical reactions for V-Ce cell followed by the electrochemical stability window for water.

oxidative V^{5+} is eliminated, ensuring the stability of cell components, especially the membrane. Initial making of the battery involved making the electrolytes with VCl_3 and $FeCl_2$ dissolved in 3 M HCl. Scaling up of this chemistry is held back due to the poor energy density associated with the limited solubility. Although V^{2+}/V^{3+} has shown a very high solubility of 4 M in HCl, the solubility of F^{2+}/Fe^{3+} is limited. The use of mixed electrolyte is the commonly encountered strategy to enhance solubility. H_2SO_4 was introduced to lower the chloride ion concentration and hence improve the solubility of Fe. With this battery operation using Fe^{2+}/Fe^{3+}, O_2 evolution as the overcharging reaction at the positive electrode leading to carbon electrode disintegration can be avoided. Because of the minimal parasitic side reactions in Fe/V cells, the percentage utilisation ratio of electrolyte is close to 100, which otherwise was around 80.

15.5 Zn/Ce-BASED FLOW BATTERY

Zn is a very attractive negative electrode material. It can be used as an anode in a wide range of pH using suitable electrolytes (–0.76 V_{SHE} for Zn^{2+}/Zn in acidic and neutral media, –1.26 V_{SHE} for $Zn(OH)_4^{2-}/Zn$ in alkaline medium). It has very low redox potential in aqueous medium and offers excellent redox kinetics requiring minimal overvoltage during charge–discharge. Zn-Ce RFB has been extensively studied since its inception in 2005 as it exhibits an OCV of 2.2 V, which is considered very high among the aqueous RFBs. The electrodes have to operate in high acidic medium and also withstand the oxidising power of Ce. Preferably platinised Ti mesh is used as a positive electrode current collector as carbon electrodes can't withstand high oxidising power of Ce^{4+} and deteriorate over time. Methanesulfonic acid (MSA) is used as electrolyte as it gives good reversibility for redox activity and solubility of 0.8 M for Ce, which is ten times higher than in sulphuric acid. Zn^{2+} is soluble up to 2 M in MSA, thereby enhancing the energy density of the cell. Zn/Zn^{2+} redox activity is comparatively sluggish as compared with Ce^{3+}/Ce^{4+} kinetics. Employing mixed electrolyte (MSA and H_2SO_4 in 1:1 ratio) at the negative side enhances the electron transfer process and the exchange current density, reducing the overvoltage during charge/discharge [32]. The use of mixed electrolyte also enjoys the advantage of improved solubility of Ce salt along with enhanced mass transfer and electron transfer parameters. Half-cell reactions of the positive electrode and negative electrode are shown in equations 15.17 and 15.18, respectively. From the redox potential values shown in equations 15.20 and 15.21, it is clear that OER and HER are thermodynamically favoured at the positive and the negative electrodes, respectively, over the Ce and Zn redox chemistry.

$$Ce^{4+} + e^- \underset{\text{Charge}}{\overset{\text{Discharge}}{\rightleftarrows}} Ce^{3+}, E^0 = 1.44 \text{ V vs SHE} \tag{15.20}$$

$$Zn^{2+} + 2e^- \underset{\text{Discharge}}{\overset{\text{Charge}}{\rightleftarrows}} Zn, E^0 = -0.76 \text{ V vs SHE} \tag{15.21}$$

Zn deposition and HER being competitive reactions lead to poor CE. During charging, there is a continuous change in pH near the electrode–electrolyte interface as concentration of H^+ varies owing to HER, and hence, non-uniform Zn deposition is expected. Redox reactions of Zn are dependent on the pH of the medium, as is depicted in the Pourbaix diagram in Figure 15.7a. Zn is in equilibrium with Zn^{2+} from pH of 0 up to 8 with a constant potential of –0.76 V_{SHE}. Upon increasing the pH further, Zn^{2+} forms $Zn(OH)_2$ and $[Zn(OH)_4]^{2-}$ with decreasing redox potential. Figure 15.7b shows the distribution of fractions of different Zn species as a function of pH of the medium. Electrolyte additives are used to minimise HER and OER and obtain efficient Zn

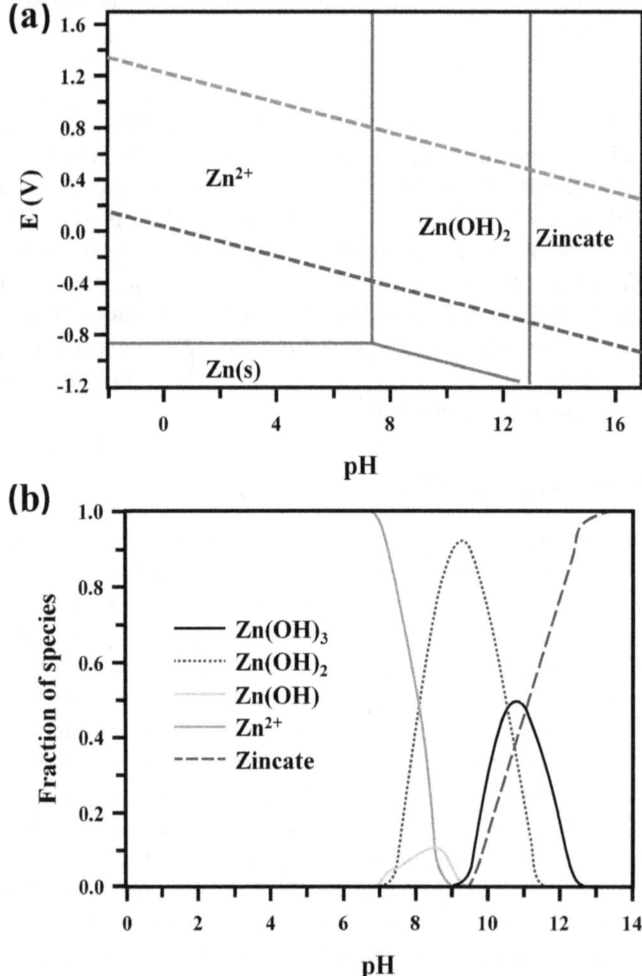

FIGURE 15.7 (a) Pourbaix diagram of Zn depicting different forms of Zn ions at different pHs. (b) Distribution of fraction of different Zn species as a function of pH of the medium.

deposit on the negative side upon charging. Classical glue, gum Arabic, nonylphenoloxylethylene, 2-butene-1,4-diol, Furfuraldehydethiosemicarbazone and salicylaldehyde thiosemicarbazone are some of the additives used to inhibit HER and get efficient Zn deposition. The dyes also influence the morphologies of the deposited Zn, giving compactness upon deposition. Adsorbed Pb(II) on the electrode surface can also strongly inhibit HER and result in Zn deposition. Care must be exercised as the excessive use of the additive in pursuit of eliminating the side reactions might block the nucleation centres on the positive side and hence deprive the electrode activity to decrease the overall EE.

15.6 ACIDIC ORGANIC RFBs

Any functional group which is easily convertible from one redox state to another and vice versa without expending much energy for the interconversion qualifies to be reversibly redox-active. A few organic functional groups like ketone, quinone and N-substituted organic core meet these requirements. Quinones have been studied in both nonaqueous and aqueous media in almost all pH ranges. And for the same reason, quinones remain the most extensively studied organic moiety so far for the rechargeable battery chemistry. Quinones have been studied for both solid electrode type and flow battery chemistry type. Large molecular weight quinones with minimum number of functional groups attached to the organic core render its poor solubility and are suitable for solid electrode type cell. Whereas quinone moiety with minimal molecular weight and appropriate functional group, depending on the medium chosen, makes it more soluble and is preferred for flow battery application. Hence, it is imperative to be clear of the chemical structure and electrochemical reduction potential and decomposition mechanism and validate how these factors are interlinked in understanding the source of instability leading to the decomposition of organic molecules in water.

15.6.1 Solubility

Functionalisation away from the ketone facilitates the efficient solvation of the substituted functional group and so higher solvation energy resulting in improved solubility. Decreasing the symmetry of the molecules also improves solubility in water. A molecule effectively solvated will exhibit higher energy of solvation and better solubility. Polar functionalities away from the ketone redox centre have more effect on the solubility than the one adjacent to ketone. Functionality adjacent to ketone will have poor solvation from the solvent due to the steric effect from the ketone leading to poor solvation energy and solubility. So, a molecule with functional group away from the redox centre will show better solubility than the molecule with similar organic core, but its functional group is adjacent to redox centre. Substituents like -OH, $-NH_2$, $-SO_3H$, $-PO_3H_2$ and -COOH being polar and hydrophilic in nature are known to improve water solubility. Precisely, acidic functional groups like phenolic -OH, $-SO_3H$, $-PO_3H_2$ and -COOH exhibit favourable acid–base interaction (as ionic interactions are stronger than dipole–dipole interaction) thereby manipulating the solubility in the basic medium further. On a similar note, basic functionalities like $-NH_2$ and $-N(CH_3)_2$ show enhanced interaction in the acidic medium to show pronounced solubility which otherwise would show mediocre value. Keynote here is if the interest is to manipulate the compound's solubility without affecting the redox potential, the hydrophilic functional group should be placed away from the aromatic core separated by alkyl chain from the core. Asymmetric distribution of charge in a molecule leads to better solubility owing to the generation of the temporary dipole moment in the molecule [33]. Capacity utilisation may also be affected due to possible dimerisation of the redox-active molecules forming a charge transfer complex in the electrolyte.

15.6.2 Redox Potential

Reduction potential is closely related to the electronic energy levels of the molecular orbitals. Expected cell voltage from a cell is the difference between the highest occupied molecular orbital

(HOMO) of the negative electrode and the lowest unoccupied molecular orbital (LUMO) of the positive electrode [34]. By introducing an electron-donating group (EDG) and electron-withdrawing group (EWG), the molecule's electronic energy levels can be altered and so is the redox potential. Experimental evidence shows that EDGs enhance the energy levels of HOMO and LUMO, while EWGs decrease the HOMO and LUMO energy levels [35]. Observations show that the presence of functionality capable of modulating electronic energy level (EDG and EWG) near to redox centre has predominant effect in varying the molecule's electronic energy level. Functionalisation of quinones near to ketone functionality affects the redox potential more due to the predominant inductive effect of the substituted group in the vicinity of the functional group [36]. An EWG (-COOH, -CHO, $-PO_3H_2$, $-COOCH_3$, $-SO_3H$, $-CF_3$, -CN and $-NO_2$) on a molecule reduces the electron density (cloud) on the molecular core, thereby decreasing LUMO energy level, making it thermodynamically facile for reduction and increasing the redox potential of the molecule. Similarly, EDG (-OH, $-CH_3$, $-NH_2$ and $-N(CH_3)_2$) on a molecule enhances the electron density (cloud), which increases the HOMO level and making it thermodynamically favourable for oxidation, and so lesser is its redox potential. But as the functionalisation moves away from the ketone functionality, its effect on the redox centre also decreases due to poor inductive effect. So, an organic molecule possessing EWG group would give a higher redox potential than its core structure. Hence, it is ideal for organic positive electrode material. Similarly, it is preferrable to use an EDG on an organic molecule chosen as the negative electrode to maximise the cell voltage within the medium's stability window. This strategy helps in getting better cell voltage, albeit has a huge impact on the stability of the molecules, thereby being the double-edged sword in the development of organic RFBs. Hence, the role of substituents on the stability of molecules takes us to the next part of discussion.

15.6.3 Stability

The stability study of the cell in its charged and discharged state is one of the crucial parameters to estimate the cell's longevity. The cost of a flow cell depends on the lifetime of the electrolyte as the cell becomes expensive with frequent replenishing of the electrolytes. So, understanding and improving the chemical stability of the electrolyte is essential. A cell may boast of good performance by virtue of excellent solubility of redox-active material and high cell voltage. But its performance can be hampered due to hydrolysis issues at the negative or positive electrodes, crossover of active material across the membrane, and most importantly, the decomposition of active material. Though the thermodynamic stability window of the solvent is limited, decomposition kinetics can be made sluggish and extend the stability window by varying few variables of the electrolyte (pH being one such variable). Crossover of active material depends on the compatibility of the active material and the functionalities of the membrane composition. The active material's temporal stability involves the combination of thermodynamic and kinetic stability of the charged active material in the medium. One of this or all put together may bring down the capacity utilisation or CE of a cell. Nucleophilic addition or substitution is one of the most commonly encountered ways of decomposition in aqueous flow batteries. Nucleophilic species like water, hydroxide and hydroperoxide anions react with electron deficient species or oxidised species. In this case, the attacking nucleophile can irreversibly replace a good leaving group present in the molecule (substitution) or attach to an unsubstituted position (addition) resulting in the manipulation of redox potential and/or solubility. EDG on the negative electrode makes the molecule nucleophilic, which also lowers the potential and makes it susceptible to electrophilic attack. Similarly, EWD has contrasting effect on the molecule making it susceptible to nucleophilic reaction.

It has been computationally estimated that quinones with higher redox potentials become severely susceptible to irreversible Michael addition/substitution reactions with water. The same has been proved experimentally that so far, no quinones have been reported with redox potential more than 0.9 V_{SHE} and stable towards Michael addition/substitution or gem/diol formation [37]. By definition, an acidic medium is one where there is deficiency of electron and hence becomes electrophilic in nature. Redox compounds with higher potential are deficient of electrons than the ones with lower redox potential. Moreover, the electrophilicity is enhanced further in the presence of acidic medium (like protonation of quinone) and it becomes unstable even in the presence of milder nucleophiles. This makes the acidic medium notorious for the cathode materials with high potential. Nonetheless, few high redox potential molecules are reported to exhibit good cycling performance as the molecule exists with stable configurations in both charged and discharged states.

15.6.4 Quinone-Based Acidic Flow Batteries

The presence of nucleophile and α,β-unsaturated carbonyl group (an electron lean centre) sets the tone for the Michael type addition reaction. The reaction involves nucleophilic attack of water molecule to an electrophilic centre present in the β-position to the electron-withdrawing ketone functionality (redox-active centre). The presence of EWGs like sulfonic or carboxylic acid leads to the development of electron deficiency in certain portion of the molecule, making it more electrophilic in nature. They easily become susceptible to the attack of nucleophilic species like water in the medium. Hence, the chances of nucleophilic attack are more in molecules with higher redox potential. For similar reasons, Michael addition has been observed in TIRON (disodium 4,5-dihydroxy-1,3-benzenedisulfonate), details of which is mentioned in the subsequent sections. A schematic of the possible reaction is shown in Scheme 15.1a. Water can be added to carbonyl functional group in an acidic or basic medium by establishing a reversible equilibrium. This gem-diol equilibrium with carbonyl group affects the reversibility of the electron transfer process (Scheme 15.1b). Giving prominence to the above said points, here are some of the systems that discuss the key aspects of RFB and issues impeding its growth to technological level.

Structure of TIRON is greatly affected by the pH of aqueous medium, so is its electrochemical behaviour [38]. At pH above 4, peak–peak separation of the TIRON is more than 100 mV and the ratio of anodic to cathodic peaks increased from 1.12 to 1.87 indicating the loss of reversibility upon increase in pH. Hence, electrochemical studies of TIRON were conducted in pH <0. Formal redox potential of TIRON is around 0.66 V vs SHE, which is within the stability window of water; therefore, during its redox activity, there would not be any O_2 evolution. Constant current charge–discharge profile reveals that the cell underwent charging at around 1.4 V with a CE of 38% in the first cycle. In the subsequent cycles, the charging voltage dropped by maintaining constant CE and EE at around 93% and 82%, respectively.

Two things that require further probing from the cycling studies of TIRON are higher charging voltage and low CE in the first cycle as in Figure 15.8a. The mechanism of electrooxidation of TIRON, which can be explained by Michael addition, is given in Scheme 15.2. It is evident from the scheme that TIRON in the charged state (ii) is susceptible to Michael addition at the second position (dotted circle). After the first Michael addition, sixth C-H proton is susceptible to further one more step of Michael addition as it will be in β-position to electronegative ketone functionality in the charged state of the molecule (iv). After the second Michael addition, it is observed that the product (vi) doesn't show any redox activity, as is depicted in Figure 15.8b [39]. The product (iii) formed at

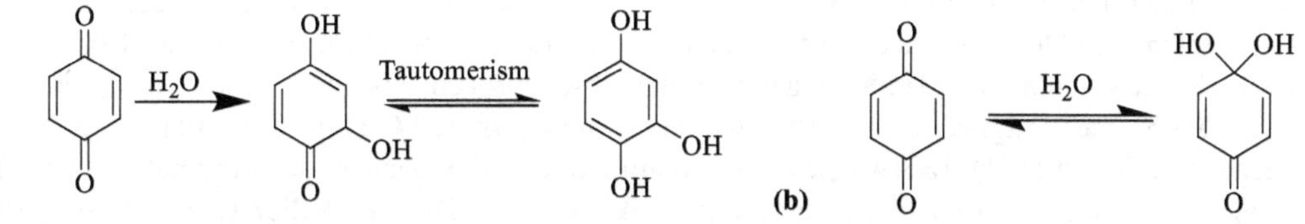

SCHEME 15.1 (a) Michael addition of water followed by keto-enol tautomerism. (b) Gem-diol formation for benzoquinone.

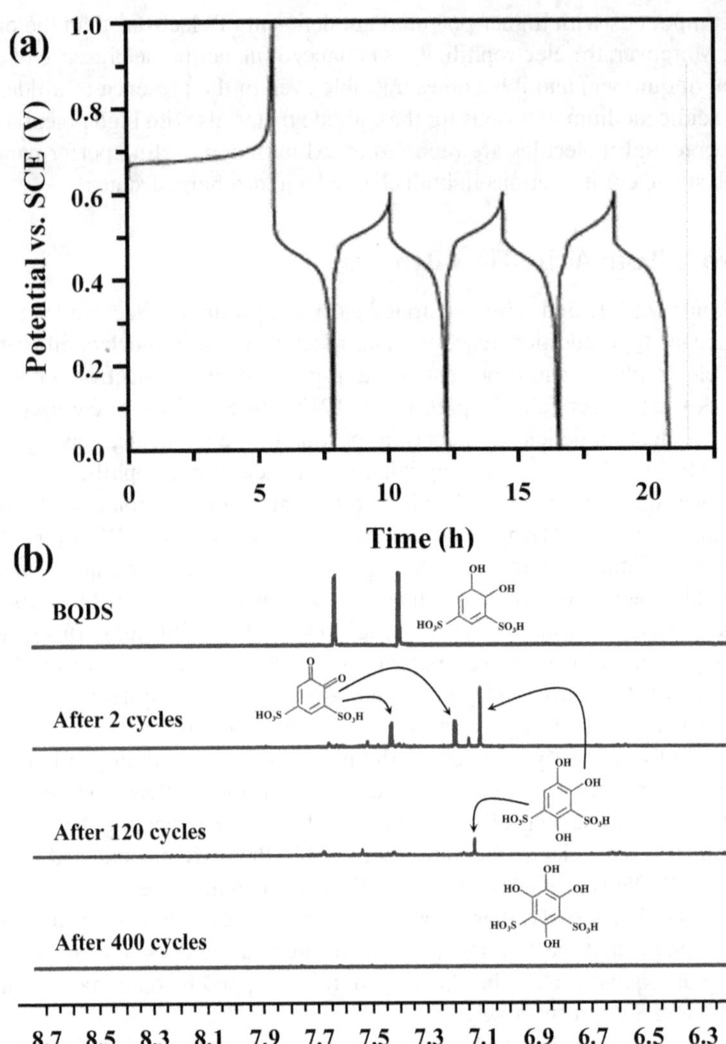

FIGURE 15.8 (a) Charge–discharge profile of RFB using 50 mM TIRON in 3 M H_2SO_4 as the positive electrode material. Adapted and reproduced from ref. [38], Copyright 2010 Elsevier. (b) Formation of side products with increasing number of charge–discharge cycles. Adapted and reproduced from ref. [39]. Copyright 2016 IOP Publishing.

the end of the first cycle has one -OH group more than in TIRON. -OH, being an EDG, reduces the reduction potential (iii) by 100 mV and hence the charging voltage of the cell.

To overcome the disadvantages of Michael addition, it is vital to design a molecule resistant to Michael addition. In an attempt to minimise the occurrence of Michael addition reaction, the free C-H protons need to be blocked. Sulfonic acid groups are desirable in the molecule as they enhance the molecule's solubility and reduction potential. In this approach, 3,6-dimethylbenzenesulfonic acid (DHDMBS) was designed with a lesser number of unsubstituted C-H positions than TIRON. The vacant position in DHDMBS was less electrophilic due to the electron-donating nature of methyl groups and hence less prone to Michael addition. As expected, DHDMBS didn't show Michael addition upon cycling and thereby exhibited better cycling performance than BQDS/TIRON.

However, due to the presence of sulfonate groups, DHDMBS exhibited proto-desulfontion with longer cycling and resulted in gradual capacity fade as in Scheme 15.3. Table 15.1 lists organic

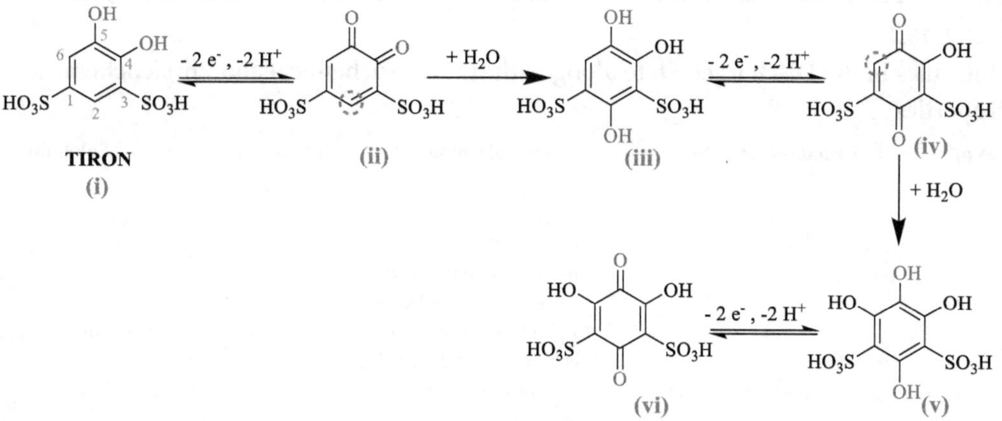

SCHEME 15.2 Redox process followed by the chemical transformation of Michael addition leading to irreversible capacity fading in the cell.

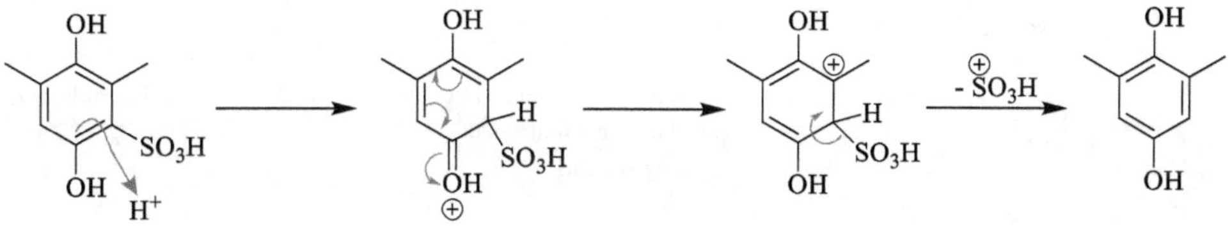

SCHEME 15.3 Proto-desulfonation of DHDMBS.

molecules explored for RFB applications in the acidic medium. Possible strategy to make molecule resistant to chemical degradation necessitates substituting unsubstituted C-H positions with alkyl chains containing terminal polar group. In this view, 1,4-hydroquinone substituted with four (dimethylamino)methyl (v) (Table 15.1) groups was used as a positive electrolyte. After cycling for a few hours, capacity fade was observed in the cell. NMR of the positive electrolyte recorded after cycling for 42 hours clearly indicates the formation of redox-inactive compound indicating instability of the electrode material despite protecting all 4 vacant C-H positions. This promoted researchers to use thio-ether substituents containing a different number of alkyl chains with terminal sulfonate groups (vi) and (vii) to assist in solubility of the molecule in acidic environment. Tetra thio-ether-substituted quinones greatly enhance stability; in addition, the thioether linkage makes the molecule resistant to substitution over prolonged cycling in aqueous acidic medium. This proposition has been proved by carrying out half-life stability tests of the molecules and compared with its contemporaries. The studies revealed that the oxidised form (charged) of (vi) and (vii) is at least two orders of magnitude more stable than that of (ii), (iii) and (iv). This clearly shows the efficacy of the strategy employed in stabilising the molecule. Upon varying the chain length, alkyl chain with 3 methylene groups showed better stability than a molecule with 2 methylene groups due to sweeping nature of the long alkyl handles reducing the probability of approach of the nucleophile or electrophile to the reactive site of the molecule. Negative electrode materials are prone to undergo electrophilic reaction. (ix), (xii) and alizarin red S (ARS) have been tried as negative electrode material, while Br_2 was used as positive electrode material. (xii) and ARS undergo bromination due to the presence of hydroxyl group which enhances electron density around the π conjugation making it susceptible to electrophiles.

TABLE 15.1
Molecules in the Discharged State along with Its Electrochemical and Physicochemical Properties

S. No.	Structure/Index No.	Special Comments	$E_{1/2}$/V vs SHE	Solubility
1	 **(1)**	One of the early quinones reported for flow battery which further led to functionalisation to improve energy and power density	0.69	0.5 M in 1 M H_2SO_4
2	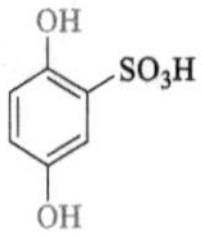 **(2)**	-SO3H is incorporated to enhance its potential and solubility in an acidic medium	0.82	0.8 M in 1 M H_2SO_4
3	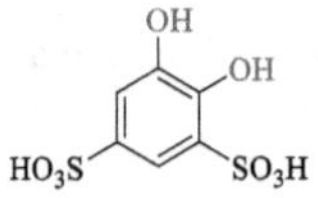 **(3)**	Vacant 2 C-H protons are susceptible to Michael addition	0.85	1.7 M in 1 M H_2SO_4
4	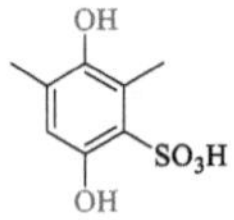 **(4)**	Relatively stable to Michael addition but undergoes proto-desulfonation	0.65	2 M in 2 M H_2SO_4
5	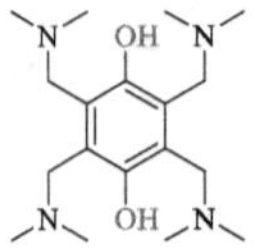 **(5)**	Resistant to Michael addition and hence exhibits good cycling stability	0.72	1.4 M in H_2SO_4
6	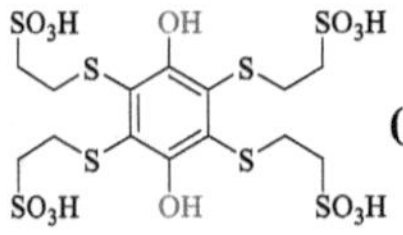 **(6)**	Resistant to Michael addition and proto-desulfonation	0.62	
7	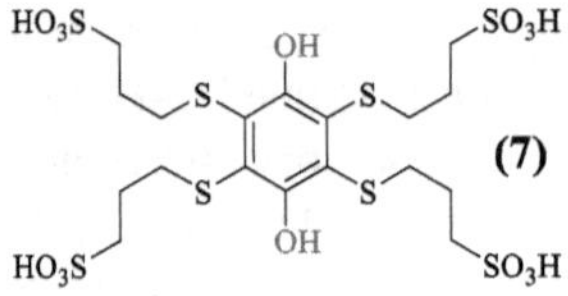 **(7)**	Resistant to Michael addition and proto-desulfonation exhibiting better stability than **(6)**	0.61	
8	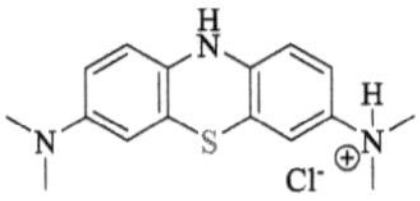 **(8)**	Redox couple is very stable due to the presence of stabilised benzenoid structure in the reduced state	0.57 in 3 M H_2SO_4	1.5 M in 3 M H_2SO_4
9	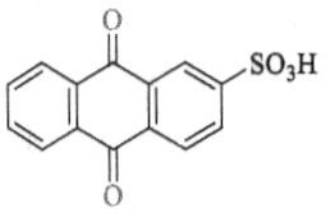 **(9)**	Lends solubility in H_2SO_4 due to the sulfonate group	0.19	0.2 M in 1 M H_2SO_4
10	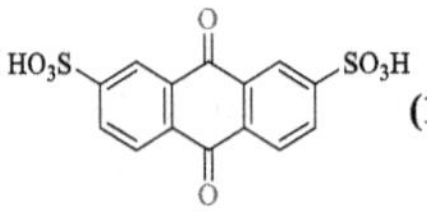 **(10)**	There is pronounced solubility due to the presence of second sulfonate group	0.29	1 M in 1 M H_2SO_4
11	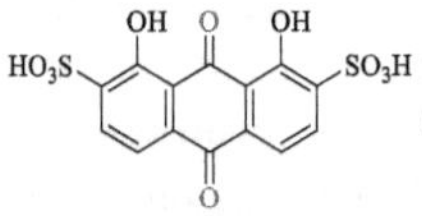 **(11)**	Incorporation of hydroxy groups in first and eighth position improves the solubility and reduces redox potential in comparison to 12	0.19	1.5 M in 1 M H_2SO_4

15.6.5 N-Heterocyclic Molecules

Nucleophilic reaction is likely to occur where Lewis basicity of the nucleophile or acidity of the substrate increases. Hence depending upon the extent of disparity, irrespective of the functional groups involved, nucleophile-initiated reactions can occur in non-quinone substrates as well as in the case of nitroxide radicals. A number of flow battery chemistries have been studied using N-heterocyclic molecules as electrode materials. (2,2,6,6-Tetramethylpiperidin-1-yl)oxyl (TEMPO) and its derivative are one such class of materials used in the positive electrode. But their stability is subjected to the meticulous maintenance of the pH of the medium during the course of the cell operation. 4-OH-TEMPO in acidic conditions undergoes disproportionation reaction to form 4-HO-TEMPO$^+$ and 1,4-dihydroxy-2,2,6,6-tetramethylpiperidine, wherein the latter is redox inactive as in Scheme 15.4. So far, methylene blue (MB), derivative of phenothiazine, has been used as positive electrode in 3 M H_2SO_4 with stable cycling up to 1200 cycles. MB can exist in 2 different forms in different pHs, as shown in Scheme 15.5. Terminal benzenoid ring structures of reduced MB (R-MB) and Leuco-MB are more aromatic than MB and are more stable. As a consequence of its stability, MB and R-MB can remain for longer time without any side reactions bestowing longer cycle life to the cell. This observation was also made in symmetric cell mode where MB and R-MB were used as electrode materials. The fabricated cell exhibited a stable cycling up to 1200 cycles with negligible capacity fade at 80 mA cm^{-2}. Table 15.1 shows the molecules with their property in addition to its $E_{1/2}$ value and solubility in medium.

15.7 CHARACTERISATION TECHNIQUES

15.7.1 Electrochemical Characterisation

Electrochemical stability estimation metrics primarily involve monitoring cycle life (capacity retention per cycle) and calendar life (capacity retention per day) of the cell. Broadly, degradation mechanism of material could be due to one or all of the following reasons: (i) crossover of active material, (ii) intrinsic instability or irreversibility of the active material, (iii) formation of dendrites on metal

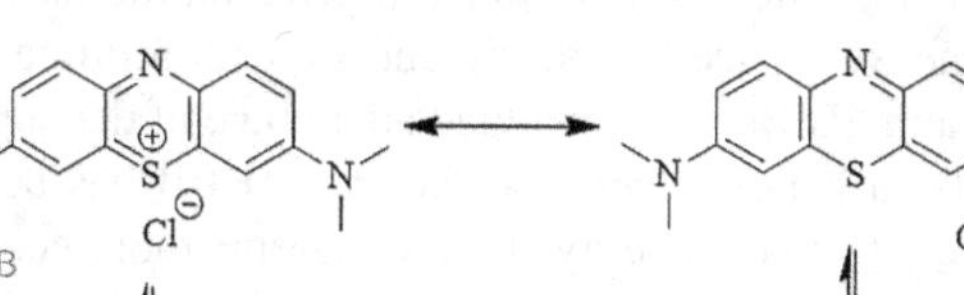

SCHEME 15.4 Disproportionation reaction of 4-OH-TEMPO in the acidic medium.

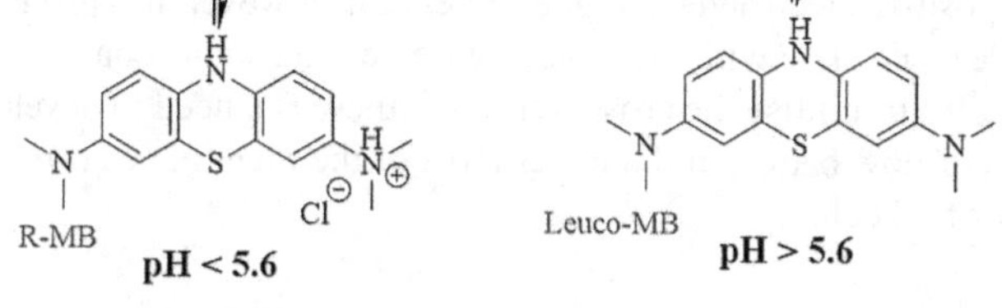

SCHEME 15.5 Different forms of MB in different pHs.

electrodes in hybrid aqueous RFBs and (iv) insufficient capacity utilisation. In the case of aqueous inorganic RFB, which use small molecules like polysulfides, halogen and vanadium, time-dependent crossover largely limits the batteries' stability, which makes calendar life an important evaluation metric. It has been revealed that capacity fading in the case of aqueous organic RFB is independent of charge–discharge cycle number but dependent on time. To evaluate the redox-active electrolyte's intrinsic stability, charge–discharge can be carried out on compositionally symmetric flow cell. In this technique, a redox-active material and its conjugate redox state is taken in two separate compartments of a cell and charge–discharge cycling is carried out. This technique exclusively monitors the stability of one of the redox-active compounds. However, this technique doesn't take into consideration crossover issue and studies only the temporal stability of the cell by virtue of monitoring capacity of the cell with time. Electrochemical techniques like cyclic voltammetry, linear sweep voltammetry, rotating disc electrode study and electrochemical impedance spectroscopy are important in evaluating redox potential and kinetics of electron transfer of the analytes.

15.7.2 Material Characterisation

Stability of the molecules dissolved in the medium is critical in establishing the longevity of the cell operation. Materials when dissolved in a different medium and subjected to an external stimulus (application of current or potential) interact differently possibly leading to various modes of disintegration with time. In order to stop/prolong this temporal disintegration, it is important to understand their mode of interaction and mechanism of operation. UV–Visible spectroscopy is suitable for active materials that absorb ultraviolet and visible light radiation with onus on quantitative studies governed by the Beer–Lambert law. Variation in the concentration of the active material during the cycling studies can easily be estimated by monitoring the intensity of the absorption maxima. Similarly, NMR spectroscopy has been widely used to explain the mechanism of material decomposition. Temporal stability of various quinones (anthraquinone-2-sulfonic acid, 1,8-dihydroxyanthraquinone-2,7-disulfonic acid and ARS) in the presence of crossed-over Br_2 is monitored using NMR. Aromatic electrophilic attack by Br_2 on 1,8-dihydroxyanthraquinone-2,7-disulfonic acid and ARS was confirmed by the disappearance of peaks for these 2 compounds, whereas the peaks for AQS remained after charge–discharge indicating the stability of AQS towards electrophilic attack by Br_2.

15.8 CONCLUDING REMARKS

Reducing the carbon footprint and greenhouse gas emissions to the atmosphere is the need of the hour. So, there is a dire need to switch to a renewable energy source for our daily needs like never before. Given the intermittency of renewable energy sources, the technological innovations are on a roll in search of the efficient and safer battery system to iron out the fluctuation in energy procurement. RFBs are a very attractive choice for storing energy with minimal safety issues and feed to the electrical grid supply chain. RFBs have an added advantage of decoupling of energy and power which lends more flexibility in battery operation. So far, VRFB has been commercialised given the minimal crossover issues. To make energy storage systems more economically viable, organic compounds have been explored. The challenge in commercialising the acidic organic RFB has been the instability of the positive electrolyte at higher potential. However, it is promising to note that it is possible to have stable electrolyte with the molecule possessing stable configurations in the charged and discharged states. To minimise the crossover issue, there is a need to develop membranes keeping the specifications of flow battery in focus, as most of the membranes used for RFB so far were initially developed for fuel cells.

REFERENCES

1. Biswas S, Senju A, Mohr R, et al (2017) Minimal architecture zinc-bromine battery for low cost electrochemical energy storage. *Energy Environ Sci* 10:114–120.
2. Lim HS, Lackner AM, Knechtli RC (1977) Zinc-bromine secondary battery. *J Electrochem Soc* 124:1154.
3. Thaller LH (1976) U.S. Patent 3996064A.
4. Ding Y, Zhang C, Zhang L, et al (2019) Pathways to widespread applications: Development of redox flow batteries based on new chemistries. *Chem* 5:1964–1987. https://doi.org/10.1016/j.chempr.2019.05.010.
5. Skyllas-Kazacos M, Cao L, Kazacos M, et al (2016) Vanadium electrolyte studies for the vanadium redox battery-A review. *ChemSusChem* 9:1521–1543. https://doi.org/10.1002/cssc.201600102.
6. Knehr KW, Kumbur EC (2011) Open circuit voltage of vanadium redox flow batteries: Discrepancy between models and experiments. *Electrochem Commun* 13:342–345. https://doi.org/10.1016/j.elecom.2011.01.020.
7. Vijayakumar M, Li L, Graff G, et al (2011) Towards understanding the poor thermal stability of V^{5+} electrolyte solution in vanadium redox flow batteries. *J Power Sources* 196:3669–3672. https://doi.org/10.1016/j.jpowsour.2010.11.126.
8. Mousa A, Skyllas-Kazacos M (2015) Effect of additives on the low-temperature stability of vanadium redox flow battery negative half-cell electrolyte. *ChemElectroChem* 2:1742–1751. https://doi.org/10.1002/celc.201500233.
9. Rahman F, Skyllas-Kazacos M (2017) Evaluation of additive formulations to inhibit precipitation of positive electrolyte in vanadium battery. *J Power Sources* 340:139–149. https://doi.org/10.1016/j.jpowsour.2016.11.071.
10. Roe S, Menictas C, Skyllas-Kazacos M (2016) A high energy density vanadium redox flow battery with 3 M vanadium electrolyte. *J Electrochem Soc* 163:A5023–A5028. https://doi.org/10.1149/2.0041601jes.
11. Skyllas-Kazacos M, Peng C, Cheng M (1999) Evaluation of precipitation inhibitors for supersaturated vanadyl electrolytes for the vanadium redox battery. *Electrochem Solid-State Lett* 2:121–122. https://doi.org/10.1149/1.1390754.
12. Sun J, Li X, Xi X, et al (2014) The transfer behavior of different ions across anion and cation exchange membranes under vanadium flow battery medium. *J Power Sources* 271:1–7. https://doi.org/10.1016/j.jpowsour.2014.07.111.
13. Prifti H, Parasuraman A, Winardi S, et al (2012) Membranes for redox flow battery applications. *Membranes (Basel)* 2:275–306. https://doi.org/10.3390/membranes2020275.
14. Wei W, Zhang H, Li X, et al (2013) Hydrophobic asymmetric ultrafiltration PVDF membranes: An alternative separator for VFB with excellent stability. *Phys Chem Phys* 15:1766–1771. https://doi.org/10.1039/c2cp43761a.
15. Qiu J, Zhao L, Zhai M, et al (2008) Pre-irradiation grafting of styrene and maleic anhydride onto PVDF membrane and subsequent sulfonation for application in vanadium redox batteries. *J Power Sources* 177:617–623. https://doi.org/10.1016/j.jpowsour.2007.11.089.
16. Sukkar T, Skyllas-Kazacos M (2003) Water transfer behaviour across cation exchange membranes in the vanadium redox battery. *J Memb Sci* 222:235–247. https://doi.org/10.1016/S0376-7388(03)00309-0.
17. Sukkar T, Skyllas-Kazacos M (2004) Membrane stability studies for vanadium redox cell applications. *J Appl Electrochem* 34:137–145. https://doi.org/10.1023/B:JACH.0000009931.83368.dc.
18. Huang X, Pu Y, Zhou Y, et al (2017) In-situ and ex-situ degradation of sulfonated polyimide membrane for vanadium redox flow battery application. *J Memb Sci* 526:281–292. https://doi.org/10.1016/j.memsci.2016.09.053.
19. Reynard D, Dennison CR, Battistel A, Girault HH (2018) Efficiency improvement of an all-vanadium redox flow battery by harvesting low-grade heat. *J Power Sources* 390:30–37. https://doi.org/10.1016/j.jpowsour.2018.03.074.
20. Reynard D, Vrubel H, Dennison CR, et al (2019) On-site purification of copper-contaminated vanadium electrolytes by using a vanadium redox flow battery. *ChemSusChem* 12:1222–1228. https://doi.org/10.1002/cssc.201802895.
21. Doan TNL, Hoang TKA, Chen P (2015) Recent development of polymer membranes as separators for all-vanadium redox flow batteries. *RSC Adv* 5:72805–72815. https://doi.org/10.1039/c5ra05914c.
22. Zhong S, Padeste C, Kazacos M, Skyllas-Kazacos M (1993) Comparison of the physical, chemical and electrochemical properties of rayon- and polyacrylonitrile-based graphite felt electrodes. *J Power Sources* 45:29–41. https://doi.org/10.1016/0378-7753(93)80006-B.

23. Skyllas-Kazacos M, Chakrabarti MH, Hajimolana SA, et al (2011) Progress in flow battery research and development. *J Electrochem Soc* 158:R55. https://doi.org/10.1149/1.3599565.
24. Kaneko H, Nozaki K, Wada Y, et al (1991) Vanadium redox reactions and carbon/a. *Electrochim Acta* 36:1191–1196.
25. Vázquez-Galván J, Flox C, Jervis JR, et al (2019) High-power nitrided TiO_2 carbon felt as the negative electrode for all-vanadium redox flow batteries. *Carbon N Y* 148:91–104. https://doi.org/10.1016/j.carbon.2019.01.067.
26. He Z, Li M, Li Y, et al (2019) Electrospun nitrogen-doped carbon nanofiber as negative electrode for vanadium redox flow battery. *Appl Surf Sci* 469:423–430. https://doi.org/10.1016/j.apsusc.2018.10.220.
27. Jiang Y, Feng X, Cheng G, et al (2019) Electrocatalytic activity of MnO_2 nanosheet array-decorated carbon paper as superior negative electrode for vanadium redox flow batteries. *Electrochim Acta* 322:134754. https://doi.org/10.1016/j.electacta.2019.134754.
28. Kabir H, Gyan IO, Francis Cheng I (2017) Electrochemical modification of a pyrolytic graphite sheet for improved negative electrode performance in the vanadium redox flow battery. *J Power Sources* 342:31–37. https://doi.org/10.1016/j.jpowsour.2016.12.045.
29. He Z, Li M, Li Y, et al (2018) Flexible electrospun carbon nanofiber embedded with TiO_2 as excellent negative electrode for vanadium redox flow battery. *Electrochim Acta* 281:601–610. https://doi.org/10.1016/j.electacta.2018.06.011.
30. Amstutz V, Toghill KE, Powlesland F, et al (2014) Renewable hydrogen generation from a dual-circuit redox flow battery. *Energy Environ Sci* 7:2350–2358. https://doi.org/10.1039/c4ee00098f.
31. Rahman F, Skyllas-Kazacos M (1998) Solubility of vanadyl sulfate in concentrated sulfuric acid solutions. *J Power Sources* 72:105–110. https://doi.org/10.1016/S0378-7753(97)02692-X.
32. Amini K, Pritzker MD (2020) In situ polarization study of zinc-cerium redox flow batteries. *J Power Sources* 471:228463. https://doi.org/10.1016/j.jpowsour.2020.228463.
33. Han KS, Rajput NN, Vijayakumar M, et al (2016) Preferential solvation of an asymmetric redox molecule. *J Phys Chem C* 120:27834–27839. https://doi.org/10.1021/acs.jpcc.6b09114.
34. Yang B, Hoober-Burkhardt L, Wang F, et al (2014) An inexpensive aqueous flow battery for large-scale electrical energy storage based on water-soluble organic redox couples. *J Electrochem Soc* 161:A1371–A1380. https://doi.org/10.1149/2.1001409jes.
35. Bulbring E, Kuriyama H (1963) Effects of changes in ionic environment on the action of acetylcholine and adrenaline on the smooth muscle cells of guinea-pig taenia coli. *J Physiol* 166:59–74.
36. Tabor DP, Gómez-Bombarelli R, Tong L, et al (2019) Mapping the frontiers of quinone stability in aqueous media: Implications for organic aqueous redox flow batteries. *J Mater Chem A* 7:12833–12841. https://doi.org/10.1039/c9ta03219c.
37. Kwabi DG, Ji Y, Aziz MJ (2020) Electrolyte lifetime in aqueous organic redox flow batteries: A critical review. *Chem Rev* 120:6467–6489. https://doi.org/10.1021/acs.chemrev.9b00599.
38. Xu Y, Wen Y-H, Cheng J, et al (2010) A study of tiron in aqueous solutions for redox flow battery application. *Electrochim Acta* 55:715–720.
39. Yang B, Hoober-Burkhardt L, Krishnamoorthy S, et al (2016) High-performance aqueous organic flow battery with quinone-based redox couples at both electrodes. *J Electrochem Soc* 163:A1442–A1449. https://doi.org/10.1149/2.1371607jes.

16 Lithium Redox Flow Battery

Jishnu N. S, Neethu T. M. Balakrishnan, Akhila Das, Jabeen Fatima M. J., Alexandru Vlad, and Prasanth Raghavan

16.1 INTRODUCTION

The increasing need of the electricity demands the generation of large proportion of the energy that can effectively meet the consumer requirement. The overreliance over the non-renewable energy sources will lead to the depletion of the fossil fuel energy, which will eventually lead to the scarcity of energy resources. The exploration of renewable resources instead of the non-renewable systems was proposed as the best alternative that can effectively reduce the dependence on the non-renewable systems. The major concern with the use of the renewable source is the difficulty to supply the generated energy when it demands. In that case, the introduction of energy storage systems is significant to ensure the storage of energy generated and the proper supply of energy on demand. Different energy storage systems like solar cells, fuel cells, super capacitors batteries, etc. were proposed in which the batteries are the primary choice. The batteries were normally categorised as primary and secondary systems in which primary batteries are means to use without recharging that is disposed once it is in completely discharged state. Unlike the primary batteries, the rechargeable batteries are the reusable systems that can be charged once it is completely discharged and can be used for a long-time span. Different rechargeable battery systems like lead acid batteries, nickel metal hydride batteries, and lithium-ion batteries were introduced, among which the lithium-ion batteries are widely accepted rechargeable systems.

Lithium-ion batteries are significantly explored in most of the portable electronic gadgets owing to its superior electrochemical performances. The superior performance of lithium-ion battery, such as high volumetric and gravimetric energy densities, enables its use in advanced applications such as electric vehicles. However, there are so many major concerns with the lithium-ion battery technology like safety, which researchers are still trying to rectify by the introduction of innovative technological advancement. Different technological competitors were developed for lithium-ion battery that could sufficiently outperform as compared to that of this established battery technology. Among these redox flow batteries are the major competitors available in markets. The safer operation and performance advancement are the key features that bring considerable attention for the redox flow battery technology. The safer chemistry and long cycle life as compared to that of rechargeable lithium-ion battery technology make redox flow batteries suitable for the energy storage applications. The chemistry behind the redox flow batteries enables it to store sufficient amount of energy which is higher than that of the commercial lithium-ion battery. The enhanced safety features and performance advancement immensely promote the research over the redox flow battery technology. This chapter gives a detailed discussion of redox flow that comprises detailed discussion on its electrochemical mechanism, performance and recent advancements.

16.2 HISTORY OF REDOX FLOW BATTERY

In the 1970s, the National Aeronautics and Space Administration (NASA) initiated the development of the redox flow battery with the insight of promoting bulk energy storage. The extensively studied redox flow system by NASA was the iron chromium redox flow battery which was fabricated and tested in late 1980s [1–5]. Owing to the poor electrochemical stability offered by chromium and the cross-contamination of the active material, NASA was forced to halt further research over this

DOI: 10.1201/9781003310167-16

system. Later in Japan, the same system was evaluated as part of the Mooney project and even now extensive research is carried out over it for employing in energy storage in wireless telecom applications. Similar to the iron–chromium system, another redox flow system parallelly studied was the zinc chloride and zinc bromide redox flow battery [6,7]. Zinc chloride system was observed to be complicated while the zinc bromide attracted much attention, and in 1990, it was installed in Japan. Currently, this system was developed by ZBB Energy Corporation, RedFlow Ltd., and Premium Power with an expandable battery modulus from 50 to 500 kWh.

Among the development of different redox flow systems, vanadium-based redox flow batteries attracted much attention. NASA carried out the evaluation of vanadium redox flow battery based on the redox couples V^{2+}/V^{3+} and V^{4+}/V^{5+} in 1977 [8–10]. Followed by it, in 1985, the successful demonstration of this redox flow battery system was done by Skyllas-Kazacos et al. [8,9,11,12] at the University of New South Wales, Australia. In 1993, vanadium redox flow batteries were commercialised and installed for various applications such as renewable coupling and power quality. It was globally accepted and its wide acceptance was assisted by the progress in electrode and membrane materials. Another redox flow system investigated was the bromide–polysulphide developed by Regenesys Technologies (UK) at Innogy plc. The most attractive features of the bromide–polysulphide redox flow system were the moderate cost of the electrodes and low-cost electrolyte that make it affordable with a combination of better performance. In 2004, different new generation flow batteries, such as lead-acid [13,14], zinc–nickel [15,16], and zinc–cerium [17,18] based, were introduced. These systems are simple since they have explored a single electrode hence do not have the requirement of an ion exchange membrane which make them low cost and simply operable [19,20].

Lithium-based redox flow batteries are the system that are developed more recently with an energy density of 397 Wh dm^{-3} and a discharge cell voltage of 3.8 V. The excellent performance exhibited by these systems enabled them to explore in the advanced energy storage applications such as electric vehicles. The advantage of lithium redox flow battery over the lithium-ion battery is that they are least expensive and can reveal a high voltage as compared to that of conventional redox flow battery systems. The lithium redox flow battery combines the advantage of both redox flow battery and lithium-ion battery to ensure the better electrochemical performance. It is considered to be a promising energy storage system owing to the modular design, which reveals a flexible operation, transportability, and affordable manufacturing cost [21,22]. The timeline of redox flow battery is depicted in Figure 16.1. The advanced performance of the lithium redox flow systems can be revealed from the battery components that will be discussed in the subsequent sections of the chapter.

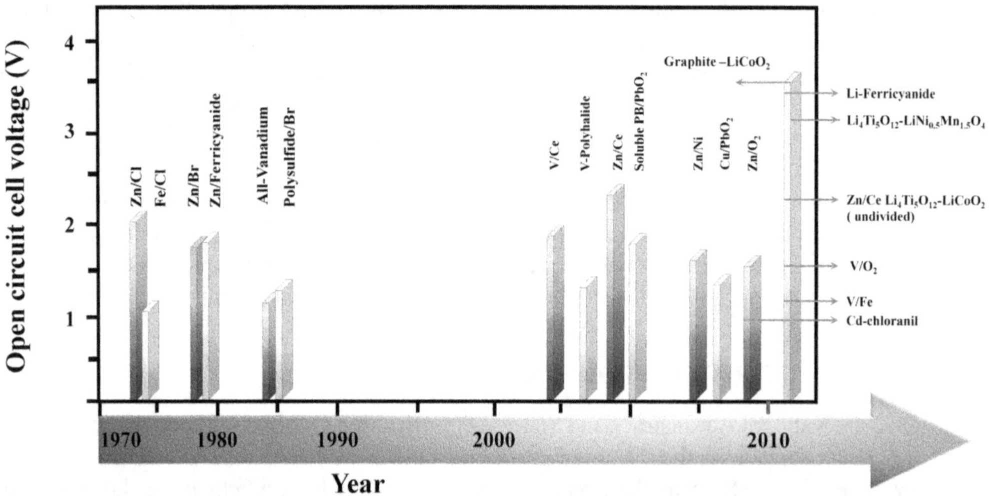

FIGURE 16.1 The timeline of the development of redox flow batteries over the past 40 years. Adapted and reproduced with permission from Ref. [3]. Copyright © 2012 The Royal Society of Chemistry.

16.3 THE BATTERY COMPONENTS AND MECHANISM OF LITHIUM REDOX FLOW BATTERY

A redox flow battery is different from the other rechargeable batteries in the mechanism of storing the energy. In the other rechargeable systems, the energy is stored in the electrodes, while in redox flow batteries the rechargeability arises as a result of the chemical components dissolved in a liquid with a reversible redox behaviour, which are stored within the system and an external reservoir. The typical structure and working principle of a lithium redox flow battery with a positive and negative reservoir separated by an ion exchange membrane is shown in Figure 16.2 [23,24].

The performance of the redox flow batteries will depend mostly on the selection of the battery components such as active cathode material and separator membrane. The cathode material opted for the lithium redox flow battery should satisfy several requirements, for example, their oxidised and reduced form should be highly soluble to ensure high mass transfer and current densities; in order to maximise the cell potential and energy density, the potential of one couple should be highly positive and the other should be highly negative; both the oxidised and reduced form should be highly stable; the material should cheap and abundant; and it should not corrosive and should be inert with the other battery components. In the case of separator, lithium conducting NASICON glass ceramics or solid polymer are employed as the separator. In the case of glass ceramic electrolyte, they exhibit a room-temperature ionic conductivity in the range of 10^{-4} S cm^{-1}. The separator employed should be stable against the aqueous and aprotic electrolyte, but in the case of low potential, they are observed to be unstable with the direct contact of metallic lithium. So usually glass ceramic separators employed for a long-term operations are coated with buffer for the safer operation.

The lithium redox flow battery constitutes a catholyte which is a soluble cathode active redox couple dissolved in a solvent and an anode lithium or anolyte that constitute an anode active redox couple dissolved in a solvent. Both the catholyte and anolyte are stored in an external reservoir and circulated by means of an external circulating subsystem. The chemistry of lithium redox flow battery during the charging and discharging enables them to provide high operating voltage which

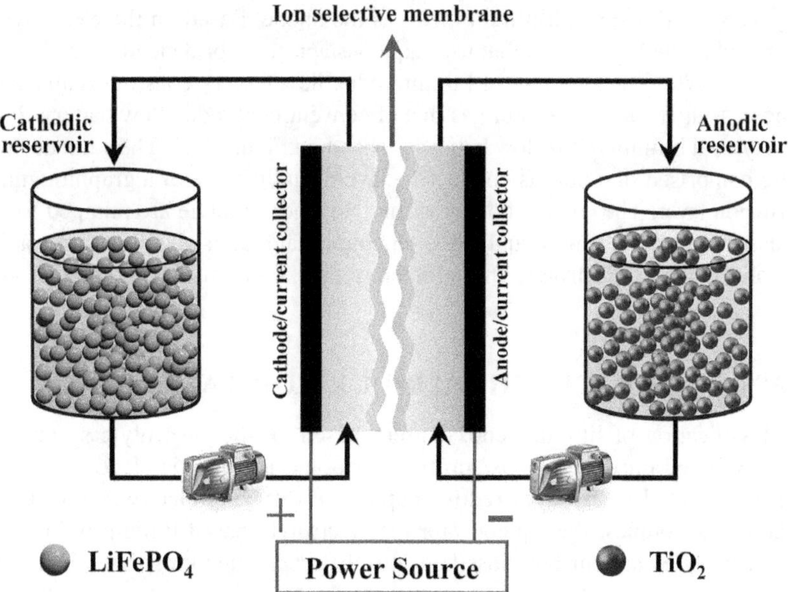

FIGURE 16.2 Structure and working principle of lithium redox-flow battery. Adapted and reproduced with permission from Ref. [25]. Copyright © 2016 The Royal Society of Chemistry.

is comparable to lithium-ion batteries [26,27]. The mechanism of the charge transfer is basically associated with the lithium-ion transport in between the anode and cathode. During the charging process, the reduced form of a redox species is oxidised to its oxidised form at the positive electrode, while at the anolyte the oxidised form of the redox couple is reduced to its reduced form or as metallic lithium at the negative electrode. The reverse process will take place during the discharge process. The lithium ions are conducted through the lithium-ion conducting membrane with the insight of balancing the charge. This reaction is observed to be continued until the redox species are consumed completely in the electrolyte. The electrochemical reactions (equations 16.1–16.5) observed are given as follows [24,28]:

$$\text{Positive electrode: } C^{x+} - ye^{-} \rightarrow C^{(x+y)+} \quad (16.1)$$

$$\text{Negative electrode: } Li^{+} + e^{-} \rightarrow Li \quad (16.2)$$

$$A^{m+} + ne^{-} \rightarrow A^{(m-n)+} \quad (16.3)$$

$$\text{Overall reactions: } C^{x+} + yLi^{+} \rightarrow C^{(x+y)+} + yLi \quad (16.4)$$

$$nC^{x+} + yA^{m+} \rightarrow nC^{(x+y)+} + yA^{(m-n)+} \quad (16.5)$$

where $C^{(x+y)+}/C^{x+}$ denotes the redox couple in the anolyte and $A^{m+}/A^{(mn)+}$ stands for the redox couple in the anolyte.

Lithium redox flow batteries are normally categorised as full flow lithium redox cell and semi flow lithium redox cell based on the anode structure used. The full flow lithium redox cell explores redox-active liquid or slurry as anolyte or catholyte. While the semi flow utilises redox-active liquid or slurry as the catholyte and lithium metal as the anode. Based on the electrolyte employed, they are further classified as the one that uses aqueous/aprotic hybrid electrolyte and the other that uses aprotic electrolyte. The structure of lithium redox flow battery constitutes an aqueous/aprotic electrolyte possessing a similar structure as that of conventional redox flow battery. The schematic structure of a typical lithium redox flow battery is shown in Figure 16.3. The cathode or anode compartments are comprised of a porous conducting layer usually made of a graphitic material which acts as a diffusion layer. The current collector, anolyte, and catholyte are pumped to a flow frame with a thickness of about 1–3 mm. Usually, the separator material used is a lithium conducting glass ceramic or solid polymer electrolyte that efficiently blocks the mixing of anolyte and catholyte [29–31].

16.4 CLASSIFICATION OF LITHIUM REDOX FLOW BATTERY

The major classification of lithium redox battery based on the electrolyte system employed is the flow battery based on the aqueous/aprotic hybrid and the aprotic electrolyte system. These systems are investigated in different redox couples with varying electrochemical performance. Based on the redox couples, the aqueous/aprotic electrolyte-based lithium redox flow batteries and aprotic electrolytes are further classified into different categories that will be discussed in detail in this section.

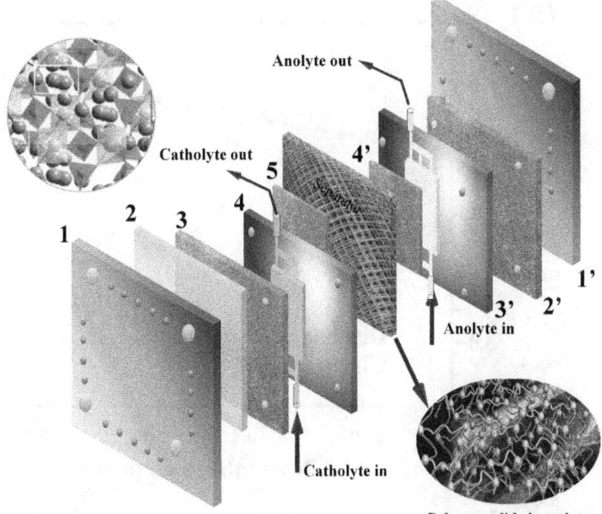

Cathode compartment:
(1) End plate
(2) Isolation plate
(3) Supporting substrate
(4) Flow frame
(5) Current collector

Anode compartment:
(1') End plate
(2') Supporting substrate
(3') Flow frame
(4') Current collector

FIGURE 16.3 Schematic structure of a Li-redox flow cell. Adapted and reproduced with permission from Ref. [25]. Copyright © 2016 The Royal Society of Chemistry.

16.5 LITHIUM REDOX FLOW BATTERY BASED ON AQUEOUS/APROTIC HYBRID ELECTROLYTE SYSTEM

16.5.1 Iron-Based Lithium Redox Flow Battery

The first rechargeable lithium redox flow battery based on aqueous/aprotic hybrid was proposed by John Goodenough and Haoshen Zhou [32,33]. The initial investigation was carried out with an iron-based redox couple using lithium metal as the anode. The abundance and high solubility of the iron in most of the solvents made it the first and prime option for the flow batteries. Goodenough developed the lithium redox system by employing an aqueous solution containing Fe^{3+} and lithium salt as the catholyte. The cell is observed to be discharged from 3.7 to 3 V in relatively low current density to reach a normalised capacity (Figure 16.4a). Even though the major issue faced with the hydrolysis of Fe^{3+} and Fe^{2+} in a neutral pH environment leads them to extend the research to another redox couple of $Fe(CN)_6{}^{3-}/Fe(CN)_6{}^{4-}$. This cell exhibited a good Coulombic efficiency in the potential rage of 3–3.75 V (Figure 16.4b). In order to prevent the hydrolysis of Fe^{2+}/Fe^{3+}, the cell constructed by Zhou et al. [32] uses an acidic aqueous solution as the catholyte. However, the sluggish kinetics of the ferric and ferrous ions, as well as the poor solubility of the iron precursor, still remains challenging [32].

16.5.2 Halogen-Based Redox Flow Battery

The low molecular weight, highly reversible redox reaction and suitable redox couple offered by the halogen redox couple such as $Br_3{}^-$/ Br and $I_3{}^-/I^-$ make them suitable for the redox flow battery. In the case of iodine system, iodine in the range of $I_2{}^-$ to $I_{29}{}^3$ has been investigated in redox systems and

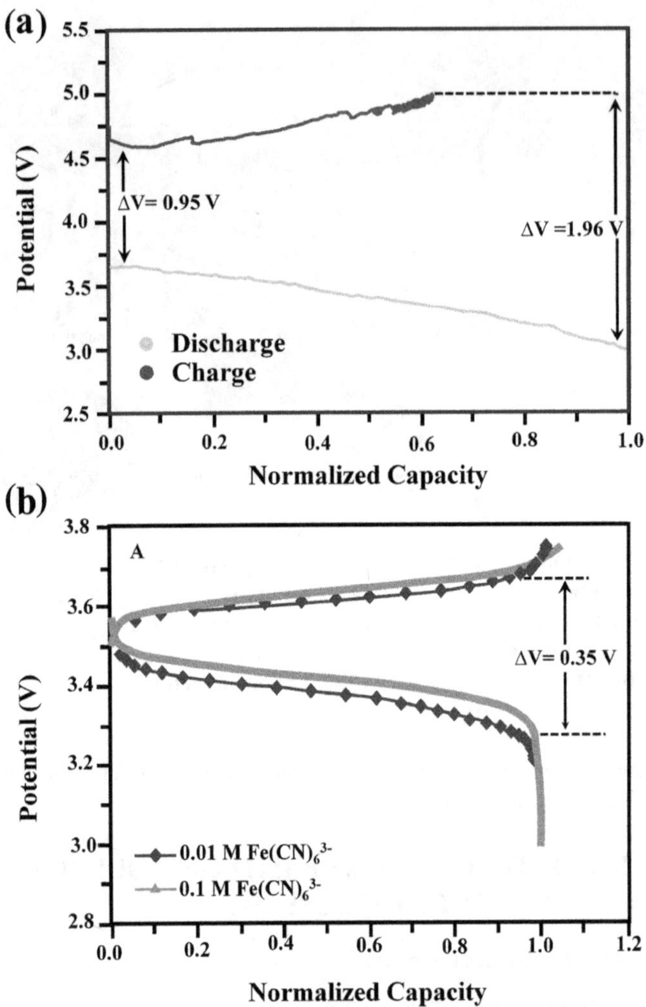

FIGURE 16.4 Electrochemical performance of iron-based Li-redox flow batteries with aqueous catholyte. (a) Charge/discharge profile of a Li-redox cell operated via Fe^{3+}/Fe^{2+} redox reaction and (b) Fe $(CN)_6^{3-}$/Fe $(CN)_6^{4}$ redox reaction under different Fe $(CN)_6^{3-}$ concentration. Adapted and reproduced with permission from Ref. [32]. Copyright © 2011 American Chemical Society.

the polyiodides reveal a high room-temperature electrical conductivity. The behaviour of bromine is observed to be similar to that of iodine in aqueous phase. The formation of tribromide ions by the combination of bromide ions with the bromine molecules being an enhanced energy density. Br_3^-/ Br redox couples are extensively applicable for the die-sensitised solar cells, alkali-ion batteries, etc. [34–36].

The work over the iodine-based catholyte was initiated by Byon et al. [37]. The 1 M solution of KI with a controlled concentration of I_3^- by the proper addition of I_2 was used as the catholyte. The cell is capable to exhibit an initial Columbic efficiency of about 80%, which is observed to be enhanced to 100% on preloading of the lithium salt or by the enhancement of iodine concentration. The flow cathode design reported by Zhao et al. [38] explored a single electrolyte system of LiI as the supporting electrolyte and the active material. The lithium redox cell fabricated by employing this electrolyte system exhibits a good energy density and power density (110 mW cm^{-2} at a current density of 40 mA cm^{-2}) [38]. Later, the same group proposed the fabrication of lithium-iodide redox

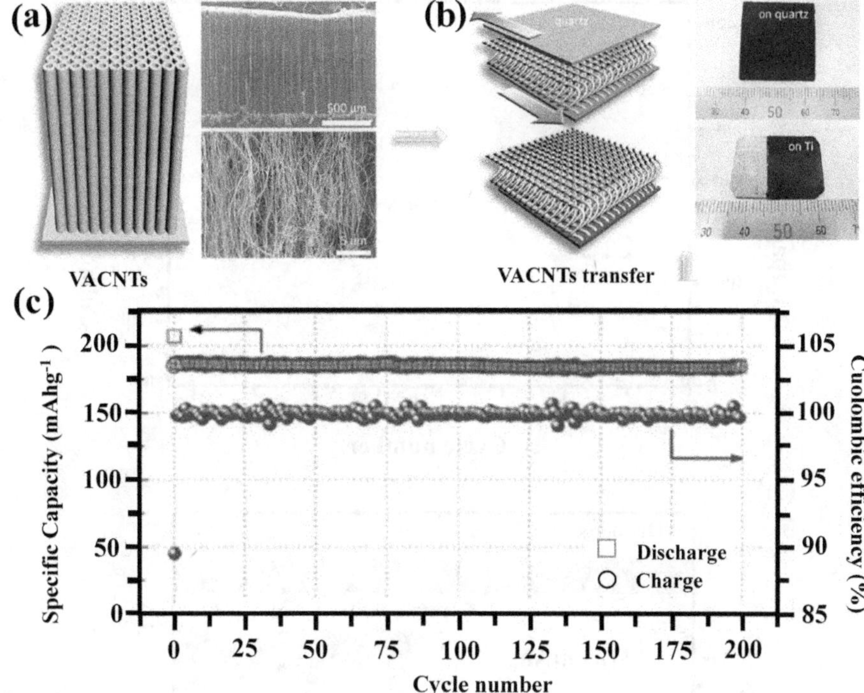

FIGURE 16.5 (a) Schematic illustrations and images of Li–I2 cell preparation process. (a) Scheme (left) and SEM images (right) of VACNTs. (b) Dry contact transfer of VACNTs from quartz to Ti film. Optical images are VACNTs before (right top) and after (right bottom) transfer. (c) Specific capacity and Coulombic efficiency for 200 cycles Li-I_2 cells with VACNT current collectors at a current rate of 2.5 mA cm^{-2} at 298 K. Adapted and reproduced with permission from Ref. [39]. Copyright © 2014 American Chemical Society.

cell, which involves the performance enhancement of a lithium-iodide hybrid battery by making use of a three-dimensional vertically aligned multiwalled carbon nanotube current collector. The current collector was fabricated by the chemical vapour deposition and dry contact transfer method without the addition of binder (Figure 16.5a). The highly dense and lightweight current collector on surface treatment yields a hydrophilic surface that can reveal the accommodation of the aqueous cathode having short I_3^-/I^- diffusion length. This cell is capable to exhibit a best electrochemical performance with about 98% of the theoretical capacity and a Columbic efficiency of about 100% (Figure 16.5b). Moreover, the aqueous cathode with the I_2 saturation results in an energy density of about 330 Wh Kg^{-1} [39].

16.5.3 Chalcogen-Based Redox Flow Battery

The chalcogen-based lithium-ion batteries are prominent by taking advantage of chalcogens such as oxygen and sulphur. The high energy density and inexpensive nature of the chalcogens are promising to explore in the application of lithium redox flow battery system. The nonaqueous system of Li-O_2 was first reported by Abraham et al. [40] and these studies are extended further. The major challenges faced while using the redox flow battery was the dendrite formation in metallic lithium, solubility of O_2, stability of the aprotic electrolyte, and more importantly, the requirement of cathode for a membrane that can selectively pass the oxygen atom without allowing the carbon dioxide and water. The lithium redox cell utilising the alkaline catholyte LiOH with lithium salts such as $LiClO_4$, $LiNO_3$, or LiCl are examined as the supporting electrolyte [41]. The system of nonaqueous electrolyte of rechargeable Li–S cell observed to yield high discharge capacity, owing to the high

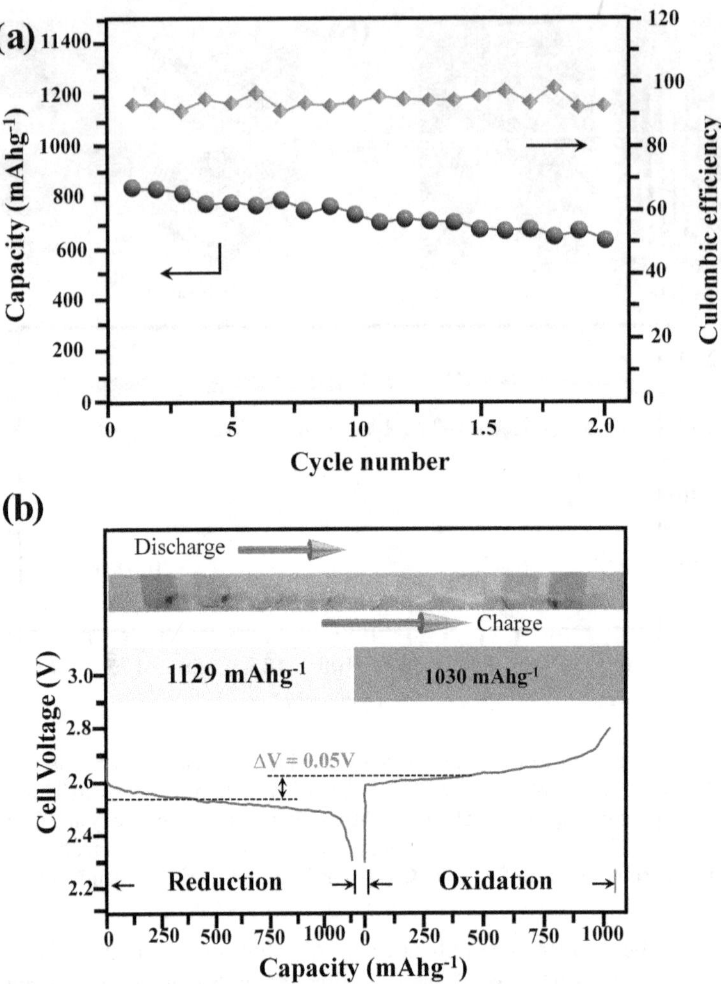

FIGURE 16.6 (a) An image of colour gradients of the Li_2S_4/Li_2S electrolyte, from yellow to transparent with increasing S^{2-} concentration. (b) Initial discharge/charge curve of the aqueous lithium–polysulphide battery using 0.1 M Li_2S_4 solution at 0.2 mA cm^2. Adapted and reproduced with permission from Ref. [44]. Copyright © 2014 The Royal Society of Chemistry.

specific capacity displayed by sulphur (1673 mAh g^{-1}) [42]. The concern with the sulphur system is the formation of soluble polysulphides that leads to a shuttle, which could reduce the efficiency and cycling stability of the system [43].

The sulphur-based lithium redox cell is reported to exhibit a high energy density of about 654 Wh Kg^{-1}, which is attributed by the redox reaction taking place between the S_2^{4-} and S^{2-} ions in aqueous phase. The redox reaction in the cell is visualised by the change in colour gradient from yellow to transparent as shown in Figure 16.6a. The cell reveals a high charge and discharge capacity of about 1129 and 1030 mAh g^{-1}. Moreover, the cell is able to exhibit a power density of about 4 mW cm^{-2} at 1.6 mA cm^{-2} with a discharge potential of 2.53 V vs Li^+/Li [44].

16.6 LITHIUM REDOX FLOW BATTERY IN APROTIC REDOX FLOW SYSTEM

The examination of aqueous catholyte for lithium redox flow battery was advantageous owing to their availability and better solubility with redox-active materials. Even though the reactivity of lithium with aqueous electrolyte can lead to severe safety issues, it is the most dangerous concern of aqueous electrolyte in lithium-based systems. The exploration of aprotic electrolytes is found to be a best choice with regard to its ability to form stable systems with a wide range of redox couples and its property to inhibit the corrosion of battery components as compared to that of aqueous electrolyte [45]. Even though the performance of the aprotic electrolytes is superior as compared to that of the aqueous electrolyte, they are facing numerous challenges such as moisture sensitivity, toxicity, and flammability. In the nonaqueous electrolyte, most of the systems explored are anion exchange systems with metal complexes or transition metals as the active materials. The lithium redox flow batteries based on aprotic solvents are categorised as metallocene, polysulphide, and organic-based lithium redox flow batteries [46,47].

16.6.1 Metallocene-Based Redox Flow Battery

Metallocenes are the organic metallic complexes that constitute a central atom attached to an aromatic ligand. Metallocenes are able to exist in various oxidation states and exhibit a sandwiched structure. Among different metallocenes, the mostly investigated metallocene in lithium-ion battery was ferrocene [48,49]. Ferrocene is observed to exhibit good redox flow kinetics, better thermal stability, and good solubility, which make them promising for the application in lithium redox flow batteries. Ferrocene- and ferrocenium-based redox couple in aprotic solvent was proposed by Zhao et al. [50]. The ferrocene/ferrocenium redox couple displayed a superior redox kinetics along with a specific capacity of about 130 mAh g^{-1}. Even after a 250 charge/discharge cycles, the cell maintained about 90% of capacity with a Coulombic efficiency in the range of 98%–100%. In order to enhance the solubility of ferrocene, another method was introduced by the functionalisation of ferrocene. The ionic-derivatised ferrocene compound on electrochemical evaluation in a nonaqueous hybrid redox flow battery reveals a cell voltage of 3.5 V with an energy density of 50 Wh L^{-1} and good cycling properties [51]. As a contradictory approach, semi-solid lithium redox flow battery examined ferrocene and ferrocene derivatives as the mediator and lithium iron phosphate as the Li^+-ions storage material. The redox couple acts as the mediator for targeting the redox to lithium iron phosphate nanoparticles [52], which are capable to enhance the performance of the cell.

16.6.2 Polysulphide-Based Lithium Redox Flow Battery

Polysulphide-based lithium redox flow batteries are promising since the sulphur can offer a good electrochemical performance of sulphur. They exhibit a better battery gravimetric capacity and low cost owing to its abundance even though they suffered from poor cyclability owing to the sluggish kinetics. The choice of polysulphide-based catholyte instead of the solid sulphur-based electrolyte is found to be fruitful to overcome this issue. The first polysulphide-based redox flow battery was proposed by Rauh et al. [53], which employs metallic lithium as the anode and the polysulphide dissolved in the catholyte comprised of tetrahydrofuran with the lithium salt $LiAsF_6$. Employing a Teflon bonded carbon as the current collector, it has been observed that the cell can achieve a best electrochemical performance. About 90% efficiency was observed with a discharge depth of 0.5 Li^+ sulphur at 50°C. They have observed a fade in performance that might arise owing to the ineffectiveness to block the shuttle effect between the polysulphide and lithium [53].

In order to enhance the performance, Zhang et al. [54] introduced a system constitute of lithium nitrate ($LiNO_3$) that can effectively form a passivation layer over the lithium anode to inhibit the movement of electrons to polysulphide and hence exhibit a good electrochemical performance and a better capacity retention over 70 cycles. The cell fabricated by Yung et al. [55] by exploring a catholyte designed that cycled only in the range between sulphur and Li_2S_2. The lithium-polysulphide battery designed constitutes a lithium foil as anode, lithium polysulphide solution as the catholyte, and carbon paper as the current collector as shown in Figure 16.7a. Since this cell configuration can effectively avoid the detrimental effect due to the volume expansion, the cell is observed to exhibit an excellent

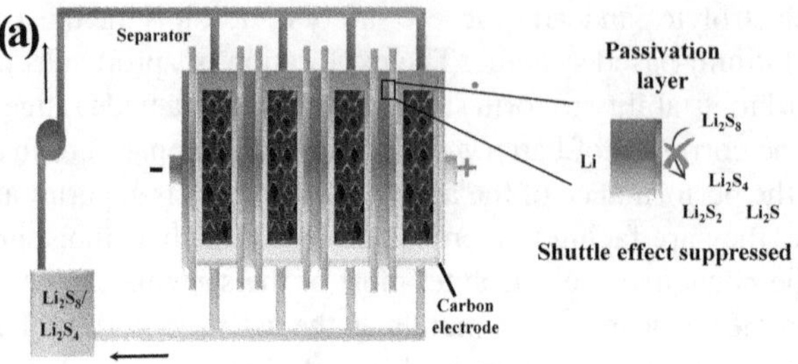

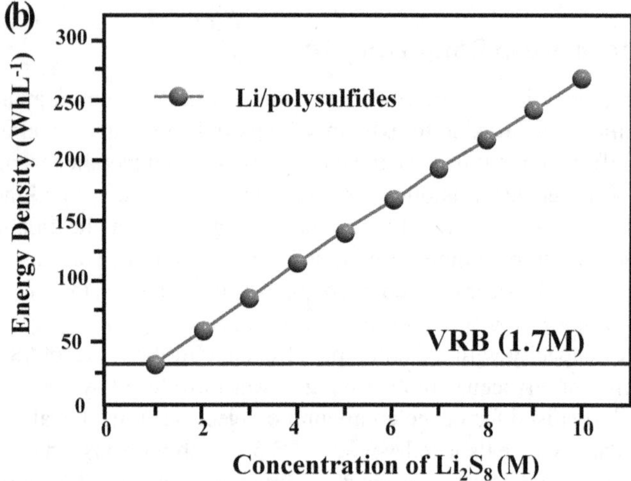

FIGURE 16.7 (a) The schematic illustrating the structure of the Li/PS battery. Polysulphides flow through the system during operation and are stored in the tank in downtime. Magnified scheme on the right: the SEI passivation layer has a high resistance towards internal reaction between PS and lithium due to the presence of $LiNO_3$. Consequently, the shuttle effect is significantly suppressed. The narrow voltage window avoids the formation of insoluble Li_2S_2/Li_2S and leads to excellent cycle life. (b) The volumetric energy density of the Li/PS system vs concentration of sulphur in the catholyte. The capacity of lithium is set to 2000 mAh g^{-1} and 1000 mAh cm^{-3} to take the extra lithium needed for practical operation into consideration. The specific capacity between sulphur and Li_2S_4 is 418 mAh g^{-1} organic-based separators. Adapted and reproduced with permission from Ref. [55]. Copyright © 2013 The Royal Society of Chemistry.

electrochemical performance. The energy density is observed to reach at 149 Wh kg^{-1} (133 Wh^{-1}), which is about five times higher than that of vanadium redox flow battery (Figure 16.7b) [55].

16.6.3 Organic-Based Lithium Redox Flow Battery

Organic redox flow batteries are normally categorised as p-type and n-type and bipolar on the basis of their reduction from the neutral state. The redox reaction taking place between the neutral state and the redox state is called n-type, while the redox reaction taking place between the neutral and positively charged state is recognised as p-type; bipolar is the state in which either oxidation or reduction can be possible (Figure 16.8) [56]. During the electrochemical rection, the corresponding cations (Li^+, Na^+) or anions (ClO_4^-, PF_6^-) should neutralise the redox product. Compared to the other redox flow batteries, the organic-based lithium redox system is observed to reveal a better dissolution, and the solubility problems significantly overcome by the addition of proper functional group. In the case of the redox-active organics, p-type redox-active organics are not yet widely explored in lithium redox batteries, and for the n-type redox organics, the carbonyl groups are widely explored [57,58].

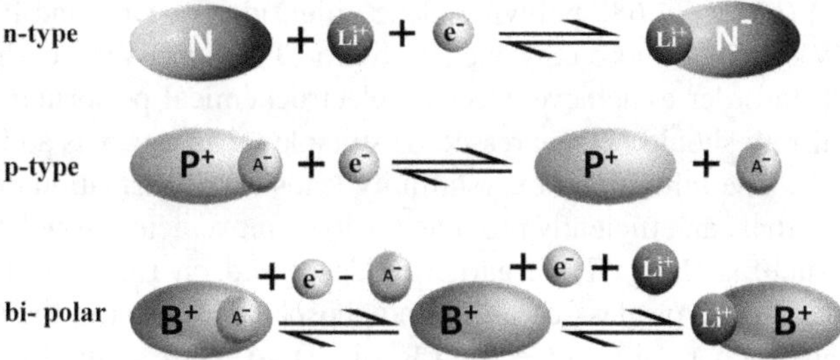

FIGURE 16.8 Redox reaction of n-type (N), p-type (P), and bipolar (B) redox-active organics and metal organics. Adapted and reproduced with permission from Ref. [56]. Copyright © 2013 The Royal Society of Chemistry.

Siroma et al. [59] proposed a benzoquinone and its derivative for the lithium redox and observed that with increasing the concentration of the catholyte the cycling property of the cell gets reduced. They reported that the concentration-dependent cycling stability arises due to the sublimation of the 1,4-benzoquinone and instability of radicle anion. Another derivative of anthraquinone, 1,5-bis(2-(2-(2-methoxyethoxy)ethoxy)ethoxy)anthracene-9,10-dione, suggested by Wang et al. [60], exhibits an improved electrochemical performance owing to its better solubility as compared to that reported by Siroma et al. [59] Another attempt to enhance the solubility is the introduction of two triethylene glycol monomethyl ether groups to anthraquinone that enable them to easily get soluble in most of the aprotic and polar solvents.

The cell designed by using 2,3,6-trimethylquinoxaline as anode-active material and 2,5-di-tert-butyl-1,4-bis(2-methoxyethoxy)benzene as cathode active material displays a low Coulombic efficiency (70%) and poor performance, which reveals the major challenge of achieving advanced battery performance [26].

16.7 SEPARATORS

Separators are the materials that will sandwich between the electrodes in batteries in order to avoid the contact of electrodes. They are the most crucial components inside the battery that equally participate to yield best electrochemical performance and to ensure better safety features in batteries [61,62]. The separator membrane used in lithium-ion battery can either be solid electrolyte, such as ceramic electrolyte, or polymer electrolyte, such as organic polymers. The major requirement of a separator is it should be electronically insulating, chemically stable, low cost, and non-toxic [63]. Different electrolytes such as NASICONs, garnets, perovskite anti-perovskite, and polymer-based electrolytes are extensively examined in lithium redox flow batteries. The NASICON-type $LiZr_2(PO_4)_3$ (LZP) is reported to reveal a better chemical stability. Similarly, garnet-type materials having cubic $A_3B_3C_2O_{12}$ garnet structures are able to exhibit an ionic conductivity of about 10^{-3} S cm^{-1} at 25°C [64,65]. Perovskite and antipervoskites constitute large number of vacant sites that are able to accommodate the Li^+ ions and establish better ionic conductivity. Certain antipervoskites are reported to exhibit better ionic conductivity similar to that of garnet-type materials that are superior in performance enhancement.

Another category of electrolytes that are extensively investigated today is the polymer electrolytes that are broadly classified as solid polymer electrolytes and gel polymer electrolytes having different electrochemical performance in rechargeable batteries. The solid polymer electrolytes are observed to exhibit a low ionic conductivity as compared to its gel counterparts. The gel polymer electrolytes that combine the characteristics of both solid and liquid electrolytes are investigated extensively for the rechargeable batteries and the most favourable polymer matrices used for these applications are polyethylene oxide (PEO) [66], poly(acrylonitrile) (PAN) [67], poly (methyl methacrylate) (PMMA) [68], polyvinylidene difluoride (PVDF), and PVDF-based copolymers [69], such as PVDF-co-trifluoro ethylene (PVdF-co-TFE) and PVDF-co-hexafluoropropylene (PVdF-co-HFP) [70]. In order to achieve excellent electrochemical performance in polymer electrolytes, their crystallinity should be suppressed by suitable means such as addition of fillers or by other suitable methods. The inhibition of crystallinity leads to the formation of proper amorphous content in the polymer that can efficiently promote the ionic movement through the polymer matrix.

Oxide materials such as (La,Li)TiO_3, garnet oxides based on Li_5La_3-Ta_2O_{12}, crystalline sulphide such as Li_2S–P_2S_5 glass or glass-ceramic, and phosphates similar to the Na+-ion conducting NASICON compounds such as $Li_{1+x}Al_xGe_{2x}(PO_4)_3$, $Li_{1+x}Ti_{2x}Al_x(PO_4)_3$, and $Li_{1+x+y}Ti_{2x}Al_xSi_y(PO_4)_{3y}$ are observed to be superior in performance than that of the polymer counterparts owing to its ability to exhibit excellent mechanical, chemical, and electrochemical stability. Apart from that, they are able to display a superior bulk ionic conductivity of about 10^{-3} S cm^{-1} [71,72].

16.8 CONCLUSION

Redox flow batteries are the emerging systems of energy storage that can ensure better performance as compared to the currently established systems. Different redox flow systems were examined in which the lithium redox flow batteries are attracting significant interest nowadays. They are the rechargeable systems that combine the advantage of both lithium-ion batteries and redox flow batteries. The performance of redox flow batteries is relying on the chemical components with reversible redox behaviour that are dissolved in the liquid contained in the system and in external reservoir. Based on the components, especially the electrolyte counterparts, they are broadly classified that exhibit characteristic performances. The presence of redox couples in lithium redox flow batteries can provide an enhanced electrochemical performance and thereby lithium-based redox systems are considered to be the promising energy storage devices for future electronics.

REFERENCES

1. Thaller L (1979) Redox flow cell energy storage systems. *Terrestrial Energy systems conference* 989, Patent Number DOE/NASA/1002-79/. https://doi.org/10.2514/6.1979-989.
2. Thaller LH (1979) Recent advances in storage systems office of energy technology. NASA Tm-79186 79186.
3. Leung P, Li X, Ponce De León C, et al (2012) Progress in redox flow batteries, remaining challenges and their applications in energy storage. *RSC Adv* 2:10125–10156. https://doi.org/10.1039/c2ra21342g.
4. Tirukkovalluri SR, Hanuman Gorthi RK (2013) Synthesis, characterization and evaluation of pb electroplated carbon felts for achieving maximum efficiency of fe-cr redox flow cell. *J New Mater Electrochem Syst* 16:287–292. https://doi.org/10.14447/jnmes.v16i4.155.
5. Anon (1983) Nasa redox storage system development project calendar year 1982. https://ntrs.nasa.gov/citations/19840003511
6. Darling RM, Gallagher KG, Kowalski JA, et al (2014) Pathways to low-cost electrochemical energy storage: A comparison of aqueous and nonaqueous flow batteries. *Energy Environ Sci* 7:3459–3477. https://doi.org/10.1039/c4ee02158d.
7. Shigematsu T (2019) The development and demonstration status of practical flow battery systems. *Curr Opin Electrochem* 18:55–60. https://doi.org/10.1016/j.coelec.2019.10.001.
8. Sarkar J, Bhattacharyya S (2012) Application of graphene and graphene-based materials in clean energy-related devices Minghui. *Arch Thermodyn* 33:23–40. https://doi.org/10.1002/er.
9. Skyllas-Kazacos M, Chakrabarti MH, Hajimolana SA, et al (2011) Progress in flow battery research and development. *J Electrochem Soc* 158:R55. https://doi.org/10.1149/1.3599565.
10. Yang Z, Zhang J, Kintner-meyer MCW, et al (2011) Electrochemical energy storage for green grid: Status and challenges. *ECS Meet Abstr* 3577–3613. https://doi.org/10.1149/ma2011-02/4/155.
11. Skyllas-Kazacos M, Kazacos M (2011) State of charge monitoring methods for vanadium redox flow battery control. *J Power Sources* 196:8822–8827. https://doi.org/10.1016/j.jpowsour.2011.06.080.
12. Sum E, Rychcik M, Skyllas-kazacos M (1985) Investigation of the V(V)/V(IV) system for use in the positive half-cell of a redox battery. *J Power Sources* 16:85–95. https://doi.org/10.1016/0378-7753(85)80082-3.

13. Mohamed MR, Sharkh SM, Walsh FC (2009) Redox flow batteries for hybrid electric vehicles: Progress and challenges. *5th IEEE Veh Power Propuls Conf VPPC'09*, Milan, Italy, 551–557. https://doi.org/10.1109/VPPC.2009.5289801.
14. Collins J, Kear G, Li X, et al (2010) A novel flow battery: A lead acid battery based on an electrolyte with soluble lead(II) Part VIII. The cycling of a 10 cm×10 cm flow cell. *J Power Sources* 195:1731–1738. https://doi.org/10.1016/j.jpowsour.2009.09.044.
15. Zhang L, Cheng J, Yang Y, et al (2008) Study of zinc electrodes for single flow zinc/nickel battery application. *J Power Sources* 179:381–387. https://doi.org/10.1016/j.jpowsour.2007.12.088.
16. Ito Y, Nyce M, Plivelich R, et al (2011) Zinc morphology in zinc-nickel flow assisted batteries and impact on performance. *J Power Sources* 196:2340–2345. https://doi.org/10.1016/j.jpowsour.2010.09.065.
17. Leung PK, Ponce De Leon C, Walsh FC (2012) The influence of operational parameters on the performance of an undivided zinc-cerium flow battery. *Electrochim Acta* 80:7–14. https://doi.org/10.1016/j.electacta.2012.06.074.
18. Leung PK, Ponce-De-León C, Recio FJ, et al (2014) Corrosion of the zinc negative electrode of zinc-cerium hybrid redox flow batteries in methanesulfonic acid batteries. *J Appl Electrochem* 44:1025–1035. https://doi.org/10.1007/s10800-014-0714-y.
19. Walsh FC (2001) Electrochemical technology for environmental treatment and clean energy conversion. *Pure Appl Chem* 73:1819–1837.
20. Zhao P, Zhang H, Zhou H, Yi B (2005) Nickel foam and carbon felt applications for sodium polysulfide/bromine redox flow battery electrodes. *Electrochim Acta* 51:1091–1098. https://doi.org/10.1016/j.electacta.2005.06.008.
21. Xu K, Von Wald Cresce A (2012) Li+-solvation/desolvation dictates interphasial processes on graphitic anode in Li ion cells. *J Mater Res* 27:2327–2341. https://doi.org/10.1557/jmr.2012.104.
22. Duduta M, Ho B, Wood VC, et al (2011) Semi-solid lithium rechargeable flow battery. *Adv Energy Mater* 1:511–516. https://doi.org/10.1002/aenm.201100152.
23. Huang Q, Yang J, Ng CB, et al (2016) Environmental science targeting reactions between LiFePO 4 and iodide. *Energy Environ Sci* 9:917–921. https://doi.org/10.1039/c5ee03764f.
24. Zhao Y, Ding Y, Li Y, et al (2015) A chemistry and material perspective on lithium redox flow batteries towards high-density electrical energy storage. 44. https://doi.org/10.1039/c5cs00289c.
25. Zhao Y, Ding Y, Li Y, et al (2015) A chemistry and material perspective on lithium redox flow batteries towards high-density electrical energy storage. *Chem Soc Rev* 44:7968–7996. https://doi.org/10.1039/c5cs00289c.
26. Brushett FR, Vaughey JT, Jansen AN (2012) An all-organic non-aqueous lithium-ion redox flow battery. *Adv Energy Mater* 2:1390–1396. https://doi.org/10.1002/aenm.201200322.
27. Wang Y, He P, Zhou H (2012) Li-redox flow batteries based on hybrid electrolytes: At the cross road between Li-ion and redox flow batteries. *Adv Energy Mater* 2:770–779. https://doi.org/10.1002/aenm.201200100.
28. Shah AA, Al-Fetlawi H, Walsh FC (2010) Dynamic modelling of hydrogen evolution effects in the all-vanadium redox flow battery. *Electrochim Acta* 55:1125–1139. https://doi.org/10.1016/j.electacta.2009.10.022.
29. Han X, Li X, White J, et al (2018) Metal-air batteries: From static to flow system. *Adv Energy Mater* 8. https://doi.org/10.1002/aenm.201801396.
30. Prifti H, Parasuraman A, Winardi S, et al (2012) Membranes for redox flow battery applications. *Membranes (Basel)* 2:275–306. https://doi.org/10.3390/membranes2020275.
31. Li J, Yang L, Yang S, Lee JY (2015) The application of redox targeting principles to the design of rechargeable Li-S flow batteries. *Adv Energy Mater* 5. https://doi.org/10.1002/aenm.201501808.
32. Lu Y, Goodenough JB, Kim Y (2011) Aqueous cathode for next-generation alkali-ion batteries. *J Am Chem Soc* 133:5756–5759. https://doi.org/10.1021/ja201118f.
33. Wang Y, Wang Y, Zhou H (2011) A Li-liquid cathode battery based on a hybrid electrolyte. *ChemSusChem* 4:1087–1090. https://doi.org/10.1002/cssc.201100201.
34. Svensson PH, Kloo L (2003) Synthesis, structure, and bonding in polyiodide and metal iodide-iodine systems. *Chem Rev* 103:1649–1684. https://doi.org/10.1021/cr0204101.
35. Ramette RW, Sandford RW (1965) Thermodynamics of iodine solubility and triiodide ion formation in water and in deuterium oxide. *J Am Chem Soc* 87:5001–5005. https://doi.org/10.1021/ja00950a005.
36. Mader MJ, White RE (1986) Mathematical model of a Zn/Br//2 cell on charge. *Proc - Electrochem Soc* 86-8:87–108. https://doi.org/10.1149/1.2108857.

37. Zhao Y, Wang L, Byon HR (2013) High-performance rechargeable lithium-iodine batteries using triiodide/iodide redox couples in an aqueous cathode. *Nat Commun* 4:1–7. https://doi.org/10.1038/ncomms2907.
38. Zhao Y, Byon HR (2013) High-performance lithium-iodine flow battery. *Adv Energy Mater* 3(12):1630–1635. https://doi.org/10.1002/aenm201300627.
39. Zhao Y, Hong M, Bonnet Mercier N, et al (2014) A 3.5 v lithium-iodine hybrid redox battery with vertically aligned carbon nanotube current collector. *Nano Lett* 14:1085–1092. https://doi.org/10.1021/nl404784d.
40. Abraham KM, Jiang Z (1996) A polymer electrolyte-based rechargeable lithium/oxygen battery. *J Electrochem Soc* 143:1–5.
41. Liang Y, Li Y, Wang H, et al (2011) Co_3O_4 nanocrystals on graphene as a synergistic catalyst for oxygen reduction reaction. *Nature* 10(10):780–786.
42. Seh ZW, Li W, Cha JJ, et al (2013) Sulphur-TiO_2 yolk-shell nanoarchitecture with internal void space for long-cycle lithium-sulphur batteries. *Nat Commun* 4. https://doi.org/10.1038/ncomms2327.
43. Zhang SS (2013) Liquid electrolyte lithium/sulfur battery: Fundamental chemistry, problems, and solutions. *J Power Sources* 231:153–162. https://doi.org/10.1016/j.jpowsour.2012.12.102.
44. Li N, Weng Z, Wang Y, et al (2014) An aqueous dissolved polysulfide cathode for lithium-sulfur batteries. *Energy Environ Sci* 7:3307–3312. https://doi.org/10.1039/c4ee01717j.
45. Xu K (2004) Nonaqueous liquid electrolytes for lithium-based rechargeable batteries. *Chem Rev* 104:4303–4417. https://doi.org/10.1021/cr030203g.
46. Chakrabarti MH, Dryfe RAW, Roberts EPL (2007) Evaluation of electrolytes for redox flow battery applications. *Electrochim Acta* 52:2189–2195. https://doi.org/10.1016/j.electacta.2006.08.052.
47. Shin SH, Yun SH, Moon SH (2013) A review of current developments in non-aqueous redox flow batteries: Characterization of their membranes for design perspective. *RSC Adv* 3:9095–9116. https://doi.org/10.1039/c3ra00115f.
48. Ding Y, Zhao Y, Yu G (2015) A membrane-free ferrocene-based high-rate semiliquid battery. *Nano Lett* 15:4108–4113. https://doi.org/10.1021/acs.nanolett.5b01224.
49. Vorotyntsev MA, Casalta M, Pousson E, et al (2001) Redox properties of titanocene-pyrrole derivative and its electropolymerization. *Electrochim Acta* 46:4017–4033. https://doi.org/10.1016/S0013-4686(01)00717-4.
50. Zhao Y, Ding Y, Song J, et al (2014) Sustainable electrical energy storage through the ferrocene/ferrocenium redox reaction in aprotic electrolyte. *Angew Chem* 126:11216–11220. https://doi.org/10.1002/ange.201406135.
51. Wei X, Cosimbescu L, Xu W, et al (2015) Towards high-performance nonaqueous redox flow electrolyte via ionic modification of active species. *Adv Energy Mater* 5:1–7. https://doi.org/10.1002/aenm.201400678.
52. Huang Q, Li H, Grätzel M, Wang Q (2013) Reversible chemical delithiation/lithiation of $LiFePO_4$: Towards a redox flow lithium-ion battery. *Phys Chem Chem Phys* 15:1793–1797. https://doi.org/10.1039/c2cp44466f.
53. Rauh RD, Abraham KM, Pearson GF, et al (1979) A lithium/dissolved sulfur battery with an organic electrolyte. *J Electrochem Soc* 126:523–527. https://doi.org/10.1149/1.2129079.
54. Zhang SS, Read JA (2012) A new direction for the performance improvement of rechargeable lithium/sulfur batteries. *J Power Sources* 200:77–82. https://doi.org/10.1016/j.jpowsour.2011.10.076.
55. Yang Y, Zheng G, Cui Y (2013) A membrane-free lithium/polysulfide semi-liquid battery for large-scale energy storage. *Energy Environ Sci* 6:1552–1558. https://doi.org/10.1039/c3ee00072a.
56. Song Z, Zhou H (2013) Towards sustainable and versatile energy storage devices: An overview of organic electrode materials. *Energy Environ Sci* 6:2280–2301. https://doi.org/10.1039/c3ee40709h.
57. Suga T, Ohshiro H, Ugita S, et al (2009) Emerging n-type redox-active radical polymer for a totally organic polymer-based rechargeable battery. *Adv Mater* 21:1627–1630. https://doi.org/10.1002/adma.200803073.
58. Liang Y, Tao Z, Chen J (2012) Organic electrode materials for rechargeable lithium batteries. *Adv Energy Mater* 2:742–769. https://doi.org/10.1002/aenm.201100795.
59. Senoh H, Yao M, Sakaebe H, et al (2011) A two-compartment cell for using soluble benzoquinone derivatives as active materials in lithium secondary batteries. *Electrochim Acta* 56:10145–10150. https://doi.org/10.1016/j.electacta.2011.08.115.
60. Wang W, Xu W, Cosimbescu L, et al (2012) Anthraquinone with tailored structure for a nonaqueous metal-organic redox flow batter. *Chem Commun* 48:6669–6671. https://doi.org/10.1039/c2cc32466k.

61. Sim LN, Sentanin FC, Pawlicka A, et al (2017) Development of polyacrylonitrile-based polymer electrolytes incorporated with lithium bis(trifluoromethane)sulfonimide for application in electrochromic device. *Electrochim Acta* 229:22–30. https://doi.org/10.1016/j.electacta.2017.01.098.
62. Meyer BWH (1998) Polymer electrolytes for lithium-ion batteries. *Rechargeable Lithium-Ion Batteries.* CRC Press, 260–288.
63. Lowe M, Gary Gereffi ST (2010) Lithium ion batteries for electric vehicles. SAE Tech. Pap. Ser. 1.
64. Li J, Yuan CF, Guo ZH, et al (2011) Limiting factors for low-temperature performance of electrolytes in $LiFePO_4$/Li and graphite/Li half cells. *Electrochim Acta* 59:69–74. https://doi.org/10.1016/j.electacta.2011.10.041.
65. Goodenough JB, Hong HY, Kafalas JA (1976) R. G. Graves and J. B. Smathers. *Mater Res Bull* 5:77843.
66. Pitawala HMJC, Dissanayake MAKL, Seneviratne VA (2007) Combined effect of Al_2O_3 nano-fillers and EC plasticizer on ionic conductivity enhancement in the solid polymer electrolyte (PEO)9LiTf. *Solid State Ionics* 178:885–888. https://doi.org/10.1016/j.ssi.2007.04.008.
67. Osman Z, Md Isa KB, Ahmad A, Othman L (2010) A comparative study of lithium and sodium salts in PAN-based ion conducting polymer electrolytes. *Ionics (Kiel)* 16:431–435. https://doi.org/10.1007/s11581-009-0410-9.
68. Kalhoff J, Eshetu GG, Bresser D, Passerini S (2015) Safer electrolytes for lithium-ion batteries: State of the art and perspectives. *ChemSusChem* 8:2154–2175. https://doi.org/10.1002/cssc.201500284.
69. Motamedi AS, Hamid Mirzadeh FH, Bagheri-Khoulenjani S, Shokrgozar M (2017) Effect of electrospinning parameters on morphological properties of PVDF nanofibrous scaffolds. *Prog Biomater* 6:113–123. https://doi.org/10.1007/s40204-017-0071-0.
70. Aravindan V, Vickraman P (2008) Characterization of SiO_2 and Al_2O_3 incorporated PVdF-HFP based composite polymer electrolytes with $LiPF_3(CF_3CF_2)_3$. *J Appl Polym Sci* 108:1314–1322. https://doi.org/10.1002/app.
71. Thokchom JS, Gupta N, Kumar B (2008) Superionic conductivity in a lithium aluminum germanium phosphate glass-ceramic. *J Electrochem Soc* 155:A915. https://doi.org/10.1149/1.2988731.
72. Barré M, Le Berre F, Crosnier-Lopez MP, et al (2009) The NASICON solid solution $Li_{1-x}La_{x/3}Zr_2(PO_4)_3$: Optimization of the sintering process and ionic conductivity measurements. *Ionics (Kiel)* 15:681–687. https://doi.org/10.1007/s11581-009-0332-6.

17 Sulfurized Polyacrylonitrile Cathodes for Advanced Lithium–Sulfur Batteries

Xiaohui Zhao, Amir Abdul Razzaq, Zhao Deng, Jou-Hyeon Ahn, and Prasanth Raghavan

17.1 INTRODUCTION

The burning of fossil fuels not only depletes natural resources but also leads to global warming which is a severe threat to mankind on the earth. Therefore, it is an urgent demand to adopt environmentally friendly and renewable energy resources. In this alarming scenario, shifting the electricity production from burning fuels to eco-friendly renewable energy sources should be advanced and implemented. No doubt, wind and solar energy are clean sources of renewable energy resources which can be effective in terms of suppressing global warming and depletion of fossil resources, but they are limited to stationary power productions [1–4]. Since the first-ever commercialized lithium-ion (Li-ion) batteries have been introduced by Sony, Li-ion batteries played a significant role in portable and electric storage devices. However, the ever-increasing portable devices, vehicles, and large-scale grids require energy storage devices of high energy density beyond the state-of-art Li-ion batteries [5–7].

The Li–S batteries, owing to their higher volumetric energy density of 2800 Wh L^{-1} and excellent gravimetric energy density of 2600 Wh kg^{-1} compared to Li-ion batteries, are one of the most promising candidates (Figure 17.1a) [8,9]. What's more, sulfur is naturally abundant with low toxicity and economic benefits [10–12]. Nonetheless, Li–S batteries are far from commercialization resulting in their low utilization of active material, low Coulombic efficiency, and poor cycle stability. Both the insulating sulfur and reduced lithium sulfides (Li_2S) significantly impede the reaction kinetics of sulfur [13–15]. Moreover, the known shuttle effect raised by the dissolution of long-chain lithium polysulfides (LiPS) in ether electrolytes leads to the loss of sulfur, self-discharge, capacity fading, volume expansion, and anode corrosion in Li–S batteries [10,13,14,16]. The electrochemical reaction in a lithium–sulfur (Li–S) battery is the reaction of sulfur and lithium to form lithium sulfide as shown in Figure 17.1b.

During the discharge process, sulfur is reduced to a series of LiPS intermediate products, and these polysulfide ions are easily dissolved in the electrolyte and diffuse through the separator between the positive and negative electrodes (Figure 17.2a). This complicated reaction process brings about a series of problems: (i) The insulation of the elemental sulfur and its final discharge product, lithium sulfide, reduces the utilization rate of active materials; (ii) the polysulfide ions are positively charged during charging and discharging. Shuttle between the negative electrodes causes loss of active material and decreased cycle performance (Figure 17.2b); (iii) LiPS disproportionate reaction on the surface of the lithium sheet corrodes the lithium negative electrode and increases the internal resistance of the battery; and (iv) the volume effect of the electrode during charge and discharge is easy to cause the structural collapse of the electrode, resulting in the degrading of cycle stability.

Various strategies including physical hosting, chemical adsorption, and catalytical promoting methods have been adopted to elevate the shuttle effect and improve the electrochemical

DOI: 10.1201/9781003310167-17

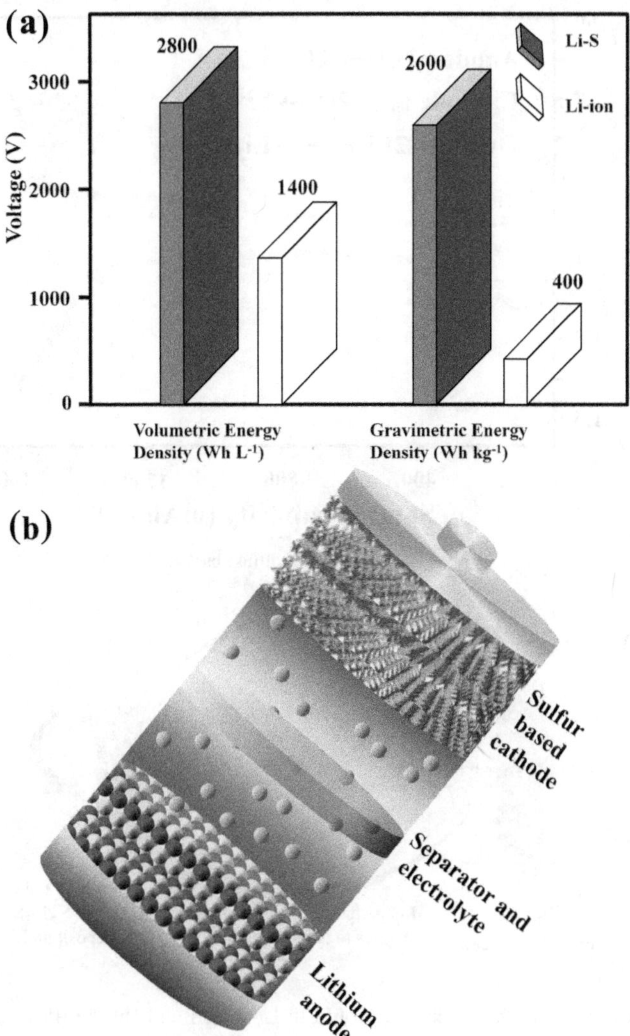

FIGURE 17.1 (a) The comparison of Li–S batteries vs Li ion (based on graphite anodes and $LiNi_{1/3}Mn_{1/3}Co_{1/3}O_2$ cathodes) in terms of specific capacity and energy density and (b) the structure of a Li–S cell.

performance of sulfur cathodes (Figure 17.3) [17,18]. The impregnation of sulfur into porous and conductive carbon matrix is proven to be a useful strategy for forming nanostructured composites. The porous carbonous matrixes provide confined spaces to inhibit the outward diffusion of polysulfides and help to accommodate the volume expansion during cycling [19]. Nazar et al. [20] first used highly order mesoporous (CMK-3) as a cathode in Li–S batteries in 2009. After that, numerous carbon materials have endeavored, including porous carbon particles [21–23], CNTs [24,25], carbon nanofibers [26–28], graphene [29,30], and so on. Carbonaceous materials were doped with heteroatoms to enhance their chemical anchoring ability of LiPS and thus catalyzed the redox reaction in Li–S batteries. Other numerous researches have been devoted, such as combining sulfur with polar mediators like metal oxides [31], metal sulfides [32,33], and metal nitrides [34]; modifying separators [35,36]; developing novel electrolyte additives [37,38]; and the protection of lithium anode [39,40]. Nevertheless, the poor physical and chemical interaction between sulfur and the carbon hosts results in low sulfur utilization. Also, the capacity decay in the first few cycles due to the loss of sulfur in ether-based electrolytes and the short-term cycle stability remain a challenge.

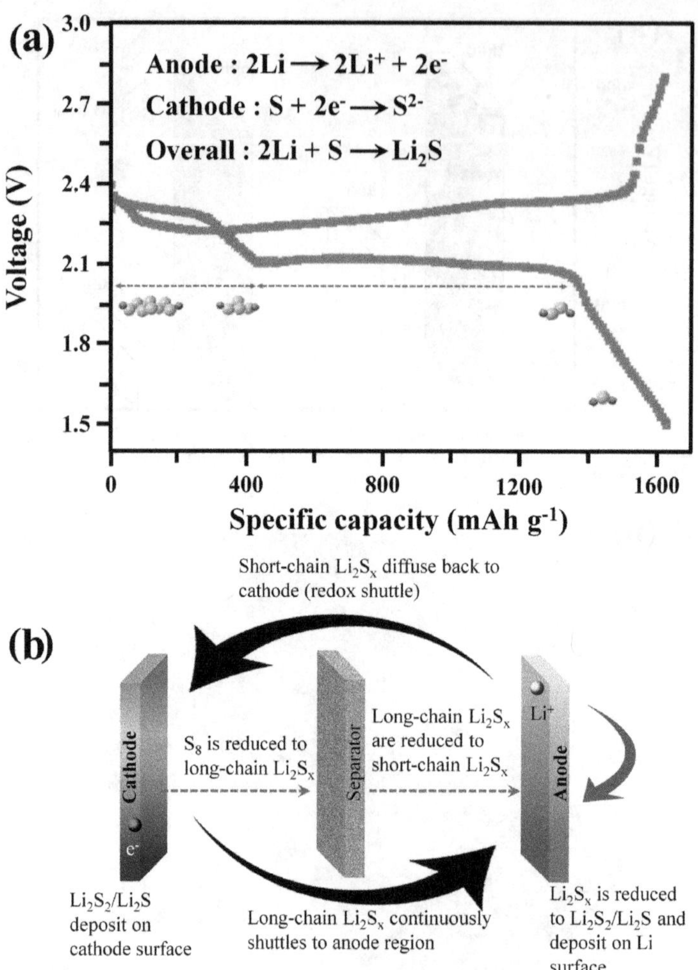

FIGURE 17.2 (a) A typical discharge/charge profile in Li–S cell and (b) the illustration of sulfur shuttle phenomenon.

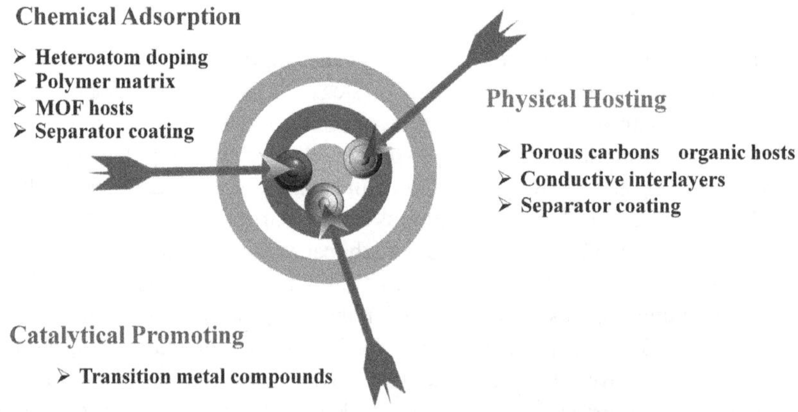

FIGURE 17.3 Summary of the strategies applied for improving the electrochemical performance of sulfur cathodes.

17.2 DEVELOPMENT OF SULFURIZED POLYACRYLONITRILE (SPAN) CATHODES

SPAN cathodes are known as an alternative cathode material of elemental sulfur for practical applications of Li–S batteries [41], thanks to their high capability without LiPS shuttling. SPAN composite with a sulfur content of around 40 wt.% can be obtained simply by heating polyacrylonitrile (PAN) and sulfur powder under the protection of an inert atmosphere, where PAN is dehydrogenated into a heterocyclic structure and sulfur is intercalated in the conductive backbone of PAN [41]. It is generally accepted that only non-soluble short-chain LiPS are involved in the redox reaction processes [42,43]. Even if calculated according to the current 40 wt.% sulfur content, the energy density of SPAN is higher than 500 Wh kg^{-1}, which exceeds that of Li-ion batteries (Table 17.1). Moreover, compared with the sulfur–carbon composite cathodes based on elemental sulfur, SPAN can use carbonate electrolytes although their energy density and working voltage are slightly lower, making it economically more feasible. It is easy to integrate with the existing Li-ion battery preparation process and implement large-scale commercial production.

17.2.1 Chemical Structure of SPAN

The exact structure of the SPAN has remained debatable since its discovery initially by Wang et al. [44], who proposed the existence of elemental sulfur in SPAN composites. However, Yu et al. [45] and He et al. [46] separately identified the Fourier-transform infrared spectroscopy (FT-IR) and the Raman characteristics of C-S peaks in SPAN along with the presence of C=C and C=N groups. In 2011, an important contribution was made by Fanous et al. [47] to understand the chemical structure of SPAN, and their analyzed results indicated that at the synthetic temperature of 330°C, sulfur is only chemically linked to carbon in the form of 2-pyridylthiolates, oligosulfide, and thioamide as shown in Figure 17.4a. Wang et al. [48] first time postulated that the two repeated and/or periodic conjugated units containing carbon and nitrogen atoms may be interconnected with sulfur atoms in the SPAN molecular structure (Figure 17.4b). Jin et al. [42] conducted the FT-IR, Raman, X-ray photoelectron spectroscopy (XPS), and solid-state nuclear magnetic resonance spectroscopy (SS-NMR) to propose the molecular structure of SPAN via reaction between PAN and sulfur, where they investigated the structure of SPAN as pyridinic ring containing C=C, C=N and 2–4 number of sulfur atoms linked with two adjacent pyridinic rings with at least 1 C-H bond in three rings as shown in Figure 17.4c. In 2020, Weret et al. [49] proposed the synthetic route of SPAN; at a temperature above 170°C, elemental sulfur converts into free radicals and covalently bonded to PAN at the nitrile carbon (C1), and when temperature further rises to 300°C, the sulfurization of the PAN to lead a conjugated heteroaromatic structure as shown in Figure 17.4d. The chemical structure of SPAN has been broadly explored, but the exact structure is still controversial. Nonetheless, it is generally recognized that the 2–4 sulfur atoms are covalently linked to the pyridinic rings of SPAN and no physically confined elemental sulfur exists.

TABLE 17.1
The Comparison of Cell Performances for SPAN, Sulfur-Carbon Composites, and Li-Ion $LiNi_{1/3}Mn_{1/3}Co_{1/3}O_2$ Cathodes

Cathode Material	Voltage (V)	Capacity (mAh/g)	Electrolyte	Energy Density (Wh/kg)	Anode
$LiNi_{1/3}Mn_{1/3}Co_{1/3}O_2$ cathodes	3.8~4.0	220–280	Carbonate	240–300	Graphite
Sulfur/Carbon composite	2.1	1000	Ether	350–600	Lithium metal
SPAN	1.8	750–1000	Carbonate	>500	Lithium metal or graphite, pre-lithiated SiO_x/C

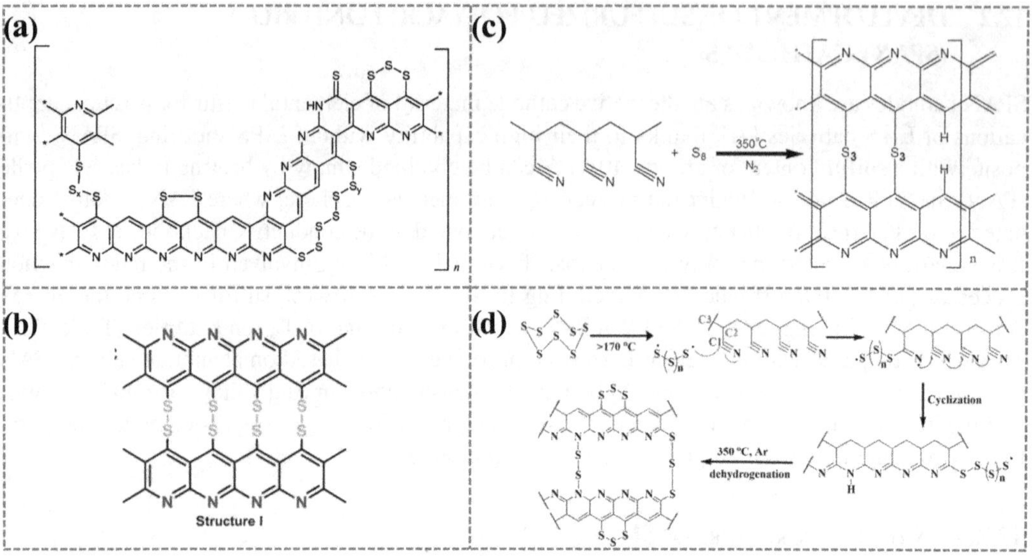

FIGURE 17.4 Proposed structure of SPAN composite. (a) The structure proposed by Fanous et al. Adapted and Reproduced from ref. [47]. Copyright 2011 American Chemical Society. (b) The proposed molecular structure of SPAN by Wang et al. Adapted and Reproduced from ref. [48]. Copyright 2018 American Chemical Society. (c) Reaction route proposed by Jin et al. Adapted and Reproduced from ref. [42]. Copyright 2018 Elsevier. (d) Weret et al. proposed synthesis reaction of SPAN. Adapted and Reproduced from ref. [49]. Copyright 2020 Elsevier.

17.2.2 Redox Reactions in SPAN

Nonetheless, the chemical structure of SPAN is still debatable, and it is generally accepted that sulfur (Sn, $2 \geq n \leq 4$) is chemically linked to the carbon atoms of the pyrolyzed skeleton. The pyridine backbone is considered a conductive framework and sulfur serves as the active material to afford the reversible capacity. Wang et al. [48] studied the involvement of breakage and reformation of C–S and S–S bonds in SPAN during the lithiation and delithiation process using solid NMR, XPS, and electron paramagnetic resonance (EPR) spectroscopy. Moreover, the irreversible bonding of lithium with nitrogen was observed in the first discharge only; the ionic coordinate thus formed accelerated the electron transfer process and improved the stability of SPAN composite in the following cycles. Jin et al. [50] used XPS and NMR to study the lithiation and delithiation behavior of the SPAN after the first discharge, first charge, and second cycles. They found the breakage of S–S and C–S bonds after reaction with lithium during first discharge, and at the same time, lithium could also react with C=C and C=N. On charging, the Li–S–S and Li–C–S bonds were supposed to be reversibly delithiated, whereas the portion of Li–C=C and Li–C–N wasn't involved in the complete delithiation process, which could be the reason for the capacity fading after the first discharge in SPAN cathodes. From the second discharge, the reversible lithiation occurred except for those lithium ions which reacted with carbon and nitrogen after the first charge. The same type of observations was also made by Wang et al. [51] during the study of freestanding. SPAN–CNT cathode was prepared by electrospinning of PAN and CNT, followed by sulfurization. In addition to NMR and XPS studies, they also computed the electron structure of the SPAN–CNT backbone during the redox process by frontier molecular orbital (FMO) investigation. The energy gap between the highest occupied molecular orbital (HOMO) and lowest unoccupied molecular orbital (LUMO) was decreased from SPAN (1.92 eV) to (SPAN-Li) 1.15 eV, and lastly to 0.5 eV for PAN-Li, which indicates the increase in conductivity of SPAN after lithiation process [52]. Weret et al. [49] observed that during the first discharge the covalent bonded C–S, N–S, and S–S bonds were broken first, and then new Li–C,

Li–N, and Li–S bonds were formed. However, in the following cycles, the covalent bonds were reversibly lithiated and delithiated.

17.2.3 SPAN Composites

17.2.3.1 Carbon Additives

Vulcanization of an insulating PAN and a sulfur with an electric conductivity of ~10^{-30} S cm^{-1} results in a semiconductor SPAN composite with a comparatively higher electric conductivity of 10^{-4} S cm^{-1}. Despite the improved electric conductivity of SPAN, the composites cast with additional conductive agents on current collectors by hand milling process exhibited comparatively poor electrochemical performance [41,45,47,53]. However, Wei et al. [54] proved that homogenous mixing of the conductive agent with SPAN attained by the ball milling method upgraded the specific capacity of 1843 mAh g^{-1} and remained over 1000 mAh g^{-1} after 1000 cycles at 0.4 C. Sulfur utilization in the redox reactions of SPAN cathodes was further enhanced by the addition of the conductive carbonaceous materials like graphene [55,56], Ketjenblack [57], and CNT [58] during vulcanization of PAN into SPAN composite (Figure 17.5). Mentbayeva et al. [59] obtained binder and current collector free sulfur-dehydrogenated polyacrylonitrile-multi-walled carbon nanotube (MWCNT) cathode (S-DPAN-MWCNT) by vacuum filtration of the aqueous suspension of sulfurized S-DPAN and functionalized MWCNT. The specific capacity of 1250 mAh g^{-1} at 0.2 C after 260 cycles was achieved by S-DPAN-MWCNT cathode.

The limited sulfur content (~40 wt%) resulting from limited chemical bonding sites of sulfur in the polymeric chains challenges SPAN to achieve a high sulfur loading as compared to elemental sulfur cathodes [60]. Moreover, a large amount of conducting agents and binders is required to cast highly conductive SPAN cathodes with a high specific capacity, which makes Li–S cells with high energy density a great challenge. Alternatively, freestanding and flexible electrodes without any extra current collectors or binders could maximize the sulfur loading and areal capacity of SPAN cells. Razzaq et al. [61] reported the fabrication of the flexible and fibrous SPAN–CNT film by co-spinning of sulfur, PAN, and CNT, followed by vulcanization in the presence of an additional sulfur atmosphere. Preloaded sulfur and CNT alignment in the fibers were found to exert a synergy in creating conductive and porous nanofibers, which in turn significantly endorsed the electrochemical performance of the as-synthesized SPAN–CNT composite.

17.2.3.2 Metal Compound Additives

Metal oxides and sulfides can also effectively enhance the performance of SPAN cathode in Li–S batteries, for example, SPAN composites with $Mg_{0.6}Ni_{0.4}O$ [62] and NiS_2 [63] stabilized the structure of the composites from irregular aggregated particles to regular smooth nanoparticles. Moreover, these oxides and sulfides have an affinity to absorb polysulfides during redox reactions and hence improve the electrochemical kinetics of electrodes. S/PAN/$Mg_{0.6}Ni_{0.4}O$ cathode delivered an initial specific capacity of 1223 mAh g^{-1} with 100% Coulombic efficiency over 100 cycles [62]. Whereas, Ni_2S-SPAN delivered a reversible specific capacity of 1533 mAh g^{-1} after 100 cycles at 0.12 C and a rate capability of 1180 mAh g^{-1} at 1.2 C [63].

Currently, the sulfur loading of SPAN cathodes is usually below 1 mg cm^{-2}, leading to a low areal capacity that is far from enough to meet the requirements of practical Li–S batteries. In our previous work [64], a freestanding CoS_2–SPAN–CNT cathode was fabricated via in situ growth of ZIF-67 on electrospun fibers of PAN–CNT containing ligand 2-methylimidazole (MeIM), followed by vulcanization process at 400°C. The schematic of the synthesis of CoS_2–SPAN–CNT fibrous composite electrodes is shown in Figure 17.6. The surface morphology of PAN–CNT–MeIM, ZIF-67@PAN-CNT, and CoS_2–SPAN–CNT is displayed in Figure 17.6b–d. A smooth surface image of PAN–CNT–MeIM was obtained with an average fiber diameter of ~430 nm (Figure 17.6b). An elevated average fiber diameter of 714 nm was obtained when ZIF-67 nanoparticles were added to

(ZIF-67@PANI–CNT) in a uniform distribution of ZIF-67 nano particle (Figure 17.6c). Average fiber diameter of CoS_2–SPAN–CNT was decreased drastically to 459 nm upon vulcanization. The surface layer of CoS_2 converting from decorated ZIF-67 on the fiber not only efficiently suppresses the swelling expansion of the SPAN film upon vulcanization but also enhances the electrical conductivity of the fiber. Thus, the ultrathin freestanding CoS_2–SPAN–CNT film (90 µm) was obtained and offered high sulfur loading up to 4.6 mg cm^{-2} with an areal capacity as high as 8.1 mAh cm^{-2}. The prototype pouch cell with CoS_2–SPAN–CNT cathode delivered a high discharge capacity of

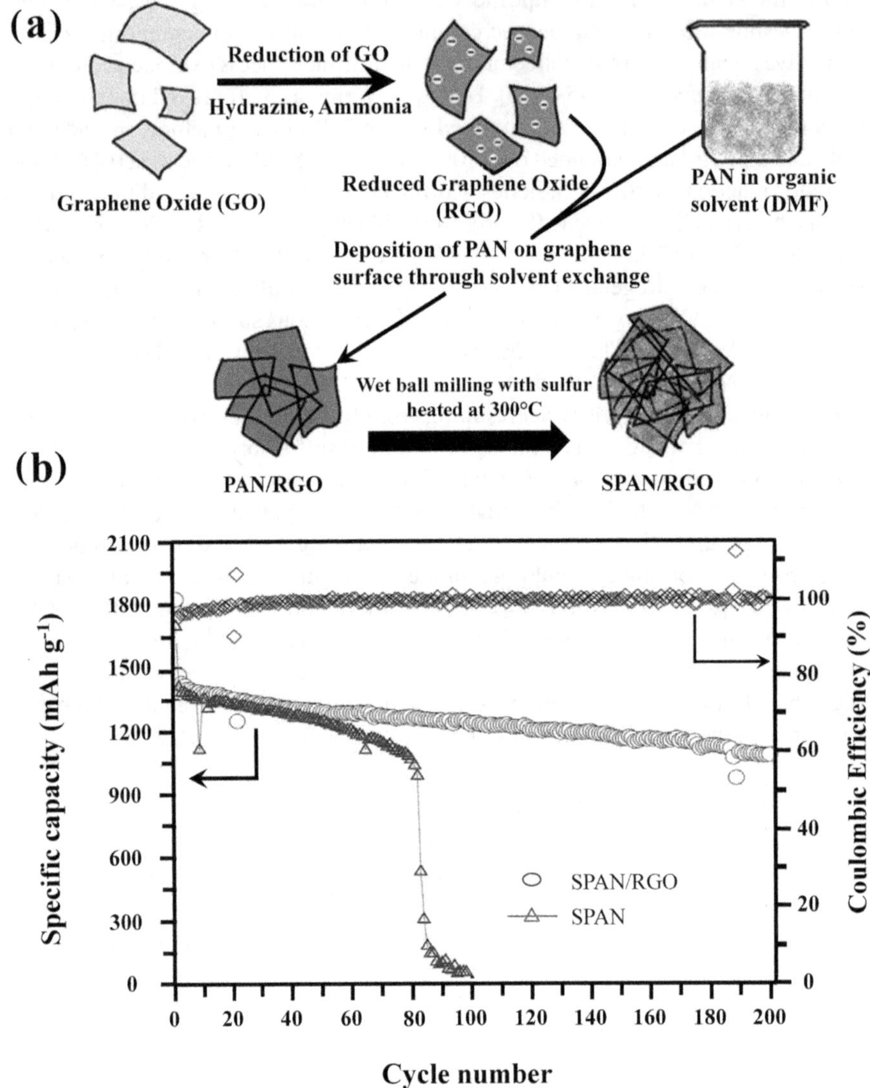

FIGURE 17.5 Synthesis of SPAN composites with conductive carbonaceous materials and their cyclic performance comparison. (a) In situ polymerization of graphene oxide (GO) and monomer acrylonitrile (AN). Adapted and Reproduced from ref. [55]. Copyright © 2012 Royal Society of Chemistry. (b) Solvent exchange process. Adapted and Reproduced from ref. [56]. Copyright © 2014 Elsevier. (c) In situ polymerization of MWCNT, AN, and itaconic acid (IA). Adapted and Reproduced from ref. [58]. Copyright © 2011 Royal Society of Chemistry. (d) Aerosol-assisted process. Adapted and Reproduced from ref. [57]. Copyright © 2016 Elsevier.

(Continued)

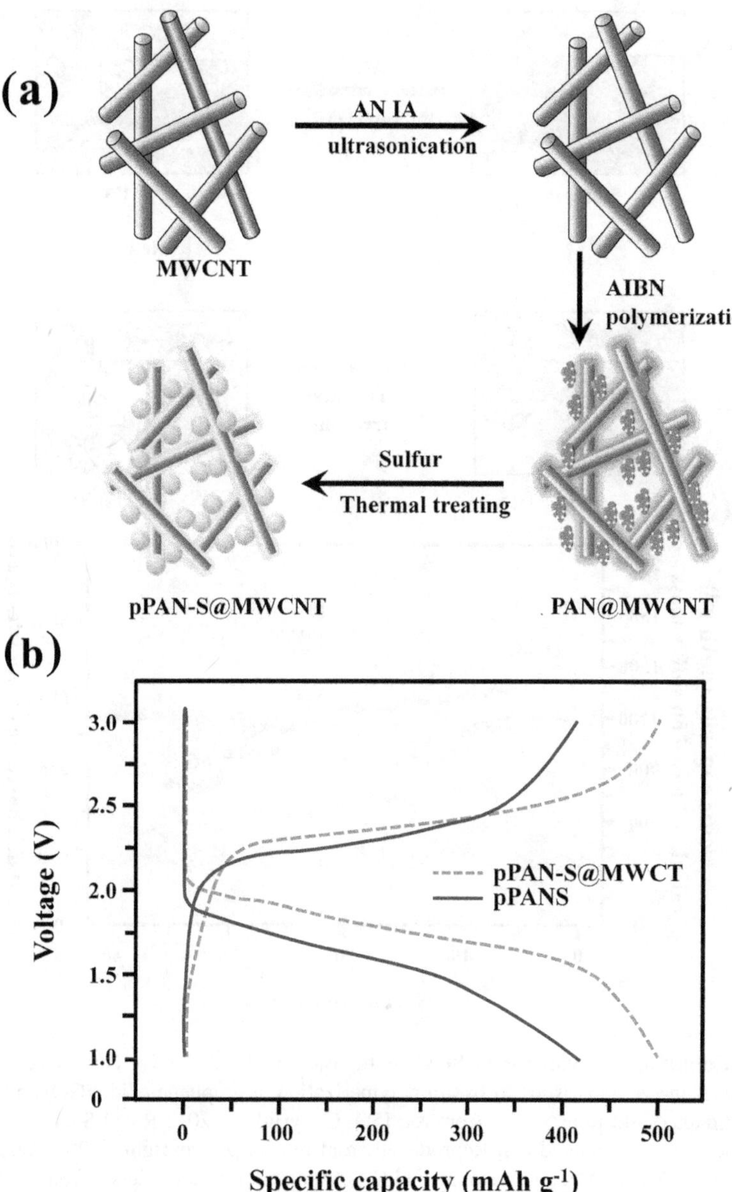

FIGURE 17.5 (*Continued*) Synthesis of SPAN composites with conductive carbonaceous materials and their cyclic performance comparison. (a) In situ polymerization of graphene oxide (GO) and monomer acrylonitrile (AN). Adapted and Reproduced from ref. [55]. Copyright © 2012 Royal Society of Chemistry. (b) Solvent exchange process. Adapted and Reproduced from ref. [56]. Copyright © 2014 Elsevier. (c) In situ polymerization of MWCNT, AN, and itaconic acid (IA). Adapted and Reproduced from ref. [58]. Copyright © 2011 Royal Society of Chemistry. (d) Aerosol-assisted process. Adapted and Reproduced from ref. [57]. Copyright © 2016 Elsevier.

(*Continued*)

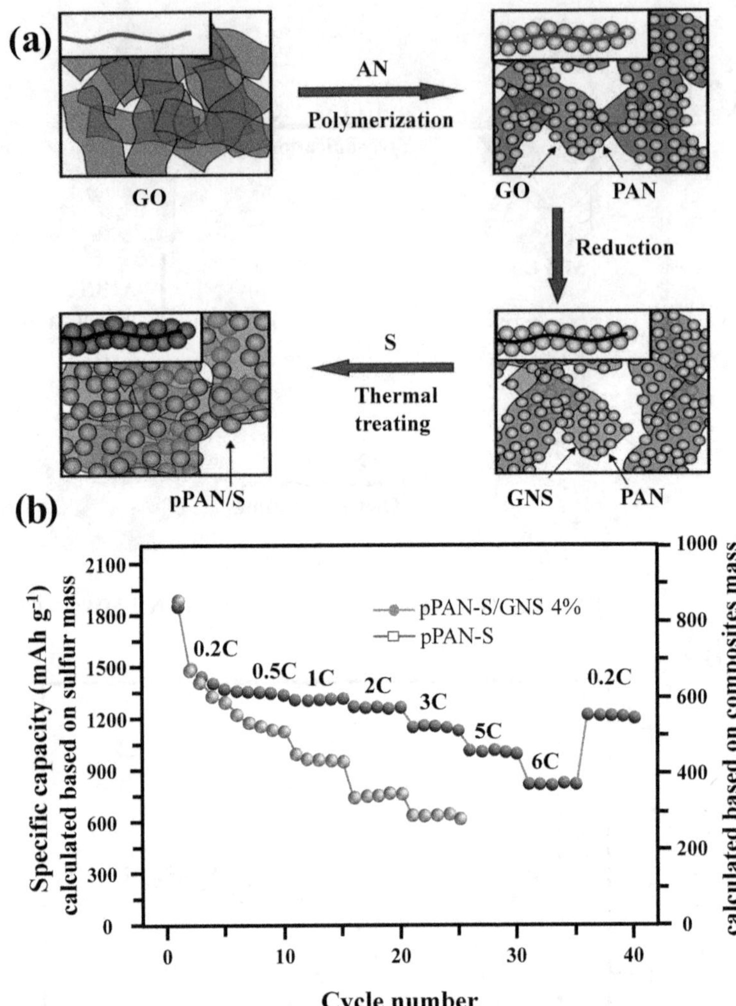

FIGURE 17.5 (*Continued*) Synthesis of SPAN composites with conductive carbonaceous materials and their cyclic performance comparison. (a) In situ polymerization of graphene oxide (GO) and monomer acrylonitrile (AN). Adapted and Reproduced from ref. [55]. Copyright © 2012 Royal Society of Chemistry. (b) Solvent exchange process. Adapted and Reproduced from ref. [56]. Copyright © 2014 Elsevier. (c) In situ polymerization of MWCNT, AN, and itaconic acid (IA). Adapted and Reproduced from ref. [58]. Copyright © 2011 Royal Society of Chemistry. (d) Aerosol-assisted process. Adapted and Reproduced from ref. [57]. Copyright © 2016 Elsevier.

(*Continued*)

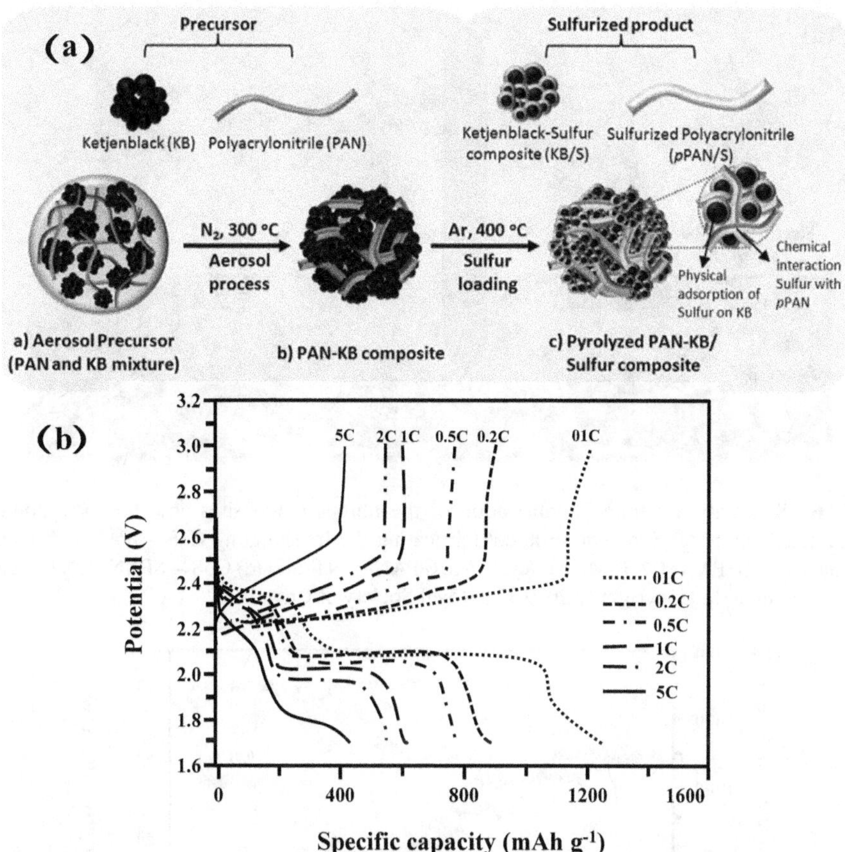

FIGURE 17.5 *(Continued)* Synthesis of SPAN composites with conductive carbonaceous materials and their cyclic performance comparison. (a) In situ polymerization of graphene oxide (GO) and monomer acrylonitrile (AN). Adapted and Reproduced from ref. [55]. Copyright © 2012 Royal Society of Chemistry. (b) Solvent exchange process. Adapted and Reproduced from ref. [56]. Copyright © 2014 Elsevier. (c) In situ polymerization of MWCNT, AN, and itaconic acid (IA). Adapted and Reproduced from ref. [58]. Copyright © 2011 Royal Society of Chemistry. (d) Aerosol-assisted process. Adapted and Reproduced from ref. [57]. Copyright © 2016 Elsevier.

1332 mAh g^{-1} [64] (Figure 17.7). This study demonstrates a promising perspective of applying SPAN–CNT fibers in flexible Li–S batteries and other bendable energy storage devices.

17.2.3.3 Heteroatom Doping

Chemical doping is considered another strategy to further constrain LiPS dissolution by immediately catalyzing into the final reducing product in the SPAN cathodes. Xie et al. [65] doped selenium in SPAN which assisted the catalyzation of a Li_2Sn ($n \leq 4$) to the final reduction product Li_2S. Importantly, Se_xSPAN composite is also a compatible ether-based electrolyte, which has more tendency to dissolute the LiPS. Subsequently, the Se_xSPAN cathode delivered a high reversible capacity (based on S and Se) of 1320 mAh g^{-1} at 0.13 C and a high-rate capability of 900 mAh g^{-1} at 6.5 C. Very similar behavior was observed again by Xie et al. [66] where tellurium was doped on SPAN; this composite also accelerated the reaction kinetics and compatible in ether-based electrolytes. Razzaq et al. [67] have developed a facile multivariate electrospinning decorum to fabricate freestanding SPAN cathodes comprising conductive CNTs and atomically dispersed Co centers to maximize the reaction kinetics in Li-SPAN batteries. CNTs not only enhanced the fiber conductivity but also generated porous structures to facilitate both the charge and mass transportation.

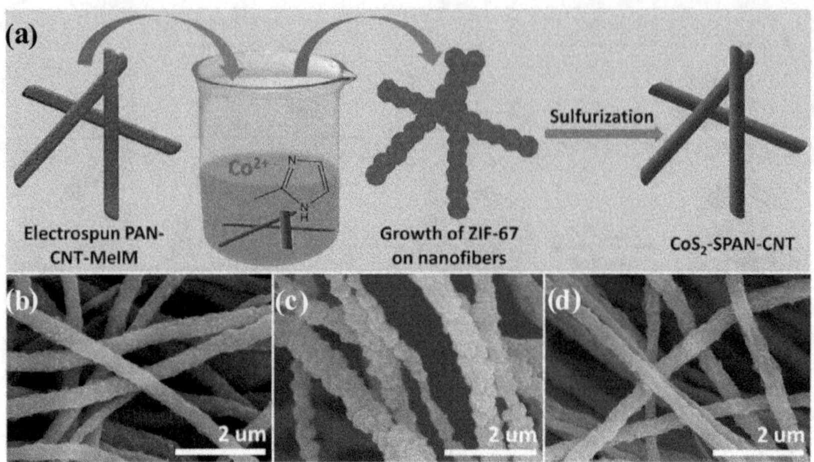

FIGURE 17.6 Schematic diagram of fabrication of the fibrous composites and the corresponding morphology characterization: (a) Synthetic protocol fabricating the freestanding CoS_2–SPAN–CNT electrodes; FE-SEM images of (b) PAN–CNT–MeIM, (c) ZIF-67@PAN–CNT, and (d) CoS_2– SPAN–CNT. Adapted and reproduced from ref. [64]. Copyright © 2019 The Royal Society of Chemistry.

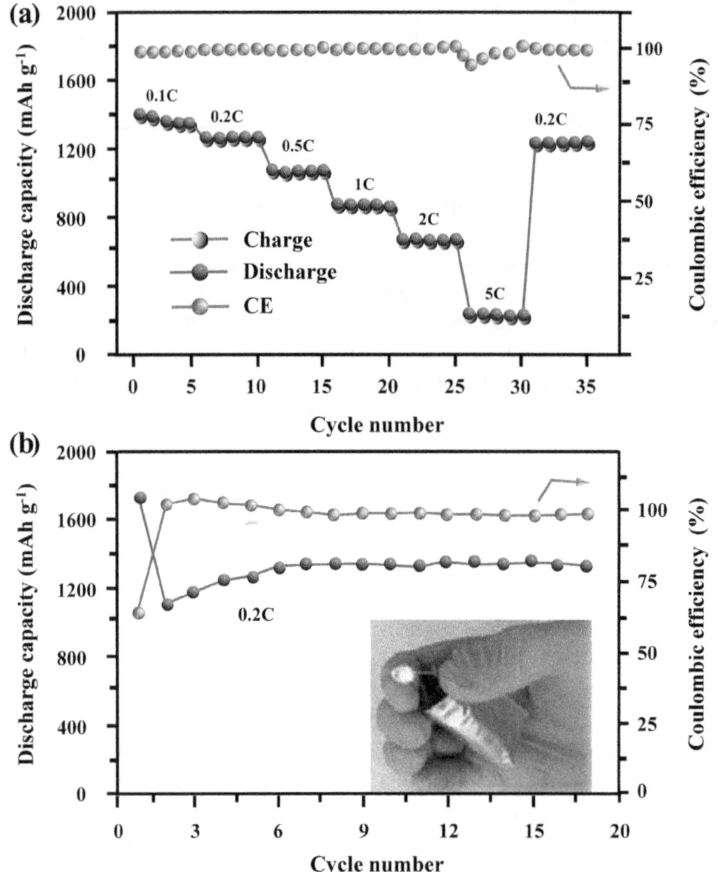

FIGURE 17.7 Electrochemical characterization of CoS_2–SPAN–CNT: (a) Rate capability from 0.1 to 5C and (b) Cycling performance of the prototype pouch cell at 0.2C. Adapted and reproduced from ref. [64]. Copyright © 2019 The Royal Society of Chemistry.

Besides, the cross-linked Co-N_4S motifs in SPAN created extra charge conduction pathways and served as the catalytic active sites for fastening the sulfur conversion. The optimized cathode of Co_{10}–SPAN–CNT delivered high discharge capacity, superb rate capability, and ultra-long battery life. Moreover, the CV results of Co_{10}–SPAN–CNT show the presence of small peaks at 1.4 V in all reduction curves, confirming the higher capacity contribution due to catalyzation of Li_2S_2/Li_2S enabled by introducing Co-N_4S species into the SPAN–CNT system.

17.3 CONCLUSIONS AND PERSPECTIVES

To clarify the working mechanism of SPAN cathode materials and improve their electrochemical performance, the current research work in the field mainly focuses on the following two aspects: material preparation and mechanism study. The research mainly includes optimizing the synthesis conditions of SPAN, increasing the sulfur content of SPAN and sulfur loading in Li–S cells, and understanding of reaction mechanisms between SPAN and lithium metal. It would take a long time to finally clarify the redox mechanism of SPAN upon repeated lithiation/delithiation due to the complex organosulfur molecular structure. While analyzing the mechanism, researchers are also actively looking for ways to improve the specific capacity and cycle stability of SPAN cathode materials in Li–S batteries, further increase the specific energy density of Li–S batteries, and finally realize the practical application of SPAN materials.

The addition of carbon materials, oxides, or sulfides can reduce the impedance in the battery and increase the kinetic rate of the redox reaction, thereby obtaining higher sulfur utilization and rate performance. Therefore, carbon additives can be used as an effective way to improve the electrochemical performance of SPAN composites by maintaining a certain sulfur content and sufficient electric conductivity for cell performance. But at present, there is no work to make in-depth discussion and research on the dosage and working mechanism of additives.

In addition, based on the analysis of molecular structure, the low sulfur content of SPAN becomes the biggest obstacle to the practical application of this material. Generally, increasing the sulfur content will lead to a decrease in the electrochemical performance of the battery. However, there is still no method that can effectively solve the contradiction between increasing the sulfur content and maintaining excellent electrochemical performance, which restricts the energy density of SPAN-based Li–S batteries.

Although the state-of-art SPAN-based Li–S batteries are still far away from the requirement for real commercial applications, the unique structure and superior electrochemical performance of SPAN will probably endow this promising cathode large breakthrough toward practical realization in the near future.

ACKNOWLEDGMENTS

This work was supported by the National Natural Science Foundation of China (No. 21805201), China Postdoctoral Science Foundation (No. 2018T110544 and 2017M611899), and the Key Technology Initiative of Suzhou Municipal Science and Technology Bureau (SYG201748). We also acknowledge the support of Suzhou Key Laboratory for Advanced Carbon Materials and Wearable Energy Technologies.

REFERENCES

1. Arico, A.S., et al., Nanostructured materials for advanced energy conversion and storage devices, in *Materials for Sustainable Energy: A Collection of Peer-Reviewed Research and Review Articles from Nature Publishing Group*, UK, 2011, World Scientific. 148–159.
2. Amponsah, N.Y., et al., Greenhouse gas emissions from renewable energy sources: A review of lifecycle considerations. *Renewable Sustainable Energy Reviews*, 2014. 39: 461–475.

3. Khare, V., S. Nema, and P. Baredar, Solar-wind hybrid renewable energy system: A review. *Renewable Sustainable Energy Reviews*, 2016. 58: 23–33.
4. Kamat, P.V., Meeting the clean energy demand: Nanostructure architectures for solar energy conversion. *The Journal of Physical Chemistry C*, 2007. 111(7): 2834–2860.
5. Seh, Z.W., et al., Designing high-energy lithium-sulfur batteries. *Chemical Society Reviews*, 2016. 45(20): 5605–5634.
6. Bruce, P.G., et al., Li-O_2 and Li-S batteries with high energy storage. *Nature Materials*, 2012. 11(1): 19–29.
7. Yin, Y.X., et al., Lithium-sulfur batteries: Electrochemistry, materials, and prospects. *Angewandte Chemie – International Edition*, 2013. 52(50): 13186–13200.
8. Fu, Y., Y.S. Su, and A. Manthiram, Li_2S-carbon sandwiched electrodes with superior performance for lithium-sulfur batteries. *Advanced Energy Materials*, 2014. 4(1): 1300655.
9. Kang, W., et al., A review of recent developments in rechargeable lithium-sulfur batteries. *Nanoscale*, 2016. 8(37): 16541–16588.
10. Manthiram, A., Y. Fu, and Y.-S. Su, Challenges and prospects of lithium-sulfur batteries. *Accounts of Chemical Research*, 2012. 46(5): 1125–1134.
11. Fu, A., et al., Recent advances in hollow porous carbon materials for lithium-sulfur batteries. *Small*, 2019. 15(10): 1804786.
12. Liang, X., et al., A highly efficient polysulfide mediator for lithium-sulfur batteries. *Nature Communications*, 2015. 6: 5682.
13. Urbonaite, S., T. Poux, and P. Novák, Progress towards commercially viable Li-S battery cells. *Advanced Energy Materials*, 2015. 5(16): 1500118.
14. He, J. and A. Manthiram, A review on the status and challenges of electrocatalysts in lithium-sulfur batteries. *Energy Storage Materials*, 2019. 20: 55–70.
15. Wild, M., et al., Lithium sulfur batteries, a mechanistic review. *Energy & Environmental Science*, 2015. 8(12): 3477–3494.
16. Manthiram, A., et al., Rechargeable lithium-sulfur batteries. *Chemical Reviews*, 2014. 114(23): 11751–11787.
17. Helen, M., et al., Single step transformation of sulphur to Li_2S_2/Li_2S in Li-S batteries. *Scientific Reports*, 2015. 5: 12146.
18. Li, Y., et al., Sulfur-infiltrated three-dimensional graphene-like material with hierarchical pores for highly stable lithium-sulfur batteries. *Journal of Materials Chemistry A*, 2014. 2(13): 4528–4533.
19. Zhou, W., et al., Yolk-shell structure of polyaniline-coated sulfur for lithium-sulfur batteries. *Journal of the American Chemical Society*, 2013. 135(44): 16736–16743.
20. Ji, X., K.T. Lee, and L.F. Nazar, A highly ordered nanostructured carbon-sulphur cathode for lithium-sulphur batteries. *Nature Materials*, 2009. 8(6): 500–506.
21. Peng, X.X., et al., Graphitized porous carbon materials with high sulfur loading for lithium-sulfur batteries. *Nano Energy*, 2017. 32: 503–510.
22. Zhang, B., et al., Novel hierarchically porous carbon materials obtained from natural biopolymer as host matrixes for lithium-sulfur battery applications. *ACS Applied Materials Interfaces*, 2014. 6(15): 13174–13182.
23. Zhang, L., et al., Advanced nanostructured carbon-based materials for rechargeable lithium-sulfur batteries. *Carbon*, 2019. 141: 400–416.
24. Guo, J., Y. Xu, and C. Wang, Sulfur-impregnated disordered carbon nanotubes cathode for lithium-sulfur batteries. *Nano Letters*, 2011. 11(10): 4288–4294.
25. Li, M., et al., Sulfur vapor-infiltrated 3D carbon nanotube foam for binder-free high areal capacity lithium-sulfur battery composite cathodes. *ACS Nano*, 2017. 11(5): 4877–4884.
26. Xu, Y.H., et al., Confined sulfur in microporous carbon renders superior cycling stability in Li/S batteries. *Advanced Functional Materials*, 2015. 25(27): 4312–4320.
27. Kang, W., et al., Sulfur-embedded porous carbon nanofiber composites for high stability lithium-sulfur batteries. *Chemical Engineering Journal*, 2018. 333: 185–190.
28. Feng, S., et al., One-step synthesis of carbon nanosheet-decorated carbon nanofibers as a 3D interconnected porous carbon scaffold for lithium-sulfur batteries. *Journal of Materials Chemistry A*, 2017. 5(45): 23737–23743.
29. Zhao, M.Q., et al., Unstacked double-layer templated graphene for high-rate lithium-sulphur batteries. *Natture Communications*, 2014. 5: 3410.
30. Pei, F., et al., Self-supporting sulfur cathodes enabled by two-dimensional carbon yolk-shell nanosheets for high-energy-density lithium-sulfur batteries. *Nature Communications*, 2017. 8(1): 482.

31. Tao, X., et al., Balancing surface adsorption and diffusion of lithium-polysulfides on nonconductive oxides for lithium-sulfur battery design. *Nature Communications*, 2016. 7: 11203.
32. Zhou, G., et al., Catalytic oxidation of Li_2S on the surface of metal sulfides for Li-S batteries. *Proceedings of the National Academy of Sciences of the United States of America*, 2017. 114(5): 840–845.
33. Liu, X., et al., Nanostructured metal oxides and sulfides for lithium-sulfur batteries. *Advanced Materials*, 2017. 29(20): 1601759.
34. Deng, D.-R., et al., Co_4N nanosheet assembled mesoporous sphere as a matrix for ultrahigh sulfur content lithium-sulfur batteries. *ACS Nano*, 2017. 11(6): 6031–6039.
35. Song, Y., et al., Enhanced sulfur redox and polysulfide regulation via porous VN-modified separator for Li-S batteries. *ACS Applied Materials Interfaces*, 2019. 11(6): 5687–5694.
36. Li, F., et al., Free-standing sulfur-polypyrrole cathode in conjunction with polypyrrole-coated separator for flexible Li-S batteries. *Energy Storage Materials*, 2018. 13: 312–322.
37. Zheng, J., et al., High-fluorinated electrolytes for Li-S batteries. *Advanced Energy Materials*, 2019. 9(16): 1803774.
38. Gupta, A., A. Bhargav, and A. Manthiram, Highly solvating electrolytes for lithium-sulfur batteries. *Advanced Energy Materials*, 2019. 9(6): 1803096.
39. Yao, Y., et al., Mosaic rGO layers on lithium metal anodes for the effective mediation of lithium plating and stripping. *Journal of Materials Chemistry A*, 2019. 7(19): 12214–12224.
40. Shen, X., et al., Lithium anode stable in air for low-cost fabrication of a dendrite-free lithium battery. *Nature Communications*, 2019. 10(1): 900.
41. Wang, J., et al., A novel conductive polymer-sulfur composite cathode material for rechargeable lithium batteries. *Advanced Materials*, 2002. 14(13-14): 963–965.
42. Jin, Z.Q., et al., A new insight into the lithium storage mechanism of sulfurized polyacrylonitrile with no soluble intermediates. *Energy Storage Materials*, 2018. 14: 272–278.
43. Peng, H., et al., Synthesis of a flexible freestanding sulfur/polyacrylonitrile/graphene oxide as the cathode for lithium/sulfur batteries. *Polymers*, 2018. 10(4): 399.
44. Wang, J., et al., Sulfur composite cathode materials for rechargeable lithium batteries. *Advanced Functional Materials*, 2003. 13(6): 487–492.
45. Yu, X.-g., et al., Lithium storage in conductive sulfur-containing polymers. *Journal of Electroanalytical Chemistry*, 2004. 573(1): 121–128.
46. Wang, L., et al., Analysis of the synthesis process of sulphur-poly (acrylonitrile)-based cathode materials for lithium batteries. *Journal of Materials Chemistry*, 2012. 22(41): 22077–22081.
47. Fanous, J., et al., Structure-related electrochemistry of sulfur-poly (acrylonitrile) composite cathode materials for rechargeable lithium batteries. *Chemistry of Materials*, 2011. 23(22): 5024–5028.
48. Wang, W., et al., Recognizing the mechanism of sulfurized polyacrylonitrile cathode materials for Li-S batteries and beyond in Al-S batteries. *ACS Energy Letters*, 2018. 3(12): 2899–2907.
49. Weret, M.A., et al., Mechanistic understanding of the Sulfurized-Poly(acrylonitrile) cathode for lithium-sulfur batteries. *Energy Storage Materials*, 2020. 26: 483–493.
50. Jin, Z.-Q., et al., A new insight into the lithium storage mechanism of sulfurized polyacrylonitrile with no soluble intermediates. *Energy Storage Materials*, 2018. 14: 272–278.
51. Wang, X., et al., Sulfurized polyacrylonitrile cathodes with high compatibility in both ether and carbonate electrolytes for ultrastable lithium-sulfur batteries. *Advanced Functional Materials*, 2019. 29(39): 1902929 (1–12).
52. Zhu, T., et al., Structural motifs for modeling sulfur-poly (acrylonitrile) composite materials in sulfur-lithium batteries. *ChemElectroChem*, 2017. 4(10): 2494–2499.
53. Yu, X., et al., Stable-cycle and high-capacity conductive sulfur-containing cathode materials for rechargeable lithium batteries. *Journal of Power Sources*, 2005. 146(1-2): 335–339.
54. Wei, S., et al., Metal-sulfur battery cathodes based on PAN-sulfur composites. *Journal of the American Chemical Society*, 2015. 137(37): 12143–12152.
55. Yin, L., et al., Polyacrylonitrile/graphene composite as a precursor to a sulfur-based cathode material for high-rate rechargeable Li-S batteries. *Energy & Environmental Science*, 2012. 5(5): 6966–6972.
56. Li, J., et al., A sulfur-polyacrylonitrile/graphene composite cathode for lithium batteries with excellent cyclability. *Journal of Power Sources*, 2014. 252: 107–112.
57. Sohn, H., et al., Porous spherical polyacrylonitrile-carbon nanocomposite with high loading of sulfur for lithium-sulfur batteries. *Journal of Power Sources*, 2016. 302: 70–78.
58. Yin, L., et al., A novel pyrolyzed polyacrylonitrile-sulfur@ MWCNT composite cathode material for high-rate rechargeable lithium/sulfur batteries. *Journal of Materials Chemistry*, 2011. 21(19): 6807–6810.

59. Mentbayeva, A., et al., High performance freestanding composite cathode for lithium-sulfur batteries. *Electrochimica Acta*, 2016. 217: 242–248.
60. Wang, L., et al., Charge/discharge characteristics of sulfurized polyacrylonitrile composite with different sulfur content in carbonate based electrolyte for lithium batteries. *Electrochimica Acta*, 2012. 72: 114–119.
61. Razzaq, A.A., et al., High-performance lithium sulfur batteries enabled by a synergy between sulfur and carbon nanotubes. *Energy Storage Materials*, 2019. 16: 194–202.
62. Zhang, Y., et al., Ternary sulfur/polyacrylonitrile/$Mg_{0.6}Ni_{0.4}O$ composite cathodes for high performance lithium/sulfur batteries. *Journal of Materials Chemistry A*, 2013. 1(2): 295–301.
63. Liu, Y., et al., A polysulfide reduction accelerator-NiS_2-modified sulfurized polyacrylonitrile as a high performance cathode material for lithium-sulfur batteries. *Journal of Materials Chemistry A*, 2017. 5(42): 22120–22124.
64. Razzaq, A.A., et al., Anchoring MOF-derived CoS_2 on sulfurized polyacrylonitrile nanofibers for high areal capacity lithium-sulfur batteries. *Journal of Materials Chemistry A*, 2020. 8(3): 1298–1306.
65. Chen, X., et al., Ether-compatible sulfurized polyacrylonitrile cathode with excellent performance enabled by fast kinetics via selenium doping. *Nature Communications*, 2019. 10(1): 1021.
66. Li, S., et al., Manipulating kinetics of sulfurized polyacrylonitrile with tellurium as eutectic accelerator to prevent polysulfide dissolution in lithium-sulfur battery under dissolution-deposition mechanism. *Nano Energy*, 2019. 60: 153–161.
67. Abdul Razzaq, A., et al., Cobalt coordination with pyridines in sulfurized polyacrylonitrile cathodes to form conductive pathways and catalytic M-N_4S sites for accelerated Li-S kinetics. *Journal of Energy Chemistry*, 2021. 61: 170–178.

18 An Overview of Magnesium–Sulfur Battery
Next-Generation Energy Storage Device

Jabeen Fatima M. J., Akhila Das, Leya Rose Raphel, Alexandru Vlad, Jou-Hyeon Ahn, Xiaohui Zhao, and Prasanth Raghavan

18.1 INTRODUCTION

The current generation relies on portable gadgets and electronic devices which are powered by lithium-ion batteries. Advancements in portable electronics and the need for greener energy have paved the way to establish the possibilities of portable rechargeable energy storage devices for powering the automobile sector. The slow charging and decreased power density have diminished the possibility of enhancement in the performance of lithium-ion batteries. Apart from the technical retardation, the increased cost of lithium and the reactivity with air and water have also led to the exploration of better alternatives. The researchers are hence focused on developing newer technologies for energy storage which is greener, cost-effective, and reduce the risk of firing. The change in the base material was the primary concern considering an alternative energy storage device which led to the introduction of multivalent ions, such as magnesium, aluminum, calcium, iron, and zinc, to the account. The abundance of magnesium and the safety to handle the material have opened an extensive area of research in magnesium-based batteries [1].

Magnesium-based batteries are hence being widely explored for future rechargeable battery technology. The elevated volumetric capacity of magnesium ions is 3837 mAh cm^{-3} compared to that of lithium (2062 mAh cm^{-3}) as well as sodium (1136 mAh cm^{-3}) [1]. The volumetric capacity of the ions dominates the size reduction of the batteries without compromising the performance of the energy storage. Mg metal possesses a reduction potential of –2.4 V vs SHE, which serves as a backbone imparting a lower reactivity of magnesium metal in the atmospheric exposure leading to an enhancement of safety features by reducing the tendency of the metal to be fired [1].

Magnesium batteries utilize magnesium as the primary ions for intercalation mechanism. A wide variety of magnesium-based secondary batteries are being investigated by the researchers, which includes magnesium-ion battery, magnesium–air battery, and magnesium–sulfur battery. Herein, the present chapter is being focused on magnesium–sulfur batteries. Sulfur has been an attractive component in batteries owing to the enhanced theoretical capacity of 1673 mAh g^{-1} [2]. The availability of the material that makes such batteries to be provided with lower cost also imparts a market appreciation for the material. Lithium–sulfur-based batteries were greatly studied wherein the theoretical energy density was predicted to be 2800 Wh L^{-1} [3]. The major drawback of the material is the formation of some chemical byproducts with sulfur which in turn enhance the tendency for dendrite growth arising with the risk of firing in the batteries which is not acceptable. Hence, there is commercialization of such high energy batteries yet being a prototype. Considering the magnesium–sulfur batteries, the theoretical capacity is expected to exceed beyond 3000 Wh L^{-1} [4]. Even

DOI: 10.1201/9781003310167-18

though magnesium–sulfur battery research has not made considerable advancements, the technology world expects a high-performance battery development. The present chapter is designed to review from the basics behind the energy storage and the chemistry of magnesium–sulfur batteries, extending toward the components investigated for the development of Mg–S batteries.

18.2 COMPONENTS AND CHEMISTRY BEHIND MAGNESIUM–SULFUR BATTERY

The magnesium–sulfur batteries are considered to be the next-generation high-capacity batteries. Apart from lithium batteries, magnesium batteries are considered to be safer with lower firing risks. The reaction of sulfur with magnesium metal anode involves a bi-electron reaction wherein magnesium ion reacts with sulfur to form MgS on release of two electrons to the external circuit as given in the following equation [5]:

$$Mg^{2+} + S + 2e^- \leftrightarrow MgS \tag{18.1}$$

Magnesium–sulfur batteries are expected to possess about 1.77 V as theoretical cell voltage. Aurbach et al. introduced the scope for the development of newer batteries based on magnesium. The magnesium–sulfur battery comprises an anode, cathode, and electrolyte (Figure 18.1). Usually, metallic magnesium is being used as anode and cathode, which consists of any sulfur-based compounds or elemental sulfur. A separator is held in between the electrodes to avoid short circuiting through contact between the electrodes. The discharging of the Mg–S battery takes place when Mg metal is converted to Mg^{2+} ion, i.e., oxidation of magnesium metal takes place where electron flows through the external circuit. The generated divalent magnesium ion moves toward the cathode through the electrolyte wherein the chemical transformation to magnesium sulfide takes place. The magnesium usually reacts with sulfur reversibly to form polysulfides like MgS_2, MgS_4, MgS_6, MgS_8, etc. These polysulfides gradually get converted into MgS [5].

The reaction of sulfur with magnesium proceeds via three stages [6]. Initially, the magnesium polysulfide formation occurs to produce MgS_8 (equation 18.2). The solubility of these polysulfides doesn't show considerable change in the reaction [6].

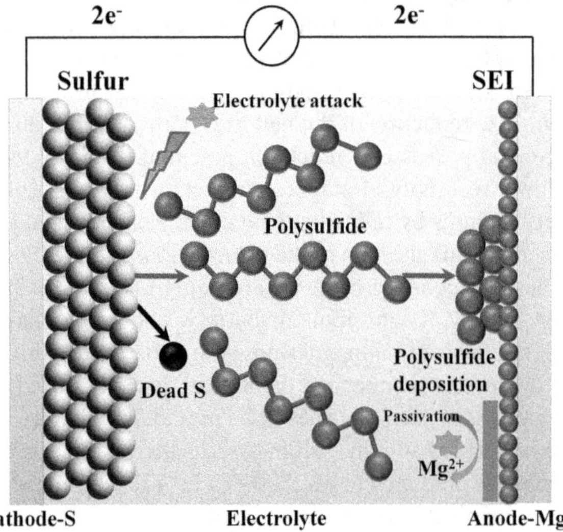

FIGURE 18.1 Schematic diagram of Mg–S batteries showing the working principle and formation of magnesium–polysulfides (Mg–PS), which passivate the anode surface. Adapted and reproduced with permission from Ref. [5]. Copyright © 2020 Elsevier.

$$S_8 + Mg^{2+} + 2e^- \rightarrow MgS_8 \qquad 2.5-1.5\ V \qquad (18.2)$$

In the second stage, the longer chain polysulfide is decomposed to form shorter chain sulfides. The experimental results reveal the formation of magnesium bisulfide as per equation 18.3, which is considered to form a modified rock salt structure [6].

$$MgS_8 + 3Mg^{2+} + 6e^- \rightarrow 4MgS_2 \qquad 1.5\ V \qquad (18.3)$$

In the following stage, the crystalline MgS_2 will be reduced to MgS. The reaction proceeds through lethargic kinetics. The chemical equation of the reaction is given in the following equation [6]:

$$MgS_2 + Mg^{2+} + 2e^- \rightarrow MgS \qquad (18.4)$$

18.3 ANODE FOR MAGNESIUM–SULFUR BATTERY

The most promising property of the Mg–S battery is the enhanced energy density, which is achieved on the usage of Mg metal-based anodes. The major drawback of the magnesium ion battery relies on the decreased cyclability as appropriate electrolyte species has yet to be discovered which could suppress the formation of passivation layer on the anode during the cycling process. A detailed outlook on to the anode electrolyte interface was explained by Hu et al. [7] wherein they systematically investigated the changes occurring at the interface of anode and the electrolyte during the cycling of the battery and formation of Mg deposition/stripping at the interface by in situ investigation using AFM technique and optical imaging. A schematic illustration on the basis of analysis is depicted in Figure 18.2, which describes the formation of Mg deposition/stripping at the interface in two different electrolytic medium diglyme (DEG) and tetraglyme (TEG). On the usage of DEG as an electrolytic medium, Mg formation is initiated as labeled in Figure 18.2a, which was subsequently proceeded to a swift deposition of spheres or particles of Mg as illustrated in Figure 18.2a3–a5 during charging process. At the time of discharging, finite morphologic crystals strip spontaneously than the other amorphous particulates as defined in Figure 18.2a5–a7. Whereas during the

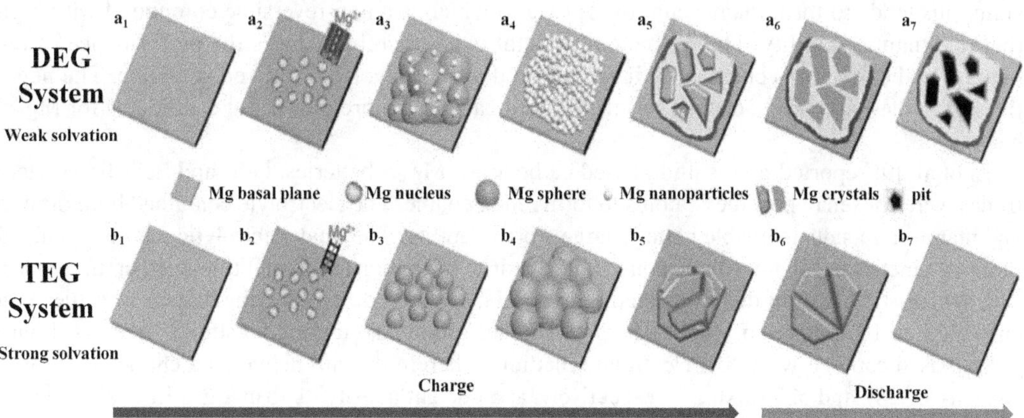

FIGURE 18.2 Schematic illustration of interfacial processes of the Mg deposition/stripping in DEG system (a1–a7) and TEG system (b1–b7). (a1) OCP, (a2) bowl-like Mg nucleation, (a3) Mg spheres and NPs emergence, (a4) NPs stack upon charge, (a5) crystals growth, (a6) crystals stripping, and (a7) pits emergence; (b1) OCP, (b2) bowl-like Mg nucleation, (b3) Mg spheres emergence, (b4) Mg spheres growth, (b5) crystals growth, (b6) crystals stripping, and (b7) the complete stripping of crystals. Adapted and reproduced with permission from Ref. [7]. Copyrights © 2018 Elsevier.

usage of TEG as electrolyte medium, Mg particles formed as displayed in Figure 18.2b2 combine together (Figure 18.2b3–b4), followed by the deposition of crystals during the charging process (Figure 18.2b5). While on discharging, the entire crystals strip out completely at slow dynamic process (Figure 18.2b5–b7). Mg ions are having weak solvation of DEG causing the non-uniform nucleation and stripping, whereas a single TEG delivers a better chelation effect enabling the formation of uniform nucleation and stripping [7]. Thus, TEG showed enhanced performance than the DEG electrolytic medium for better cycling of the Mg–S batteries.

Friedrich et al. [8] investigated the role of powder-based magnesium anodes for Mg–S batteries. The magnesium metal powder was ball milled with graphite and pelletized in a hydraulic press at 75 and 350 MPa to be used as anode. The performance of the metal powder anodes was compared with Mg foil anode. The cell was analyzed with sulfur cathode and bis-hexamethyldisilazide magnesium-based electrolyte in a Swagelok type cell. The electrochemical cycle studies were conducted galvanostatically within a voltage span of 2.8–0.5 V at a current density of 167 A kg^{-1}. During initial cycling, low-pressure pelletized powder showed a capacity of 600 Ah kg^{-1}, whereas high-pressure pellet showed a capacity of almost half. A constant degradation in the performance of the cells was observed in prolonged cycling. 88% decrease in performance was observed on cycling from 1 to 100 cycles of low-pressure pelletized anode. The enhanced performance of the low-pressure pelletized anode may be considered to be the voids or inter-grain space between the Mg powders enabling the intercalation of the Mg^{2+} ions.

18.4 CATHODE FOR MAGNESIUM–SULFUR BATTERY

The energy density of a battery is relied on the cathode material, and the material enables intercalation of the active ions electrochemically. In Mg–S batteries, sulfur-based compounds are used as cathodes especially including metal sulfides like FeS_2, carbon–sulfur compounds, etc. In some trials, metal oxides are also used which are capable of replacing the oxide with sulfur. Apart from these chalcogenides, organic-based cathodes are also been studied, 2,5-dimethoxy-1,4-benzoquinone. Considering the sulfur-based cathodes involves in the energy storage by chemical redox reaction involving sulfur. But in the case of oxygen-based cathodes, the replacement of oxygen occurs which in turn supports the sulfur-based redox reactions. Hence, the chalcogenide-based cathodes may be categorized within the same section. In the presence of oxygen, the magnesium ions tends to form magnesium oxide (MgO) which is a non-reversible compound with high thermodynamic stability. The formation of metal oxide thereby reduces the performance of the battery as the system is chemically inactive and does not involve in the redox couple. The negatives of oxide-based cathodes make sulfur-based cathodes more promising candidates for Mg–S batteries.

Li et al. [9] reported dual sulfide-based cathode for Mg–S batteries. FeS- and FeS_2-based electrodes were chosen to host the S-atoms to form redox couple. The electrolyte contained both lithium and magnesium salts to enable better charge storage ability. The dual electrolytic medium enabled the slow kinetics of Mg ion which reduced the dendrite formation ability of lithium ion at the anodes. The synergistic effect of dual ion electrolyte and FeS_2 electrode enabled the battery to achieve an enhanced energy density of ~400 Wh kg^{-1}. Another group also reported a similar study [10] with $FeSe_2$-based cathode with dual electrolyte medium wherein the cell delivered a charge–discharge capacity of 287 and 525 mAh g^{-1}, respectively, at a current density of 15 mA g^{-1}. Li et al. [11] also reported carbon structure from metal–organic framework (ZIF-67), which was further doped with nitrogen and cobalt forming the cathodes which supports the redox reaction of sulfur atoms. The synthesis of carbonaceous compound from ZIF-67 was achieved by the pyrolysis of the material in an inert condition (Ar atmosphere) followed by acid leaching (0.5 M H_2SO_4) for purification. The resultant product was heated in the presence of sulfur to infiltrate the atoms into the interstitial space and void of the carbon framework. The schematic illustration of the same has been depicted in Figure 18.3a. The XRD pattern of the samples before carbonization, carbonization and after sulfur

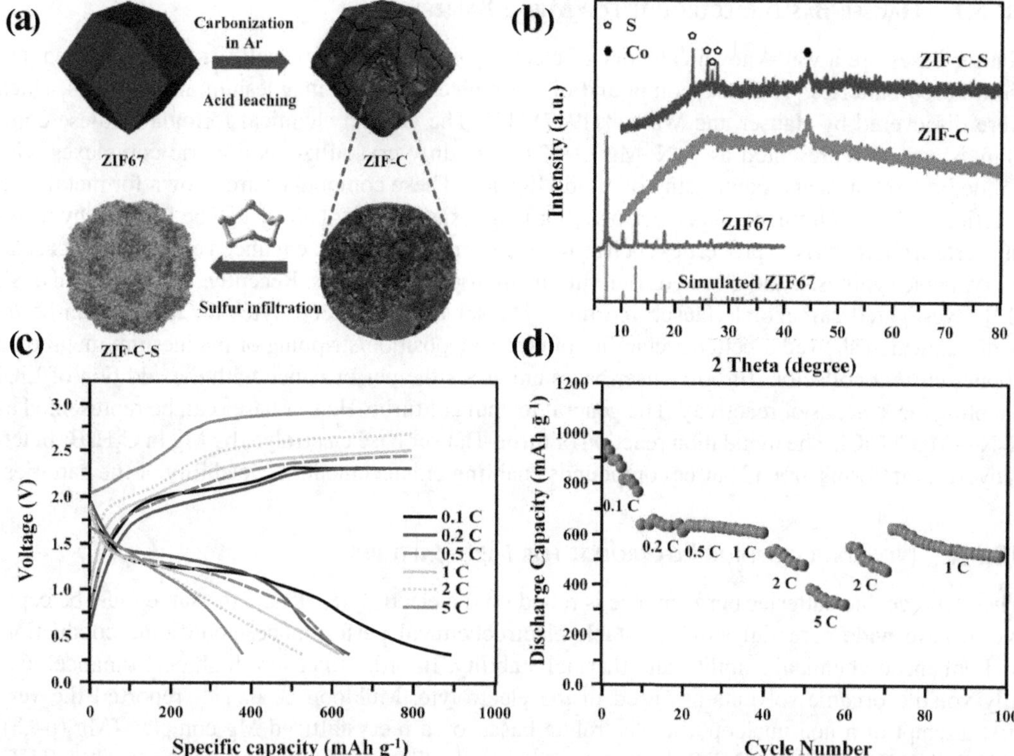

FIGURE 18.3 (a) Schematic illustration to prepare ZIF-C and ZIF-C-S. (b) XRD patterns of ZIF-67, ZIF-C, and ZIF-C-S. The simulated pattern of ZIF-67 is also listed as a reference. ZIF-C is highly disordered with crystalline Co particles embedded. The infiltration of crystalline S in ZIF-C-S is indicated. Galvanostatic discharge/charge curves of ZIF-C-S under a protocol of charge cutoff capacity to 800 mAh g^{-1} and discharge voltage to 0.3 V. (d) At different rates from 0.1 to 5 C in $(HMDS)_2Mg$-$AlCl_3$-LiTFSI electrolyte. (e) Rate performance for LiTFSI contained Mg–S system from 0.1 to 5 C. Adapted and reproduced with permission from Ref. [11]. Copyrights © 2018 Wiley.

infiltration revealed the presence of metallic cobalt particles in the samples which is attributed to the conversion of Co from ZIF-67 to nanoparticles on treatment. The electrochemical charge storage analysis of ZIF-C-S at different C rates is depicted in Figure 18.3c and d, which clearly imparts a discharge capacity of 900 mAh g^{-1} at 0.1 C which was gradually decreased to 400 mAh g^{-1} on change in C rate to 5 C. The remarkable advantage of the cell was observed when the cell was cycled at 0.2, 0.5, and 1 C, where the performance was ~600 mAh g^{-1} where no considerable change was observed on change in C rate.

18.5 ELECTROLYTES FOR MAGNESIUM–SULFUR BATTERY

Electrolytes are considered to be the heart of the battery. In the case of magnesium-based batteries, the performance of the energy storage is relied on the electrolyte efficiency and compatibility with the electrolyte medium [5]. Magnesium-based batteries on cycling form a passivation layer on the anode, which suppress the performance of the battery. The electrolytes for Mg–S battery can be divided on the basis of constituents of the electrolyte: Hauser-based electrolyte, Grignard reagent-based electrolyte, halogen-based electrolyte, boron-based electrolyte, imide-based electrolyte, etc.

18.5.1 Hauser-Based Electrolyte for Mg–S Batteries

Hauser bases are a well-known category of electrolytes which are specially employed for magnesium-based batteries. The base compound of these electrolytes is magnesium amide bases which were discovered by Hauser and Walker (1947) [12]. The general chemical formula of these compounds can be represented as $R_2N–Mg(L)–X$ that mainly crystallize as dimeric complexes with halide bridge bonds and comprising of amido ligands. These compounds are known for metalation reactions. The requirement of an electrolyte for magnesium–sulfur battery can be fulfilled by meeting certain characteristic properties such as reduced corrosive nature, enhanced electrochemical stability, liable synthesis, and elevated solubility in appropriate solvents. Recently, Westerhausen et al. [13] investigated carbazolyl Hauser and turbo-Hauser bases as electrolytes for magnesium-based batteries which showed a better cyclability of electrodeposition/stripping of magnesium metal on a copper current collector. Turbo-Hauser bases enhances the performance with the addition of LiCl enabling the kinetics of reactivity. The general formula of turbo-Hauser bases can be represented as $R_2N—MgCl{\cdot}LiCl$. The metalation reaction of turbo-Hauser base electrolyte by Mg in C_2H_5Br in tetrahydrofuran forms green Hauser compounds enabling enhancement in cyclability of the batteries.

18.5.2 Non-Nucleophilic Electrolyte for MgS Batteries

The rechargeable batteries performance is relied on the electrolyte. Hence, the same must be capable to have wide potential window, stable electrochemical performance, good ionic conduction, and enhanced chemical stability and thermal stability. In order to ensure high performance, usually volatile organic solvents are used in the electrolyte. Muldoon et al. [14] reported the very first attempt of a non-nucleophilic electrolyte based on a recrystallized Mg complex $[Mg_2(\mu\text{-}Cl)_3(THF)_6][HMDS{\cdot}AlCl_3]$ (HMDS=hexamethyldisilazide). The crystal structure $[Mg_2(\mu\text{-}Cl)_3{\cdot}6THF][HMDS{\cdot}AlCl_3]$ was elucidated by single-crystal X-ray diffraction depicted in Figure 18.4. The structure explains the presence of a cation with two octahedrally coordinated magnesium ions bridged together by three chlorine atoms. The remaining sites of the magnesium are connected with tetrahydrofuran through the oxygen atoms. The counter anion is an aluminum atom tetrahedrally coordinated by one HMDS group and three chlorine atoms.

In 2020, Huang et al. [15] reported magnesium trifluoromethanesulfonate (MTB-Mg$(CF_3SO_3)_2$)-based electrolyte for magnesium–sulfur batteries. The electrolyte was synthesized by dissolving $Mg(CF_3SO_3)_2$ in DME followed by the addition of almost double the quantity of $MgCl_2$, and the samples were mixed thoroughly by magnetic stirring for 6 hours subsequently adding equimolar concentration of $AlCl_3$ and stirring for 12 hours at 60°C. The electrolyte concentration was varied and the performance was studied, which revealed that the equimolar concentration of $MgCl_2$ and $AlCl_3$ delivered the maximum performance. On cycling, the electrolyte formed stable ionic

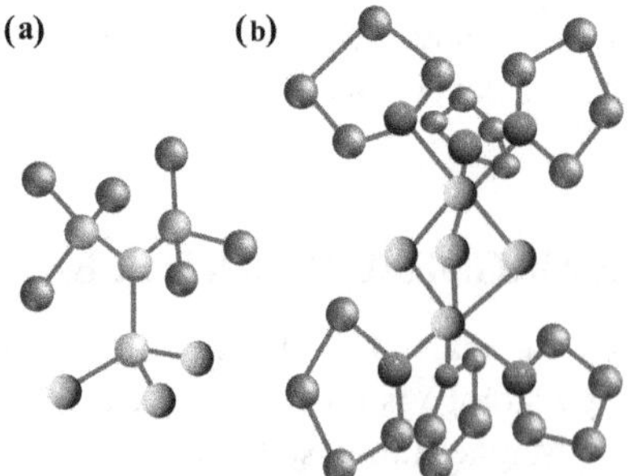

FIGURE 18.4 ORTEP plot (25% thermal probability ellipsoids) of $[Mg_2Cl_3\text{-}6THF][HMDS{\cdot}AlCl_3]$. Hydrogen atoms, THF of crystallization, and second component of disorder are omitted for clarity. Adapted and reproduced from Ref. [14]. Creative common licence.

species like $[Mg_2(\mu\text{-}Cl)_2(DME)_4]^{2+}$ and tetracoordinated anions $[AlCl_3(CF_3SO_3)]^-$ which were irreversible. The cells also showed a Coulombic efficiency of ~99%. The electrolyte favored the shuttle effect of sulfur ions delivering a specific capacity of 866 mAh g^{-1} at 200 mA g^{-1}. Also, the cell offered an enhanced power density of 550 W kg^{-1}.

Non-nucleophilic halogen-based electrolytes are also a commonly explored category of electrolyte in Mg–S batteries. Among halogens, mostly fluorine- or chlorine-based electrolytes are explored. These electrolyte species are mainly synthesized by Lewis acid–base reactions. In 2013, Aurbach et al. [16] reported the first inorganic solution-based electrolyte based on acid base reaction of organomagnesium compound (R_2Mg) which corresponds to a Lewis base, with a Lewis acid like $AlCl_3$, $AlCl_2Et$, BPh_3 to form $(Mg(AlCl_{4n}R_n)_2)$, wherein R represents an organic moiety (e.g., alkyl and aryl). The electrolyte delivered an anodic stability of 3.1 V against Mg and is expected to deliver a Coulombic efficiency of 99%: ROMgCl [17], RSMgCl [17], and $MgCl_2$ [16].

Considering various electrolyte species, the crystallographic investigation reveals that the electrolytic mechanism is achieved by the formation of an intermediate ionic structure with the chemical formula $[Mg_2(\mu Cl)_3(THF)_6]^{2+}$. This structure is considered to be highly reactive and acts as a source for the corrosion of the electrodes. Hence, Li et al. [18] demonstrated yet another simple electrolyte omitting the chloride ions to form $[Mg(THF)_6][AlCl_4]_2$. The investigation of the electrochemical behavior of the electrolyte was conducted by CV technique. The platinum disk was used as the working electrode and magnesium foil was used as counter as well as reference electrodes. The CV was performed at a scan rate of 25 mV s^{-1} and swept between 1 and 2 V. The cyclic voltammogram is depicted in Figure 18.5a in which a clear peak observed at −0.6 V represented the metallic deposition of magnesium metal. The oxidation peak observed at 0.5 V is considered to be due to the electrochemical dissolution of the Mg metal. Figure 18.5b represents the performance of the cell using copper metal as the current collector, wherein it clearly displays the reversibility of the magnesiation reaction. The LSV analysis of the electrolyte indicates narrow potential window of 2.5–3 V (Figure 18.5c). The samples showed excellent Coulombic efficiency and performance at various current densities as indicated in Figure 18.5d. The cycling on 15 min interval scanned from 1 to 500 mA cm^{-2}.

18.5.3 Grignard Reagent-Based Electrolyte for Mg–S Batteries

Grignard reagent-based electrolytes are chemically reversible (i.e., Mg deposition/dissolution) [19] and are commonly represented as RMgX, where R is alkyl or aryl and X is the halide anion. Grignard reagent-based electrolytes are usually corrosive for non-noble metals [20] and sensitive to the humidity. The major advantage of using Grignard reagent is the ability for cycling by reversible magnesium stripping and deposition. Considering the counterpart, the potential stability of the electrolyte is comparatively low. The initial study on similar electrolyte was reported by Gregory et al. [21] wherein the electrolyte was synthesized by reacting an equimolar quantity of di(N-methyl aniline)magnesium and Li in a mixture of solvents THF and diethyl ether (6.5:3.5). The testing of Mg plating/dissolution ability revealed better results. Later, Yagi et al. [20] investigated the role of combination of phenyl magnesium bromide and aluminum chloride in THF solvent. The report extensively investigated the electrochemical stability of the electrolytic medium with various electrode materials. The metals such as platinum, nickel, copper, stainless steel, titanium, and aluminum were noted and compared with glassy carbon electrode. Among the electrodes, Pt and glassy carbon electrodes were stable in the potential range from −1 to 4 V, whereas corrosion of the electrode surfaces was observed for Cu, Ni, stainless steel, Ti, and Al electrodes after the first anodic polarization up to 4 V. Lee et al. [22] modified the conventional Grignard reagent-based electrolyte by using the same as a mixture in ionic liquid electrolyte. N-allyl N-methyl pyrrolidinium chloride was employed as the ionic liquid and phenyl magnesium chloride was used as Grignard reagent in THF with varying concentrations, which has been used as electrolyte to study the behavior of the

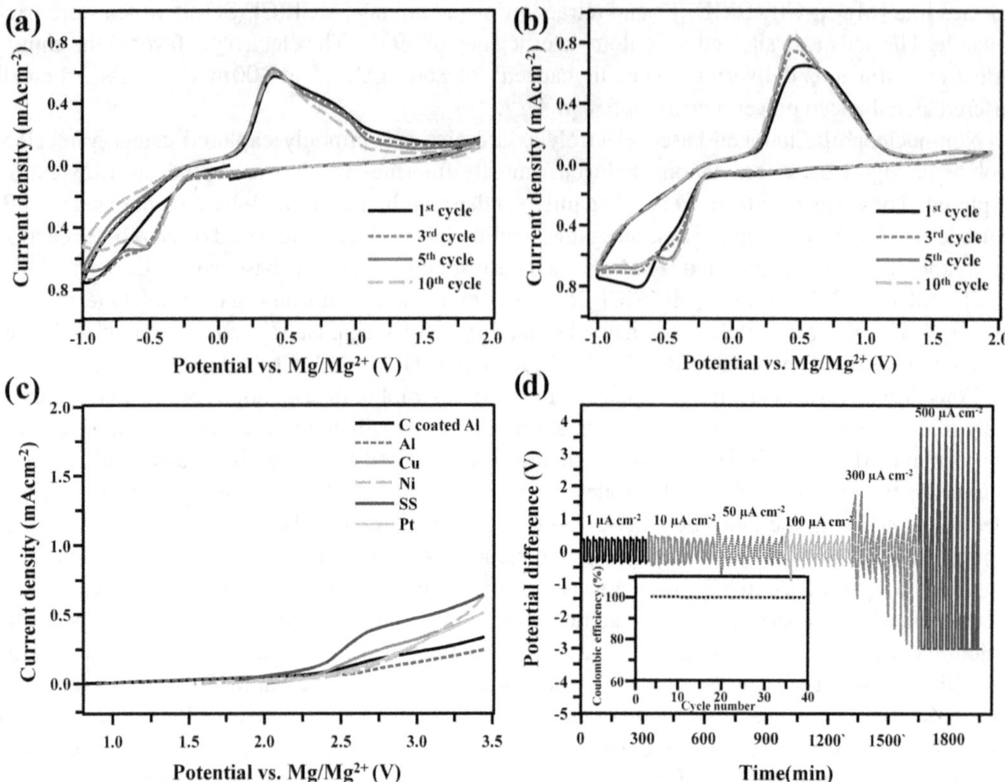

FIGURE 18.5 Cyclic voltammogram of electrolyte $[Mg(THF)_6][AlCl_4]_2$ in $PYR_{14}TFSI/THF$ (1:1 v/v). The working electrodes are (a) Pt disk and (b) Cu foil while, the counter- and reference electrodes are Mg metal. Measurements obtained at 25 mV s^{-1} and under ambient conditions. (c) The LSV of electrolyte $[Mg(THF)_6][AlCl_4]_2$ in $PYR_{14}TFSI/THF$ (1:1 v/v) on different working electrodes while the counter- and reference electrodes are Mg metal. Measurements obtained at 25 mV s^{-1} and ambient conditions. (d) Cycling behavior of a symmetrical cell with electrolyte $[Mg(THF)_6][AlCl_4]_2$ in $PYR_{14}TFSI/THF$ (1:1 v/v) at different current densities of 1 to 500 mA cm^{-2}. The cycle time was 30 min per cycle (15 min charging and 15 min discharging). Adapted and reproduced from Ref. [18]. Copyright ©2016 Wiley International.

same in rechargeable magnesium batteries. 0.1–0.4 M concentration of the ionic liquid was dissolved in the 0.4 M Grignard reagent. The performance evaluation of the electrolyte was conducted with Chevrel phase cathode and magnesium metal anode. Galvanostatic cycling study was conducted with a potential sweep between 0.5 and 1.8 V at 25°C. The cell delivered a discharge capacity of 90 mAh g^{-1} at 0.1 C, and on increase in the C rate to 0.25 C, the reversible capacity was decreased to 66 mAh g^{-1}. The cell delivered a stable charge–discharge performance even after 100 cycles. The charge–discharge profiles and cycling stability of the cell are depicted in Figure 18.6.

18.5.4 Boron-Based Electrolyte for Mg–S Batteries

The first boron-based electrolytes for Mg–S batteries were reported by Gregory et al [23]. $Mg(BPh_2Bu_2)_2$ (where Ph – Phenyl groups, Bu – Butyl groups) solution was used as an electrolyte in that work. These batteries are observed with efficient anodic stability. Lewis acids containing boron, such as BPh_2Cl, $BPhCl_2$, $B[(CH_3)2N]$, BPh_3, BEt_3, BBr_3, BCl_3 and BF_3, were used with Lewis bases such as THF solutions of Bu_2Mg by Aurbach et al. [24]. In 2012, Guo et al. [25] synthesized a new class on boron-based electrolyte for rechargeable magnesium batteries by reacting

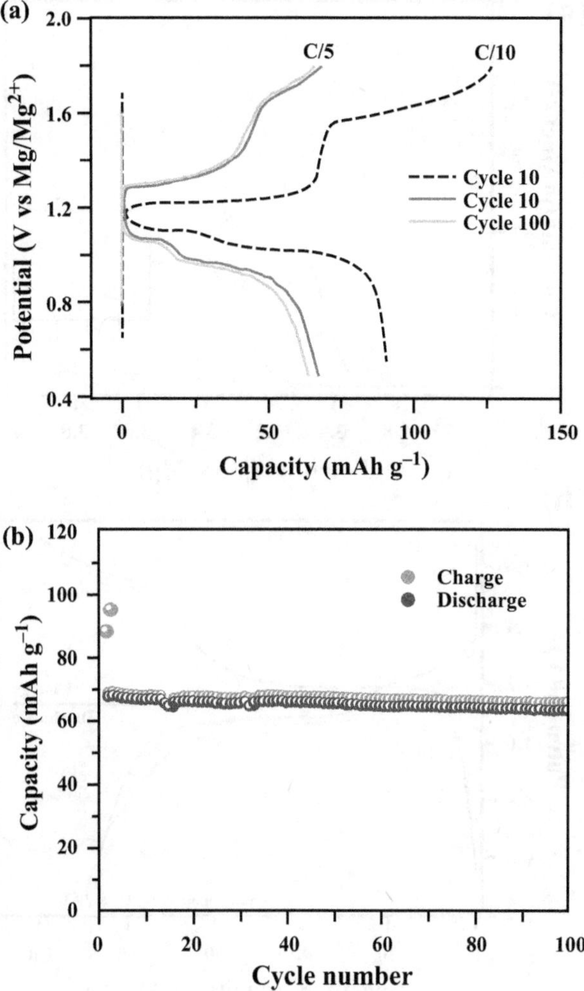

FIGURE 18.6 (a) Galvanostatic discharge–charge profiles of Mo6S8/Mg cells with the (PMC)(AMPC)1.0 electrolyte at different current rates of C/5 (solid line) and C/10 (dotted line) and (b) the cycling performance at the C/5 rate up to 100 cycles. Adapted and reproduced with permission from Ref. [22]. Copyright © 2018 The Royal Society of Chemistry.

tri(3,5-dimethylphenyl)borane (Mes_3B) and Grignard reagent (PhMgCl) in THF for rechargeable magnesium batteries in the medium of THF. The electrolyte delivered a charge–discharge capacity of approximately 85 and 80 mAh g^{-1} at 0.1 C for the third cycle, which gradually decreased the performance on prolonged cycling. The overall performance of the electrolyte and the lesser affinity of reaction of boron-based compounds with the current collectors were the most attractive phenomena for the exploration of these kinds of electrolytes in Mg-based secondary batteries.

Carter et al. [26] proposed a boron clusters as stable electrolyte for secondary magnesium batteries. Among the boron clusters, *closo* boranes were considered for the study with the general formula $B_nH_n^{2-}$. The boron compound was reacted with Grignard reagent in the THF medium to form 1-(1,7-carboranyl) magnesium chloride. The LSV measurements revealed an onset potential of 3.2 V with Pt, Al, and SS grade as electrodes (Figure 18.7a). The charge–discharge performance of the electrolyte at 0.05 C in full cell with Chevrel phase Mo_6S_8 as cathode and magnesium metal as anode delivered a reversible specific capacity of 90 mAh g^{-1}, which was cyclable up with about 90% Coulombic efficiency (Figure 18.7b).

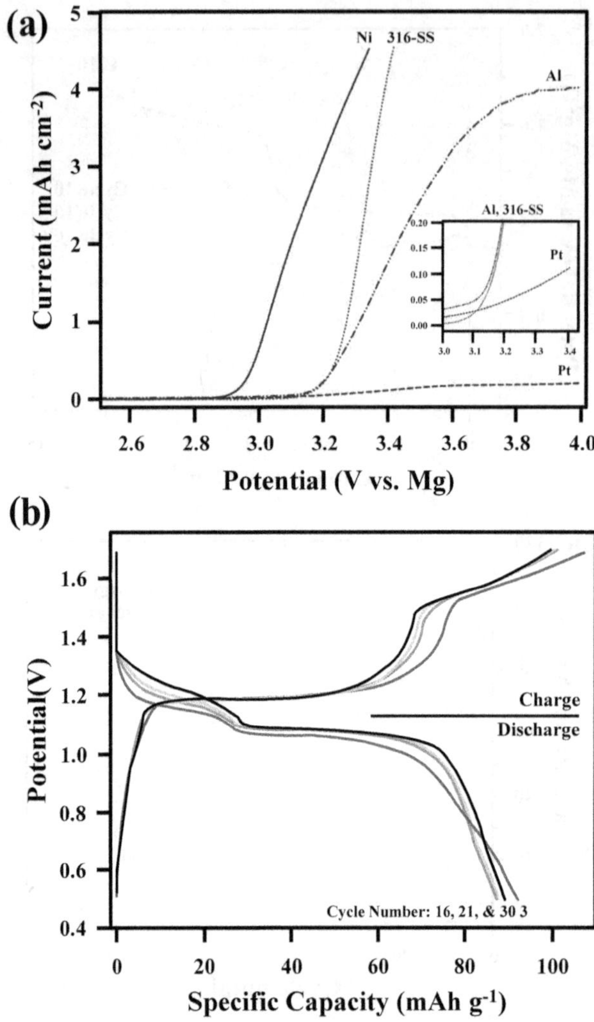

FIGURE 18.7 (a) Linear sweep voltammograms of 2 on Pt, SS, Ni, and Al electrodes (inset: expanded view of the oxidation onset). (b) Typical charge/discharge profiles for a rechargeable battery with 2 as the electrolyte, a Mg anode, and a Chevrel phase cathode at 0.05 C. Adapted and reproduced with permission from Ref. [26]. Copyright ©2014 Wiley International.

18.6 CONCLUSION

Magnesium-based batteries are the next-generation batteries with high capacity. The abundance of magnesium and sulfur as well as high theoretical capacity makes the magnesium–sulfur battery most attractive among other secondary batteries. Magnesium forms divalent ion compared to that of lithium which forms monovalent ion, hence the volumetric capacity is much higher. Considering the safety and firing issues faced in lithium-based batteries as commercialized portable energy storage devices relied on lithium-ion batteries, the need for more safer energy storage device with enhanced performance has been aroused as a necessity. Herein, the relevance of the investigations in magnesium–sulfur battery becomes prominent. The magnesium ions are even safer and less reactive in atmospheric condition with better properties. The abundance and cost of the batteries are another major drawback of the lithium-based batteries, in which also a revolutionary progress can be made on the commercialization of magnesium-based batteries. The major drawback of these battery is the cyclability, which can be improved by proper tuning and investigation of compatible electrode–electrolyte combinations. The bright future of secondary batteries is also dependent on the magnesium–sulfur batteries.

REFERENCES

[1] Z. Zhao-Karger and M. Fichtner, Magnesium-sulfur battery: Its beginning and recent progress, *MRS Commun.* 7, 770 (2017).

[2] F. Jin, S. Xiao, L. Lu, and Y. Wang, Efficient activation of high-loading sulfur by small CNTs confined inside a large CNT for high-capacity and high-rate lithium-sulfur batteries, *Nano Lett.* 16, 440 (2016).

[3] G. Li, Z. Li, B. Zhang, and Z. Lin, Developments of electrolyte systems for lithium-sulfur batteries: A review, *Front. Energy Res.* 3, 1 (2015).

[4] Z. Zhao-Karger and M. Fichtner, Beyond intercalation chemistry for rechargeable Mg batteries: A short review and perspective, *Front. Chem.* 7, 1 (2019).

[5] M. Rashad, M. Asif, and Z. Ali, Quest for magnesium-sulfur batteries: Current challenges in electrolytes and cathode materials developments, *Coord. Chem. Rev.* 415, 213312 (2020).

[6] T. Gao, X. Ji, S. Hou, X. Fan, X. Li, C. Yang, F. Han, F. Wang, J. Jiang, K. Xu, and C. Wang, Thermodynamics and kinetics of sulfur cathode during discharge in $MgTFSI_2$-DME electrolyte, *Adv. Mater.* 30, 2 (2018).

[7] X. C. Hu, Y. Shi, S. Y. Lang, X. Zhang, L. Gu, Y. G. Guo, R. Wen, and L. J. Wan, Direct insights into the electrochemical processes at anode/electrolyte interfaces in magnesium-sulfur batteries, *Nano Energy* 49, 453 (2018).

[8] B. Sievert, J. Häcker, F. Bienen, N. Wagner, and K. A. Friedrich, Magnesium sulfur battery with a new magnesium powder anode, *ECS Trans.* 77, 413 (2017).

[9] Y. Zhang, J. Xie, Y. Han, and C. Li, Dual-salt Mg-based batteries with conversion cathodes, *Adv. Funct. Mater.* 25, 7300 (2015).

[10] H. Batteries, Studies of $FeSe_2$ cathode materials for Mg-Li hybrid batteries, *Energies* 13, 4375 (2020).

[11] X. Zhou, J. Tian, J. Hu, and C. Li, High rate magnesium - sulfur battery with improved cyclability based on metal - organic framework derivative carbon host, *Adv. Mater.* 1704166, 1 (2018).

[12] C. R. Hauser and H. G. Walker, Condensation of certain esters by means of diethylaminomagnesium bromide, *J. Am. Chem. Soc.* 69, 295 (1947).

[13] P. Schüler, S. Sengupta, S. Zaubitzer, F. Fiesinger, S. Dongmo, H. Görls, M. Wohlfahrt-Mehrens, M. Borg, D. Gaissmaier, S. Krieck, M. Marinaro, T. Jacob, and M. Westerhausen, Suitability of carbazolyl hauser and turbo-hauser bases as magnesium-based electrolytes, *Eur. J. Inorg. Chem.* 2022(12), e202200149 (1–6) (2022).

[14] H. S. Kim, T. S. Arthur, G. D. Allred, J. Zajicek, J. G. Newman, A. E. Rodnyansky, A. G. Oliver, W. C. Boggess, and J. Muldoon, Structure and compatibility of a magnesium electrolyte with a sulphur cathode, *Nat. Commun.* 2, 427 (2011).

[15] D. Huang, S. Tan, M. Li, D. Wang, C. Han, Q. An, and L. Mai, Highly efficient non-nucleophilic $Mg(CF_3SO_3)_2$-based electrolyte for high-power Mg/S battery, *ACS Appl. Mater. Interfaces* 12, 17474 (2020).

[16] R. E. Doe, R. Han, J. Hwang, A. J. Gmitter, I. Shterenberg, H. D. Yoo, N. Pour, and D. Aurbach, Novel, electrolyte solutions comprising fully inorganic salts with high anodic stability for rechargeable magnesium batteries, *Chem. Commun.* 50, 243 (2014).

[17] P. Bian, Y. Nuli, Z. Abudoureyimu, J. Yang, and J. Wang, A novel thiolate-based electrolyte system for rechargeable magnesium batteries, *Electrochim. Acta* 121, 258 (2014).

[18] W. Li, S. Cheng, J. Wang, Y. Qiu, Z. Zheng, H. Lin, S. Nanda, Q. Ma, Y. Xu, F. Ye, M. Liu, L. Zhou, and Y. Zhang, Synthesis, crystal structure, and electrochemical properties of a simple magnesium electrolyte for magnesium/sulfur batteries, *Angew. Chemie – Int. Ed.* 55, 6406 (2016).

[19] O. Tutusaus, R. Mohtadi, T. S. Arthur, F. Mizuno, E. G. Nelson, and Y. V. Sevryugina, An efficient halogen-free electrolyte for use in rechargeable magnesium batteries, *Angew. Chemie – Int. Ed.* 54, 7900 (2015).

[20] S. Yagi, A. Tanaka, Y. Ichikawa, T. Ichitsubo, and E. Matsubara, Electrochemical stability of magnesium battery current collectors in a grignard reagent-based electrolyte, *J. Electrochem. Soc.* 160, C83 (2013).

[21] John Muldoon, Claudiu B. Bucur, and Thomas Gregory,, Quest for nonaqueous multivalent secondary batteries: Magnesium and beyond, *Chem. Rev.* 114, 23, 11683–11720 (2014).

[22] B. Lee, J. H. Cho, H. R. Seo, S. Bin Na, J. H. Kim, B. W. Cho, T. Yim, and S. H. Oh, Strategic combination of grignard reagents and allyl-functionalized ionic liquids as an advanced electrolyte for rechargeable magnesium batteries, *J. Mater. Chem. A* 6, 3126 (2018).

[23] T. D. Gregory, R. J. Hoffman, and R. C. Winterton, Nonaqueous electrochemistry of magnesium: applications to energy storage. *J Electrochem Soc.* 137, 775 (1990).

[24] D. Aurbach, H. Gizbar, A. Schechter, O. Chusid, H. E. Gottlieb, Y. Gofer, and I. Goldberg, Electrolyte solutions for rechargeable magnesium batteries based on organomagnesium chloroaluminate complexes, *J. Electrochem. Soc.* 149, A115 (2002).

[25] Y. S. Guo, F. Zhang, J. Yang, F. F. Wang, Y. Nuli, and S. I. Hirano, Boron-based electrolyte solutions with wide electrochemical windows for rechargeable magnesium batteries, *Energy Environ. Sci.* 5, 9100 (2012).

[26] T. J. Carter, R. Mohtadi, T. S. Arthur, F. Mizuno, R. Zhang, S. Shirai, and J. W. Kampf, Boron clusters as highly stable magnesium-battery electrolytes, *Angew. Chemie – Int. Ed.* 53, 3173 (2014).

19 Metal–Sulfur Battery
A Promising Candidate for Next-Generation Battery

Roshny Joy, Akhila Das, Devika S. Lal, Xiaohui Zhao, Jabeen Fatima M. J., Jou-Hyeon Ahn, and Prasanth Raghavan

19.1 INTRODUCTION

With ever increasing energy needs and challenging environmental issues, there is a lot of pressure on society to find a sustainable energy alternative that depletes fossil fuel resources [1]. Over the past 150 years, the battery technology has tremendously developed from lead acid battery to lithium-ion batteries (LIBs) [2]. LIBs have a very huge role in advancing battery technologies toward portable electronic devices due to their high storage capacity and promising efficiency. But the higher cost is always been a major challenge in terms of long-term grid-scale applications as well as the limited resources of lithium metal and poor cyclic stability of LIBs [3]. Nevertheless, there is a requirement to develop an intelligent battery with high energy density, long lifespan and efficiency, which leads to the idea of metal–sulfur battery with sulfur being an environmentally friendly, inexpensive, earth-abundant element that is useful as a cathode material in metal–sulfur batteries. Among all solid-state cathodes, elemental sulfur is a promising candidate with reversibly storing large amount of electrical energy up to 2.8 k Wh L^{-1}. Mechanistically, sulfur undergoes a two-electron transfer process, but pairing with the sodium anode promotes multi-electron transfer conversion and exhibits the theoretical energy density of 1274 Wh kg^{-1}, which is about five times that of LIBs [4]. The studies based on Li–S system have been initiated in the early 1960s and yet not reached to the marketplace because there are few scientific hurdles that remain to be cleared. LIBs mainly follow the intercalation chemistry, which yields one or less than one electrons to the transition metal ion, but the combination Li–S is open up to an easy way for energy storage with an impulsive reversible reaction [5]. This combination shows a theoretical specific energy density of 2500 Whkg^{-1} and a theoretical specific capacity of 1675 mAh g^{-1}, which have received a lot of attention [6]. Based on the mechanism, the solution is needed for the main problems such as the poor cyclability and redox shuttle mechanisms [7]. The researchers started to focus on sulfur as a cathode and parallel studies were conducted on different metal anodes to couple with sulfur electrodes. There are different combinations of theoretical metal–sulfur batteries such as calcium, sodium, potassium and magnesium, and aluminum as metal anodes [8]. Figure 19.1 depicts the abundance of these metals along with gravimetric energy density of the metal–sulfur battery with few metal anodes.

Comparing these metal anode combination with sulfur cathode, Li–S shows a higher density. In addition, non-lithium batteries are far less expensive than lithium batteries, and these are promising battery systems with right choice of electrolytes and combinations. These can be very widely used in a variety of applications, including electric vehicles and portable devices, but there is a conflict during the ongoing research of non-lithium metal batteries as they could not meet all demands like energy density and longer cycle life. However, the gravimetric or volumetric energy densities are higher than the LIBs. But it can surely provide a very low cost as well as safety guarantees than LIBs [2]. The advantages of various battery combinations with sulfur batteries are listed in

DOI: 10.1201/9781003310167-19

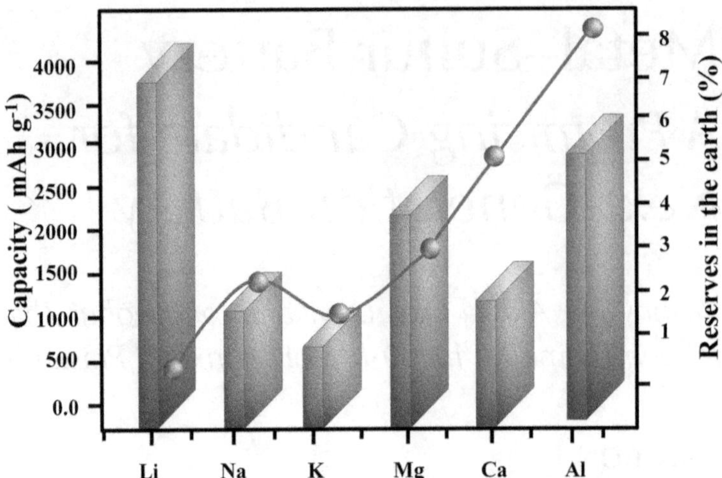

FIGURE 19.1 Abundance of Li, Na, K, Mg, Ca, and Al elements in the earth's crust and gravimetric energy density of MSBs with various metal anodes. Adapted and reproduced with permission from Ref. [8]. Copyright © 2021, American Chemical Society.

TABLE 19.1
Comparison of Advantages of Different MSBs

Metal–Sulfur Batteries	Advantages
Li–S	• High gravimetric density
Na-S	• High natural abundance • Low cost • High charge and discharge efficiency
Mg–S	• High natural abundance • Free dendrites • High volumetric energy density • High security
Al–S	• High natural abundance • Low cost • High energy density
K–S	• High natural abundance • Low reduction potential • High specific capacity • Energy density
Ca–S	• High natural abundance • Free dendrites • High security

Source: Adapted and reproduced with permission from Ref. [8].

Table 19.1, and the schematic diagram of the development of non-lithium metal–sulfur batteries is shown in Figure 19.2.

The research based on metal batteries and their developments will face large difficulties. The fundamental concepts, electrochemical mechanisms, and intrinsic performance of the Na–S, Mg–S, Al–S, K–S, and Ca–S chemistries are included in this chapter, followed by an assessment with the Li–S batteries and rundown of how far Li–S batteries have progressed [9]. Along with the history of

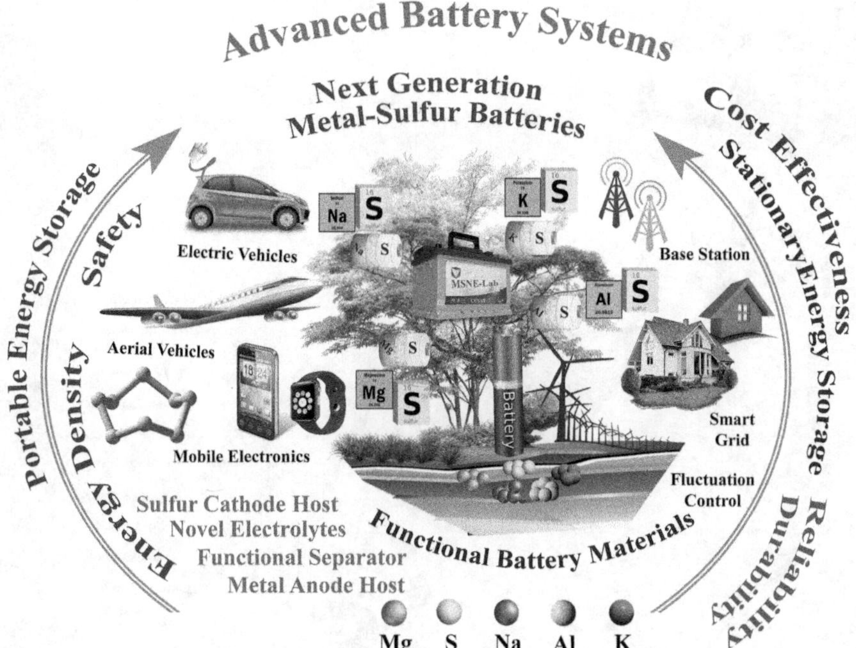

FIGURE 19.2 Schematic illustration of the development of non-Li metal–sulfur batteries. Adapted and reproduced with permission from Ref. [2]. Copyright © 2018, Wiley.

progress, research trends, and technical advancements, Na–S, Ca–S, Mg–S, Al–S, and K–S progress are summarized. Timeline of important developments is shown in Figure 19.3 [10].

19.2 ALUMINUM–SULFUR BATTERIES

In metals, aluminum is the most abundant element and the third most element in the earth's crust. It shows three-electron redox properties, which will enhance high capacity while also offering low cost, low reactivity, easy handling as well as low flammability [11]. Researchers and academicians believed that the history of aluminum batteries started in 1970s, but it is still unknown who came up with the idea. The pioneer work of Hulot paved a new trend in the electrochemistry of aluminum. There are a lot of examples for aluminum batteries such as Al–MnO^{-}_2, Al–H_2O_2, Al–S, Al–Ni, Al–$KMnO_4$, and Al–air batteries, but these failed to recharge [12].

The development of various aluminum–sulfur systems began in the 1980s. A nonaqueous aluminum–sulfur system was investigated by Marassi et al. [13–15] in 1977, while the aqueous aluminum–sulfur battery was initially proposed in 1993 by Licht et al. [16–18]. Recently, Cohn et al. [15] proposed the first aluminum–sulfur cell fabricated using an aluminum metal anode and room-temperature ionic liquid electrolyte of $AlCl_3$ in 1-ethyl-3-methylimidazolium chloride and a sulfur/carbon composite cathode. This system can reach a high capacity of 1600 mAh g^{-1} and shows the potential to be electrochemically reversible. A few follow-up studies have since begun improving the aluminum–sulfur cell performance with ionic liquid electrolyte. This progress report discusses aluminum–sulfur cells that were developed after 2015, as well as their electrochemical conversion battery chemistry. Similar to lithium–sulfur and other metal–sulfur batteries, aluminum–sulfur cells generate electricity by electrochemical conversion reactions among sulfur, polysulfides, and sulfides. The use of an aluminum-metal anode forms aluminum polysulfides with different chain lengths and Al during discharging, which end as aluminum sulfides [19].

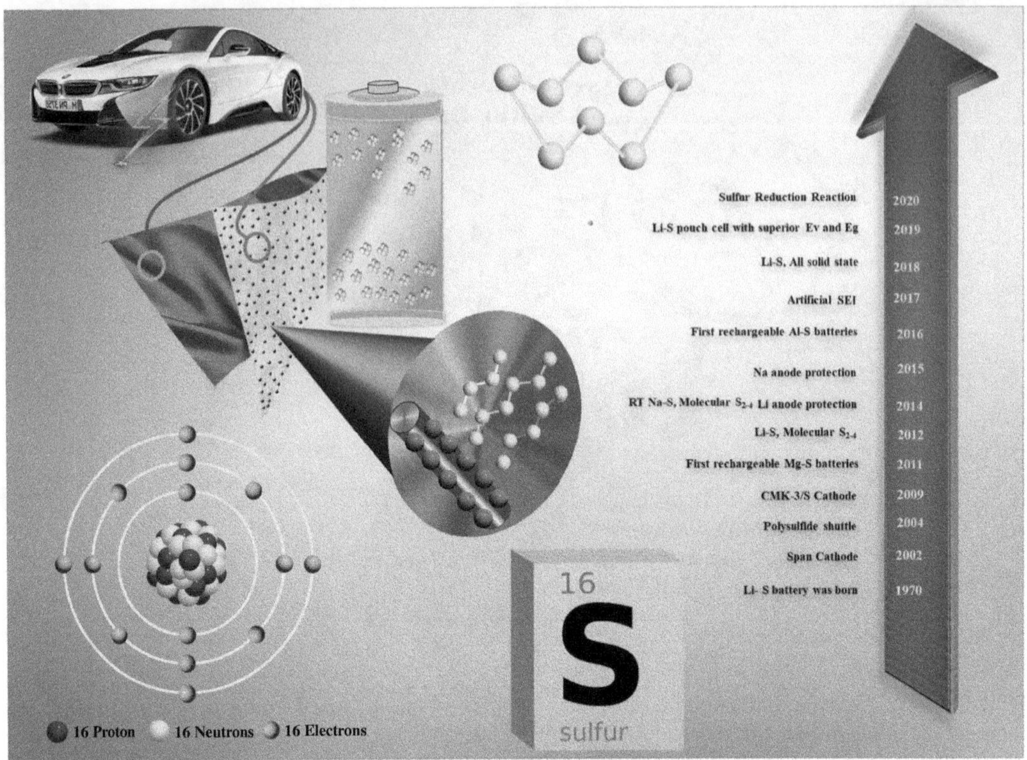

FIGURE 19.3 Principle and mechanisms of metal–sulfur chemistries. Adapted and reproduced with permission from Ref. [10]. Copyright © 2021, Elsevier.

The overall electrochemical reaction is expected to follow the conversion reaction (equation 19.1):

$$3S + 2Al \rightarrow Al_2S_3 \tag{19.1}$$

Another reported work has been done by Manthiram et al. in which they added lithium ions to the electrolyte to improve performance since the lithium ions have a greater affinity toward sulfur than Al^{3+} and the lithium polysulfides have much better solubility in electrolyte than the Al polysulfides. The lifespan of the Al–S battery increased up to 50 cycles with a capacity retention of >400 mAh g^{-1} under a cycling rate of C/10 though the voltage hysteresis (1.2 V). Similar to Li–S batteries, the high energy barrier of sulfur conversion in the cathode of Al–S batteries is the main problem for the low reaction kinetics, which is associated with poor sulfur utility and large voltage hysteresis. Later, Yue Guo et al. [20] contributed another observation to develop the aluminum–sulfur batteries. They use Co (II, III) as an electrochemical catalyst, which enhance the activity of polysulfides thereby reducing the voltage hysteresis. When Co (II, III) was immobilized by carbon matrix, it shows a low-voltage hysteresis of 0.8 V and a specific capacity retention of 500 mAh g^{-1} after 200 times of discharge–charge cycling at 1 Ag^{-1} and of 300 mAh g^{-1} when cycled at 3 A g^{-1}; this indicates the improved reaction kinetics and sulfur utilization in the cathode, as shown in Figure 19.4 [20]. The development of Al–S batteries with high electrochemical cyclic stability can be explored and carried forward in to the next-generation battery system. Despite all of these efforts toward this system, the research on Al–S still has a long way to reach its practicality [21] (Figure 19.5).

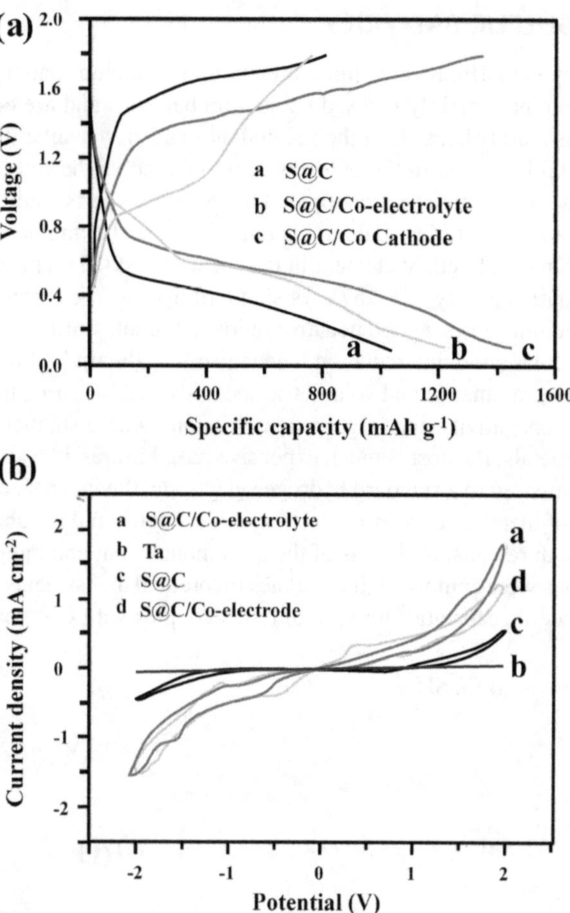

FIGURE 19.4 (a) CV profiles of symmetric cells. (b) Galvanostatic discharge–charge curves measured under a current density of 1 A g^{-1} at the first cycle of Al–S batteries consisting of different types of cathode materials. (c) Voltage hysteresis of the Al–S batteries that are the values of the difference between the average voltages of the charge and discharge curves. Adapted and reproduced with permission from Ref. [20]. Copyright © 2020, Wiley.

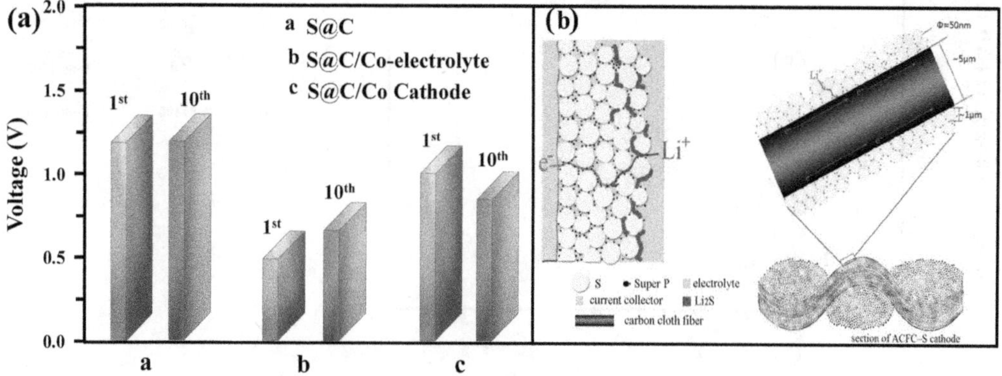

FIGURE 19.5 (a) Voltage hysteresis of the Al–S batteries that are the values of the difference between the average voltages of the charge and discharge curves. (b) Schematic of section of ACFC-S cathode. Adapted and reproduced with permission from Ref. [20]. Copyright © 2020, Wiley.

19.3 CALCIUM–SULFUR BATTERIES

Battery techniques beyond LIBs receive more attention and calcium battery is one among them. In the beginning, researchers mainly focused on sodium batteries and are considered as the finest alternative for the LIBs mainly because of their abundance and easy availability, which give another scope to the other metal batteries. In the 1960s, sulfur emerged as the cathode, and the important chemistry paved a new interesting research topic to the battery experts leading to the breakthrough of non-lithium metal–sulfur batteries. Calcium is one of the good combination with sulfur for batteries and it is the fifth most abundant element in the earth's crust [22]. This divalent alkaline earth metal has a great oxidative ability of –2.87V vs standard hydrogen electrode. When compared to other metal atoms, calcium shows a good negative redox potential, similar ionic radii with Na^+, and easily undergoes intercalation–deintercalation mechanism. The main highlight of calcium-ion batteries is that theoretical gravimetric and volumetric specific capacities are almost same as the LIBs. Calcium has low electronegativity and large ionic radii along with a smaller charge density. When compared with other metals, the cost is much expensive [23]. Figures 19.6 and 19.7 show the metal/ion redox potential compared to a standard hydrogen electrode, the mean size of the ion species, the theoretical capacities of metal electrodes in electrochemical cells, and the abundance on the earth's crust as well as the mean raw material costs of the used monovalent and multivalent atomic species beyond lithium batteries were compared. It combines theoretical assessments of the thermodynamic redox potential and specific capacities for several redox couples with Ca^{2+} ions as well as estimates

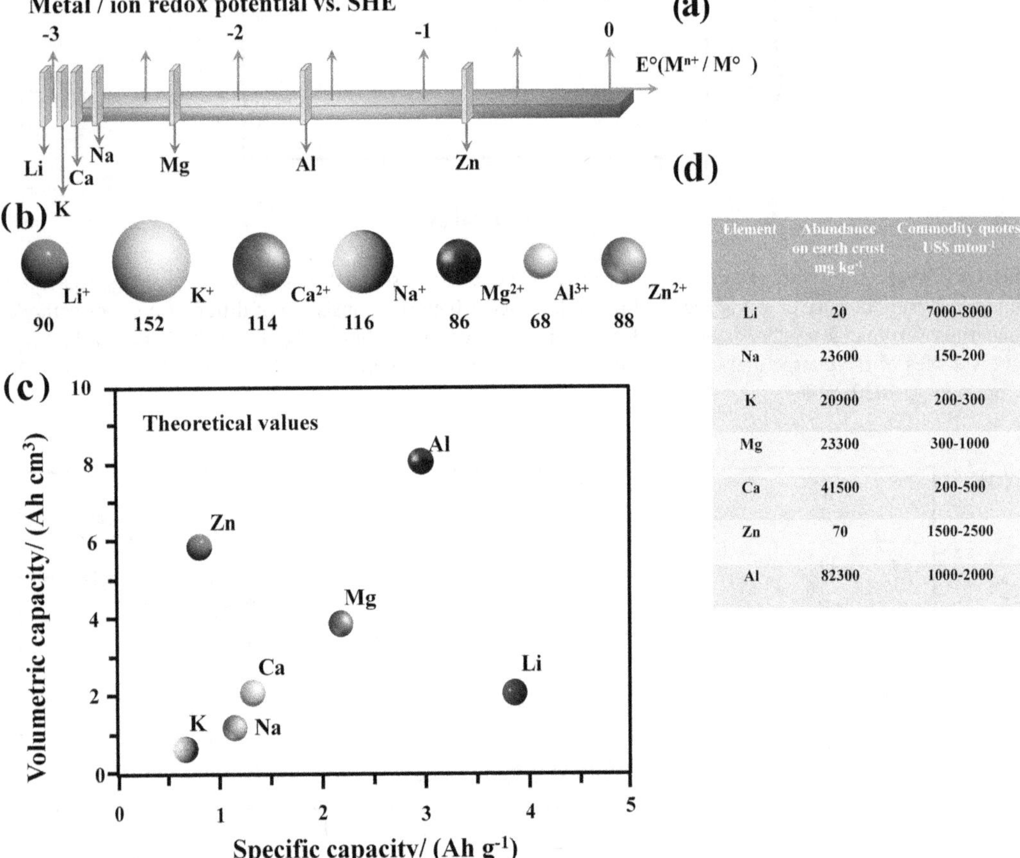

Element	Abundance on earth crust mg kg⁻¹	Commodity quotes US$ mton⁻¹
Li	20	7000-8000
Na	23600	150-200
K	20900	200-300
Mg	23300	300-1000
Ca	41500	200-500
Zn	70	1500-2500
Al	82300	1000-2000

FIGURE 19.6 (a and b) Redox-potential and ionic radius of different elements. (c) volumetric capacity vs theoretical capacity of different elements (d) tabulation of different elements of earth abundance and commodity quotes. Adapted and reproduced with permission from Ref. [23]. Copyright © 2021, Elsevier.

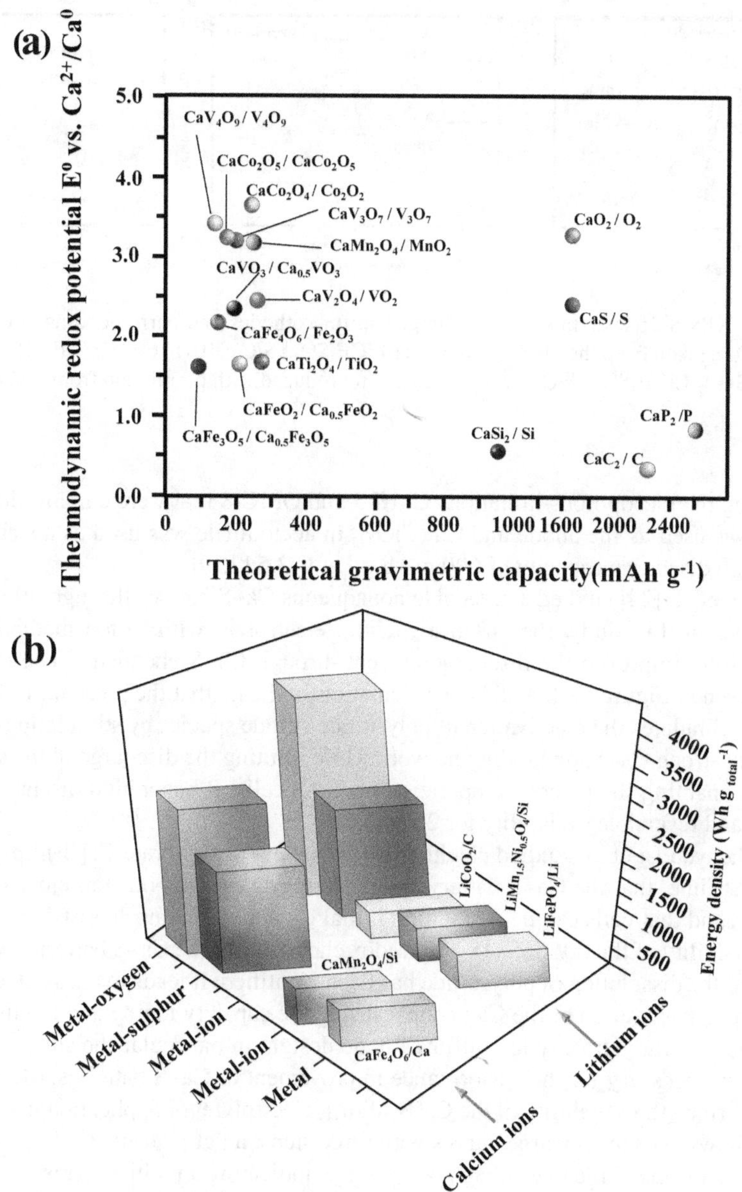

FIGURE 19.7 (a and b) Combined theoretical evaluations of the thermodynamic redox potential and specific capacities for several redox couples involving Ca^{2+} ions. Adapted and reproduced with permission from Ref. [23]. Copyright © 2021, Elsevier.

the theoretical energy density for calcium-based batteries compared to lithium-based analogs. The estimates of the standard redox potentials of various hypothetical pairs were obtained through the available literature; thermodynamic Gibbs energies of formation or semi-empirical evaluations, theoretical capacities, and estimated energy densities are normalized to the sum of the masses of the active materials [23].

The current status of calcium batteries and a plenty of elementary research is ongoing, and from the data, the reversible calcium sulfur batteries have not yet been reported. The lack of suitable electrolytes is the main challenge. For calcium anodes, sulfur is considered as the best cathode combination in terms of reversibility and higher utilization. In 2013, primary calcium–sulfur cell was

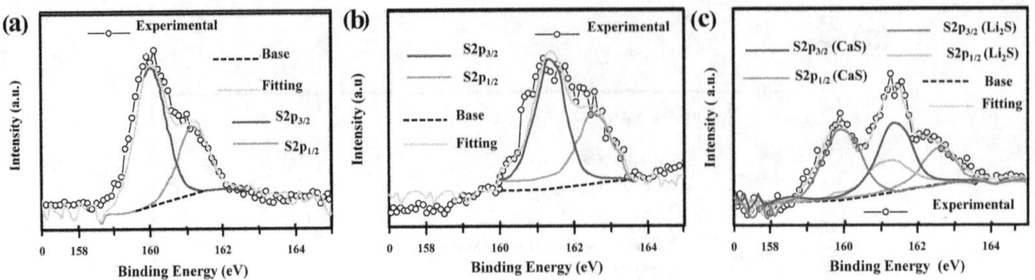

FIGURE 19.8 XPS S 2p spectra of the discharged sulfur cathodes and corresponding curve fittings. The sulfur cathodes are taken from the discharged (a) Li ‖ $LiCF_3SO_3$ ‖ S/C cell, (b) Ca ‖ $Ca(CF_3SO_3)_2$ ‖ S/C cell, and (c) Ca ‖ $Ca(CF_3SO_3)_2$-$LiCF_3SO_3$ ‖ S/C cell. Adapted and reproduced with permission from Ref. [24]. Copyright © 2019, Wiley.

fabricated using the mesoporous imitations CMK-3 and OMC-8 that were employed to host sulfur. Ca antipode was used as the anode and Ca $(ClO_4)_2$ in acetonitrile was used as a stable electrolyte achieving a high discharge capacity of 600 mAh g^{-1} at C/3.5 [2,19].

Xingwen Yu et al. [24] studied a reversible nonaqueous Ca–S battery through lithium-ion meditation approach. In this study, they used a calcium electrolyte with Li-ion mediation approach, which successfully improved the discharge current through Ca–S chemistry. The results of XPS analysis are given in Figure 19.8. And DFT calculations indicate that the presence of Li^+ ions in the Ca electrolyte stimulates the reactivation of polysulfide/sulfide species by stimulating the formation of the relatively strong Ca–S ion binding network while limiting the discharge of the sulfur cathode to a low level, enabling the reversible operation of Ca–S cells. It has a high discharge capacity of 800 mA h g^{-1} and reversible cyclability for 20 cycles.

In 2020, Zhenyou Li et al. studied calcium–sulfur battery with borate $Ca[B(hfip)_4]_2$ electrolyte, and for the first time, that the Ca–S battery can deliver a discharge potential close to the thermodynamic value and that cells can deliver a high initial capacity. The mechanistic study verifies the chemical reversibility of the S0/ CaSx (x=28) redox chemistry in the Ca–S batteries with the borate electrolyte, and the dissolution of polysulfide has been identified. It leads to a loss of active material and causes strong passivation of the Ca anode, causing the capacity fading and the limited lifetime of the batteries. Another strategy for sulfur cathode design, in particular the surface protection of the Ca anode, is necessary for the performance improvement of Ca–S batteries. Most importantly, this study confirmed the suitability of the $Ca[B(hfip)_4]_2$ electrolyte for application in Ca–S batteries. This system shows discharge/charge curves with three steps: a flat plateau at 2.1 V, a slope at about 1.5 V, and another slope in the low voltage range at 1 V, indicating a multi-step response. In terms of voltage, a discharge capacity of about 760 mAh g^{-1} was delivered in the window between the open circuit voltage (OCV) of 2.5 V and the cutoff at 1 V, while the charge capacity far exceeds the discharge capacity, implying that a polysulfide shuttle effect is common for the calcium–sulfur battery system, and cycling stability of the Ca/S capacity of about 120 mAh g^{-1} remained after 15 cycles, as depicted in Figure 19.9 [25]. For Ca–S batteries to operate better, more research into sulfur cathode design and, in particular, surface protection of the Ca anode is required.

19.4 MAGNESIUM–SULFUR BATTERIES

This driving force of cost reduction inspired the application of earth-abundant Mg metal as the anode and S as the cathode material in Mg/S batteries, which shows high density (1.738 g cm^3) and divalent nature of Mg^{2+}, Mg^- metal has a higher theoretical volumetric capacity of 3832 mAh cm^3 than Li (2062 mAh cm^3). Moreover, Mg has a reduction potential of 2.4 V vs SHE and can be handled in air. A typical Mg/S battery consists of an S-carbon-based composite cathode, an organic

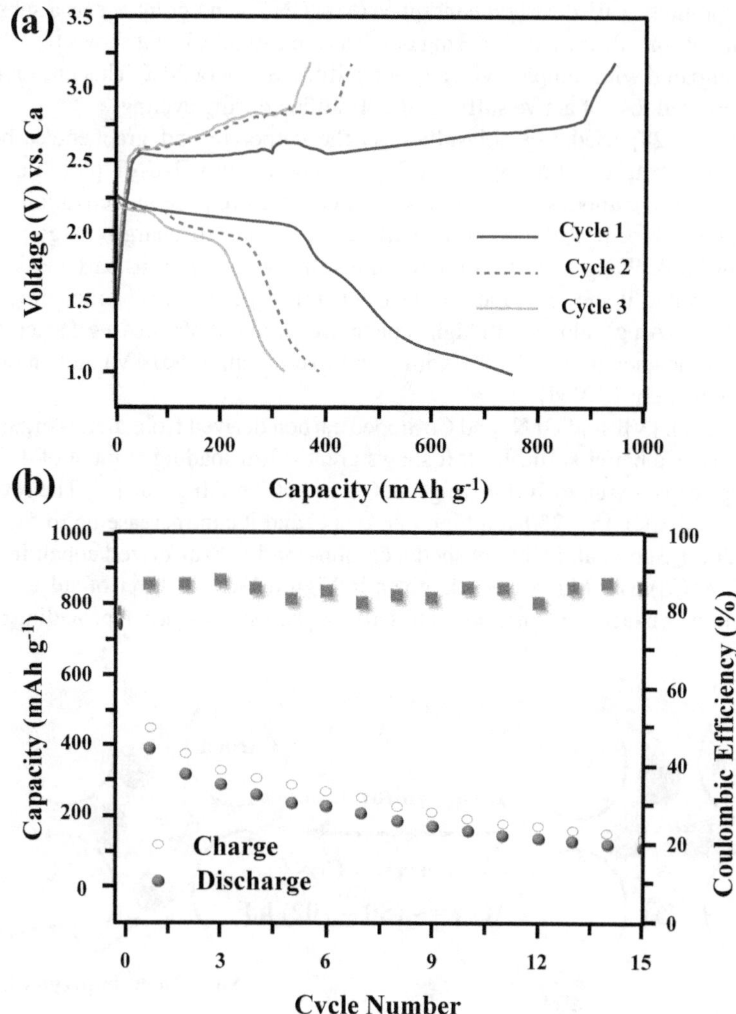

FIGURE 19.9 Electrochemical performance of the Ca–S batteries with the $Ca[B(hfip)_4]_2$/DME electrolyte in a voltage window of 1.0–3.2 V: (a) charge–discharge profiles and (b) cycling performance. Adapted and reproduced with permission from Ref. [25]. Copyright © 2020, Wiley.

electrolyte, and an Mg metal anode. The Mg metal is oxidized to produce Mg^{2+}, which is separated by the organic electrolyte, and the separator to the S-Cathode migrates as electrons arrive at active S material via an external circuit (1684 Wh kg^{-1} or 3221 Wh L^{-1}). The theoretical volumetric energy density of Mg/S batteries exceeds that of Li/S batteries (2856 Wh L^{-1}). However, the theoretical gravimetric energy density of Mg/S batteries is lower than that (2600 Wh kg^{-1}) of Li/S batteries. The main problem with magnesium–sulfur batteries is controlling the polysulfides dissolution and diffusion from cathode to electrolyte. To overcome this, carbonaceous materials were incorporated with Mg–S cathodes to act as the conductive matrix. Otherwise, the sulfur cathode may undergo rapid capacity fading from the very first cycle. The mainly used conductive matrices are carbon cloth (ACC), graphene-based matrices, carbon nanotubes (CNTs), carbon nanofibers (CNFs), sulfur graphdiyne (SGDY), and nanoporous carbon (CMK-3).

Kim et al. [26] prepared a cathode by simply coating a mixture of elemental sulfur, carbon black, and poly(tetrafluoroethylene) adhesive on porous carbon substrate, which was the first reported Mg–S cell work. Vinayan et al. [27] prepared another one with hybrid nitrogen-doped nanocomposite of multi-walled carbon nanotubes (MWCNTs) and graphene as a substrate containing varying amounts of sulfur (with 0.5–3 mg cm^{-2}) as the cathode. But it shows that the capacitance fades rapidly compared with samples with a lower sulfur loading of 20%. This unusual phenomenon is due to the increased loss of active sulfur and polysulfide during cycling.

Later, Xu et al. [28] used Mg polysulfide as the catholyte and graphene/carbon nanotubes (GCNTs) as carbon matrix and dropping the MgS_8 solution onto G-CNT pasted pyrolytic graphite substrate to prepare cathodes. These polysulfide cathodes require an activation process, which results in a slightly lower probability in the early cycles. Their discharge/charge plateaus were at 1.2 and 2.2 V. In 2018, Wang et al. [29] proposed a sulfur@microporous carbon (S@MC) cathode for Mg–S batteries and it showed superior electrochemical performances including high capacity approximately 979 mAh g^{-1} along with high capacity retention of 369 mAh g^{-1} after 200 cycles and high Coulombic efficiency (CE) of 100% with small overpotential (<0.4 V) and an obvious voltage plateau at approximately 1.1 V [1].

Xuejun Zhou et al. [30] studied N- and Co-doped carbon derived from metal–organic framework ZIF-67, which act as a better sulfur host. It shows great sulfur loading amount of 47% and it could release discharge capacity up to 700 mAh g^{-1} at 0.1 and 600 mAh g^{-1} at 1 C. The reversible capacity is still preserved after 150–250 at a high rate of 1 C and it can increase up to 5 C, which is still achievable. In 2020, Sun et al. [31] published a carbon-confined sulfurized cobalt in a mesoporous matrix (MesoCo@CS) as a Mg–S cathode given in Figure 19.10. A layer of sulfurized CoSx species develops on the surface of Co nanoparticles after in situ sulfurization, providing robust binding

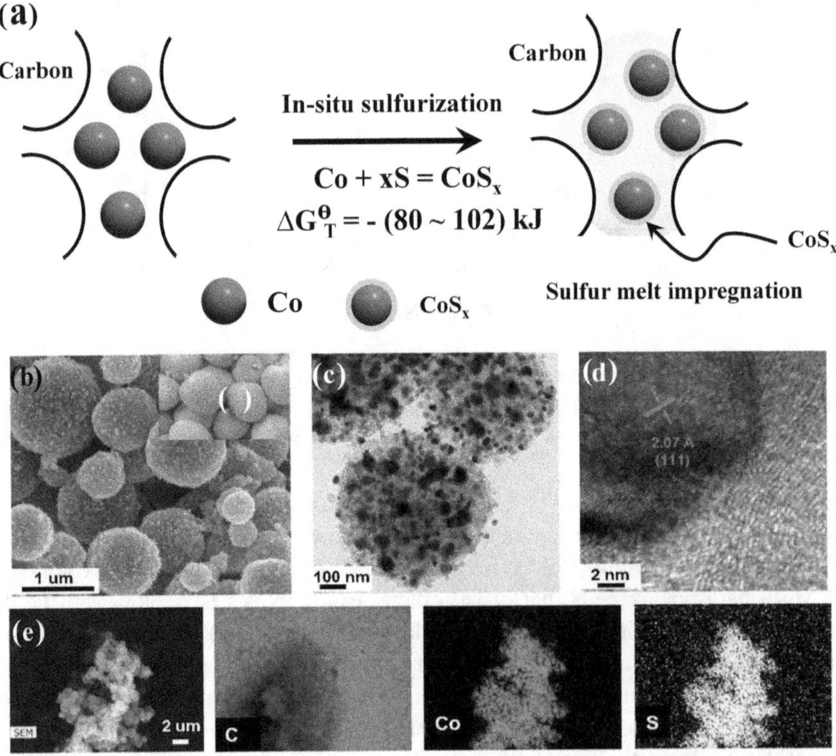

FIGURE 19.10 A carbon-confined sulfurized cobalt in a mesoporous matrix (MesoCo@C–S) as Mg–S cathode. (a) Schematics of the in situ sulfurization of MesoCo@C precursor. (b–d) Characterization of MesoCo@C precursor: (b) SEM, (c) TEM, (d) HRTEM, and (e) EDS images. The inset picture in (b) is a digital photo of lychee fruits for comparison. Adapted and reproduced with permission from Ref. [31]. Copyright © 2020, American Chemical Society.

to polysulfides. The carbon matrix and conserved Co cores both have high electrical conductivity, which aids charge transfer to the sulfurized CoSx species and hence enhances cathode kinetics. MesoCo@CS cathodes exhibited a high initial capacity of 830 mAh g^{-1} and a capacity of 280 mAh g^{-1} after 400 cycles, according to the researchers.

Zhao-Karger et al. [32] reported a S/CMK400PEG cathode using CMK-3 as the substrate and a CMK-3 to sulfur weight ratio of 3:7 which is exactly like the mesoporous carbon substrate for sulfur in Li–S batteries. Du et al. [33] created a sulfur cathode based on sulfur-based graphdiyne in 2017. Sulfur-based Graphdiyne (SGDY) is used as a cathode with all phenyl complex (APC) electrolytes. Graphdiyne is a layered structure composed of benzene rings and connected by butadiyne linkage that works with nucleophilic electrolytes in 2017 [1].

From the studies, the fluorinated alkoxyborate-based electrolyte used in Mg–S batteries has constant cyclability, near-theoretical cell voltage, and high Coulomb efficiency. Electrolyte solvents or solvent mixtures along with cathode designing are challenging. Most of the basic understanding is still needed. There is a lack of understanding of the interaction between electrodes and electrolytes. It is still unclear whether the formation of soluble polysulfides or the formation of soluble sluggish electrochemical reactivity of MgSx species is responsible for the rapid capacity degradation in the first cycles [34]. Studies reveal that the future is looking forward to magnesium–sulfur cell and the challenges will be successfully addressed.

19.5 POTASSIUM–SULFUR BATTERIES

Aside from Li and Na, potassium has the ability to build a K–S battery when combined with sulfur. Due to the higher ionic size of potassium, it is difficult to generate K_2S in a K–S battery as it is in Li–S and Na–S batteries. Instead, as described, the usual discharge product is $KS_{0.67-0.85}$ [35,36]. Furthermore, the greater reactivity of K than that of Na worsens the anode difficulties of the K–S battery, necessitating careful design of each battery component to minimize unwanted anode reactions in a K–S battery. The same principles that apply to the design of cathode materials for Na–S and Li–S batteries apply to K–S batteries. As a result, sulfur is still present in carbonaceous materials, resulting in a composite cathode. Ordered mesoporous carbon, for example, CMK-3, was utilized as the sulfur host for RT K–S batteries. To prepare, it was covered with conductive polyaniline (PANI) and named as PANI@CMK-3/S cathode. This composite cathode with an organic liquid electrolyte shows two reduction peaks at 2.1 and 1.8 V and oxidation peak at 2.2 V with discharge product K_2S_3 and charge products S and K metal. In another reported work, SPAN cathode was made up by a mixture of sulfur and PAN by heating. SPAN cathode showing that the sulfur species can be covalently bound to PAN in these K–S batteries, and it performs a decent cycling stability of 54% capacity retention after 200 cycles and worthy rate performance were achieved, even though its actual capacity was relatively low because of the slow reaction kinetics related with the large ionic size of potassium [37]. Later, "K metal-free" K–S battery was reported, which is derived from the active material of Li/Na polysulfides, a cell with a potassium polysulfide active material and a hard carbon anode. In this battery, the redox reactions take place in K_2S_3 and K_2S_{5-6} at the cathode, and K ions were reversibly stored in the hard carbon at the anode. And the latest reported work is HT K–S battery, which shows excellent cycling stability with very small capacity decay even for 1000 with the final discharge capacity over 300 mAh g^{-1} cycles, as shown in Figure 19.11. This novel HT K–S battery effectively overcomes the cycling stability issues of the RT K–S batteries and has potential prospects for stationary energy storage [2,38]. Potassium–sulfur battery is at their early stages of development, hence it has a number of obstacles that must be addressed before it can become a reality. Similar problems can be seen in Li–S and Na–S batteries; however, some are unique to K–S batteries. They are all crucial and must be overcome in order to produce a workable K–S battery with a high energy density and long cycling life.

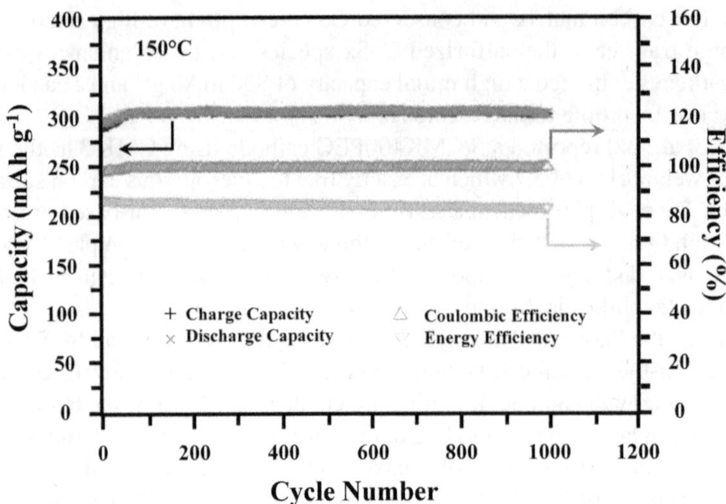

FIGURE 19.11 Discharge/charge capacity, energy efficiency, and Coulombic efficiency of the HT K–S battery. Adapted and reproduced with permission from Ref. [39]. Copyright © 2015, Wiley.

19.6 SODIUM–SULFUR BATTERIES

Sodium–sulfur batteries have received more attention in the research field like lithium–sulfur batteries only because of the high temperature withstanding capacity. It is widely used in large scale because of the low cost. The working temperature is in between 300°C and 350°C with the voltage window in between 1.78 and 2.06 V. Nowadays, room temperature Na–S batteries are more studied due to the higher theoretical specific density of 1274 Wh kg^{-1} compared to HT Na–S batteries.

In 2016, Park et al. proposed a RT Na–S battery with Na-ion conductivity 5.1×10^{-4} S cm^{-1} at 25°C with a gel polymer electrolyte. Since then, a huge progress has been listed in the development of Na–S batteries. Significant efforts have been made to develop new cathode materials to immobilize the soluble polysulfides, novel electrolytes, and functional separators to scavenge the polysulfides, conduct Na ions and/or protect the Na metal surface, and novel anode hosts to regulate stripping/Plating of Na metal to avoid the formation of unwanted Na dendrites and to buffer the variations in Na metal volume during cycling, which has made progress in RT Na–S batteries. For Na–S batteries, sulfur is studied extensively as a cathode material because chalcogenide insulators have a conductivity of only 5×10^{-3} S cm^{-1} at room temperature, resulting in low utilization of sulfur in the electrode. Therefore, conductive materials are essential to coat sulfur particles to increase the conductivity of the cathode material. Carbon in the form of Sulfur–Carbon hollow nanospheres is used as host in the first industrialized cathode material for Na–S battery after which Sulfur–Graphene composite cathode, Sulfur-porous Carbon composite cathode, Sulfur-Organic Polymer cathode, covalent Sulfur-based carbonaceous cathode (CSCM), and metal sulfide cathode are established. Figure 19.12 provides the challenges linked with electrolytes for RT Na–S along with an overview.

19.7 LITHIUM–SULFUR BATTERIES

Li–S batteries, which have a theoretical specific gravimetric capacity of 1675 mAh g^{-1}, are considered one of the most prominent competitors for renewable energy storage. However, the development of the Li–S system is burdened by several problems distributed by cathodes, separators, electrolytes, and anodes. Carbonate electrolytes, which are widely used in LIBs, are not applicable for Li–S batteries. Ether electrolytes have been used extensively in Li–S systems. Polysulfide shuttle effect related to the complicated electrochemical reactions of sulfur cathode is the most difficult

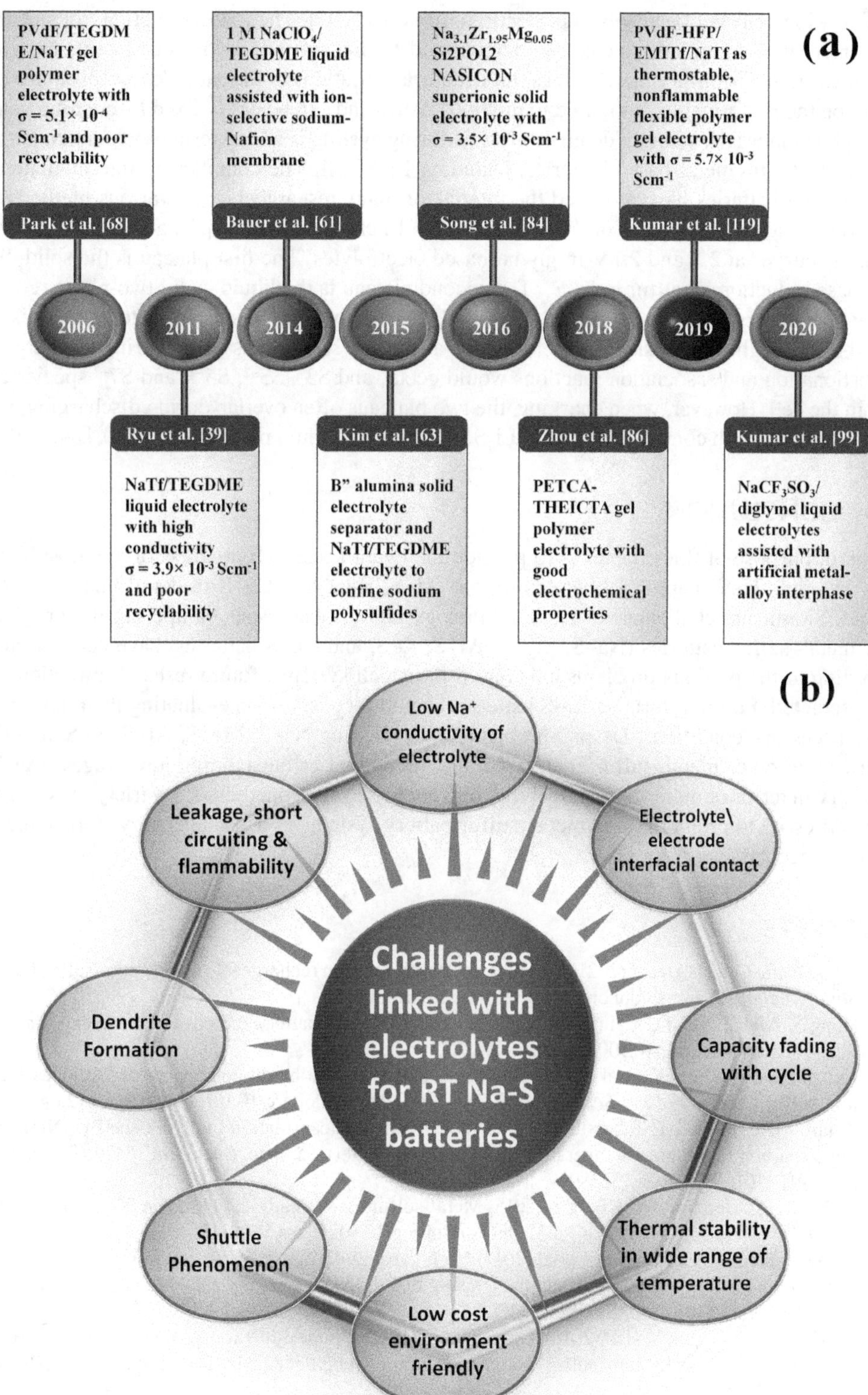

FIGURE 19.12 (a) The challenges linked with electrolytes for RT Na–S batteries' cycling performance. (b) The key electrolyte architectures development/achievements for RT Na–S cell. Adapted and reproduced with permission from Ref. [40]. Copyright © 2020, Elsevier.

problem which has not been solved perfectly until now [41,42]. The low conductivity of sulfur and its counterpart lithium sulfide can be compensated by integrating various conductive nanostructured materials. The minor flaw, such as the structural variations of the cathodes, seems to have less impact on the performance compared to previous issues and can be easily fixed by creating voids in the cathode materials. Lithium dendrite growth during cycling is a long-term problem that hampers the use of lithium metal-based batteries, including Li–S cell. The complicated intermediate reactions of Li–S batteries have attracted the interest of many researchers, and various methods have been tried to study the essence of the mechanism of Li chemistry [43,44]. In general, there are two discharge curves at 2.3 and 2.1 V in glyme-based electrolytes. The first plateau is the solid–liquid two-phase reduction of sulfur to Li_2S_8. The second plateau is the liquid–solid two-phase reduction of low-order polysulfides to solid Li_2S_2 and Li_2S. Usually before the second plateau, there is a dip caused by the highest viscosity of the electrolyte as the polysulfides dissolve. During discharge, disproportionation and association reactions would occur, and $S3^-$, $S3^{-2}$, $S5^{-2}$, and $S7^{-2}$ species could exist in the cell. However, when charging, the two plateaus often overlap during discharging, which is due to the sluggish conversion of solid Li_2S into soluble lithium polysulfides [9,10,41,45,46].

19.8 CONCLUSION

The main purpose of this chapter is to provide the overall research progress of metal–sulfur batteries in brief. This chapter deliberates Mg–S, Al–S, K–S, and Ca–S in detail and Li–S, Na–S in brief. Significant challenges in Li–S technology and critical issues hampering the progress of other metal–sulfur batteries (Na–S, Mg–S, Al–S, K–S, and Ca–S batteries) have been discussed. According to the perilous problems existing in these cell systems, future research directions have been predicted. Future efforts on Li–S batteries should focus more on evaluating them under practically necessary conditions. Despite problems in developing Na–S, Mg–S, Al–S, K–S, and Ca–S batteries, the other metal–sulfur energy storage systems offer outstanding advantages over Li–S chemistry in terms of material sustainability and energy costs. Nonetheless, a variety of issues need to be addressed to promote these metal–sulfur battery systems as a viable energy storage technology with acceptable performance.

REFERENCES

1. Lu Y, Wang C, Liu Q, et al (2021) Progress and perspective on rechargeable magnesium-sulfur batteries. *Small Methods* 5. https://doi.org/10.1002/smtd.202001303.
2. Hong X, Mei J, Wen L, et al (2019) Nonlithium metal-sulfur batteries: Steps toward a leap. *Adv Mater* 31:1–30. https://doi.org/10.1002/adma.201802822.
3. Luo J, Guan K, Lei W, et al (2022) One dimensional carbon-based composites as cathodes for lithium-sulfur battery. *J Mater Sci Technol* 122:101–120. https://doi.org/10.1016/j.jmst.2021.12.048.
4. Aslam MK, Hussain T, Tabassum H, et al (2022) Sulfur encapsulation into yolk-shell Fe_2N@nitrogen doped carbon for ambient-temperature sodium-sulfur battery cathode. *Chem Eng J* 429:132389. https://doi.org/10.1016/j.cej.2021.132389.
5. Wei S, Ma L, Hendrickson KE, et al (2015) Metal-sulfur battery cathodes based on PAN-sulfur composites. *J Am Chem Soc* 137:12143–12152. https://doi.org/10.1021/jacs.5b08113.
6. Wang Y, Sahadeo E, Rubloff G, et al (2019) High-capacity lithium sulfur battery and beyond: A review of metal anode protection layers and perspective of solid-state electrolytes. *J Mater Sci* 54:3671–3693. https://doi.org/10.1007/s10853-018-3093-7.
7. Bao W, Zhang Z, Qua Y, et al (2014) Confine sulfur in mesoporous metal-organic framework @ reduced graphene oxide for lithium sulfur battery. *J Alloys Compd* 582:334–340. https://doi.org/10.1016/j.jallcom.2013.08.056.
8. Mu P, Dong T, Jiang H, et al (2021) Crucial challenges and recent optimization progress of metal-sulfur battery electrolytes. *Energy and Fuels* 35:1966–1988. https://doi.org/10.1021/acs.energyfuels.0c04264.
9. Yu X, Manthiram A (2020) A progress report on metal-sulfur batteries. *Adv Funct Mater* 30:1–27. https://doi.org/10.1002/adfm.202004084.

10. Liu X, Li Y, Xu X, et al (2021) Rechargeable metal (Li, Na, Mg, Al)-sulfur batteries: Materials and advances. *J Energy Chem* 61:104–134. https://doi.org/10.1016/j.jechem.2021.02.028.
11. Ambroz F, Macdonald TJ, Nann T (2017) Trends in aluminium-based intercalation batteries. *Adv Energy Mater* 7. https://doi.org/10.1002/aenm.201602093.
12. Das SK, Mahapatra S, Lahan H (2017) Aluminium-ion batteries: Developments and challenges. *J Mater Chem A* 5:6347–6367. https://doi.org/10.1039/c7ta00228a.
13. Mamantov G, Hvistendahl J (1984) Rechargeable high voltage low temperature molten, salt cell Na/β''-alumina/SCl_3+ in $AlCl_3$-NaCl. *J Electroanal Chem* 168:451–466. https://doi.org/10.1016/0368-1874(84)87115-4.
14. Manan NSA, Aldous L, Alias Y, et al (2011) Electrochemistry of sulfur and polysulfides in ionic liquids. *J Phys Chem B* 115:13873–13879. https://doi.org/10.1021/jp208159v.
15. Elia GA, Marquardt K, Hoeppner K, et al (2016) An overview and future perspectives of aluminum batteries. *Adv Mater* 28:7564–7579. https://doi.org/10.1002/adma.201601357.
16. Dillon R, Licht S, October R (1993) Investigation of a novel aqueous aluminum/sulfur battery. *J Power Sources* 45:311–323.
17. Licht S, Jeitler JR, Hwang JH (1997) Aluminum anodic behavior in aqueous sulfur electrolytes. *J Phys Chem B* 101:4959–4965. https://doi.org/10.1021/jp9700490.
18. Peramunage D, Licht S (2008) A solid sulfur cathode for aqueous batteries. *Science* 261:1029–1032.
19. Chung SH, Manthiram A (2019) Current status and future prospects of metal-sulfur batteries. *Adv Mater* 31:39–42. https://doi.org/10.1002/adma.201901125.
20. Guo Y, Hu Z, Wang J, et al (2020) Rechargeable aluminium-sulfur battery with improved electrochemical performance by cobalt-containing electrocatalyst. *Angew Chemie* 132:23163–23167. https://doi.org/10.1002/ange.202008481.
21. Soni CB, Kumar V (2021) Recent advances in cathode engineering to enable reversible room-temperature aluminium-sulfur batteries. *Nanoscale Adv* 3:1569–1581. https://doi.org/10.1039/d0na01019g.
22. Yang Y, Yang H, Wang X, et al (2022) Multivalent metal-sulfur batteries for green and cost-effective energy storage: Current status and challenges. *J Energy Chem* 64:144–165. https://doi.org/10.1016/j.jechem.2021.04.054.
23. Stievano L, de Meatza I, Bitenc J, et al (2021) Emerging calcium batteries. *J Power Sources* 482:228875. https://doi.org/10.1016/j.jpowsour.2020.228875.
24. Yu X, Boyer MJ, Hwang GS, Manthiram A (2019) Toward a reversible calcium-sulfur battery with a lithium-ion mediation approach. *Adv Energy Mater* 9:1–10. https://doi.org/10.1002/aenm.201803794.
25. Li Z, Vinayan BP, Diemant T, et al (2020) Rechargeable calcium-sulfur batteries enabled by an efficient borate-based electrolyte. *Small* 16:1–6. https://doi.org/10.1002/smll.202001806.
26. Kim HS, Arthur TS, Allred GD, et al (2011) Structure and compatibility of a magnesium electrolyte with a sulphur cathode. *Nat Commun* 2:426–427. https://doi.org/10.1038/ncomms1435.
27. Vinayan BP, Euchner H, Zhao-Karger Z, et al (2019) Insights into the electrochemical processes of rechargeable magnesium-sulfur batteries with a new cathode design. *J Mater Chem A* 7:25490–25502. https://doi.org/10.1039/c9ta09155f.
28. Xu Y, Zhou G, Zhao S, et al (2019) Improving a Mg/S battery with YCl3 additive and magnesium polysulfide. *Adv Sci* 6:6–11. https://doi.org/10.1002/advs.201800981.
29. Wang W, Yuan H, Nuli Y, et al (2018) Sulfur@microporous carbon cathode with a high sulfur content for magnesium-sulfur batteries with nucleophilic electrolytes. *J Phys Chem C* 122:26764–26776. https://doi.org/10.1021/acs.jpcc.8b09003.
30. Zhou X, Tian J, Hu J, Li C (2018) High rate magnesium-sulfur battery with improved cyclability based on metal-organic framework derivative carbon host. *Adv Mater* 30:1–8. https://doi.org/10.1002/adma.201704166.
31. Sun J, Deng C, Bi Y, et al (2020) In-situ sulfurized carbon-confined cobalt for long-life Mg/S. *ACS Appl Energy Mater* 3:2516–2525.
32. Zhao-Karger Z, Zhao X, Wang D, et al (2015) Performance improvement of magnesium sulfur batteries with modified non-nucleophilic electrolytes. *Adv Energy Mater* 5:1–9. https://doi.org/10.1002/aenm.201401155.
33. Du H, Zhang Z, He J, et al (2017) A delicately designed sulfide graphdiyne compatible cathode for high-performance lithium/magnesium-sulfur batteries. *Small* 13:1–9. https://doi.org/10.1002/smll.201702277.
34. Zhao-Karger Z, Fichtner M (2017) Magnesium-sulfur battery: Its beginning and recent progress. *MRS Commun* 7:770–784. https://doi.org/10.1557/mrc.2017.101.

35. Eftekhari A, Jian Z, Ji X (2017) Potassium secondary batteries. *ACS Appl Mater Interfaces* 9:4404–4419. https://doi.org/10.1021/acsami.6b07989.
36. Hwang JY, Kim HM, Yoon CS, Sun YK (2018) Toward high-safety potassium-sulfur batteries using a potassium polysulfide catholyte and metal-free anode. *ACS Energy Lett* 3:540–541. https://doi.org/10.1021/acsenergylett.8b00037.
37. Liu Y, Wang W, Wang J, et al (2018) Sulfur nanocomposite as a positive electrode material for rechargeable potassium-sulfur batteries. *Chem Commun* 54:2288–2291. https://doi.org/10.1039/c7cc09913d.
38. Zhao Q, Hu Y, Zhang K, Chen J (2014) Potassium–sulfur batteries: A new member of room-temperature rechargeable metal–sulfur batteries. *Inorg Chem* 53:9000–9005.
39. Lu X, Bowden ME, Sprenkle VL, Liu J (2015) A low cost, high energy density, and long cycle life potassium-sulfur battery for grid-scale energy storage. *Adv Mater* 27:5915–5922. https://doi.org/10.1002/adma.201502343.
40. Syali MS, Kumar D, Mishra K, Kanchan DK (2020) Recent advances in electrolytes for room-temperature sodium-sulfur batteries: A review. *Energy Storage Mater* 31:352–372. https://doi.org/10.1016/j.ensm.2020.06.023.
41. Zhang Y, Zhang X, Silva SRP, et al (2022) Lithium-sulfur batteries meet electrospinning: recent advances and the key parameters for high gravimetric and volume energy density. *Adv Sci* 9:1–34. https://doi.org/10.1002/advs.202103879.
42. Wang J, Han WQ (2022) A review of heteroatom doped materials for advanced lithium-sulfur batteries. *Adv Funct Mater* 32:1–31. https://doi.org/10.1002/adfm.202107166.
43. Park K, Gil B, Jiwan Yun A, et al (2022) Mixed-valence iron phosphate as an effective catalytic host for the high-rate lithium-sulfur battery. *Chem Eng J* 435:134814. https://doi.org/10.1016/j.cej.2022.134814.
44. Yang W, Huang R, Ni Z, et al (2022) Application and research of current collector for lithium-sulfur battery. *Ionics (Kiel)* 28:1713–1738. https://doi.org/10.1007/s11581-022-04461-2.
45. Kim SJ, Kim K, Park J, Sung YE (2019) Role and potential of metal sulfide catalysts in lithium-sulfur battery applications. *ChemCatChem* 11:2373–2387. https://doi.org/10.1002/cctc.201900184.
46. Zhang Q, Huang Q, Hao SM, et al (2022) Polymers in lithium-sulfur batteries. *Adv Sci* 9:1–30. https://doi.org/10.1002/advs.202103798.

20 Electrolyte for Metal–Sulfur Batteries

Anandu Nair, Devika S. Lal, S. K. Vineeth Akhila Das, Xiaohui Zhao, Vijayamohanan Pillai, Jou-Hyeon Ahn, and Prasanth Raghavan

20.1 INTRODUCTION

The sudden rise of the international economy and infrastructure necessitates the development of urgently sustainable energy systems. The rise of renewable energy technologies such as solar and wind presents a significant challenge in terms of managing their intermittent nature. For our technologically advanced civilization, energy storage is becoming extremely important. In this sense, secondary battery technologies play a critical role [1]. Among numerous available technologies, batteries show a vital role for providing power to the rechargeable world, from portable devices to electric vehicles (Figure 20.1) [2], and, on a bigger scale, in grid-scale energy storage for renewable energy resources like tidal and wind power [3]. In association with power grids, photovoltaic, and wind power generation systems, batteries can be employed as stationary energy storage devices (ESD). Batteries using ESS can considerably benefit mankind in distributed generation systems, whether they reside in huge cities or in remote mountain villages. Electrochemical batteries are a crucial technology that keeps the global civilization flowing efficiently [4]. High-energy Li-ion batteries, in fact, have aided the rapid growth of smart phones, electric vehicles, and commercialized remotely operated aircraft in latest days. They also shrink the size and weight of personal computers. High-energy batteries are transforming not only existing industries, but also making previously inconceivable technologies possible. Battery systems with high energy density, high safety, and extended life are required by the rapid development of high-tech devices. Composite technologies are being discovered, and our basic understanding of their structure is improving. Relationships between composition, property, and performance have aided in the advancement of the profession. Battery technology has evolved over the last half-century as a result of a focused effort on basic solid-state chemistry of materials. Due to its attractive qualities such as great single cell voltage, low memory effect, extended cycle life, and high energy density, the lithium-ion battery is without a question one of the great accomplishments of modern electrochemistry.

Over the last three decades, the spectacular success of lithium-ion (Li^+-ion) batteries has clearly driven Li^+-ion technology to the forefront for large-scale energy storage [5]. The carbonaceous anode and a lithium transition metal-oxide cathode are differentiated by a polyolefine porous separator soaking in a nonaqueous liquid electrolyte in today's lithium-ion batteries. Traditional Li^+-ion batteries, on the other hand, are unable to meet the stringent demands of grid-scale energy storage in terms of energy density and pricing due to a number of inherent constraints [6]. Traditional insertion-type cathodes, for instance, have a storage capacity of <200 mAh g^{-1} [7]. Resources of some elements in the traditional metal-oxide cathodes are less abundant and are geographically concentrated. These concerns usually induce short-term or long-term price fluctuations and raise multifarious ethical and social concerns [8]. Furthermore, completely eliminating the hazardous components in typical metal-oxide cathodes is unattainable [9]. As a result, when considering the wide-scale use of Li+-ion batteries, environmental impacts are unavoidable. As a result, future storage technologies for numerous applications must investigate abundant, high-capacity, environmentally friendly, and

DOI: 10.1201/9781003310167-20

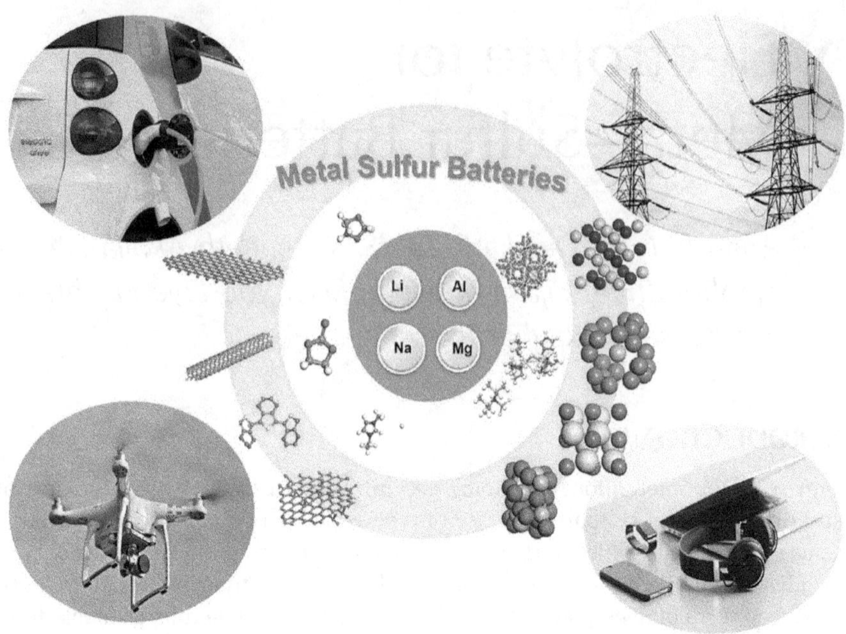

FIGURE 20.1 Electric vehicles, grid energy storage, autonomous drones, consumer electronics, and other possibilities of metal–sulfur batteries. Adapted and reproduced from Ref. [2]. Copyright © 2021 Elsevier.

low-cost materials [10]. The engineering hurdles of this kind of battery system, on the other hand, create an extremely significant reward ratio for pursuing accomplishment in this scenario.

The research interests in Li–air science have gradually decreased over the second decade of this century. Indeed, since the invention of Li^+-ion batteries, researchers have been looking for cathode materials that may deliver higher capacity than standard insertion-type metal oxides. Despite lithium-ion (Li-ion) batteries' enormous success, there is a potential incentive to investigate different electrochemical systems in order to boost energy and power densities, improve security features, and cut costs. As a cathode material for lithium batteries, element sulfur has a high specific capacity of 1672 mAh g^{-1} and is common, cheap, and non-toxic. Metal–sulfur batteries (MSBs) with sulfur cathodes and lightweight metal anodes are gaining popularity in this area [11]. Sulfur has a large theoretical capacity (1675 mAh g^{-1}), cheap cost, and abundant [12–14]. Li–S batteries are considered as one of most projecting players for renewable energy storage with charge–discharge curves shown in Figure 20. 2 [15]. Since sulfur is electrochemically active and can take up to two electrons per atom at 2.1 V vs Li/Li^+, Li–S batteries have recently caught the interest of numerous researchers. Furthermore, there are concerns about the lack of availability of metal supplies for Li-ion cathodes, like Co and Ni, in order to satisfy the massive requirements arising from automobile electrification and renewable energy adoption. Li, Na, K, Mg, and Al constitute lightweight anodes having lower potentials and large capacities. Sulfur is an excellent cathode material for the generation of high energy density, relatively inexpensive, and ecofriendly battery systems, as evidenced by the above characteristics. However, while cycling, sulfur comprises a sequence of chemical and structural transformations which include soluble polysulfides and insoluble sulfides. However, fully releasing the theoretical capacities of MSBs and ensuring consistent cycling remain practical constraints.

Bulk sulfur (S_8) insulation is indeed a priority to address since it will obstruct electron transfer (5.0×10^{-30} S cm^{-1}), resulting in slower sulfur redox reactions and low sulfur use [16]. The S–S bond would be disrupted, forming long-chain lithium polysulfide (LiPS) molecules if the S_8 molecules were lucky enough to acquire electrons (Li_2S_n, n = 8, 7, 6, 5, 4) [17].

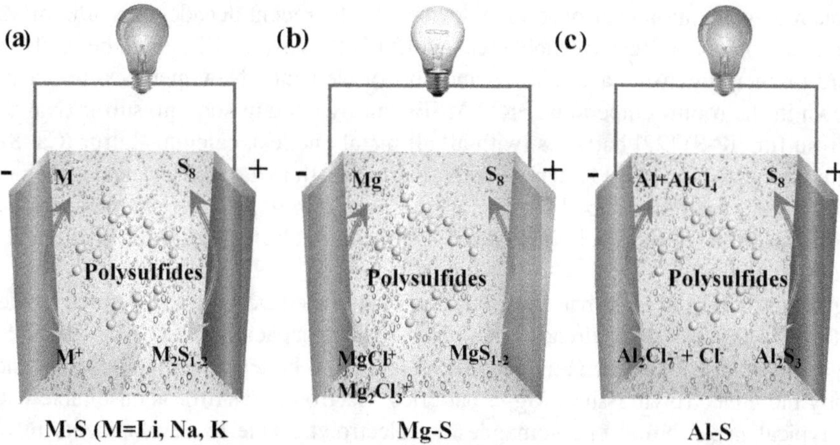

FIGURE 20.2 A conventional Li–S battery's galvanostatic charge–discharge patterns with the associated products at each voltage stage. Adapted and reproduced from Ref. [15]. Copyright © 2021 Elsevier.

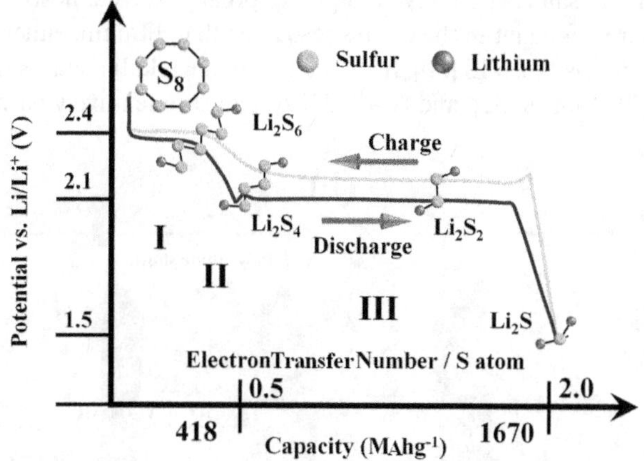

FIGURE 20.3 Schematic representation of (a) M–S (M = Li, Na, K), (b) Mg–S, and (c) Al–S batteries with polysulfides consist in the electrolyte. Adapted and reproduced from Ref. [20]. Copyright © 2020 Wiley.

This one as LiPSs are soluble in typical ether electrolyte and can diffuse to the lithium (Li) anode to create solid Li_2S_2/Li_2S precipitations (as known as the "shuttle effect"), resulting in continuous S consumption and low Coulombic efficiency.

Furthermore, in this case, the mass density differences between S (2.07 g cm^{-3}) and Li_2S (1.66 g cm^{-3}) create substantial volume shifts of up to 80%, compromising the rationally intended arrangements of cathode [18]. Last but not least, even during repetitive charge/discharge process, lithium ions (Li^+) are likely to deposit irregularly on the surface of the Li anode, resulting in the production of Li dendrites and raising possible security problems [19]. Since their beginnings, the aforementioned four elements have been the main bottlenecks of Li–S batteries (Figure 20.3) [20]. As a result, the search for alternative (non-lithium) anode materials based on abundant earth elements has begun to gain impetus in the development of next-generation sulfur-based batteries.

20.2 OVERLOOK ON DIFFERENT METAL–SULFUR BATTERIES

Because of their higher specific capacities, massive energy densities, high abundance, and relatively low cost of the electroactive materials, MSBs with sulfur cathodes and lightweight metal anodes are possible next-generation energy storage devices. In the recent decade, non-lithium MSBs have made significant progress. Rechargeable metal–sulfur batteries (RMSB) could be really interesting with regard to energy density, availability, and cost of elements. New members (other than Li–S) make known in the room-temperature (RT) MSBs family contain sodium–sulfur (Na–S) [21] and potassium–sulfur (K–S) [22] batteries (with alkali metal anodes), calcium–sulfur (Ca–S) [23] and magnesium–sulfur (Mg–S) [24] batteries (with alkaline earth metal anodes), and aluminum–sulfur (Al–S) [25] battery (with a group-III-met anode). Figure 20.4 depicts the characteristic prototypes of four different MSBs. In the field of electrolytes, the core electrolytes utilized in these four systems are also explained [2].

Owing to the success of Li–S batteries, RT Na–S batteries have grown in popularity; dependable cycling of over 1000 cycles has already been reported. The capacity of 100-cycle RT K–S batteries exceeding 1000 mAh g^{-1} has been attained. Mg–S and Al–S batteries with multivalent anodes have significantly more electrolyte issues. Mg–S batteries' electrophilic sulfur is unsuitable with Mg-ion batteries' typical nucleophilic organomagnesium electrolyte. Due to developments in electrolyte technology (Figure 20.5), high-capacity 100-cycle Mg–S and Al–S batteries have been recorded [26]. Alkali metal–sulfur (Li–S, Na–S, and K–S) systems have very similar chemistry; electrochemical conversion reactions in these cells are opened with liquid electrolytes including either a Li-salt [27], a Na-salt [28], or a K-salt [29] dissolved in proper organic solvent hosts. Although sodium and potassium are even more abundant in the earth's resources than lithium, either Na or K is extremely active and offers a greater risk than Li [30]. In distinction to the alkali metal–sulfur (Li–S, Na–S, and K–S) systems, Ca–S [31], Mg–S [32], and Al–S [33] systems are relatively harmless.

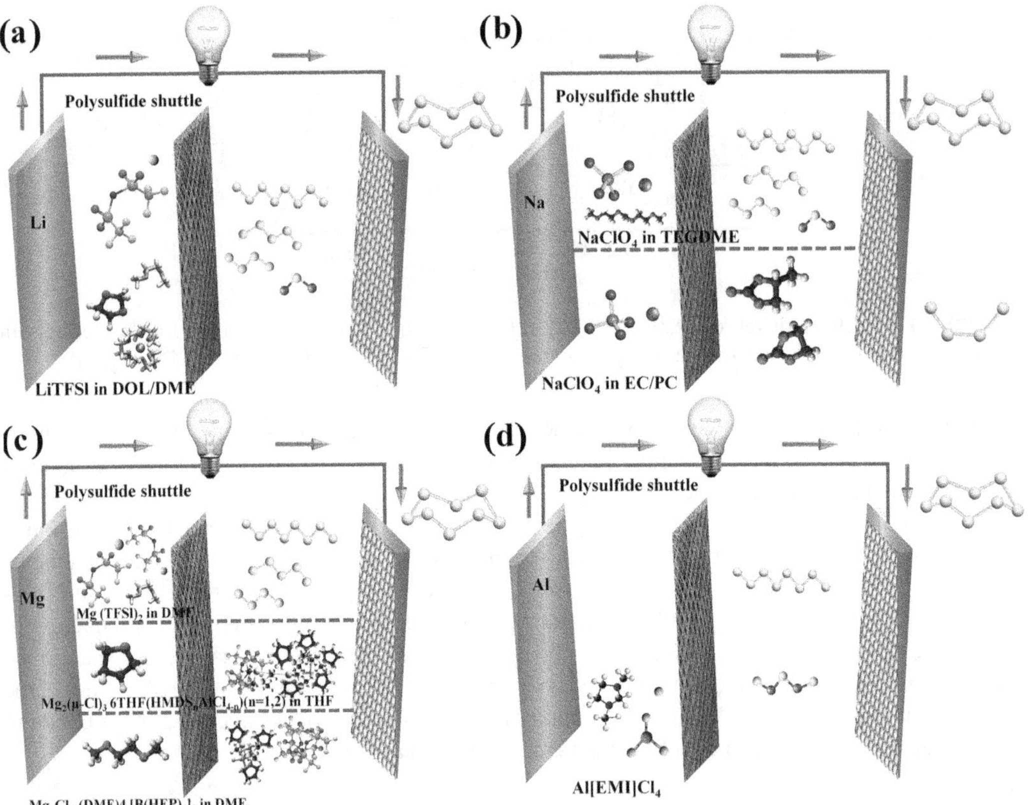

FIGURE 20.4 Multiple kinds of rechargeable metal–sulfur batteries: (a) Li–S batteries, (b) RT Na–S batteries, (c) Mg–S batteries, and (d) Al–S batteries. A metal anode, a sulfur cathode, an electrolyte, and a separator are all used in these devices. Adapted and reproduced from Ref. [2]. Copyright © 2021 Elsevier.

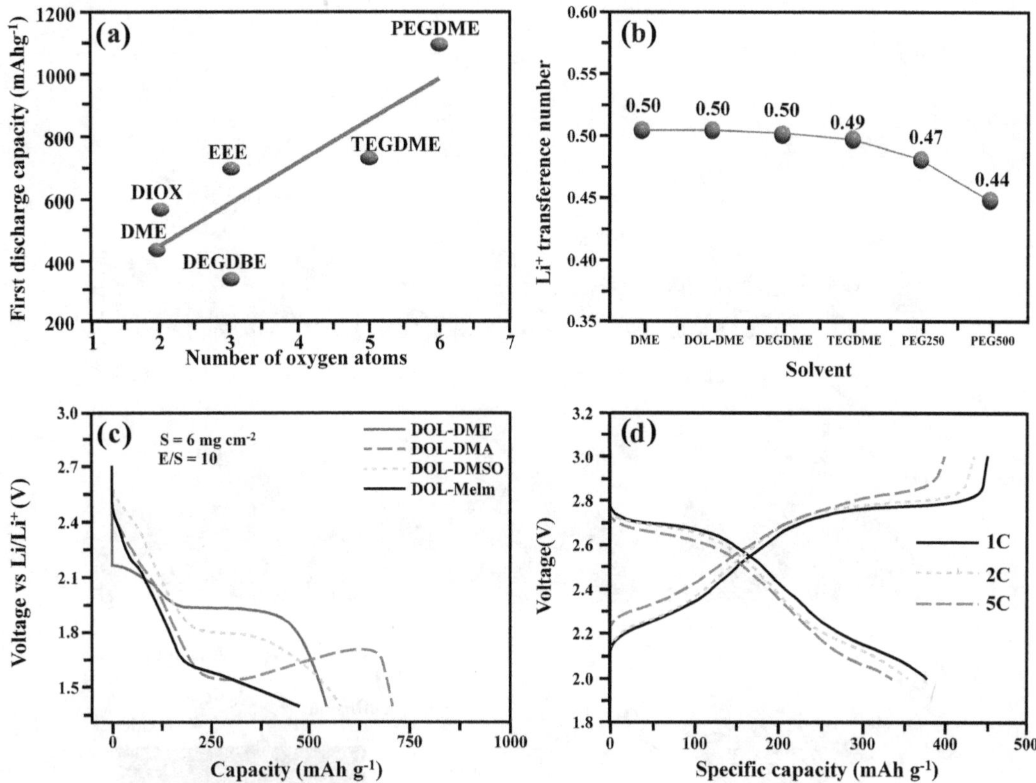

FIGURE 20.5 Electrolytes used in the development of RT Na–S batteries, including their properties, benefits and drawbacks, present status, and future prospects. Adapted and reproduced from Ref. [26]. Copyright © 2020 Wiley.

As a result, for the function of high-energy MSBs, Na, Mg, Al, K, and Ca are preferable possibilities to Li in terms of elemental abundance (sustainability), material cost (economical consideration), and volumetric capacity (electrochemical performance). Ca, Mg, and Al anodes have the added benefit of being operationally safe. Despite the outstanding research in Li–S batteries over the last decade, which might be partially translated to other MSBs, research and development in Na–S, Mg–S, Al–S, K–S, and Ca–S batteries confront even greater difficulties (Figure 20.6) [34].

Metal oxides exhibiting specialized catalytic activities, such as MnO_2, can be effective oxidizers of polysulfides [35,36]. Metal oxides typically usually found in metal ions and O^{2-} ions that have a certain crystal structure. Magnéli phase oxides, which already have oxygen vacancies, have a high conductivity [37]. Sulfides have a stronger sulfiphilic property to polysulfides and lower lithiation voltages than metal oxides. Because metal sulfides are generally found in metallic or half-metallic phases, they have a high electrical conductivity. MnO_2 accepted a position in polysulfides mediator for Li–S batteries as an effective electrocatalyst.

According to Liang et al. [37], d-MnO_2 can serve as a catalyst, oxidizing polysulfides to generate thiosulfate and assisting in the conversion of long-chain polysulfides to short-chain polysulfides and lithium sulfide as shown in Figure 20.7 [38]. In situ produced core–shell hollow sulfur-MnO_2 was the subject of a successional study [39]. It established a sign of success for making a bifunctional d-MnO_2–sulfur composite (physical protection and chemical adsorption, and catalysis). Lou and co-authors built a sulfur-infiltrated hollow carbon nanofiber with embedded d-MnO_2 nanosheets composite in this vein [40]. At 0.5 C, it could still deliver an initial specific capacity of over 900 mAh g^{-1} and sustain 662 mAh g^{-1} after 300 cycles with a 71% sulfur loading and areal sulfur

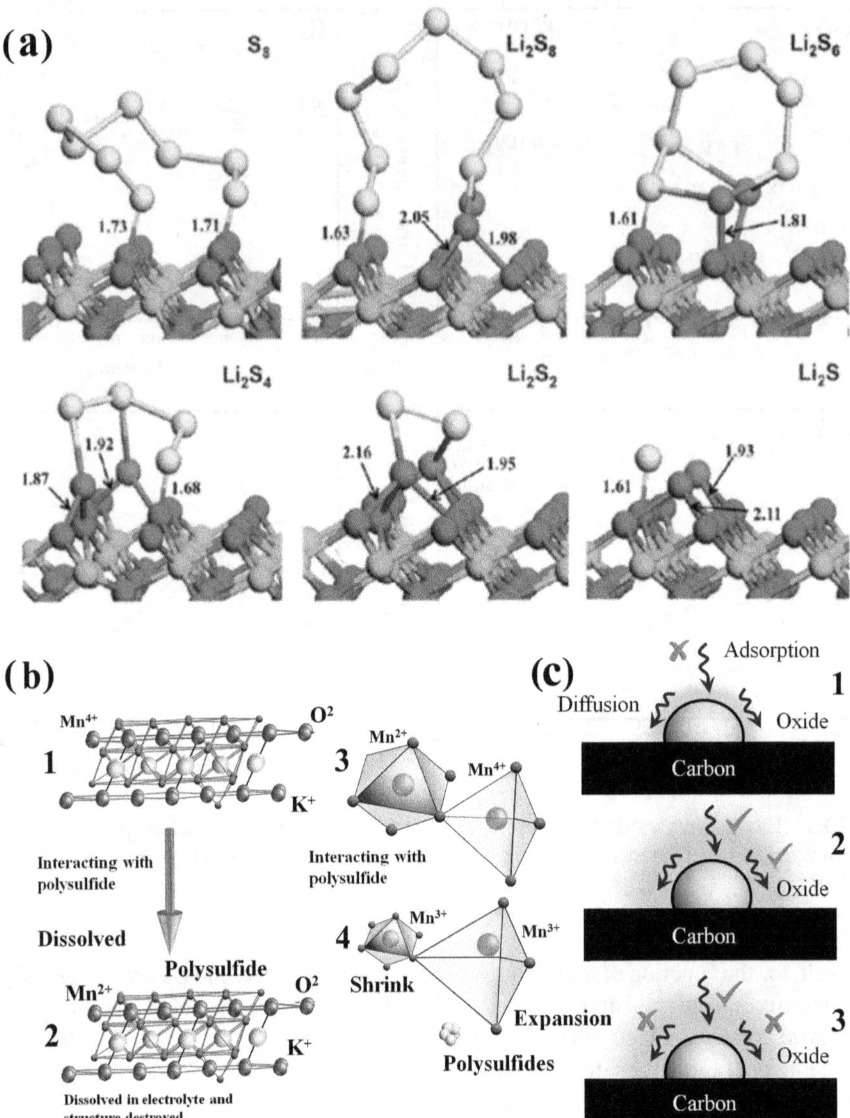

FIGURE 20.6 (a) Amount of oxygen atoms in various electrolyte solvents vs first discharge capacity. (b) At ambient temperature, the lithium transference number of ether-based electrolytes. (c) The first cycle discharge profiles of a sulfur cathode with a 2 mg cm^{-2} sulfur cathode were produced in four distinct electrolytes, with the solvents of interest employed with 50% DOL and 1 M LiTFSI. (d) Charge/discharge patterns of lithium polysulfide cells cycled with sulfur/long-chain polysulfide redox pair using N-Methyl-2-Pyrrolidone (NMP) solvent. The cell was run at various C-rates. Adapted and reproduced from Ref [34]. Copyright © 2021 Wiley.

level of 3.5 mg cm^{-2}. The coaxial PPy-MnO_2 nanotubes were designed for sulfur composite cathodes employing MnO_2 nanowires as the template and oxidation initiator. Carbon nanoboxes packed with d-MnO_2 were recently reported by Hou and co-workers [41] to empower Li–S batteries. DFT simulations demonstrated that MnO_2 may bond with S_8, Li_2S_8, Li_2S_6, and Li_2S_4 with energies of 1.60, 4.68, 3.86, and 5.15 eV, respectively, using the (100) crystal surface d-MnO_2 as an attaching support [38]. Li_2S could disintegrate into S=O and Li=O bonds due to the high interactions between MnO_2 and Li_2S (around 5.15 eV). An ultrathin interlayer composed of MnO_2/graphene oxide/carbon nanotube has been shown to be effective in capturing polysulfides in addition to being employed as

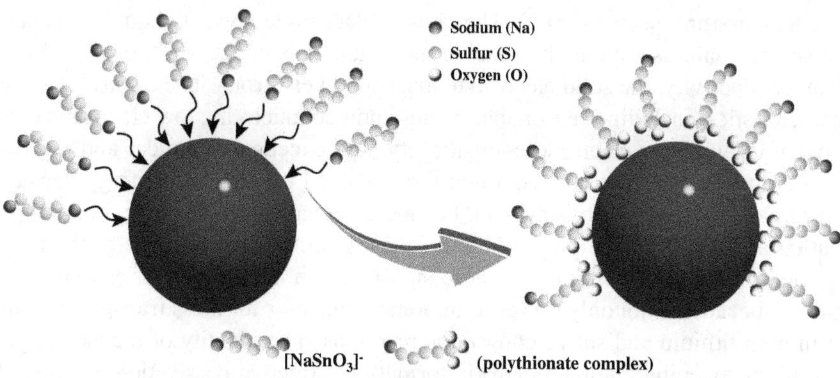

FIGURE 20.7 Schematic representation of the Mn_xO_y adsorption and possible redox reaction. Adapted and reproduced from Ref. [38]. Copyright © 2021 Wiley.

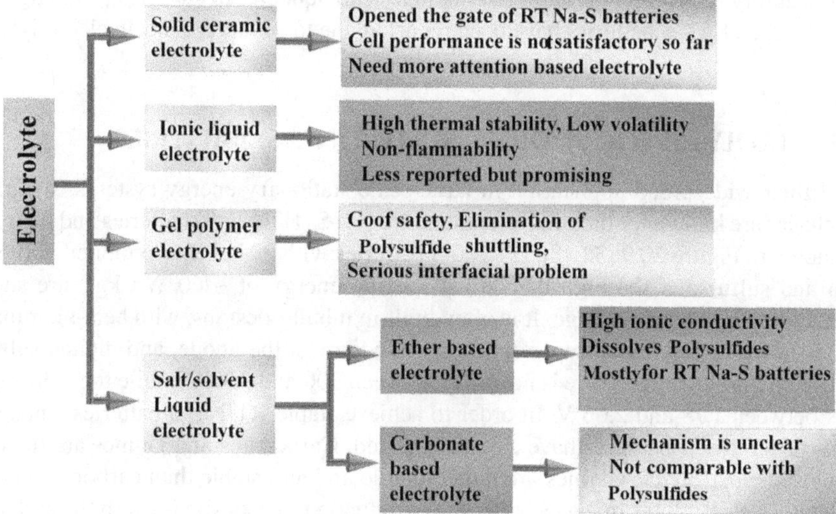

FIGURE 20.8 Classification of electrolytes in lithium–sulfur battery.

composite cathode host materials [42]. Furthermore, new studies [43] found that Mn_3O_4 was more stable than MnO_2 in Li–S batteries. When MnO_2 interacts with polysulfides, Mn^{4+} is reduced to Mn^{2+}, which then dissolves into the electrolyte, as seen in Figure 20.8. The structural equilibrium related to the shrink of MnO_6 tetrahedra and the expansion of MnO_4 tetrahedra keeps the crystal structure of Mn_3O_4 constant, even when Mn^{4+} is reduced to Mn^{3+} and Mn^{2+} is reduced to Mn^{3+} in Mn_3O_4. Overall, catalytic oxidation of polysulfides and appropriate binding of polysulfides by manganese oxides can help achieve high-performance Li–S batteries.

The electrolyte is a critical technology for building MSBs with high capacity and long cycle life. The following considerations should be made while creating an electrolyte for a specific metal–sulfur system. The electrolyte's primary function is to transport the metal ion and insulating electrons. The basic considerations are these two functions. The electrolyte, whether in gel, liquid, or solid form, must be capable of transporting the appropriate metal ion as well as insulating electrons. The conductivity of an electrolyte must be at least 5~15 mS/cm for commercial Li-ion batteries. The usual $LiPF_6$ concentration in the carbonate solvent to achieve this criterion is roughly 1 M. The minimum criterion for conductivity in modern solid-state lithium batteries is 0.1 mS cm^{-1}

at operating temperature, such as 60°C. The chosen electrolyte solvent should be capable of dissolving the specific salt, as well as the specific metal ion of Na^+, K^+, Mg^{2+}, or Al^{3+}. In addition to regulated ion conductivity, the solid electrolyte must shield electrons. The electrolytes must satisfy certain characteristics, including reasonable strong ionic conductivity, low electronic conductivity, sufficient potential window, electrode compatibility, and excellent chemical and electrochemical stability. More importantly, it is now commonly recognized that for high energy density, limiting the amount of electrolyte solution is critical. For metal–S batteries, choosing an appropriate electrolyte solution and solvating the polysulfide moieties within is arguably one of the most difficult tasks. The composition of the solution is more significant in S-based batteries than in traditional Li-ion batteries, because it not only serves as an ionic conductor for mass transport but also plays a major part in both lithium and sulfur conversion reactions. The identity of the electrolyte solution influences lithium stripping/plating, dendritic formation, effective passivation by solid electrolyte interphase (SEI), and interference by side reactions. Polysulfide solvation is an important factor in determining the sulfur reaction route. The electrolyte problems for Mg–S and Al–S batteries with multivalent anodes are substantially bigger [44,45]. The electrophilic sulfur in Mg–S batteries is incompatible with the traditional nucleophilic organomagnesium electrolyte used in Mg-ion batteries. Rechargeability of Al–S batteries is not possible with aqueous electrolytes [46]. High-capacity 100-cycle Mg–S and Al–S batteries have been reported due to advancements in electrolyte technology [47].

20.3 ELECTROLYTES FOR SODIUM–SULFUR (NA–S) BATTERIES

Because of their widespread application in large-scale stationary energy systems, sodium–sulfur (Na–S) batteries are known for their high-temperature Na–S (HT Na–S) batteries and their advancements as shown in Figure 20.9 [38,48]. HT Na–S batteries with its benefits exhibited in the low cost of sodium and sulfur and the high theoretical specific energy of ~760 Wh kg^{-1} are suitable for large-scale stationary energy storage. It is often built-in tubular designs, with beta-alumina serving as the solid-state electrolyte and separator, molten sodium as the anode, and molten sulfur as the cathode. The operating temperature is normally between 300 and 350°C, while the working voltage window is between 1.78 and 2.06 V. In order to achieve stable RT Na–S batteries, changes to the electrolytes of RT Na–S batteries have also been tested. Carbonates and glymes are the most normally employed electrolytes. Glymes are more volatile and less stable than carbonate electrolytes. When utilized as electrolytes in Na–S batteries, the situation differs significantly from that of Li–S batteries. Archer and colleagues [49] used an ionic liquid 1-methyl-3-propylimidazolium-chlorate tethered silica nanoparticle (SiO_2-IL-ClO_4) additive in a liquid carbonate electrolyte (1 M $NaClO_4$ in a mixture of ethylene carbonate (EC) and propylene carbonate (PC) as a stabilizing agent for RT Na–S batteries. Figure 20.10a shows the Na–S cell. It employs the previously mentioned electrolyte, and the SEM image on the left depicts the surface of cycled sodium metal. They discovered that sulfur currently undergoing a solid-state electrochemical reaction when using ZIF-8 derived microporous carbon polyhedron–sulfur composites as the composite cathode for RT Na–S batteries, and that this type of featured carbonate-based electrolyte was compatible with the microporous carbon–sulfur composite cathode. For 50 cycles at 0.1 C rate, assembled batteries with this electrolyte (with 5 vol % SiO_2-IL-ClO_4) and the composite cathode exhibited a reversible capacity of more than 860 mAh g^{-1}. The discharge capacity of over 600 mAh g^{-1} for 100 cycles was attained at a higher rate, 0.5 C, in the potential range of 0.6–2.6 V vs Na^+/Na (Figure 20.10b). The clear variation in sulfur electrochemical reaction processes between microporous carbons and other carbons is critical for researchers to study further if significant progress is to be achieved in this area.

Because of the changes in pore diameters of carbon materials, sulfur can exist in distinct allotropes (for example, S_4, S_3, and S_2). As a result, the processes will differ based on the porous carbon–sulfur materials used. Carbonate electrolytes have been discovered to be reactive to polysulfides in Li–S batteries; however, because the reactivity of Na^+–polysulfide ion combinations is lower in

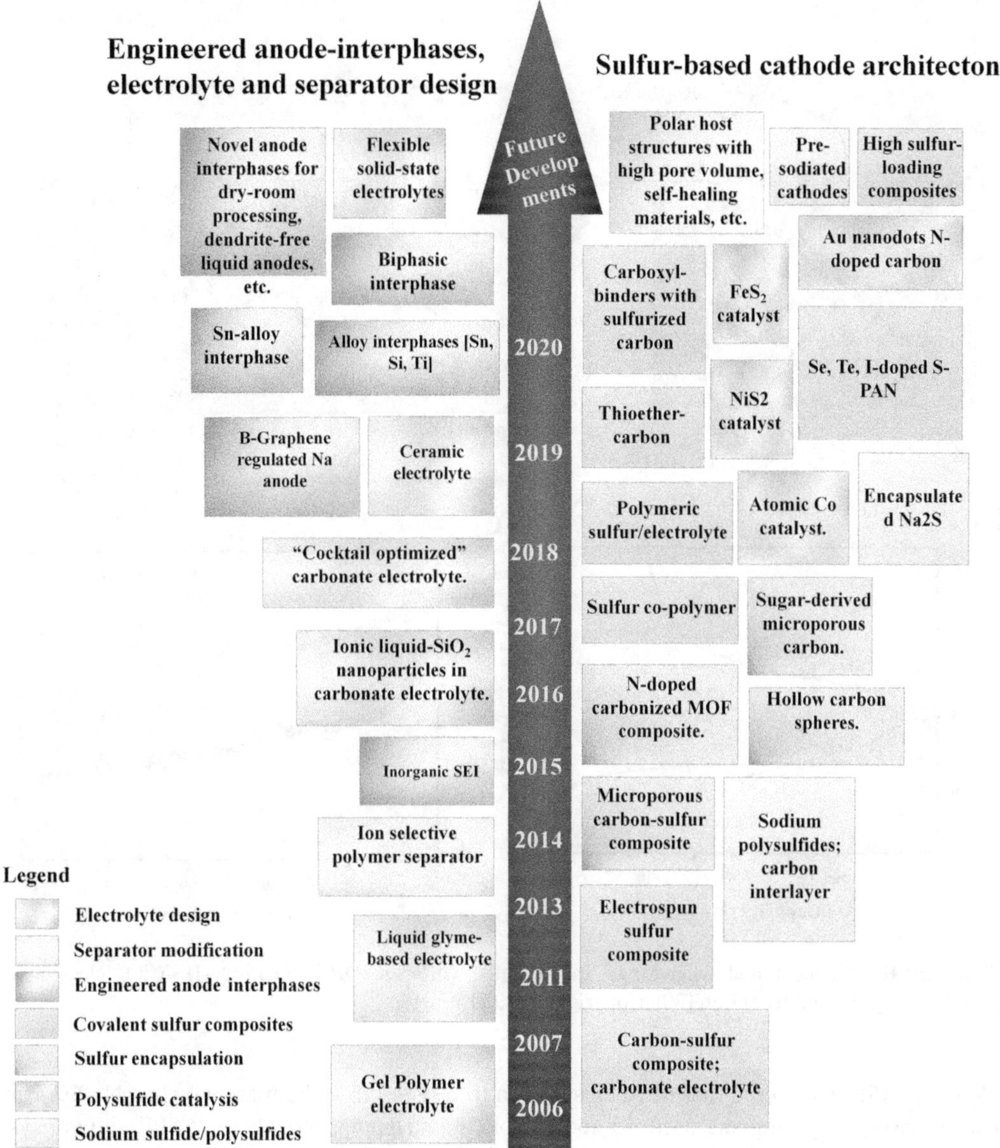

FIGURE 20.9 From 2006 to the present, an outline of room-temperature sodium–sulfur battery research is presented, covering significant advances in anode interphases, electrolyte design, and sulfur composites. Adapted and reproduced from Ref. [38]. Copyright © 2021 Wiley.

Na–S batteries, they can be used. The Wang research group [50] developed a novel electrolyte for RT Na–S batteries consisting of 2 M sodium trifluoromethane sulfonimide (NaTFSI) in polycarbonate (PC)/fluoroethylene carbonate (FEC) (v:v = 1) with InI_3 as an addition. The impedance measurement of Na–S cells utilizing two types of electrolytes is shown in Figure 20.11. Lower charge transfer resistance and higher ionic conductivity were observed in the cell enabled by 2 M NaTFSI in PC/FEC with InI_3. Based on the studies above, carbonate electrolyte appears to be a good fit for RT Na–S batteries. However, a recent investigation on the effect of the electrolyte on the reversibility of the sodium metal anode found that an electrolyte containing sodium hexafluorophosphate in glymes was better for RT Na–S batteries.

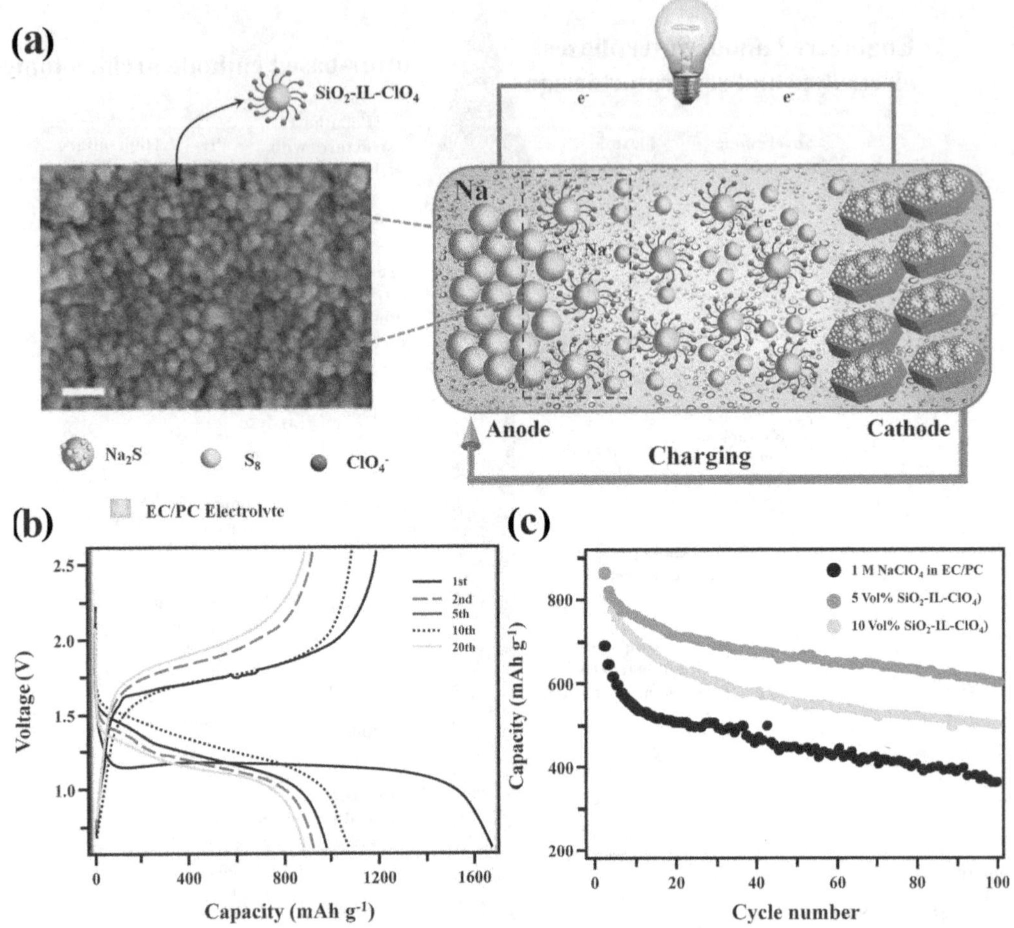

FIGURE 20.10 Concept and behavior of Na–S cells with $NaClO_4$/EC/PC/SiO_2–IL–ClO_4 electrolyte. Adapted and reproduced from Ref. [20]. Copyright © 2020 Wiley.

Seh et al. [51] discovered that there existed dendrite issues in sodium anode when $NaPF_6$–carbonate electrolyte system was used. In sodium plating stripping, though the $NaPF_6$–glyme system achieved an average Coulombic efficiency of 99.9% over 300 cycles. It was connected to the growth of inorganic SEI (Na_2O and NaF) that was impermeable to liquid electrolyte. Furthermore, sodium dendrites were efficiently inhibited by the high shear moduli of Na_2O and NaF. Carbonate electrolyte was weaker to glymes because it was quicker to break down due to its larger reduction potentials.

Batteries built up of Na anode, $NaPF_6$ in tetraglyme, and sulfur-infiltrated carbon nanofibers composite cathode were evaluated at 0.1 C from 1.2 to 2.8 V vs Na^+/Na to illustrate its application in RT Na–S batteries (Figure 20.11). It had a high specific capacity of 776 mAh g^{-1} after 20 cycles and retained 500 mAh g^{-1}. This discovery could spark new research into high-energy-density RT Na–S batteries and other sodium-based energy storage systems. Although in situ generated SEI was successful in inhibiting dendrite growth, its mechanical strength was unable to withstand the substantial volume change experienced during sodium plating/stripping. ALD was utilized to deposit an Al_2O_3 thin film on a Na metal anode; for the same reason, it was used to preserve lithium metal anodes. A low-temperature plasma-enhanced ALD approach was used by Hu and colleagues [52] to build an artificial SEI on a Na metal anode. As predicted, symmetric cells with an Al_2O_3 coating on

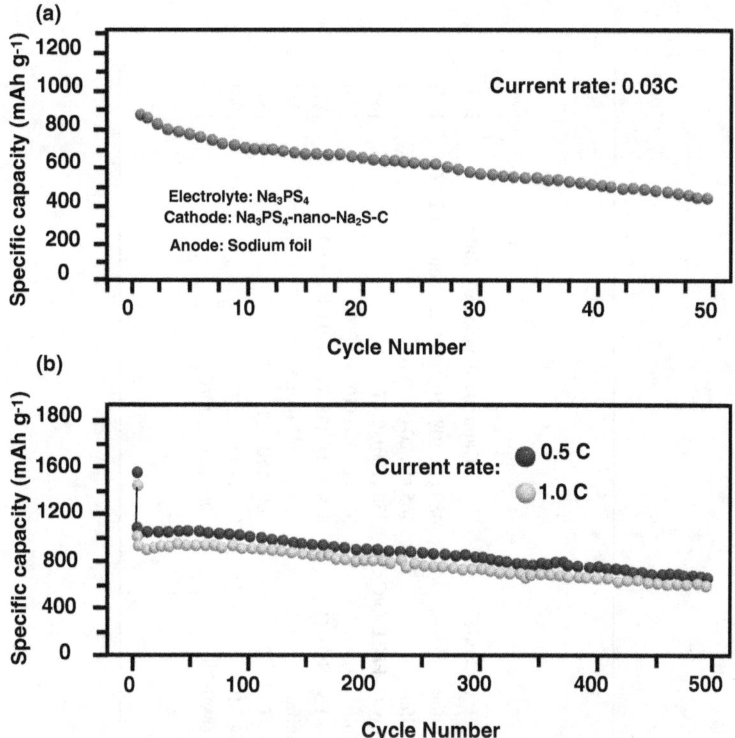

FIGURE 20.11 Electrochemical performances of a solid-state Na–S battery with electrolyte of (a) Na_3PS_4 and cathode of Na_3PS_4-nano-Na_2S-C and (b) NaTFSI/PC/FEC/InI_3. Adapted and reproduced from Ref. [20]. Copyright © 2020 Wiley.

the sodium metal could be cycled over 450 times without losing polarization, whereas the naked cell exhibited a higher polarization all the time. The morphological analysis revealed that symmetric cells covered with Al_2O_3 had a steady impedance, whereas the impedance of bare cells varied with cycle time. This meant that on the Na metal anode, a persistent artificial SEI had emerged. Sun and co-workers [53] used Rutherford backscattering spectrometry (RBS) measurements to find no cavities in the Al_2O_3 layer, confirming the Al_2O_3 layer's uniform and fine development.

A collection of recent developments in RT Na–S batteries, focusing on several electrolyte-based prototypes, was also created. Table 20.1 summarizes some significant advancements. Microporous carbon–sulfur composites were shown to be potential cathodes for RT Na–S batteries, which could contribute to lower large-scale stationary load-leveling energy storage. Electrolytes derived on carbonates are an excellent choice for RT Na–S batteries when linked with microporous carbon–sulfur composite cathodes. RT Na–S batteries are not necessarily analogous to Li–S batteries due to their very distinct application. We can infer that more research into the reaction mechanisms of RT Na–S batteries is required in diverse settings (various electrolytes). In contrast to other carbon hosts, sulfur may undertake a solid-state reaction in microporous carbon. In terms of glymes- and carbonates-based electrolytes, it was too early to declare which ones are really not suited for RT Na–S batteries.

Figure 20.12 presents the across-the-board mechanism of electrochemical reactions during discharging a Na–S cell. All solid-state, sodium–sulfur batteries have a lot of promise for future RT Na–S battery development. It uses Na metal anodes, which are considered the holy grail of Na-ion batteries. Developments on all-solid-state Na–S batteries centered on Na_3SbS_4 [54] and Na_3PS_4 [55] open the way for the development of solid-state Na–S batteries. The research of solid-state electrolytes for Na–S batteries is still in its early stages. Incorporating Ge or Sn into Li–S batteries could

TABLE 20.1
Electrolytes for Na–S Batteries Reported in the Literature

Electrolyte		Conductivity ($mS\ cm^{-1}$)	Cathode	Capacity (Cycle no), Rate	Ref
Gel electrolyte	PVDF-$NaCF_3SO_3$-G4	0.51 at 25°C	Sulfur	489 (1), 108 (10), 0.144 $mA\ cm^{-2}$ at 25°C	[83]
Solid electrolyte	PVDF-HFP-SiO_2-$NaCF_3SO_3$-EC-PC	4.1	Sulfur	165 (1), 20 (8), 0.1C	[84]
Liquid electrolyte-ether	THEICTA-PETE-NaTFSI	3.85 at 25°C	Poly (S-PETEA)/C	736 (100), 0.1 C	[85]
	PEO-$NaCF_3SO_3$	0.34 at 90°C	Sulfur	505 (1), 166 (10) at 90°C	[86]
	Na_3PS_4	No data	Na_3PS_4-Na_2S-C	869.2 (1), 438.4 (50), 0.03C at 60°C	[87]
	G4-$NaCF_3SO_3$	3.9 at 25°C	Sulfur	538 (1), 0.144 $mA\ cm^{-2}$	[88]
	G4-$NaPF_6$	No data	S@C	300 (1500), 1C	[89]
Liquid electrolyte	EC-PC-SiO_2-IL-ClO_4	0.6	S@MCP	600 (100), 0.5 C	[90]
	PC-FEC-NaTFSI-InI_3	1.95 at 25°C	S@MPCF	1000 (150), 0.1 C	[91]

Source: Adapted and reproduced from Ref. 20. Copyright © 2020 Wiley.

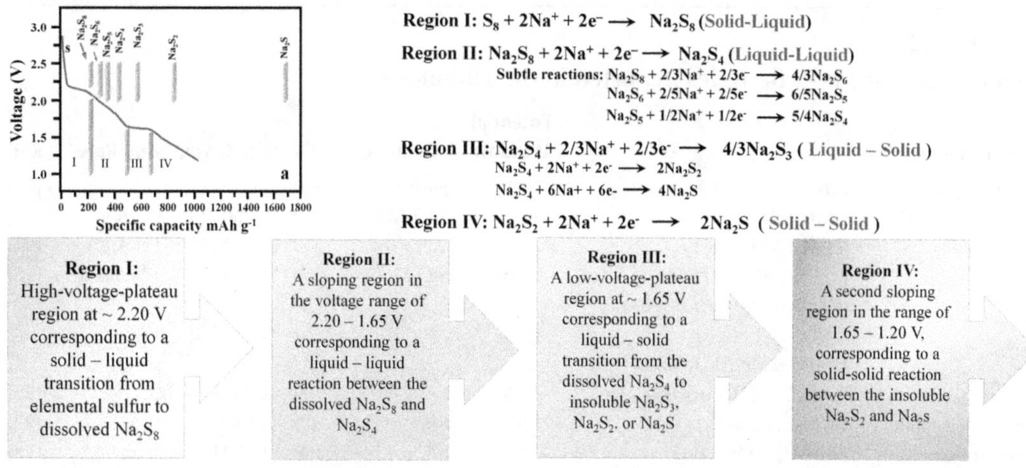

FIGURE 20.12 Discharging an RT Na–S battery using an all-encompassing process. Adapted and reproduced from Ref. [26]. Copyright © 2020 Wiley.

have been beneficial. Furthermore, the question of whether to utilize either electrolytes or carbonates electrolytes is still a hot topic. It is expected that academic and industry growth will continue in this field.

20.4 ELECTROLYTES FOR Mg–S BATTERIES

The lack of suitable electrolytes has delayed the development of multivalent metal–sulfur batteries (Mg–S and Al–S batteries) [56–70]. Electrolytes created for Mg–S batteries can now be categorized into three types: organic non-nucleophilic magnesium electrolytes, magnesium borate-based electrolytes, and $Mg(TFSI)_2$-$MgCl_2$ ether electrolytes. Nucleophilic magnesium organo-halo aluminate electrolyte has earlier been shown to react with electrophilic sulfur cathodes. The development of a non-nucleophilic electrolyte for Mg–S batteries is critical. Kim et al. [57] used an in situ reaction of Lewis acid $AlCl_3$ and hexamethyldisilazide magnesium chloride (HMDSMgCl) in THF to create a $(Mg_2(l\text{-}Cl)_{36}THF)(HMDSnAlCl_{4n})$ (n = 1, 2) crystal for the first rechargeable Mg–S batteries (tetrahydrofuran). Significant progress has been achieved in understanding the mechanism of Mg–S batteries owing to the use of this electrolyte. Zhao et al. [58] looked into the Mg-HMDS discharge mechanism of electrolytes-based Mg–S batteries. The Mg–S cell reached its first upper flat voltage plateau about 1.6 V.

All of the observed Mg-based battery systems are ether-based electrolyte solutions with a complicated structure (Table 20.2). Electrolyte solutions can be divided into three categories: those containing organometallic complexes, those having a traditional Mg salt, and those containing boron-based electrolytes. With low overpotentials and good Coulombic efficiency, all of these electrolyte solutions permit reversible magnesium deposition/dissolution processes. Because ethereal solvents are innately stable in contact with Mg metal, all of the electrolyte solutions discussed above rely on a passivation-free condition. Magnesium deposition is not dendritic, unlike that of monovalent metal anodes. This fact, combined with the availability of several types of passivation-free (thus reversible) Mg anodes, led the community to believe that using multivalent metal anodes in high-capacity rechargeable batteries would be relatively simple compared to using intercalation or conversion reaction cathodes.

Chen and colleagues [59] created a well-defined boron centered anion-based magnesium electrolyte (BCM electrolyte) and used it in Mg–S batteries to achieve high-performance Mg–S batteries. The cell showed a first discharge capacity of 1081 mAh g^{-1} and a retention of 86.4% after 30

TABLE 20.2
Electrolytes for Mg–S Batteries Reported in the Literature

Electrolyte		Potential Window	Cathode	Capacity (Cycle No.), Rate	Ref.
HMDS-based	$[Mg_2(\mu\text{-}Cl)_3\text{-}6THF]$	3.2 V	Sulfur	1200 (1), 394 (2)	[92]
non-nucleophilic	$[HMDSAlCl_3]$	3 V	S/CMK400	800 (1), 260 (20), 20 mA g^{-1}	[93]
electrolyte	$(HMDS)_2Mg/AlCl_3/$	2.7 V	Sulfur/ACC	300 (1), 900 (30), 71 mA g^{-1}	[94]
	$MgCl_2/G4/PP14TFSI$	No data	S/C-MOF	850 (1), 450 (250), 176 mA g^{-1}	[95]
	$(HMDS)_2Mg/AlCl_3/MgCl_2/$				
	G4/LiTFSI				
	$HMDS)_2$ $Mg/AlCl_3/G2/LiTFSI$				
Borate-based	$[Mg_4Cl_6(DME)_6][B(HFP)_4]_2$	3.3 V	S/CNT	700 (1), 1019 (100),	[96]
non-nucleophilic	BCM electrolyte	3.5 V	S/C	160 mA g^{-1}	[97]
electrolyte	$Mg(BH_4)_2/THFPB/G$	2.8 V	S/C	1081 (1), 934 (30), 50 mAg^{-1}	[98]
				956 (1), 526 (30), 50 mA g^{-1}	

Source: Adapted and reproduced from Ref. [20]. Copyright © 2020 Wiley.

cycles using this BCM electrolyte and a sulfur/carbon composite cathode at 0.05 A g^{-1}. 1.1 V was the flat voltage plateau vs Mg^{2+}/Mg. Du et al. [60] just developed a successful organic magnesium borate-based electrolyte using tetrakis(hexafluoroisopropyl)borate anions as the effective cation and $[Mg_4Cl_6(DME)_6]^{2+}$ as the solvated cation. It exhibited some promising characteristics, including strong oxidation stability up to 3.0 V (vs Mg^{2+}/Mg) and a high ionic conductivity of 5.58 mS cm^{-1}. When combined with an 80 wt % sulfur-carbon nanotubes (S-CNT) composite cathode in a Mg–S battery, the battery demonstrated a strong specific capacity of up to 1200 mAh g^{-1} and over 1000 mAh g^{-1} for more than 100 cycles at 0.16 A g^{-1}. The overpotential could also be as low as 0.3 V. In comparison to prior Mg electrolytes based on HMDS and monocarborane, this represents a significant advancement in electrolyte research. Because the aforementioned electrolytes are so complicated, researchers have begun to investigate the potential of the widely used bis(trifluoromethane) sulfonimide (TFSI)–glyme electrolyte in Li–S batteries. A novel TFSI–glyme electrolyte could be used in Mg–S batteries, according to the Wang research group [61]. At 0.1 A g^{-1}, a further study [62] on the $Mg(TFSI)_2$–$MgCl_2$–DME electrolyte for Mg–S batteries revealed a 69% capacity retention after 110 cycles. Around 800 mAh g^{-1} was the initial specific capacity.

20.5 ECTROLYTES FOR Al–S BATTERIES

To date, there has been minimal worry about rechargeable aluminum–sulfur batteries. In nonaqueous Al–S batteries, room-temperature ionic liquid (RTIL) was commonly employed. Al–S batteries have a more complicated charge–discharge operation than other metal–sulfur cell technologies. Al–S cells are typically operated using an ionic-liquid electrolyte rather than a salt–solvent electrolyte. The first reversible Al–S battery using an ionic-liquid electrolyte was reported by Wang and colleagues [63]. $AlCl_3$ and 1-ethyl-3-methylimidazolium chloride were combined to make the electrolyte (EMIC). The electrochemical stability of the electrolyte was assessed (Figure 20.13).

The electrolyte should determine the rate-determining step of Al–S batteries, according to the current research. A schematic of the charge–discharge operations of an Al–S cell with an ionic-liquid electrolyte is shown in Figure 20.14. In Al–S batteries, Li and colleagues [64] used a novel ionic liquid made of NBMPBr (N-butyl-N-methylpiperidinium) and $AlCl_3$ as the electrolyte. They used the $EMIBr/AlCl_3$ electrolyte at first, despite the fact that Al–S batteries utilizing this electrolyte had a large upfront specific capacity. However, the electrophilic polysulfide intermediates reacted

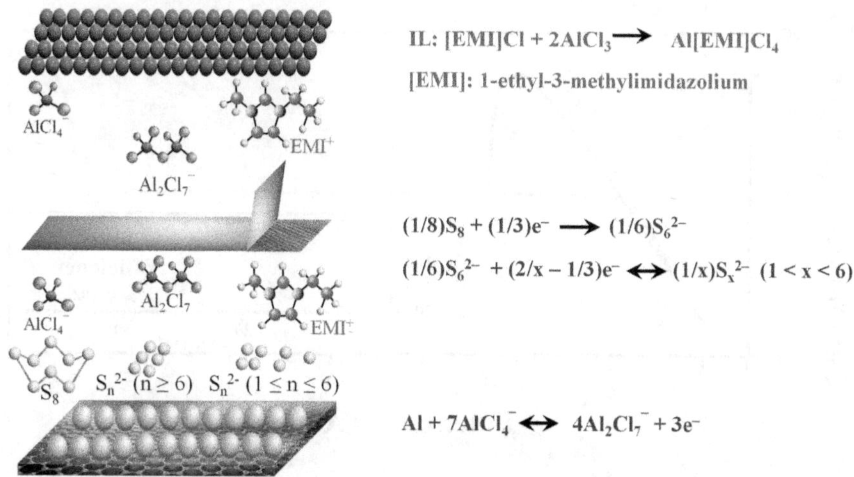

FIGURE 20.13 Schematic representation of the electrochemical discharge–charge mechanism of the Al–S battery with an ionic-liquid electrolyte developed by a reaction of $AlCl_3$ with 1-ethyl-3-ethylimidazolium chloride ([EMI]Cl). Adapted and reproduced from Ref [26]. Copyright © 2020 Wiley.

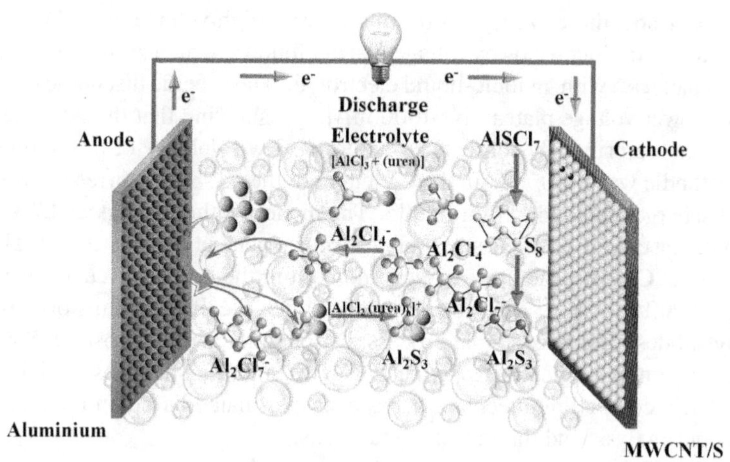

FIGURE 20.14 Schematic of an Al–S cell with the electrolyte of $AlCl_3$/urea. Adapted and reproduced from Ref. [20]. Copyright © 2020 Wiley.

with the nucleophilic imidazole ring of EMI+, resulting in rapid degradation. Instead of EMIBr, they attempted NBMPBr. The EMIBr/$AlCl_3$ electrolyte was easier to dissociate and had a stronger reactivity toward sulfur cathodes than the ionic liquid made up of EMIC and $AlCl_3$. This was due to the fact that Al_2Cl_6-Br had a larger formation energy and a narrower LUMO–HOMO gap than Al_2Cl_6Cl, which DFT simulations supported. The NBMPBr/$AlCl_3$ electrolyte and the EMIBr/$AlCl_3$ electrolyte were identical at this point. The initial discharge capacity of Al–S batteries using this electrolyte and the CMK-3/S composite cathode was 1390 mAh g^{-1}, with more than 400 mAh g^{-1} remaining after 20 cycles.

The usual charge–discharge patterns and cyclic voltammograms of an Al–S cell operated with an ionic-liquid electrolyte are shown in Figure 20.15. The upper-voltage region (the little voltage plateau in Figure 20.15) has been reported to be visible only during the initial discharge and to vanish after the first cycle (Figure 20.15). The change of elemental sulfur to a soluble S_6 2 species accounts

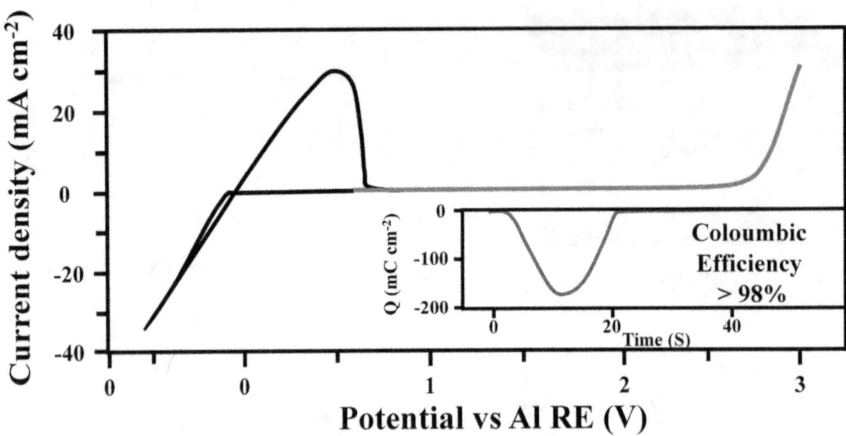

FIGURE 20.15 Al–S battery advancements. (a) Aluminum deposition/stripping in the EMIC: $AlCl_3$ (1:1.3) electrolyte (black) and electrochemical stability of the electrolyte (pink); the inset is the aluminum deposition/stripping charge-time curve; glassy carbon was the working electrode and aluminum was both the counter and reference electrode. Adapted and reproduced from Ref. [2]. Copyright © 2021 Elsevier.

for 200 mAh g^{-1} of specific capacity inside this narrow high-voltage area. Following this brief upper-voltage zone, a continuous lower-voltage plateau follows, which corresponds to the discharge behavior of Li–S batteries with an ionic-liquid electrolyte. The specific discharge capacity produced by this extended lower-voltage plateau is >1000 mAh g^{-1}, showing that the Al–S cell's polysulfide to sulfide processes are primarily solid state (due to the low solubility of polysulfides in ionic liquids). The big cathodic wave at 0.75 V in the CV curve of the Al–S cell corresponds to the extended lower-voltage discharge plateau in Figure 20.15. The modest cathodic peak at 1.7 V in Figure 20.3 corresponds to the discharge profile's small upper-voltage plateau (Figure 20.15). The high-voltage cathodic wave in the CV, like the upper-voltage domain in the discharge curve, vanishes after the first cycle (Figure 20.15). The research suggests that in Al–S cells, the transformation of sulfur to long-chain polysulfides is an irreversible process. Despite the fact that most Al–S battery research has concentrated on improving cathodes and electrolytes, mechanistic investigations have been carried out, and a fairly compelling mechanism has been postulated based on the conclusions gained from spectroscopic analysis and theoretical simulations.

Manthiram et al. [65] found that Li^+ ion mediation could improve the reversibility of Al–S batteries, similar to how lithium ions regulate the electrochemical performance of Mg–S batteries. Polysulfides shuttle vanishes during a solid-state electrochemical process in traditional Al–S batteries that use ionic liquids as the electrolyte. The solubility of intermediate polysulfides in this ionic liquid was improved following the addition of LiTFSI to the Al[EMI]Cl_4 electrolyte, and polysulfides shuttle returned. The solubility of S_4 species increased in the Li ion–mediated electrolyte, as evidenced by the UV–visible absorption experiment. The short-chain Al polysulfides were activated by the presence of Li ions, lowering the kinetic barrier in the reduction/oxidation of the discharged products. Their Al–S batteries had an initial specific capacity of more than 1000 mAh g^{-1} and a sustained capacity of up to 600 mAh g^{-1} (Figure 20.16).

20.6 ELECTROLYTES FOR K–S BATTERIES

K–S batteries, like HT Na–S batteries, have been known to work at high temperatures [66]. The electrolyte used in the HT K–S batteries was a novel beta-Al_2O_3 solid electrolyte that conducts K^+. Similar to liquid sodium, liquid potassium had better wettability on the surface of this unique solid electrolyte. Because of this beneficial feature, the K–S battery could be operated at 150°C,

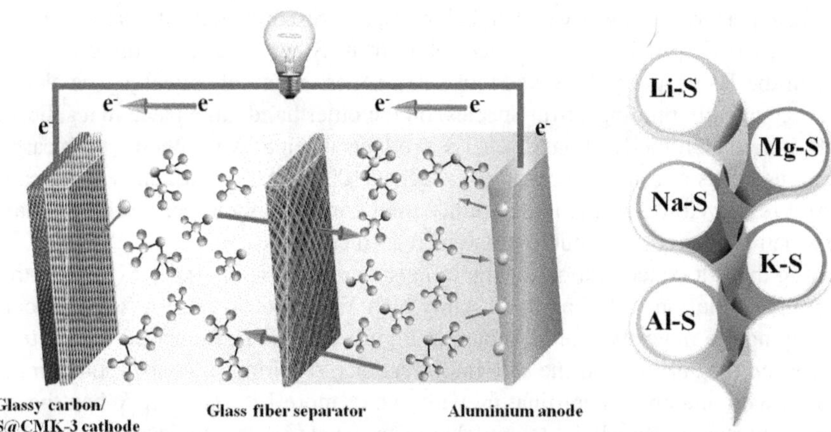

FIGURE 20.16 Al–S battery advancements. The Al–S battery discharge procedure with the NBMPBr/$AlCl_3$ electrolyte. Adapted and reproduced from Ref. [2]. Copyright © 2021 Elsevier.

compared to 300°C for the Na–S battery companion. The capacity was roughly 400 mAh g^{-1} at C/40 at this operating temperature. The capacity remained 300 mAh g^{-1} at a higher rate of C/4.2, with a life of 1000 cycles. However, one of the primary drawbacks of HT K–S batteries is their high temperature. As a result, the investigation of RT K–S batteries is a hot topic [67].

The first investigation on the RT K–S battery was established in 2014 by Zhao et al. (Figure 20.6) [68]. The electrolyte was $KClO_4/G_4$, the anode was metallic potassium, and the cathode was PANI plated CMK-3/S composite in this experiment. K_2S_3, which had a sulfur valency of −2/3, was revealed to be the major discharge product. This explains why the capacity is only 513 mAh g^{-1}, or about a third of the theoretical specific capacity. Similar to Li–S and RT Na–S batteries, ether-based electrolytes should be utilized with sulfur or mesoporous carbon/sulfur composite cathodes, while ester-based electrolytes should be used with microporous carbon/sulfur or PAN/S composites.

For a complete K–S cell, Hwang et al. produced a catholyte of K_2S_5/DEGDME and an anode of potassium/hard carbon composite [69]. The liquid state of K_2S_x with x = 1,2,3,4, and 5 in DEGDME was discovered to be K_2S_5/DEGDME. Furthermore, the DEGDME solvent was found to be non-flammable, in contrast to the DOL/DME solvent, which has a significant tendency to catch fire. The lack of potassium foil enhanced cell safety. Other than a glass fiber and Celgard 2400 as a separator to prevent short circuit, no other electrolyte was used in this K–S cell as shown in Figure 20.17. CNT film absorbed the catholyte, providing catholyte materials with reaction sites. The complete cell showed only the initial cycle with a discharge capacity of 235 mAh g^{-1} and no further cycling with a voltage window of 0.71–0.85V and a current density of 56 mA g^{-1}. Despite the fact that the

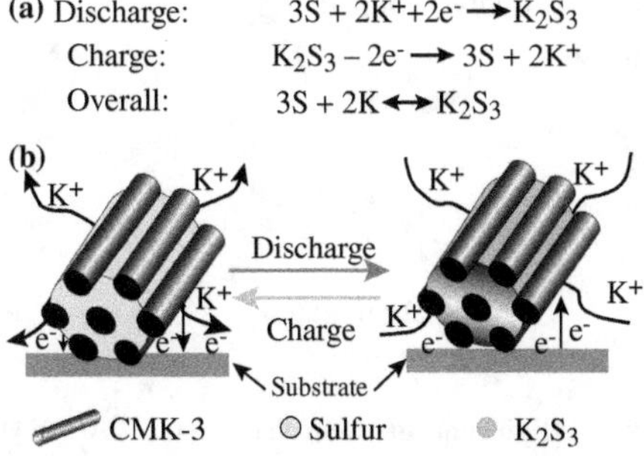

FIGURE 20.17 Electrochemical behavior of a K–S battery. Adapted and reproduced from Ref. [20]. Copyright © 2020 Wiley.

electrochemical performance was worse to half cells, this research added to the K–S battery family. Gu et al. [70] used KTFSI/DME as an electrolyte in K–S batteries, combining it with Nafion-K^+ separators. In the K–S battery, K_2S was found to be inert electrochemically. The electrochemical reactions for covalently binding sulfur species, on the other hand, take place in a solid state. $K_{0.85}S$ and K_2S_2/K_2S were identified as final discharge products when SPAN cathodes with carbonate electrolytes were used. As a result, the capacity exceeded 1000 mAh g^{-1}, which is significantly greater than the 2014 result. However, the latter studies used considerably larger discharge/charge voltage windows, which may have contributed to the increased capacities.

Controlling the salt concentration, which in most situations is around 1 M, allows the structure of the electrolyte solution to be modified. Only a small proportion of the solvent coordinates to salt ions, keeping the majority free as shown in Figure 20.18 [71]. Increased levels strengthen the contacts between the solvent and the salt anions/cations, resulting in contact ion pairs and aggregates (AGGs), with one anion coordinating with two or more Li^+ ions [72]. When the salt–solvent ratio exceeds a particular threshold, the number of free-state solvent molecules drops dramatically, perhaps to zero. The lack of free-state solvent molecules in such a high-concentration electrolyte (HCE) system limits polysulfide dissolution. HCEs have some unique features due to the strong solvent–ion interactions, such as high Li^+ transference numbers, large electrochemical windows, flame retardance, lower volatility, and thermal stability [73]. HCEs, on the other hand, have limited ionic conductivity and high viscosity. Diluents with low viscosity and poor solvating power are typically integrated to generate diluted electrolytes or localized high-concentration electrolyte (LHCE) systems to overcome this difficulty without destroying the solvation structure [74].

Solid electrolyte solutions provide a strong benefit over liquid electrolyte solutions in allowing the operation of reversible active metal electrodes [75]. Chemical, thermal, and mechanical stabilities are all characteristics of well-designed SE. Furthermore, these electrolytes serve as a separator in the cells. As a result, developing such electrolytes is of tremendous interest. By combining active metal anodes with SE, many of the problems that hamper their practical application could be avoided. Because the transference number of the solid electrolyte for the active metal cations is projected to be close to 1, SE is meant to avoid any undesired impurities from reaching the anode side. Polymer electrolytes based on PEO [76], PEGDME, and PVDF-HFP (gel type SE) [77] are now being studied, as are inorganic SE such as sulfide glass $Li_2SP_2S_5$ [78], beta-alumina [79], and NASICON [80].

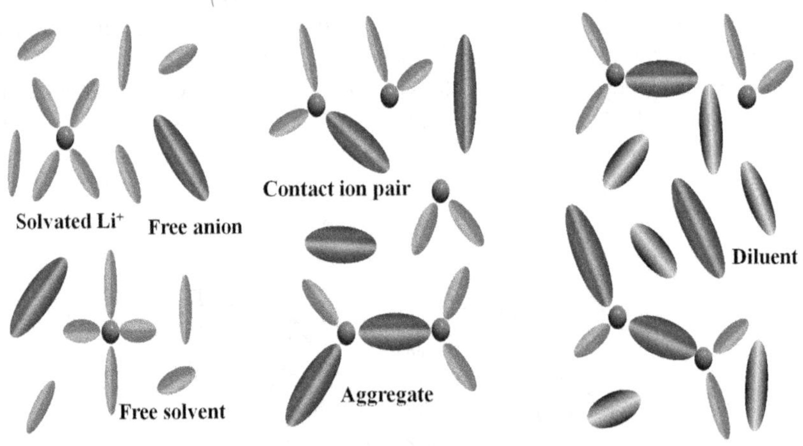

FIGURE 20.18 Schematic illustration of conventional (~1 M regime), concentrated, and diluted concentrated electrolyte solution structures. Adapted and reproduced from Ref. [92]. Copyright © 2021 Elsevier.

Ethers are the most common solvents in sulfur-based electrochemical systems. The reason for this decision is that ethers can operate as relatively good solvents for metal-polysulfide species generated as intermediate products of sulfur reduction. In ethereal electrolyte solutions, considerable usage of sulfur cathodes is thus possible. In sulfur-based batteries, however, the high solubility of polysulfides in electrolyte solutions is a key issue. The soluble polysulfides diffuse to the anode side and are repeatedly lowered therein. Because a constant electron flow from active metal anodes in these systems to the polysulfide moieties in solution avoids their re-oxidation to elemental sulfur during the charging process, this scenario, known as the "shuttle mechanism," produces a capacity loss on the cathode side. Furthermore, polysulfide moieties react on the anode side, corroding the active metal and forming dead sites. Metal–sulfur cells have a weak Coulombic efficiency as a result of these circumstances. Because of their capacity to dissolve polysulfide and their relatively high cathodic stability, 1,2-dimethoxyethane (DME) and tetra(ethylene glycol) dimethyl ether (TEGDME) are the most appealing ethereal solvents for M–S batteries. To decrease the shuttle effect while keeping the benefits of employing ethereal solvents, extensive research was carried out [81,82]. The elemental sulfur is encapsulated in conductive carbonaceous material, which keeps the dissolved polysulfide as semi-liquid cathode material trapped in a position that prevents the cathode from dissolving into the solution.

20.7 SUMMARY

The major goal of this chapter is to have an up-to-date overview on electrolytes for metal–sulfur battery research. The present state of electrolytes and reaction mechanism for Li–S battery scientific research was briefly covered. The progress of RT electrolytes for Na–S, Mg–S, Al–S, and K–S batteries was then addressed in depth. Key hurdles in Li-based battery technology, as well as crucial issues impeding the development of other MSBs, were objectively examined. Despite significant progress in tackling the issues with metal–sulfur batteries, there are still difficulties to overcome in order to improve overall performance. A few prospective routes can be suggested to enhance reaction kinetics and boost sulfur conversion efficiency, as well as to elucidate the regulatory mechanisms by electrocatalysts, which have still yet to receive merited attention but are deserving of in-depth research due to their tremendous perspectives. It will be fascinating to follow the development of Na–S batteries as humans strive to develop ecologically sustainable and efficient practical solutions to fulfill our pace with the fast future energy demands.

REFERENCES

1. Manthiram, A. (2020). A reflection on lithium-ion battery cathode chemistry. *Nature Communications*, 11(1), 1–9.
2. Liu, X., Li, Y., Xu, X., Zhou, L., & Mai, L. (2021). Rechargeable metal (Li, Na, Mg, Al)-sulfur batteries: materials and advances. *Journal of Energy Chemistry*, 61, 104–134.
3. Dunn, B., Kamath, H., & Tarascon, J. M. (2011). Electrical energy storage for the grid: a battery of choices. *Science*, 334(6058), 928–935.
4. Goodenough, J. B., & Park, K. S. (2013). The Li-ion rechargeable battery: a perspective. *Journal of the American Chemical Society*, 135(4), 1167–1176.
5. Li, W., Erickson, E. M., & Manthiram, A. (2020). High-nickel layered oxide cathodes for lithium-based automotive batteries. *Nature Energy*, 5(1), 26–34.
6. Yu, X., & Manthiram, A. (2021). Sustainable battery materials for next-generation electrical energy storage. *Advanced Energy and Sustainability Research*, 2(5), 2000102.
7. Manthiram, A. (2011). Materials challenges and opportunities of lithium-ion batteries. *The Journal of Physical Chemistry Letters*, 2(3), 176–184.
8. Whittingham, M. S. (2004). Lithium batteries and cathode materials. *Chemical Reviews*, 104(10), 4271–4302.
9. Harper, G., Sommerville, R., Kendrick, E., Driscoll, L., Slater, P., Stolkin, R., ... & Anderson, P. (2019). Recycling lithium-ion batteries from electric vehicles. *Nature*, 575(7781), 75–86.

10. Schmuch, R., Wagner, R., Hörpel, G., Placke, T., & Winter, M. (2018). Performance and cost of materials for lithium-based rechargeable automotive batteries. *Nature Energy*, 3(4), 267–278.
11. Hong, X., Mei, J., Wen, L., Tong, Y., Vasileff, A. J., Wang, L., ... & Dou, S. X. (2019). Nonlithium metal–sulfur batteries: steps toward a leap. *Advanced Materials*, 31(5), 1802822.
12. Ji, X., Lee, K. T., & Nazar, L. F. (2009). A highly ordered nanostructured carbon–sulphur cathode for lithium–sulphur batteries. *Nature Materials*, 8(6), 500–506.
13. Lee, J. S., Kim, W., Jang, J., & Manthiram, A. (2017). Sulfur-embedded activated multichannel carbon nanofiber composites for long-life, high-rate lithium–sulfur batteries. *Advanced Energy Materials*, 7(5), 1601943.
14. Seh, Z. W., Sun, Y., Zhang, Q., & Cui, Y. (2016). Designing high-energy lithium–sulfur batteries. *Chemical Society Reviews*, 45(20), 5605–5634.
15. Dong, Y., Li, T., Cai, D., Yang, S., Zhou, X., Nie, H., & Yang, Z. (2021). Progress and Prospect of organic electrocatalysts in lithium–sulfur batteries. *Frontiers in Chemistry*, 515, 703354 (1–27).
16. Li, C., Wang, Z. B., Wang, Q., & Gu, D. M. (2017). Recent advances in cathode materials for Li–S battery: structure and performance. *Rare Metals*, 36(5), 365–380.
17. Hong, X., Wang, R., Liu, Y., Fu, J., Liang, J., and Dou, S. (2020). Recent advances in chemical adsorption and catalytic conversion materials for Li-S batteries. *Journal of Energy Chemistry*, 42, 144–168.
18. Zhang, M., Chen, W., Xue, L., Jiao, Y., Lei, T., Chu, J., ... & Xiong, J. (2020a). Adsorption-catalysis design in the lithium-sulfur battery. *Advanced Energy Materials*, 10 (2), 1903008.
19. Guo, J., Pei, H., Dou, Y., Zhao, S., Shao, G., & Liu, J. (2021). Rational designs for lithium-sulfur batteries with low electrolyte/sulfur ratio. *Advanced Functional Materials*, 31(18), 2010499.
20. Pan, Y., Li, S., Yin, M., & Li, J. (2019). Electrolyte evolution propelling the development of nonlithium metal–sulfur batteries. *Energy Technology*, 7(12), 1900164.
21. Adelhelm, P., Hartmann, P., Bender, C. L., Busche, M., Eufinger, C., & Janek, J. (2015). From lithium to sodium: cell chemistry of room temperature sodium–air and sodium–sulfur batteries. *Beilstein Journal of Nanotechnology*, 6(1), 1016–1055.
22. Yu, X., & Manthiram, A. (2018). A reversible nonaqueous room-temperature potassium-sulfur chemistry for electrochemical energy storage. *Energy Storage Materials*, 15, 368–373.
23. Yu, X., & Manthiram, A. (2021). Sustainable battery materials for next-generation electrical energy storage. *Advanced Energy and Sustainability Research*, 2(5), 2000102.
24. Yu, X., & Manthiram, A. (2016). Performance enhancement and mechanistic studies of magnesium–sulfur cells with an advanced cathode structure. *ACS Energy Letters*, 1(2), 431–437.
25. Yu, X., Boyer, M. J., Hwang, G. S., & Manthiram, A. (2018). Room-temperature aluminum-sulfur batteries with a lithium-ion-mediated ionic liquid electrolyte. *Chem*, 4(3), 586–598.
26. Yu, X., & Manthiram, A. (2020). A progress report on metal–sulfur batteries. *Advanced Functional Materials*, 30(39), 2004084.
27. Yu, X., & Manthiram, A. (2015). A class of polysulfide catholytes for lithium–sulfur batteries: energy density, cyclability, and voltage enhancement. *Physical Chemistry Chemical Physics*, 17(3), 2127–2136.
28. Yu, X., & Manthiram, A. (2014). Room-temperature sodium–sulfur batteries with liquid-phase sodium polysulfide catholytes and binder-free multiwall carbon nanotube fabric electrodes. *The Journal of Physical Chemistry C*, 118(40), 22952–22959.
29. Zhao, Q., Hu, Y., Zhang, K., & Chen, J. (2014). Potassium–sulfur batteries: a new member of room-temperature rechargeable metal–sulfur batteries. *Inorganic Chemistry*, 53(17), 9000–9005.
30. Braga, M. H., Grundish, N. S., Murchison, A. J., & Goodenough, J. B. (2017). Alternative strategy for a safe rechargeable battery. *Energy & Environmental Science*, 10, 331–336.
31. Monti, D., Ponrouch, A., Araujo, R. B., Barde, F., Johansson, P., & Palacín, M. R. (2019). Multivalent batteries—prospects for high energy density: Ca batteries. *Frontiers in Chemistry*, 79, 1–6.
32. Zhao-Karger, Z., Liu, R., Dai, W., Li, Z., Diemant, T., Vinayan, B. P., ... & Fichtner, M. (2018). Toward highly reversible magnesium–sulfur batteries with efficient and practical Mg [B (hfip) 4] 2 electrolyte. *ACS Energy Letters*, 3(8), 2005–2013.
33. Yu, X., & Manthiram, A. (2017). Electrochemical energy storage with a reversible nonaqueous room-temperature aluminum–sulfur chemistry. *Advanced Energy Materials*, 7(18), 1700561.
34. Zou, Q., & Lu, Y. C. (2021). Liquid electrolyte design for metal-sulfur batteries: mechanistic understanding and perspective. *EcoMat*, 3(4), e12115.
35. Wang, X., Li, G., Li, J., Zhang, Y., Wook, A., Yu, A., & Chen, Z. (2016). Structural and chemical synergistic encapsulation of polysulfides enables ultralong-life lithium–sulfur batteries. *Energy & Environmental Science*, 9(8), 2533–2538.

36. Tang, C., Zhang, Q., Zhao, M. Q., Huang, J. Q., Cheng, X. B., Tian, G. L., ... & Wei, F. (2014). Nitrogen-doped aligned carbon nanotube/graphene sandwiches: facile catalytic growth on bifunctional natural catalysts and their applications as scaffolds for high-rate lithium-sulfur batteries. *Advanced Materials*, 26(35), 6100–6105.
37. Liang, X., Hart, C., Pang, Q., Garsuch, A., Weiss, T., & Nazar, L. F. (2015). A highly efficient polysulfide mediator for lithium–sulfur batteries. *Nature Communications*, 6(1), 1–8.
38. Eng, A. Y. S., Kumar, V., Zhang, Y., Luo, J., Wang, W., Sun, Y., ... & Seh, Z. W. (2021). Room-temperature sodium–sulfur batteries and beyond: realizing practical high energy systems through anode, cathode, and electrolyte engineering. *Advanced Energy Materials*, 11(14), 2003493.
39. Liang, X., & Nazar, L. F. (2016). In situ reactive assembly of scalable core–shell sulfur–MnO_2 composite cathodes. *ACS Nano*, 10(4), 4192–4198.
40. Li, Z., Zhang, J., & Lou, X. W. (2015). Hollow carbon nanofibers filled with MnO_2 nanosheets as efficient sulfur hosts for lithium–sulfur batteries. *Angewandte Chemie*, 127(44), 13078–13082.
41. Rehman, S., Tang, T., Ali, Z., Huang, X., & Hou, Y. (2017). Integrated design of MnO_2@ carbon hollow nanoboxes to synergistically encapsulate polysulfides for empowering lithium sulfur batteries. *Small*, 13(20), 1700087.
42 Kong, W., Yan, L., Luo, Y., Wang, D., Jiang, K., Li, Q., ... & Wang, J. (2017). Ultrathin MnO_2/graphene oxide/carbon nanotube interlayer as efficient polysulfide-trapping shield for high-performance Li–S batteries. Advanced Functional Materials, 27(18), 1606663.
43. Guo, J., Zhang, X., Du, X., & Zhang, F. (2017). A Mn_3O_4 nano-wall array based binder-free cathode for high performance lithium–sulfur batteries. *Journal of Materials Chemistry A*, 5(14), 6447–6454.
44. Li, X. Y., Qu, J. K., & Yin, H. Y. (2021). Electrolytic alloy-type anodes for metal-ion batteries. *Rare Metals*, 40(2), 329–352.
45. Xia, S., Zhang, X. M., Huang, K., Chen, Y. L., & Wu, Y. T. (2015). Ionic liquid electrolytes for aluminium secondary battery: influence of organic solvents. *Journal of Electroanalytical Chemistry*, 757, 167–175.
46. Licht, S., Hwang, J., Light, T. S., & Dillon, R. (1997). The low current domain of the aluminum/sulfur battery. *Journal of the Electrochemical Society*, 144(3), 948.
47. Gao, T., Hou, S., Wang, F., Ma, Z., Li, X., Xu, K., & Wang, C. (2017). Reversible S0/MgSx redox chemistry in a $MgTFSI_2$/$MgCl_2$/DME electrolyte for rechargeable Mg/S batteries. *Angewandte Chemie*, 129(43), 13711–13715.
48. Wen, Z., Hu, Y., Wu, X., Han, J., & Gu, Z. (2013). Main challenges for high performance NAS battery: materials and interfaces. *Advanced Functional Materials*, 23(8), 1005–1018.
49. Wei, S., Xu, S., Agrawral, A., Choudhury, S., Lu, Y., Tu, Z., ... & Archer, L. A. (2016). A stable room-temperature sodium–sulfur battery. *Nature Communications*, 7(1), 1–10.
50 Xu, X., Zhou, D., Qin, X., Lin, K., Kang, F., Li, B., ... & Wang, G. (2018). A room-temperature sodium–sulfur battery with high capacity and stable cycling performance. *Nature Communications*, 9(1), 1–12.
51. Seh, Z. W., Sun, J., Sun, Y., & Cui, Y. (2015). A highly reversible room-temperature sodium metal anode. *ACS Central Science*, 1(8), 449–455.
52. Luo, W., Lin, C. F., Zhao, O., Noked, M., Zhang, Y., Rubloff, G. W., & Hu, L. (2017). Ultrathin surface coating enables the stable sodium metal anode. *Advanced Energy Materials*, 7(2), 1601526.
53. Zhao, Y., Goncharova, L. V., Lushington, A., Sun, Q., Yadegari, H., Wang, B., ... & Sun, X. (2017). Superior stable and long life sodium metal anodes achieved by atomic layer deposition. *Advanced Materials*, 29(18), 1606663.
54. Wan, H., Weng, W., Han, F., Cai, L., Wang, C., & Yao, X. (2020). Bio-inspired nanoscaled electronic/ionic conduction networks for room-temperature all-solid-state sodium-sulfur battery. *Nano Today*, 33, 100860.
55. Yue, J., Han, F., Fan, X., Zhu, X., Ma, Z., Yang, J., & Wang, C. (2017). High-performance all-inorganic solid-state sodium–sulfur battery. *ACS Nano*, 11(5), 4885–4891.
56. Das, C., Ananthoju, B., Dhara, A. K., Aslam, M., Sarkar, S. K., & Balasubramaniam, K. R. (2017). Electron-selective TiO_2/CVD-graphene layers for photocorrosion inhibition in Cu_2O photocathodes. *Advanced Materials Interfaces*, 4(17), 1700271.
57. Kim, H. S., Arthur, T. S., Allred, G. D., Zajicek, J., Newman, J. G., Rodnyansky, A. E., ... & Muldoon, J. (2011). Structure and compatibility of a magnesium electrolyte with a sulphur cathode. *Nature Communications*, 2(1), 1–6.

58. Zhao-Karger, Z., Zhao, X., Wang, D., Diemant, T., Behm, R. J., & Fichtner, M. (2015). Performance improvement of magnesium sulfur batteries with modified non-nucleophilic electrolytes. *Advanced Energy Materials*, 5(3), 1401155.
59. Zhang, Z., Cui, Z., Qiao, L., Guan, J., Xu, H., Wang, X., ... & Chen, L. (2017). Novel design concepts of efficient Mg-ion electrolytes toward high-performance magnesium–selenium and magnesium–sulfur batteries. *Advanced Energy Materials*, 7(11), 1602055.
60. Du, A., Zhang, Z., Qu, H., Cui, Z., Qiao, L., Wang, L., ... & Cui, G. (2017). An efficient organic magnesium borate-based electrolyte with non-nucleophilic characteristics for magnesium–sulfur battery. *Energy & Environmental Science*, 10(12), 2616–2625.
61. Gao, T., Hou, S., Wang, F., Ma, Z., Li, X., Xu, K., & Wang, C. (2017). Reversible S0/MgSx redox chemistry in a $MgTFSI_2/MgCl_2$/DME electrolyte for rechargeable Mg/S batteries. *Angewandte Chemie*, 129(43), 13711–13715.
62. Ho, Y. C., & Chung, S. H. (2021). A design of the cathode substrate for high-loading polysulfide cathodes in lean-electrolyte lithium-sulfur cells. *Chemical Engineering Journal*, 422, 130363.
63. Bian, Y., Li, Y., Yu, Z., Chen, H., Du, K., Qiu, C., ... & Lin, M. C. (2018). Using an $AlCl_3$/urea ionic liquid analog electrolyte for improving the lifetime of aluminum-sulfur batteries. *ChemElectroChem*, 5(23), 3607–3611.
64. Yang, H., Yin, L., Liang, J., Sun, Z., Wang, Y., Li, H., ... & Li, F. (2018). An aluminum–sulfur battery with a fast kinetic response. *Angewandte Chemie*, 130(7), 1916–1920.
65. Yu, X., Boyer, M. J., Hwang, G. S., & Manthiram, A. (2018). Room-temperature aluminum-sulfur batteries with a lithium-ion-mediated ionic liquid electrolyte. *Chem*, 4(3), 586–598.
66. Liu, Y., Wang, W., Wang, J., Zhang, Y., Zhu, Y., Chen, Y., ... & Wu, Y. (2018). Sulfur nanocomposite as a positive electrode material for rechargeable potassium–sulfur batteries. *Chemical Communications*, 54(18), 2288–2291.
67. Ma, R., Fan, L., Wang, J., & Lu, B. (2019). Confined and covalent sulfur for stable room temperature potassium-sulfur battery. *Electrochimica Acta*, 293, 191–198.
68. Hwang, J. Y., Kim, H. M., Yoon, C. S., & Sun, Y. K. (2018). Toward high-safety potassium–sulfur batteries using a potassium polysulfide catholyte and metal-free anode. *ACS Energy Letters*, 3(3), 540–541.
69. Xiong, P., Han, X., Zhao, X., Bai, P., Liu, Y., Sun, J., & Xu, Y. (2019). Room-temperature potassium–sulfur batteries enabled by microporous carbon stabilized small-molecule sulfur cathodes. *ACS Nano*, 13(2), 2536–2543.
70. Gu, S., Xiao, N., Wu, F., Bai, Y., Wu, C., & Wu, Y. (2018). Chemical synthesis of K2S2 and K2S3 for probing electrochemical mechanisms in K–S batteries. *ACS Energy Letters*, 3(12), 2858–2864.
71. Liu, Y., Elias, Y., Meng, J., Aurbach, D., Zou, R., Xia, D., & Pang, Q. (2021). Electrolyte solutions design for lithium-sulfur batteries. *Joule*, 5(9), 2323–2364.
72. Yamada, Y., & Yamada, A. (2015). Superconcentrated electrolytes for lithium batteries. *Journal of the Electrochemical Society*, 162(14), A2406.
73. Yamada, Y., Wang, J., Ko, S., Watanabe, E., & Yamada, A. (2019). Advances and issues in developing salt-concentrated battery electrolytes. *Nature Energy*, 4(4), 269–280.
74. Peng, Z., Cao, X., Gao, P., Jia, H., Ren, X., Roy, S., ... & Zhang, J. G. (2020). High-power lithium metal batteries enabled by high-concentration acetonitrile-based electrolytes with vinylene carbonate additive. *Advanced Functional Materials*, 30(24), 2001285.
75. Gao, Z., Sun, H., Fu, L., Ye, F., Zhang, Y., Luo, W., & Huang, Y. (2018). Promises, challenges, and recent progress of inorganic solid-state electrolytes for all-solid-state lithium batteries. *Advanced Materials*, 30(17), 1705702.
76. Pfaffenhuber, C., Göbel, M., Popovic, J., & Maier, J. (2013). Soggy-sand electrolytes: status and perspectives. *Physical Chemistry Chemical Physics*, 15(42), 18318–18335.
77. Appetecchi, G. B., Dautzenberg, G., & Scrosati, B. (1996). A new class of advanced polymer electrolytes and their relevance in plastic-like, rechargeable lithium batteries. *Journal of the Electrochemical Society*, 143(1), 6.
78. Lin, Z., Liu, Z., Dudney, N. J., & Liang, C. (2013). Lithium superionic sulfide cathode for all-solid lithium–sulfur batteries. *ACS Nano*, 7(3), 2829–2833.
79. Wen, Z., Hu, Y., Wu, X., Han, J., & Gu, Z. (2013). Main challenges for high performance NAS battery: materials and interfaces. *Advanced Functional Materials*, 23(8), 1005–1018.

80. Kobayashi, E., Plashnitsa, L. S., Doi, T., Okada, S., & Yamaki, J. I. (2010). Electrochemical properties of Li symmetric solid-state cell with NASICON-type solid electrolyte and electrodes. *Electrochemistry Communications*, 12(7), 894–896.
81. Shen, C., Xie, J., Zhang, M., Andrei, P., Hendrickson, M., Plichta, E. J., & Zheng, J. P. (2017). Understanding the role of lithium polysulfide solubility in limiting lithium-sulfur cell capacity. *Electrochimica Acta*, 248, 90–97.
82. Moy, D., Manivannan, A., & Narayanan, S. R. (2014). Direct measurement of polysulfide shuttle current: a window into understanding the performance of lithium-sulfur cells. *Journal of the Electrochemical Society*, 162(1), A.
83. Kim, E. J., & Kim, D. H. (2006). Highly conductive HfN x films prepared by plasma-assisted atomic layer deposition. *Electrochemical and Solid-State Letters*, 9(8), C123.
84. Kumar, D., Suleman, M., & Hashmi, S. A. (2011). Studies on poly (vinylidene fluoride-co-hexafluoropropylene) based gel electrolyte nanocomposite for sodium–sulfur batteries. *Solid State Ionics*, 202(1), 45–53.
85. Zhou, D., Chen, Y., Li, B., Fan, H., Cheng, F., Shanmukaraj, D., ... & Wang, G. (2018). A stable quasi-solid-state sodium–sulfur battery. *Angewandte Chemie – International Edition*, 57(32), 10168–10172.
86. Park, C. W., Ryu, H. S., Kim, K. W., Ahn, J. H., Lee, J. Y., & Ahn, H. J. (2007). Discharge properties of all-solid sodium–sulfur battery using poly (ethylene oxide) electrolyte. *Journal of Power Sources*, 165(1), 450–454.
87. Yue, J., Han, F., Fan, X., Zhu, X., Ma, Z., Yang, J., & Wang, C. (2017). High-performance all-inorganic solid-state sodium–sulfur battery. *ACS Nano*, 11(5), 4885–4891.
88. Ryu, H., Kim, T., Kim, K., Ahn, J. H., Nam, T., Wang, G., & Ahn, H. J. (2011). Discharge reaction mechanism of room-temperature sodium–sulfur battery with tetra ethylene glycol dimethyl ether liquid electrolyte. *Journal of Power Sources*, 196(11), 5186–5190.
89. Carter, R., Oakes, L., Douglas, A., Muralidharan, N., Cohn, A. P., & Pint, C. L. (2017). A sugar-derived room-temperature sodium sulfur battery with long term cycling stability. *Nano Letters*, 17(3), 1863–1869.
90. Wei, S., Xu, S., Agrawral, A., Choudhury, S., Lu, Y., Tu, Z., ... & Archer, L. A. (2016). A stable room-temperature sodium–sulfur battery. *Nature Communications*, 7(1), 1–10.
91. Xu, X., Zhou, D., Qin, X., Lin, K., Kang, F., Li, B., ... & Wang, G. (2018). A room-temperature sodium–sulfur battery with high capacity and stable cycling performance. *Nature Communications*, 9(1), 1–12.
92. Kim, H. S., Arthur, T. S., Allred, G. D., Zajicek, J., Newman, J. G., Rodnyansky, A. E., ... & Muldoon, J. (2011). Structure and compatibility of a magnesium electrolyte with a sulphur cathode. *Nature Communications*, 2(1), 1–6.
93. Aurbach, D., Lu, Z., Schechter, A., Gofer, Y., Gizbar, H., Turgeman, R., ... & Levi, E. (2000). Prototype systems for rechargeable magnesium batteries. *Nature*, 407(6805), 724–727.
94. Tian, J., Cao, D., Zhou, X., Hu, J., Huang, M., & Li, C. (2018). High-capacity Mg–organic batteries based on nanostructured rhodizonate salts activated by Mg–Li dual-salt electrolyte. *ACS Nano*, 12(4), 3424–3435.
95. Hu, X. C., Shi, Y., Lang, S. Y., Zhang, X., Gu, L., Guo, Y. G., ... & Wan, L. J. (2018). Direct insights into the electrochemical processes at anode/electrolyte interfaces in magnesium-sulfur batteries. *Nano Energy*, 49, 453–459.
96. Wu, C., Hu, J., Tian, J., Chu, F., Yao, Z., Zheng, Y., ... & Li, C. (2019). Stacking of tailored chalcogenide nanosheets around MoO_2-C conductive stakes modulated by a hybrid POM⊂ MOF precursor template: composite conversion–insertion cathodes for rechargeable Mg–Li dual-salt batteries. *ACS Applied Materials & Interfaces*, 11(6), 5966–5977.
97. Zhang, Z., Cui, Z., Qiao, L., Guan, J., Xu, H., Wang, X., ... & Chen, L. (2017). Novel design concepts of efficient Mg-Ion electrolytes toward high-performance magnesium–selenium and magnesium–sulfur batteries. *Advanced Energy Materials*, 7(11), 1602055.
98. Xu, H., Zhang, Z., Li, J., Qiao, L., Lu, C., Tang, K., ... & Cui, G. (2018). Multifunctional additives improve the electrolyte properties of magnesium borohydride toward magnesium–sulfur batteries. *ACS Applied Materials & Interfaces*, 10(28), 23757–23765.

Index

For Product Safety Concerns and Information please contact our EU representative GPSR@taylorandfrancis.com
Taylor & Francis Verlag GmbH, Kaufingerstraße 24, 80331 München, Germany

www.ingramcontent.com/pod-product-compliance
Lightning Source LLC
LaVergne TN
LVHW081314110826
845149LV00006B/1503

* 9 7 8 1 0 3 2 3 1 5 3 5 5 *